ENGINEERING MATHEMATICS

VOLUME III

(Third Semester)

(As per Anna University Syllabus)

ENGINEERING MATHEMATICS

(TRANSFORMS AND PARTIAL DIFFERENTIAL EQUATIONS)

VOLUME-III

(Third Semester)

[For B.E., B.Tech. & B.Sc. (Applied Sciences) As per Anna University Syllabus and Deemed Universities]

P. KANDASAMY
M.A. Ph.D.
Professor of Mathematics (Retd.)
P.S.G. College of Technology
COIMBATORE
&
Visiting Professor
Amrita Institute of Technology and Science (Deemed University)
COIMBATORE

K. THILAGAVATHY
M.Sc., M.Phil, Ph.D
Reader in Mathematics
Kongunadu Arts and Science College
COIMBATORE

K. GUNAVATHY
M.E., Ph.D.
Professor, Department of
Electronics and Communication Engineering
P.S.G. College of Technology
COIMBATORE

S. CHAND
PUBLISHING

S Chand And Company Limited
(ISO 9001 Certified Company)

S Chand And Company Limited
(ISO 9001 Certified Company)
Head Office: D-92, Sector–2, Noida – 201301, U.P. (India), Ph. 91-120-4682700
Registered Office: A-27, 2nd Floor, Mohan Co-operative Industrial Estate, New Delhi – 110 044 | Phone: 011-49731800
www.**schandpublishing.com**; e-mail: **info@schandpublishing.com**

Marketing Offices:

Chennai : Ph: 23632120; chennai@schandpublishing.com
Guwahati : Ph: 2738811, 2735640; guwahati@schandpublishing.com
Hyderabad : Ph: 40186018; hyderabad@schandpublishing.com
Jalandhar : Ph: 4645630; jalandhar@schandpublishing.com
Kolkata : Ph: 23357458, 23353914; kolkata@schandpublishing.com
Lucknow : Ph: 4003633; lucknow@schandpublishing.com
Mumbai : Ph: 25000297; mumbai@schandpublishing.com
Patna : Ph: 2260011; patna@schandpublishing.com

First Edition 1996
Reprints 1997, 1998, 1999, 2001, 2002, 2003, 2004, 2006, 2007, 2008, 2009, 2010, 2012 (Twice), 2013, 2014, 2016, 2020, 2021, 2022, 2023, 2024

Reprint 2025

ISBN: 978-81-219-1143-6 **Product Code:** H3IEM61MATH10ENAD08S

PRINTED IN INDIA

By Vikas Publishing House Private Limited, Plot 20/4, Site-IV, Industrial Area Sahibabad, Ghaziabad – 201 010 and Published by S Chand And Company Limited, A-27, 2nd Floor, Mohan Co-operative Industrial Estate, New Delhi – 110 044.

PREFACE TO THE FOURTH EDITION

The chapter on 'LINEAR ALGEBRA' is added to suit the syllabus for Electronics and Communication Engineering branch of Anna University, Coimbatore from 2008 July.

AUTHORS

PREFACE TO THE THIRD EDITION

In this edition the chapter on '*z*-Transform' is included and the chapter on 'Laplace Transform' is shifted to the Volume II, to suit the syllabus of Anna University from 2004.

AUTHORS

PREFACE TO THE FIRST EDITION

The existing Third Volume of our series of textbooks on Engineeri Mathematics for students of B.E., B.Tech and B.Sc. (Applied Sciences) h been now split into two volumes, to cater to the needs of the syllabus s mester-wise. This volume caters to the syllabus of fourth semester.

Many worked examples are added in each chapter and a large numb of problems are included in the Exercises. In addition, short Answer Que tions, Model and University Question Papers have been added at the en We are confident that with all these modifications/additions etc., this prese edition with its new format will prove to be even more useful and our e teemed readers will receive it well.

We request the readers to feel free to write to us for the improvement the book.

AUTHOR

Author's address for contact :

66, Thirty Feet Road,
Krishnasamy Nagar,
Ramanathapuram
Coimbatore-641045

NEW SYLLABUS
ANNA UNIVERSITY

TRANSFORMS AND PARTIAL DIFFERENTIAL EQUATIONS

(Common to all B.E. / B.Tech. Degree Programmes)

OBJECTIVES

The course objective is to develop the skills of the students in the areas of Transforms and Partial Differtial Equations. This will be necessary for their effective studies in a large number of engineering subjects like heat conduction, communication systems, electro-optics and electromagnetic theory. The course will also serve as a prerequisite for post graduate and specialized studies and research.

FOURIER SERIES **9 + 3**

Dirichlet's conditions – General Fourier series – Odd and even functions – Half range sine series – Half range cosine series – Complex form of Fourier Series – Parseval's identify – Harmonic Analysis.

FOURIER TRANSFORMS **9 + 3**

Fourier integral theorem (without proof) – Fourier transform pair – Sine and Cosine transforms – Properties – Transforms of simple functions – Convolution theorem – Parseval's identity.

PARTIAL DIFFERENTIAL EQUATIONS **9 +3**

Formation of partial differential equations – Lagrange's linear equation – Solutions of standard types of first order partial differential equations - Linear partial differential equations of second and higher order with constant coefficients.

APPLICATIONS OF PARTIAL DIFFERENTIAL EQUATIONS

9 + 3

Solutions of one dimensional wave equation – One dimensional equation of heat conduction – Steady state solution of two-dimensional equation of heat conduction (Insulated edges excluded) – Fourier series solutions in cartesian coordinates.

5. **Z-TRANSFORMS AND DIFFERENCE EQUATIONS** **9**

Z-transforms - Elementary properties – Inverse Z-transform – Convoluti theorem -Formation of difference equations – Solution of differen equations using Z-transform.

TOTAL (L : 45 + T : 15) : 60 PERIO

TEXT BOOKS

1. Grewal, B.S, "*Higher Engineering Mathematics*", 40th Edition, Khan Publishers, Delhi, (2007).

REFERENCES

1. Bali.N.P and Manish Goyal, "A Textbook of Engineering Mathematic 7th Edition, Laxmi Publications(P) Ltd. (2007).
2. Ramana. B.V., "Higher Engineering Mathematics", Tata Mc-GrawH Publishing Company Limited, New Delhi (2007).
3. Glyn James, "Advanced Modern Engineering Mathematics", 3rd Editio Pearson Education (2007).
4. Erwin Kreyszig, "Advanced Engineering Mathematics", 8th editio Wiley India (2007).

CONTENTS

FOURIER SERIES AND ITS APPLICATIONS

CHAPTER I

FOURIER SERIES

Periodic functions. Let $f(x)$ be a function of x, defined as follows. For each positive integer x, $f(x)$ is the remainder obtained when x is divided by 3. We can easily make a list of the first few values of $f(x)$: $f(1) = 1$, $f(2) = 2$, $f(3) = 0$; $f(4) = 1$, $f(5) = 2$, $f(6) = 0$; $f(7) = 1$, $f(8) = 2$, $f(9) = 0$; $f(10) = 1$, $f(11) = 2$, $f(12) = 0$.

From an examination of this table, it is found that for any positive integer x, $f(x+3) = f(x)$; further, if y is any positive integer less than 3, $f(x+y) \neq f(x)$. We say that $f(x)$ is a periodic function with the period 3.

Definition. A function $f(x)$ is said to be periodic, if and only if **$f(x+p) = f(x)$ is true for some value of p and every value of x. The Smallest** value of p for which this equation is true for every value of x will be called the *period of the function.*

When n is any integer, $\sin(x + 2n\pi) = \sin x$, for all x; hence $\sin x$ is periodic. Taking $n = 1$, $\sin(x + 2\pi) = \sin x$ for every x. However, if a is any positive number smaller than 2π, it can be shown that $\sin(x + a) = \sin x$ is *not* true for *all* values of x. Therefore, 2π is the smallest number p such that $\sin(x+p) = \sin x$ for every x. Accordingly, 2π is the period of $\sin x$.

Similarly, $\cos x$ is a periodic function with the period 2π. Now $\tan(x + \pi) = \tan x$, for all x. Hence $\tan x$ is periodic; its period is π.

The graph of a periodic function of period p is obtained by periodic repetition of its graph in any interval of length p.

Note that $f(x) = k$, where k is a constant, is a periodic function in the sense of the definition, since $f(x+p) = f(x)$. Again, if $f(x)$ and $F(x)$ have the same period p, then the function $\phi(x) = af(x) + bF(x)$ has the period p.

Exercises 1(a)

1. Find the smallest period p of the following functions:

   (*i*) $\sin 2x$, (*ii*) $\cos 3x$, (*iii*) $\sin nx$, (*iv*) $\cos 2\pi x$, (*v*) $\sin(2\pi x/k)$.

2. Plot graphs of the following functions:

(*i*) $\sin x$, (*ii*) $\frac{\pi}{4} + \cos x$, (*iii*) $\sin x + \frac{1}{3}\sin 3x$, (*iv*) $\sin 2\pi x$,

(*v*) $f(x) = x$, when $-\pi < x < \pi$ and $f(x+2\pi) = f(x)$,

(*vi*) $f(x) = x^2$, when $-\pi < x < \pi$ and $f(x+2\pi) = f(x)$,

(*vii*) $f(x) = \begin{cases} -1 \text{ when } -\pi < x < 0 \\ +1 \text{ when } 0 < x < \pi \end{cases}$ and $f(x+2\pi) = f(x)$,

(*viii*) $f(x) = \begin{cases} 0 \text{ when } -\pi < x < 0 \\ x \text{ when } 0 < x < \pi \end{cases}$ and $f(x+2\pi) = f(x)$.

3. If p is a period of $f(x)$, show that np is also a period, where n is any integer, (positive or negative)

4. Show that the function $f(x) =$ constant, is a periodic function of period p, for any value of p.

5. If $f(x)$ and $F(x)$ have the period p, show that $\phi = af + bF$ where a and b are constants, has the same period p.

Bounds of a function. If for all values of x in a given interval, $f(x)$ is never greater than some fixed number M, the number M is called an *upper bound* for $f(x)$ in that interval; if $f(x)$ is never less than some number m, then m is called a *lower bound*. One or both of these bounds may not exist. Thus, in the interval $(0, 2\pi)$, $y = \sin x$ has an upper bound $+1$ and a lower bound -1. $y = \frac{1}{2} + \cos x$, has an upper bound $\frac{3}{2}$ and a lower bound $-\frac{1}{2}$. The function $y = \frac{1}{x}$ in the interval $(0, 1)$ has a lower bound unity, but has no upper bound. The function $y = x - \frac{1}{x}$ has neither an upper nor a lower bound in the interval $(0, \infty)$.

If a function has an upper bound and a lower bound in a given interval, then that function is said to be *bounded*. We can put the idea differently: for all values of x in a given interval, if $|f(x)| \le a$, where a is some fixed number, then the function is bounded in that interval.

Continuity of a function. It is sometimes important to know the behaviour of a function $f(x)$ when $x \to a$ in such a way that x always remains greater than a, that is when x is allowed to approach a from the right side only. In such a case, the following notation is adopted: we write $x \to a+$, and the limit is called the *right-hand limit*. If the limit is b, we give the symbol

$$\lim_{x \to a+} f(x) = b, \text{ or } f(a+) = b.$$

The left-hand limit is defined similarly, and is denoted by either

$$\lim_{x \to a-} f(x), \text{ or } f(a-).$$

Let $f(x) = \frac{3}{2} + \frac{x-a}{2\,|x-a|}$.

Here $f(a-) = 1$, and $f(a+) = 2$. The symbols $f(a-)$ and $f(a+)$ should not be confused with the symbol $f(a)$, which means the value of $f(x)$ calculated at the point $x = a$. In the present example, $f(a) = \frac{3}{2} + \frac{1}{2} \times \frac{0}{0}$; hence $f(a)$ *does not exist.*

Or again, if $f(x) = 2^{\frac{1}{x-1}}$, $f(1-) = 0$, $f(1+) = \infty$, but $f(1)$ is not defined.

If $f(a-) = f(a+) = b$, then we call b the limit of $f(x)$ as $x \to a$ and write $\lim_{x \to a} f(x) = b$.

If $f(x) = \frac{x^2 - a^2}{x - a}$, $f(a-) = 2a$ and $f(a+) = 2a$; thus, $\lim_{x \to a} f(x) = 2a$; note that $f(a) = \frac{0}{0}$ and hence is not defined.

We have already studied that a function $f(x)$ is said to be continuous at $x = a$, if $\lim_{x \to a} f(x) = f(a)$. With our present knowledge, we can say that $f(x)$ is continuous at $x = a$, if $f(a-) = f(a) = f(a+)$. Thus, $f(x) = \frac{x^2 - a^2}{x - a}$, when $x \neq a$, and $f(x) = 2a$, when $x = a$ is continuous at $x = a$, even though the function $f(x) = \frac{x^2 - a^2}{x - a}$ as such is not continuous at $x = a$.

From the above, it is clear that $f(x)$ is discontinuous at $x = a$, if $f(a-) \neq f(a+)$. The discontinuity at $x = a$ is called *infinite* whenever $f(x)$ becomes infinite as $x \to a$. If the function is discontinuous at point $x = a$ but is bounded in its vicinity, the discontinuity is called *finite* or *ordinary.* If $f(a+)$ and $f(a-)$ exist and are unequal, their difference $f(a+) - f(a-)$ is called the *jump of the function* at $x = a$. Thus the function $\frac{3}{2} + \frac{x-a}{2\,|x-a|}$ has a jump of one unit at $x = a$. On the other hand, the function $2^{\frac{1}{x-1}}$ has an infinite discontinuity at $x = 1$. A function which is continuous everywhere except for a finite number of jumps in a given interval is called *piece-wise continuous or sectionally continuous*, in that interval.

Exercises 1(b)

1. If $f(x) = x^2$, find $f(3-)$, $f(3+)$. What is $\lim_{x \to 3} x^3$?

2. What is $\lim_{x \to \pi/3} \cos x$? $\cos\left(\frac{\pi}{3} +\right)$? $\cos\left(\frac{\pi}{3} -\right)$?

3. Given that $f(x) = \dfrac{1}{x-1}$, find $f(1+)$ and $f(1-)$

4. Find $f(0+)$ and $f(0-)$, if $f(x) = \dfrac{e^{1/x}}{1+e^{1/x}}$; what is $\lim_{x \to 0} f(x)$?

5. If $f(x) = \dfrac{1}{1-e^{-1/x}}$, find $f(0+)$ and $f(0-)$. What is $f(0)$?

Is the function continuous at $x = 0$? What is the jump of the function at $x = 0$? Find $f(0)$.

6. $f(x) = 1,\ 0 < x < \pi$

$f(x) = 2,\ \pi < x < 2\pi$. Find $f(\pi -)$ and $f(\pi +)$. Evaluate the jumps of the function at $x = \pi$.

Fourier Series. Periodic functions occur frequently in engineering problems. Such periodic functions are often complicated; it is therefore desirable to represent these in terms of the simple periodic functions of sine and cosine. A development of a given periodic function into a series of sines and cosines was effected by the French physicist and mathematician Joseph Fourier (1768–1830); the series of sines and cosines is known after him.

Below we prove a few results which we need in deriving Fourier series. Given that m and n are positive integers or zero:

1. If $n \neq 0$,

$$\int_c^{c+2\pi} \sin nx\, dx = \left[-\frac{\cos nx}{n}\right]_c^{c+2\pi} = -\frac{\cos n(c+2\pi)}{n} + \frac{\cos nc}{n}$$

$$= \frac{\cos nc - \cos(nc + 2n\pi)}{n} = \frac{\cos nc - \cos nc}{n} = 0.$$

Note. if $n = 0$, $\displaystyle\int_c^{c+2\pi} \sin nx\, dx = \int_c^{c+2\pi} 0 \cdot dx = 0.$

2. If $n \neq 0$, $\displaystyle\int_c^{c+2\pi} \cos nx\, dx = \left[\frac{\sin nx}{n}\right]_c^{c+2\pi} = 0.$

3. $\displaystyle\int_c^{c+2\pi} \sin mx \cos nx\, dx$

$$= \frac{1}{2}\int_c^{c+2\pi} [\sin(m+n)x + \sin(m-n)x]\, dx = 0, \text{ as in 1.}$$

4. If $m \neq n, \int_c^{c+2\pi} \sin mx \sin nx \, dx$

$$= \frac{1}{2}\int_c^{c+2\pi} [\cos (m-n)x - \cos (m+n)x] \, dx = 0, \text{ as in (2).}$$

5. If $m \neq n, \int_c^{c+2\pi} \cos mx \cos nx \, dx$

$$= \frac{1}{2}\int_c^{c+2\pi} [\cos (m+n)x + \cos (m-n)x] \, dx = 0, \text{ as in (2)}$$

6. If $n \neq 0, \int_c^{c+2\pi} \sin^2 nx \, dx = \frac{1}{2}\int_c^{c+2\pi} (1 - \cos 2nx) \, dx$

$$= \frac{1}{2}\int_c^{c+2\pi} dx - \frac{1}{2}\int_c^{c+2\pi} \cos 2nx \, dx = \frac{1}{2}[x]_c^{c+2\pi} - \frac{1}{2} \times 0$$

$$= \frac{1}{2}\{(c+2\pi) - c\} = \frac{1}{2} \times 2\pi = \pi.$$

7. If $n \neq 0, \int_c^{c+2\pi} \cos^2 nx \, dx = \frac{1}{2}\int_c^{c+2\pi} (1 + \cos 2nx) \, dx = \pi.$

Theorem 1. *If $f(x)$ is a periodic function with period 2π and if $f(x)$ can be represented by a trigonometric series in $(-\infty, \infty)$.*

$$f(x) = \frac{a_o}{2} + \sum_{n=1}^{\infty} (a_n \cos nx + b_n \sin nx), \qquad ...(i)$$

then
$$a_n = \frac{1}{\pi}\int_c^{c+2\pi} f(x) \cos nx \, dx, \text{ and } b_n = \frac{1}{\pi}\int_c^{c+2\pi} f(x) \sin nx \, dx.$$

Proof. First of all, let us determine a_0. Integrating both the sides of (i) from c to $c + 2\pi$, we get

$$\int_c^{c+2\pi} f(x) \, dx = \int_c^{c+2\pi} \left[\frac{a_0}{2} + \sum_{n=1}^{\infty} (a_n \cos nx + b_n \sin nx)\right] dx.$$

We assume that term by term integration of the series is allowed. [This is possible, if the series is uniformly convergent.] Then we have

$$\int_c^{c+2\pi} f(x)\,dx = \frac{a_0}{2}\int_c^{c+2\pi} dx + \sum_{n=1}^{\infty}\left\{a_n \int_c^{c+2\pi} \cos nx\,dx + b_n \int_c^{c+2\pi} \sin nx\,dx\right\}.$$

The first term on the right is $\frac{a_0}{2}[x]_c^{c+2\pi} = \frac{a_0}{2} \times 2\pi = a_0\pi$. Using formulae 2 and 1 on p. 4, the other integrals on the right become zero each. Thus,

$$\int_c^{c+2\pi} f(x)\,dx = a_0\pi, \quad \text{giving} \quad a_0 = \frac{1}{\pi}\int_c^{c+2\pi} f(x)\,dx.$$

We now determine $a_1, a_2, a_3, \ldots$ by a similar method. Multiplying (*i*) by $\cos mx$, where m is any fixed positive integer, and integrating from c to $c + 2\pi$, we get

$$\int_c^{c+2\pi} f(x)\cos mx\,dx = \int_c^{c+2\pi}\left\{\frac{a_0}{2} + \sum_{n=1}^{\infty}(a_n \cos nx + b_n \sin nx)\right\}\cos mx\,dx \quad \text{..(ii)}$$

Using term by term integration, the right hand side becomes

$$\frac{a_0}{2}\int_c^{c+2\pi}\cos mx\,dx + \sum_{n=1}^{\infty} a_n \int_c^{c+2\pi}\cos nx \cos mx\,dx + \sum_{n=1}^{\infty} b_n \int_c^{c+2\pi}\sin nx \cos mx\,dx.$$

By formula 2 on p. 4, the first of these integrals is zero. Again by formula 3 on p. 4, all the integrals of the last Σ are zero each. The remaining integrals are

$$a_1\int_c^{c+2\pi}\cos x \cos mx\,dx + a_2\int_c^{c+2\pi}\cos 2x\cos mx\,dx + \cdots + a_m\int_c^{c+2\pi}\cos^2 mx\,dx$$

$$+ \cdots + a_n\int_c^{c+2\pi}\cos nx\cos mx\,dx + \cdots$$

Of these, $a_m \int_c^{c+2\pi}\cos^2 mx\,dx = a_m \cdot \pi$, using formula 7, p. 5.

Lastly by formula 5, p. 5 each of the remaining integrals is zero. Substituting the above results in (*ii*) above, we have

$$\int_c^{c+2\pi} f(x)\cos mx\,dx = a_m\cdot\pi, \text{ giving } a_m = \frac{1}{\pi}\int_c^{c+2\pi} f(x)\cos mx\,dx.$$

Writing n in the place of m, we get $a_n = \frac{1}{\pi}\int_c^{c+2\pi} f(x)\cos nx\,dx,\ n = 1, 2, 3, \ldots$.

Finally multiplying (*i*) by $\sin mx$, where m is any fixed positive integer, and integrating from c to $c + 2\pi$, we have

$$\int_c^{c+2\pi} f(x) \sin mx \, dx = \int_c^{c+2\pi} \left\{ \frac{a_0}{2} + \sum_{n=1}^{\infty} (a_n \cos nx + b_n \sin nx) \right\} \sin mx \, dx \quad \ldots(iii)$$

Integrating the R.H.S. term by term, we find that every one of the integrals except $\int_c^{c+2\pi} b_m \sin^2 mx \, dx$ is zero; and, by formula 6 on p. 5, this integral is $b_m \pi$.

Hence from (*iii*) above

$$\int_c^{c+2\pi} f(x) \sin mx \, dx = b_m \pi, \text{ giving } b_m = \frac{1}{\pi} \int_c^{c+2\pi} f(x) \sin mx \, dx$$

Writing n in the place of m, $b_n = \frac{1}{\pi} \int_c^{c+2\pi} f(x) \sin nx \, dx, \; n = 1, 2, 3, \ldots$

Thus we have found the values of the coefficients a_0, a_n and b_n. These relations are known as Euler formulas. Thus, the **Euler formulas** are :

$$a_0 = \frac{1}{\pi} \int_c^{c+2\pi} f(x) \, dx; \quad a_n = \frac{1}{\pi} \int_c^{c+2\pi} f(x) \cos nx \, dx, \text{ and}$$

$$b_n = \frac{1}{\pi} \int_c^{c+2\pi} f(x) \sin nx \, dx, \; n = 1, 2, \ldots.$$

They are sometimes called *Euler-Fourier* formulas. The R.H.S. of (i), when the coefficients there are given by the Euler formulas, is known as the **Fourier series** of $f(x)$. In most applications, the interval over which $f(x)$ is to be expanded is either $(-\pi, \pi)$ or $(0, 2\pi)$, so that the value of c in the Euler formulas is either $-\pi$ or 0. Actually, the formula for a_0 need not be listed, for it can be obtained from the general expression for a_n by putting $n = 0$; it was to achieve this that we wrote the constant term in the original expansion as $\frac{a_0}{2}$, (and not a_0 as usual).

Note that we have not proved that any given function $f(x)$ has a Fourier expansion that converges to $f(x)$. What we have shown is that *if* a function $f(x)$ has an expansion of the form (i) for which term by term integration is valid, *then* the coefficients in that series will be given by the Euler formulas. Questions concerning the convergence of the Fourier series, and if they converge, the conditions under which they will represent the functions which generated them are many and difficult. But, all the functions that you will have to deal with will be covered by the famous theorem of Dirichlet, (the great German mathematician 1805–1859).

Theorem 2. *If $f(x)$ is a bounded function of period 2π which in any one period has at most a finite number of maxima and minima and a finite number of points of discontinuity, then the Fourier series of $f(x)$ converges to $f(x)$ at all points where $f(x)$ is continuous; also, the series converges to the average value of the right and left-hand limits of $f(x)$ at each point where $f(x)$ is discontinuous.* The conditions given in this theorem are known as the **Dirichlet conditions**. (The proof of this theorem is not expected of you; you have merely to memorise the statement). These conditions are *not necessary*.

Dirichlet's conditions

If a function $f(x)$ is defined in $c \le x \le c + 2\pi$, it can be expanded as a Fourier series of the form $\frac{a_0}{2} + \sum_{n=1}^{\infty} (a_n \cos nx + b_n \sin nx)$, provided the following DIRICHLET'S conditions are satisfied.

(*i*) $f(x)$ is single valued and finite in $(c,\ c + 2\pi)$

(*ii*) $f(x)$ is continuous or piece-wise continuous with finite number of finite discontinuities in $(c,\ c + 2\pi)$

(*iii*) $f(x)$ has a finite number of maxima or minima in $(c,\ c + 2\pi)$

Note 1. These conditions are *not necessary* but *only sufficient* for the existence of Fourier series.

Note 2. If $f(x)$ satisfies Dirichlet's conditions and $f(x)$ is defined in $(-\infty,\ \infty)$, then $f(x)$ has to be periodic of periodicity 2π for the existence of Fourier series of period 2π.

Note 3. If $f(x)$ satisfies Dirichlet's conditions and $f(x)$ is defined in $(c,\ c + 2\pi)$, then $f(x)$ *need not be periodic* for the existence of Fourier series of period 2π

Note 4. If $x = a$ is a point of discontinuity of $f(x)$, then the value of the Fourier series at $x = a$ is $\frac{1}{2}\ [f(a+) + f(a-)]$

The Dirichlet conditions make it clear that a function need not be continuous in order to possess a valid Fourier expansion. This means that the graph of a function may consist of a number of dis-jointed arcs of different curves, each defined by a different equation, and still be representable by a Fourier series. In using the Euler formulas to find the coefficients in the expansion of such a function, it will be necessary to break up the range of integration $(c, c + 2\pi)$ to correspond to the various segments of the function. Thus, in the above

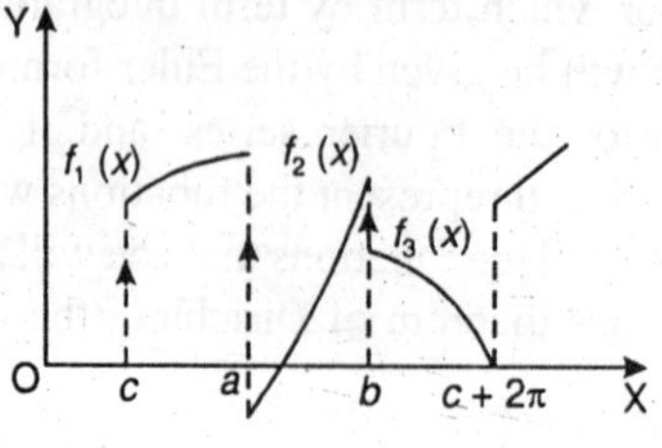

function, $f(x)$ is defined by three expressions $f_1(x), f_2(x)$ and $f_3(x)$ on successive portions of the period interval $c \le x \le c + 2\pi$. Hence it is necessary to write the Euler formulas as

$$a_n = \frac{1}{\pi} \int_c^{c+2\pi} f(x) \cos nx \, dx$$

$$= \frac{1}{\pi} \left\{ \int_c^a f(x) \cos nx \, dx + \int_a^b f(x) \cos nx \, dx + \int_b^{c+2\pi} f(x) \cos nx \, dx \right\}.$$

$$= \frac{1}{\pi} \left\{ \int_c^a f_1(x) \cos nx \, dx + \int_a^b f_2(x) \cos nx \, dx + \int_b^{c+2\pi} f_3(x) \cos nx \, dx \right\}.$$

A similar break up is necessary for evaluating b_n.

According to the above theorem, the Fourier series of the function shown in the figure will converge to the average values, indicated by dots, at the discontinuities at c, a and b, regardless of the definition (or lack of definition) of the function at these points.

The following two standard integrals which you have studied in lower classes occur in almost all problems in this chapter. It is better to remember these results to apply them in problems.

1. $\int e^{ax} \sin bx \, dx = \frac{e^{ax}}{a^2 + b^2} [a \sin bx - b \cos bx].$

2. $\int e^{ax} \cos bx \, dx = \frac{e^{ax}}{a^2 + b^2} [a \cos bx + b \sin bx].$

Bernoulli's generalised formula of integration by parts

This formula is only an extension of the formula of integration by parts.

We know that

$$\int u \, dv = uv - \int v \, du.$$

But in many problems, the integral will be of the form $\int uv \, dx$ and not $\int u \, dv$. Now, we will find a formula to evaluate $\int uv dx$.

In what follows, we write

$$\frac{du}{dx} = u', \quad \frac{d^2u}{dx^2} = u'', \text{ etc.}$$

and $\quad \int v \, dx = v_1, \int v_1 dx = v_2$ etc.

Hence,

$$\int uv \, dx = \int u d(v_1)$$

$$= uv_1 - \int v_1\, du, \text{ using integration by parts}$$
$$= uv_1 - \int v_1 u' dx$$
$$= uv_1 - \int u'\, d(v_2)$$
$$= uv_1 - u'v_2 + \int v_2 u''\, dx$$
$$= uv_1 - u'v_2 + \int u''\, d(v_3)$$
$$= uv_1 - u'v_2 + u''v_3 - \int v_3 u'''\, dx$$

....

....

$$= uv_1 - u'\,v_2 + u''v_3 - u'''v_4 + \cdots\cdots$$

Thus,

$\int uv\, dx = uv_1 - u'\,v_2 + u''v_3 - \cdots$ is called Bernoulli's extented formula of integration by parts.

Note 1. Whenever there is a polynomial as a factor in the integrand, take that polynomial as u and the remaining factor as v, since after a certain stage, the derivatives of u will vanish.

To illustrate the *advantage* of this formula, consider

$$\int (x^3 + 2x^2 + 5) \cos 2x\, dx$$

Here take $x^3 + 2x^2 + 5 = u$ and $\cos 2x = v$.

$$\int (x^3 + 2x^2 + 5) \cos 2x\, dx = \int uv\, dx$$

$$= (x^3 + 2x^2 + 5)\left(\frac{\sin 2x}{2}\right) - (3x^2 + 4x)\left(-\frac{\cos 2x}{4}\right)$$
$$+ (6x + 4)\left(-\frac{\sin 2x}{8}\right) - (6)\left(\frac{\cos 2x}{16}\right) + C$$

Note 2. In the above problem, we take the given integral as $\int uv\, dx$ and not as $\int u\, dv$.

Example 1. *Obtain the Fourier series of periodicity* 2π *for* $f(x) = e^{-x}$ *in the interval* $0 < x < 2\pi$. *Hence deduce the value of* $\sum_{n=2}^{\infty} \frac{(-1)^n}{1+n^2}$. *Further, derive a series for* $\operatorname{cosech} \pi$.

$$\text{Let } e^{-x} = \frac{a_0}{2} + \sum_{n=1}^{\infty} (a_n \cos nx + b_n \sin nx) \qquad ...(i)$$

Taking $c = 0$ in the Euler formulas, we have,

$$a_0 = \frac{1}{\pi}\int_0^{2\pi} f(x)\,dx = \frac{1}{\pi}\int_0^{2\pi} e^{-x}\,dx = \frac{1}{\pi}\left[-e^{-x}\right]_0^{2\pi}$$

$$= \frac{1}{\pi}\left[-e^{-2\pi}+1\right] = \frac{1-e^{-2\pi}}{\pi}$$

For $n = 1, 2, 3,......$

$$a_n = \frac{1}{\pi}\int_0^{2\pi} f(x)\cos nx\,dx$$

$$= \frac{1}{\pi}\int_0^{2\pi} e^{-x}\cos nx\,dx$$

$$= \frac{1}{\pi}\left[\frac{e^{-x}}{(-1)^2+n^2}\{(-1)\cos nx + n\sin nx\}\right]_0^{2\pi},$$

using the result of the standard integral.

$$= \frac{1}{\pi(1+n^2)}\left[e^{-2\pi}(-1)-(-1)\right]$$

$$= \frac{1}{\pi(1+n^2)}(1-e^{-2\pi}).$$

Similarly,

$$b_n = \frac{1}{\pi}\int_0^{2\pi} e^{-x}\sin nx\,dx$$

$$= \frac{1}{\pi}\left[\frac{e^{-x}}{1+n^2}\{(-1)\sin nx - n\cos nx\}\right]_0^{2\pi}$$

$$= \frac{1}{\pi(1+n^2)}\left[-ne^{-2\pi}+n\right]$$

$$= \frac{n}{\pi(1+n^2)}(1-e^{-2\pi})$$

Substituting the values of a_0, a_n, b_n in (i), and remembering that the constant term is $\frac{a_0}{2}$, (and not a_0), we get that in the range $(0, 2\pi)$,

$$e^{-x} = \frac{1-e^{-2\pi}}{\pi}\left[\frac{1}{2} + \sum_{n=1}^{\infty} \frac{1}{1+n^2}(\cos nx + n \sin nx)\right] \quad ...(ii)$$

In $(0, 2\pi)$, e^{-x} is continuous. Therefore, at $x = \pi$, the value of the Fourier series equals the value of the function e^{-x}.

Hence replacing x by π in (*ii*),

$$e^{-\pi} = \frac{1-e^{-2\pi}}{\pi}\left[\frac{1}{2} + \sum_{n=1}^{\infty} \frac{\cos n\pi}{1+n^2}\right]$$

$$\therefore \quad \frac{1}{2} + \sum_{n=1}^{\infty} \frac{\cos n\pi}{1+n^2} = \frac{\pi e^{-\pi}}{1-e^{-2\pi}} = \frac{\pi}{e^{\pi}-e^{-\pi}} = \frac{\pi}{2\sinh\pi}$$

When $n = 1$, $\frac{\cos n\pi}{1+n^2} = -\frac{1}{2}$; further $\cos n\pi = (-1)^n$

Hence $\sum_{n=2}^{\infty} \frac{(-1)^n}{1+n^2} = \frac{\pi \operatorname{cosech} \pi}{2}$

$$\therefore \quad \operatorname{cosech} \pi = \frac{2}{\pi} \sum_{\pi=2}^{\infty} \frac{(-1)^n}{1+n^2}.$$

Example 2. *If* $f(x) = \begin{cases} -k, \text{ when } -\pi < x < 0 \\ k, \text{ when } 0 < x < \pi \end{cases}$

and $f(x+2\pi) = f(x)$ *for all x, derive the Fourier series for* $f(x)$. *Deduce that*

$$\frac{\pi}{4} = 1 - \frac{1}{3} + \frac{1}{5} - \frac{1}{7} + \cdots \text{ to infinity.}$$

Let $f(x) = \frac{a_0}{2} + \sum_{n=1}^{\infty} (a_n \cos nx + b_n \sin nx)$

Taking $c = -\pi$ in the Euler formulas we have

$$a_0 = \frac{1}{\pi}\int_{-\pi}^{\pi} f(x)\,dx = \frac{1}{\pi}\left\{\int_{-\pi}^{0} f(x))\,dx + \int_{0}^{\pi} f(x)\,dx\right\}$$

Now using the hypothesis for the value of $f(x)$, we get

$$a_0 = \frac{1}{\pi}\left\{\int_{-\pi}^{0}(-k)\,dx + \int_{0}^{\pi} k dx\right\} = \frac{1}{\pi}\left\{\left(-kx\right)_{-\pi}^{0} + \left(kx\right)_{0}^{\pi}\right\}$$

$$= \frac{1}{\pi}\{(0 - k\pi) + (k\pi - 0)\}$$

Thus $a_0 = 0$. Again for $n = 1, 2, 3, \ldots$

$$a_n = \frac{1}{\pi}\int_{-\pi}^{\pi} f(x)\cos nx\,dx$$

$$= \frac{1}{\pi}\left\{\int_{-\pi}^{0} f(x)\cos nx\,dx + \int_{0}^{\pi} f(x)\cos nx\,dx\right\}.$$

Substituting the values supplied for $f(x)$, we have

$$a_n = \frac{1}{\pi}\left\{\int_{-\pi}^{0}(-k)\cos nx\,dx + \int_{0}^{\pi} k\cos nx\,dx\right\}$$

$$= \frac{1}{\pi}\left\{\left(-k\frac{\sin nx}{n}\right)_{-\pi}^{0} + \left(k\frac{\sin nx}{n}\right)_{0}^{\pi}\right\}$$

Since $\sin 0$, $\sin(-n\pi)$ and $\sin n\pi$ are all zero, we get $a_n = 0$.

$$b_n = \frac{1}{\pi}\int_{-\pi}^{\pi} f(x)\sin nx\,dx$$

$$= \frac{1}{\pi}\left\{\int_{-\pi}^{0} f(x)\sin nx\,dx + \int_{0}^{\pi} f(x)\sin nx\,dx\right\}$$

$$= \frac{1}{\pi}\left\{\int_{-\pi}^{0}(-k)\sin nx\,dx + \int_{0}^{\pi} k\sin nx\,dx\right\}$$

$$= \frac{1}{\pi}\left\{\left[k\frac{\cos nx}{n}\right]_{-\pi}^{0} + \left[-k\frac{\cos nx}{n}\right]_{0}^{\pi}\right\}$$

$$= \frac{1}{\pi}\left[\left\{\frac{k}{n}\cos 0 - \frac{k}{n}\cos(-n\pi)\right\} + \left\{-\frac{k}{n}\cos n\pi + \frac{k}{n}\cos 0\right\}\right]$$

But $\cos(-\alpha) = \cos\alpha$, giving $\cos(-n\pi) = \cos n\pi$; further, $\cos 0 = 1$.

Hence $b_n = \frac{k}{n\pi}[\{1 - \cos n\pi\} + \{-\cos n\pi + 1\}] = \frac{k}{n\pi}(2 - 2\cos n\pi)$

$$\therefore\ b_n = \frac{2k}{n\pi}(1 - \cos n\pi). \text{ Now } \cos n\pi = \begin{cases} -1, \text{ for odd } n \\ +1, \text{ for even } n \\ (-1)^n, \text{ for any integer } n. \end{cases}$$

Hence $b_1 = \frac{4k}{\pi}$; $b_2 = 0$; $b_3 = \frac{4k}{3\pi}$; $b_4 = 0$; $b_5 = \frac{4k}{5\pi}$; $b_6 = 0$;

$$b_7 = \frac{4k}{7\pi}\ldots..$$

Using the values of a_n and b_n in (*i*) we obtain

$$f(x) = \frac{4k}{\pi}\{\sin x + \tfrac{1}{3}\sin 3x + \tfrac{1}{5}\sin 5x + \tfrac{1}{7}\sin 7x + \cdots \text{ to } \infty\}$$

In the above equation putting $x = \pi/2$, we get

$$f\left(\frac{\pi}{2}\right) = \frac{4k}{\pi}\{1 - \tfrac{1}{3} + \tfrac{1}{5} - \tfrac{1}{7} + \cdots \text{ to } \infty\}$$

But, by hypothesis, $f\left(\frac{\pi}{2}\right) = k.$

$$\text{Hence } k = \frac{4k}{\pi}\{1 - \tfrac{1}{3} + \tfrac{1}{5} - \tfrac{1}{7} + \cdots \text{ to } \infty\}$$

Multiplying both the sides by $\frac{\pi}{4k}$, we have

$$\frac{\pi}{4} = 1 - \tfrac{1}{3} + \tfrac{1}{5} - \tfrac{1}{7} + \cdots \text{ to } \infty.$$

Note. Functions of the type given in this example occur as external forces acting on mechanical systems, electromotive forces in electric circuits etc.

Example 3. *Obtain the Fourier series of the periodic function defined by* $f(x) = -\pi$, if $-\pi < x < 0$ *and* $f(x) = x$, if $0 < x < \pi$. *Deduce that*

$$\frac{1}{1^2} + \frac{1}{3^2} + \frac{1}{5^2} + \cdots \text{ to } \infty = \frac{\pi^2}{8}.$$ **(*MS. 1986A*)**

$$\text{Let } f(x) = \frac{a_0}{2} + \sum_{n=1}^{\infty}(a_n \cos nx + b_n \sin nx) \qquad \ldots(i)$$

$$\text{Then } a_0 = \frac{1}{\pi}\int_{-\pi}^{\pi} f(x)\,dx = \frac{1}{\pi}\left[\int_{-\pi}^{0} f(x)\,dx + \int_{0}^{\pi} f(x)\,dx\right]$$

$$a_0 = \frac{1}{\pi}\left[\int_{-\pi}^{0}(-\pi)\,dx + \int_{0}^{\pi} x\,dx\right]$$

$$= \frac{1}{\pi}\left[-\pi^2 + \frac{\pi^2}{2}\right] = -\pi/2$$

$$a_n = \frac{1}{\pi}\int_{-\pi}^{\pi} f(x)\cos nx\, dx$$

$$= \frac{1}{\pi}\left[\int_{-\pi}^{0}(-\pi)\cos nx\, dx + \int_{0}^{\pi} x\cos nx\, dx\right]$$

$$= \frac{1}{\pi}\left[\left\{(-\pi)\left(\frac{\sin nx}{n}\right)\right\}_{-\pi}^{0} + \left\{(x)\left(\frac{\sin nx}{n}\right) - (1)\left(-\frac{\cos nx}{n^2}\right)\right\}_{0}^{\pi}\right]$$

$$= \frac{1}{\pi}\left[\frac{1}{n^2}(\cos n\pi - 1)\right]$$

$$= \frac{1}{\pi n^2}[(-1)^n - 1]$$

$$b_n = \frac{1}{\pi}\int_{-\pi}^{\pi} f(x)\sin nx\, dx$$

$$= \frac{1}{\pi}\left[\int_{-\pi}^{0}(-\pi)\sin nx\, dx + \int_{0}^{\pi} x\sin nx\, dx\right]$$

$$= \frac{1}{\pi}\left[(-\pi)\left(-\frac{\cos nx}{n}\right)_{-\pi}^{0} + \left\{(x)\left(-\frac{\cos nx}{n}\right) - (1)\left(-\frac{\sin nx}{n^2}\right)\right\}_{0}^{\pi}\right]$$

$$= \frac{1}{\pi}\left[\frac{\pi}{n}(1 - \cos n\pi) - \frac{\pi}{n}\cos n\pi\right]$$

$$= \frac{1}{n}(1 - 2\cos n\pi)$$

Substituting the values of the coefficients in (*i*),

$$f(x) = -\frac{\pi}{4} - \frac{2}{\pi}\left(\cos x + \frac{\cos 3x}{3^2} + \frac{\cos 5x}{5^2} + \cdots\right)$$

$$+ 3\sin x - \frac{\sin 2x}{2} + \frac{3\sin 3x}{3} - \frac{\sin 4x}{4} + \frac{3\sin 5x}{5} - \quad \ldots(ii)$$

Zero is a point of discontinuity of $f(x)$. But $f(0-) = -\pi$, and $f(0+) = 0$.

Hence $\dfrac{f(0-) + f(0+)}{2} = -\dfrac{\pi}{2}$.

By theorem 2 on p. 8, this is the sum of the Fourier series in (*ii*) for the value 0 of x. Hence

$$-\frac{\pi}{2} = -\frac{\pi}{4} - \frac{2}{\pi}\left(1 + \frac{1}{3^2} + \frac{1}{5^2} + \cdots\cdots\right)$$

$$\therefore \quad -\frac{\pi}{4} = -\frac{2}{\pi}\left(1 + \frac{1}{3^2} + \frac{1}{5^2} + \dots\right)$$

Hence $\dfrac{\pi^2}{8} = 1 + \dfrac{1}{3^2} + \dfrac{1}{5^2} + \dots$

Example 4. *What is the Fourier expansion of the periodic function, whose definition in one period is* $f(x) = \begin{cases} 0, & \text{when } -\pi < x < 0 \\ \sin x, & \text{when } 0 < x < \pi \end{cases}$?

Evaluate $\dfrac{1}{1\cdot 3} - \dfrac{1}{3\cdot 5} + \dfrac{1}{5\cdot 7} - \dfrac{1}{7\cdot 9} + \dots$ to ∞.

$$\text{Let } f(x) = \frac{a_0}{2} + \sum_{n=1}^{\infty} (a_n \cos nx + b_n \sin nx) \qquad \dots(i)$$

[In the great majority of the examples, we can calculate a_n and b_n for $n = 1$ to ∞. But, this is a special example, where a_1 and b_1 are to be calculated separately; but a_0 can be got from a_n.]

$$a_1 = \frac{1}{\pi}\left\{\int_{-\pi}^{0} 0 \,.\, \cos x\, dx + \int_{0}^{\pi} \sin x \cos x\, dx\right\}$$

$$= \frac{1}{2\pi}\Big[\sin^2 x\Big]_0^{\pi} = \frac{1}{\pi} \times 0 = 0.$$

For $\quad n = 0, 2, 3, 4, \dots$

$$a_n = \frac{1}{\pi}\left\{\int_{-\pi}^{0} 0 \cdot \cos nx\, dx + \int_{0}^{\pi} \sin x \cos nx\, dx\right\}$$

$$= \frac{1}{2\pi}\int_0^{\pi} \{\sin (1+n)\, x + \sin (1-n)\, x\}\, dx$$

$$= \frac{1}{2\pi}\left[-\frac{\cos (1+n)\, x}{1+n} - \frac{\cos (1-n) x}{1-n}\right]_0^{\pi}, \qquad \text{for } n \neq 1$$

$$= \frac{1}{2\pi}\left[\left\{-\frac{\cos (\pi + n\pi)}{1+n} - \frac{\cos (\pi - n\pi)}{1-n}\right\} + \left\{\frac{1}{1+n} + \frac{1}{1-n}\right\}\right]$$

$$= \frac{1}{2\pi}\left[\frac{\cos n\pi}{1+n} + \frac{\cos n\pi}{1-n} + \frac{1}{1+n} + \frac{1}{1-n}\right]$$

$$= \frac{1}{2\pi}\left[\frac{1+\cos n\pi}{1+n} + \frac{1+\cos n\pi}{1-n}\right] = \frac{1+\cos n\pi}{\pi(1-n^2)}$$

i.e., $$a_n = -\frac{1+\cos n\pi}{\pi(n-1)(n+1)}$$

$$b_1 = \frac{1}{\pi}\left\{\int_{-\pi}^{0} 0 \cdot \sin x\, dx + \int_0^{\pi} \sin x \sin x\, dx\right\}$$

$$= \frac{1}{\pi}\int_0^{\pi} \sin^2 x\, dx = \frac{1}{\pi}\int_0^{\pi}\left\{\tfrac{1}{2} - \tfrac{1}{2}\cos 2x\right\} dx = \frac{1}{\pi}\times\frac{\pi}{2} = \frac{1}{2}$$

For $n = 2, 3, 4, 5....$

$$b_n = \frac{1}{\pi}\left\{\int_{-\pi}^{0} 0 . \sin nx\, dx + \int_0^{\pi} \sin x \sin nx\, dx\right\}$$

$$= \frac{1}{2\pi}\int_0^{\pi} \{\cos(n-1)x - \cos(n+1)x\}\, dx$$

$$= \frac{1}{2\pi}\left[\frac{\sin(n-1)x}{n-1} - \frac{\sin(n+1)x}{n+1}\right]_0^{\pi} = \frac{1}{2\pi}\times 0 = 0.$$

Substituting the values of the coefficients in (*i*),

$$f(x) = \frac{1}{\pi} + \frac{\sin x}{2} - \frac{2}{\pi}\left\{\frac{\cos 2x}{3} + \frac{\cos 4x}{15} + \frac{\cos 6x}{35} + \frac{\cos 8x}{63} + \cdots \text{ to } \infty\right\}$$

Putting $x = \frac{\pi}{2}$,

$$f\left(\frac{\pi}{2}\right) = \frac{1}{\pi} + \frac{1}{2} - \frac{2}{\pi}\left\{-\frac{1}{3} + \frac{1}{15} - \frac{1}{35} + \frac{1}{63} - \cdots \text{ to } \infty\right\}$$

But, by hypothesis $f\left(\frac{\pi}{2}\right) = \sin\frac{\pi}{2} = 1$. Hence

$$1 = \frac{1}{\pi} + \frac{1}{2} - \frac{2}{\pi}\left\{-\frac{1}{1\cdot 3} + \frac{1}{3\cdot 5} - \frac{1}{5\cdot 7} + \frac{1}{7\cdot 9} - \cdots \text{ to } \infty\right\}$$

$$\therefore\ 1 - \frac{1}{\pi} - \frac{1}{2} = \frac{2}{\pi}\left\{\frac{1}{1\cdot 3} - \frac{1}{3\cdot 5} + \frac{1}{5\cdot 7} - \frac{1}{7\cdot 9} + \cdots \text{ to } \infty\right\}$$

i.e., $$\frac{1}{2} - \frac{1}{\pi} = \frac{2}{\pi}\left\{\frac{1}{1\cdot 3} - \frac{1}{3\cdot 5} + \frac{1}{5\cdot 7} - \frac{1}{7\cdot 9} + \cdots \text{ to } \infty\right\}$$

$$\therefore\ \frac{\pi-2}{4} = \frac{1}{1\cdot 3} - \frac{1}{3\cdot 5} + \frac{1}{5\cdot 7} - \frac{1}{7\cdot 9} + \cdots \text{ to } \infty.$$

Note. When k is a positive integer, in the Fourier expansion of any function containing either $\sin kx$ or $\cos kx$, the coefficients a_k and b_k are to be calculated separately. This is necessary in ex. 5 also.

Example 5. *Expand in Fourier series of periodicity* 2π *of* $f(x) = x \sin x$, *for* $0 < x < 2\pi$. **(MS. 1991 A, N)**

Let $f(x) = \frac{a_0}{2} + \sum_{n=1}^{\infty} (a_n \cos nx + b_n \sin nx)$ ...(i)

Then $a_0 = \frac{1}{\pi}\int_0^{2\pi} f(x)\, dx = \frac{1}{\pi}\int_0^{2\pi} x \sin x\, dx$ ...(ii)

Let us evaluate $\int_0^{2\pi} x \sin rx\, dx$, when r is an integer;

$$\int_0^{2\pi} x \sin rx\, dx = \left[(x)\left(-\frac{\cos rx}{r}\right) - (1)\left(-\frac{\sin rx}{r^2}\right)\right]_0^{2\pi}$$

$$= -\frac{2\pi}{r}\cos 2\pi r = -\frac{2\pi}{r} \quad \text{...(iii)}$$

Using this result in (ii), we get $a_0 = \frac{1}{\pi}\left\{-\frac{2\pi}{1}\right\} = -2$

$$a_1 = \frac{1}{\pi}\int_0^{2\pi} f(x) \cos x\, dx = \frac{1}{\pi}\int_0^{2\pi} x \sin x \cos x\, dx = \frac{1}{2\pi}\int_0^{2\pi} x \sin 2x\, dx$$

Hence by (iii), $a_1 = \frac{1}{2\pi}\left\{-\frac{2\pi}{2}\right\} = -\frac{1}{2}$.

For $n = 2, 3, 4, \ldots$

$$a_n = \frac{1}{\pi}\int_0^{2\pi} x \sin x \cos nx\, dx = \frac{1}{2\pi}\int_0^{2\pi} x(2 \sin x \cos nx)\, dx$$

$$\therefore\ a_n = \frac{1}{2\pi}\int_0^{2\pi} x\{\sin (n+1)x - \sin (n-1)x\}\, dx.$$

Now using (iii), $a_n = \frac{1}{2\pi}\left\{-\frac{2\pi}{n+1} + \frac{2\pi}{n-1}\right\} = -\frac{1}{n+1} + \frac{1}{n-1}$

$$= \frac{-(n-1)+(n+1)}{(n+1)(n-1)} = \frac{2}{n^2-1}$$

$$b_1 = \frac{1}{\pi}\int_0^{2\pi} f(x) \sin x\, dx = \frac{1}{\pi}\int_0^{2\pi} x \sin^2 x\, dx = \frac{1}{2\pi}\int_0^{2\pi} x(2\sin^2 x)\, dx$$

But $\cos 2x = 1 - 2\sin^2 x$, giving $2\sin^2 x = 1 - \cos 2x$.

Substituting this, $b_1 = \frac{1}{2\pi} \int_0^{2\pi} x(1 - \cos 2x)dx$ ...(*iv*)

When r is an integer,

$$\int_0^{2\pi} x \cos rx \, dx = \left[x \frac{\sin rx}{r} \right]_0^{2\pi} - \int_0^{2\pi} \frac{\sin rx}{r} dx = 0 + \left[\frac{\cos rx}{r^2} \right]_0^{2\pi} = 0 \quad ...(v)$$

Using this in (*iv*),

$$b_1 = \frac{1}{2\pi} \left\{ \left[\frac{x^2}{2} \right]_0^{2\pi} - 0 \right\} = \frac{1}{2\pi} \times \frac{(2\pi)^2}{2} = \frac{2\pi}{2} = \pi.$$

Lastly for $n = 2, 3, 4, ...$

$$b_n = \frac{1}{\pi} \int_0^{2\pi} x \sin x \sin nx \, dx = \frac{1}{2\pi} \int_0^{2\pi} x(2 \sin x \sin nx) \, dx$$

$$= \frac{1}{2\pi} \int_0^{2\pi} x\{\cos (n-1)\,x - \cos (n+1)x\} \, dx = \frac{1}{2\pi} \{0 - 0\}, \text{ using } (v).$$

Thus, for $n = 2, 3, 4, \ldots, b_n = 0$.

Substituting the values of the coefficients in (*i*), we get

$$f(x) = -1 - \tfrac{1}{2} \cos x + 2 \left\{ \sum_{n=2}^{\infty} \frac{\cos nx}{n^2 - 1} \right\} + \pi \sin x.$$

Note. Putting $x = \frac{\pi}{2}$, we derive that

$$\frac{1}{1.3} - \frac{1}{3.5} + \frac{1}{5.7} - \cdots \text{ to } \infty = \frac{\pi - 2}{4}.$$

Example 6. *Derive the Fourier series of $f(x) = k$, where k is a constant, the periodicity being 2π.*

$$a_0 = \frac{1}{\pi} \int_c^{c+2\pi} f(x)dx = \frac{1}{\pi} \int_c^{c+2\pi} k \, dx = \frac{k}{\pi} \Big[x \Big]_c^{c+2\pi} = \frac{k}{\pi} \times 2\pi = 2k.$$

$$a_n = \frac{1}{\pi} \int_c^{c+2\pi} f(x) \cos nx \, dx = \frac{1}{\pi} \int_c^{c+2\pi} k \cos nx \, dx = \frac{k}{\pi} \left[\frac{\sin nx}{n} \right]_c^{c+2\pi}$$

$$= \frac{k}{\pi} \times 0 = 0.$$

$$b_n = \frac{1}{\pi}\int_c^{c+2\pi} f(x)\sin nx\, dx = \frac{1}{\pi}\int_c^{c+2\pi} k\sin nx\, dx = \frac{k}{\pi}\left[-\frac{\cos nx}{n}\right]_c^{c+2\pi}$$

$$= \frac{k}{\pi}\times 0 = 0$$

Thus the first term (constant term) in the Fourier expansion is k; *all* the other coefficients are zero each. So, no Fourier expansion, (other than what is supplied), is available for k.

Apparently, the labour involved in this example is a waste.

Note. k being any constant, the result of this example is valid for any constant number like π.

Theorem 3. *The Fourier coefficients of a sum $f_1 + f_2$ are the sums of the corresponding Fourier coefficients of f_1 and f_2.*

Theorem 4. *If c is a constant, the Fourier coefficients of cf are the corresponding Fourier coefficients of f multiplied by c.*

Example 7. *Derive the Fourier expansion for*

$$F(x) = \begin{cases} 0, \text{ when } -\pi < x < 0 \\ 2k, \text{ when } 0 < x < \pi \end{cases} \text{ and } f(x+2\pi) = f(x), \text{ for all } x.$$

$$\text{Let } f_1(x) = \begin{cases} -k, \text{ when } -\pi < x < 0 \\ +k, \text{ when } 0 < x < \pi \end{cases}$$

and $f_2(x) = k$, for all x.

Then we have developed the Fourier expansions of f_1 and f_2 in worked examples 2 and 6 respectively. But the given $F(x) = f_1(x) + f_2(x)$.
Hence by theorem 3,

$$f(x) = k + \frac{4k}{\pi}\{\sin x + \tfrac{1}{3}\sin 3x + \tfrac{1}{5}\sin 5x + \tfrac{1}{7}\sin 7x + \cdots \text{ to } \infty\}.$$

Note. One can directly work out this problem as in example 2 without splitting into f_1 and f_2.

Example 8. *Derive the Fourier series for the periodic function, whose definition in one period is*

$$F(x) = \begin{cases} -\tfrac{1}{4}\pi, \text{ when } -\pi < x < 0 \\ \tfrac{1}{4}\pi + \sin x, \text{ when } 0 < x < \pi \end{cases}$$

$$\text{Let } f_1(x) = \begin{cases} -\tfrac{1}{4}\pi, \text{ when } -\pi < x < 0 \\ \tfrac{1}{4}\pi, \text{ when } 0 < x < \pi. \end{cases}$$

Then working exactly as in ex. 2,

$$f_1(x) = \sin x + \tfrac{1}{3}\sin 3x + \tfrac{1}{5}\sin 5x + \cdots \text{ to } \infty.$$

Now $F(x) = f(x) + f_1(x)$, where $f(x)$ is the one supplied in ex. 4.

Therefore using Th. 3,

$$F(x) = \frac{1}{\pi} - \frac{2}{\pi}\left\{\frac{\cos 2x}{3} + \frac{\cos 4x}{15} + \frac{\cos 6x}{35} + \frac{\cos 8x}{63} + \ldots \text{ to } \infty\right\}$$

$$+ \frac{3}{2}\sin x + \left\{\frac{1}{3}\sin 3x + \frac{1}{5}\sin 5x + \frac{1}{7}\sin 7x + \ldots \text{ to } \infty\right\}.$$

Note. Work out this example directly as in example 2.

Even and odd functions. $f(x)$ is said to be even, if

$f(-x) = f(x)$; x^2, $\cos kx$, 2, $\cosh x$, $x^2 - 3\cos x + 1$ are all even functions. The graph of an even function will be symmetric about the y-axis. (Refer Fig. 1.)

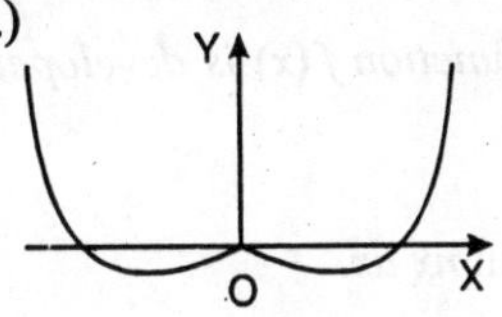

Fig. 1

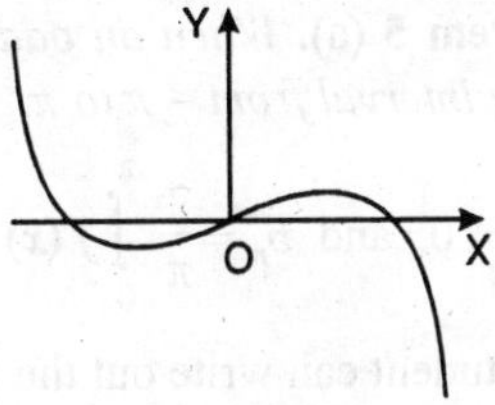

Fig. 2

The function $f(x)$ is odd, if $f(-x) = -f(x)$. For example, x, $\sin x$, $3x - 4\sin x$ are odd functions of x. The graph of an odd function will be symmetric about the origin. Refer Fig. 2.

There are functions which are neither odd nor even; $1 + x$ and e^x are such functions.

Results to Remember. (from integral calculus)

$$\int_{-a}^{a} f(x)\,dx = 2\int_{0}^{a} f(x)\,dx \text{ if } f(x) \text{ is even}$$

and $$\int_{-a}^{a} f(x)\,dx = 0 \text{ if } f(x) \text{ is odd.}$$

Theorem 5. *When an even function $f(x)$ is expanded in a Fourier series over the interval $-\pi$ to π, the series will have cosine terms only, and the coefficients of the series are given by* $a_n = \dfrac{2}{\pi}\displaystyle\int_0^{\pi} f(x)\cos nx\,dx$; *and* $b_n = 0$

Proof. We have already seen that,

$$a_n = \frac{1}{\pi}\int_{-\pi}^{\pi} f(x)\cos nx\, dx$$

$$= \frac{2}{\pi}\int_{0}^{\pi} f(x)\cos nx\, dx \text{ since the integrand is even}$$

$$b_n = \frac{1}{\pi}\int_{-\pi}^{\pi} f(x)\sin nx\, dx = 0 \because \text{ integrand is odd}$$

{$f(x)$ is even and $\sin nx$ is odd}

$$a_0 = \frac{2}{\pi}\int_{0}^{\pi} f(x)\, dx$$

Theorem 5 (a). *When an odd function $f(x)$ is developed in a Fourier series in the interval from $-\pi$ to π,*

$$a_n = 0, \text{ and } b_n = \frac{2}{\pi}\int_{0}^{\pi} f(x)\sin nx\, dx.$$

(The student can write out the proof.)

Note that the Fourier series of an odd function will contain terms in sines only. The function in worked example 2 is an odd function; and, the expansion had terms in sines only.

Example 9. *Find the Fourier series of $f(x) = x + x^2$ in $(-\pi, \pi)$ of periodicity 2π. Hence deduce $\sum \frac{1}{n^2} = \frac{\pi^2}{6}$* ***(MS. 1993 April)***

Let $f(x) = \frac{a_0}{2} + \sum_{n=1}^{\infty} (a_n \cos nx + b_n \sin nx)$ ...(1)

Then $a_0 = \frac{1}{\pi}\int_{-\pi}^{\pi} (x + x^2)dx$

$$= \frac{1}{\pi}\int_{-\pi}^{\pi} x dx + \frac{1}{\pi}\int_{-\pi}^{\pi} x^2 dx$$

$$= 0 + \frac{2}{\pi}\int_{0}^{\pi} x^2 dx$$

Since the first integral vanishes because integrand x is odd and the second integral is $\frac{2}{\pi}\int_{0}^{\pi} x^2 dx$ since x^2 is even.

we use the theorem, $\int_{-a}^{a} f(x)\, dx = 0$ if $f(x)$ is odd

and $\int_{-a}^{a} f(x)\, dx = 2\int_{0}^{a} f(x)\, dx$ if $f(x)$ is even.

$$\text{Hence } a_0 = \frac{2}{\pi}\left[\frac{x^3}{3}\right]_0^{\pi} = \frac{2}{3}\pi^2$$

$$a_n = \frac{1}{\pi}\int_{-\pi}^{\pi} (x + x^2) \cos nx\, dx$$

$$= \frac{1}{\pi}\int_{-\pi}^{\pi} x \cos nx\, dx + \frac{1}{\pi}\int_{-\pi}^{\pi} x^2 \cos nx\, dx$$

$$= 0 + \frac{2}{\pi}\int_{0}^{\pi} x^2 \cos nx\, dx, \qquad \text{since } x \cos nx \text{ is odd}$$

$$= \frac{2}{\pi}\left[(x^2)\left(\frac{\sin nx}{n}\right) - (2x)\left(-\frac{\cos nx}{n^2}\right) + (2)\left(-\frac{\sin nx}{n^3}\right)\right]_0^{\pi}$$

$$= \frac{2}{\pi}\left[\frac{2\pi}{n^2} \cos n\pi\right] = \frac{4}{n^2}(-1)^n$$

$$b_n = \frac{1}{\pi}\int_{-\pi}^{\pi} (x + x^2) \sin nx\, dx$$

$$= \frac{1}{\pi}\int_{-\pi}^{\pi} x \sin nx\, dx + \frac{1}{\pi}\int_{-\pi}^{\pi} x^2 \sin nx\, dx.$$

$$= \frac{2}{\pi}\int_{0}^{\pi} x \sin nx\, dx + 0$$

$$= \frac{2}{\pi}\left[(x)\left(\frac{-\cos nx}{n}\right) - (1)\left(\frac{-\sin nx}{n^2}\right)\right]_0^{\pi}$$

$$= \frac{2}{\pi}\left[\frac{-\pi}{n} \cos n\pi\right] = -\frac{2}{n}(-1)^n$$

Substituting the values of a_0, a_n, b_n, in (1), we get

$$x + x^2 = \frac{\pi^2}{3} + \sum_{n=1}^{\infty} (-1)^n \left[\frac{4}{n^2} \cos nx - \frac{2}{n} \sin nx\right]$$

Deduction: $x = -\pi$ and $x = \pi$ are end points of the range.

Therefore the value of Fourier series at $x = \pi$ is the average of the values of $f(x)$ at $x = \pi$ and $x = -\pi$

Hence, put $x = \pi$ in Fourier series

$$\frac{\pi^2}{3} + \sum_{n=1}^{\infty} (-1)^n \frac{4}{n^2} \cos n\pi = \frac{f(\pi) + f(-\pi)}{2} = \frac{(\pi + \pi^2) + (-\pi + \pi^2)}{2}$$

$$\frac{\pi^2}{3} + 4 \sum_{n=1}^{\infty} \frac{1}{n^2} = \pi^2 \qquad \therefore \quad 4 \sum_{1}^{\infty} \frac{1}{n^2} = \frac{2}{3}\pi^2$$

$$\therefore \qquad \sum_{1}^{\infty} \frac{1}{n^2} = \frac{1}{1^2} + \frac{1}{2^2} + \frac{1}{3^2} + \cdots\text{to } \infty = \frac{\pi^2}{6}$$

Example 10. *In $-\pi < x < \pi$, Express sinh ax and cosh ax in Fourier series of periodicity 2π.* **(*MS. 1989A; 1988 Nov.*)**

Let $f(x) = \sinh ax = \frac{a_0}{2} + \sum_{n=1}^{\infty} (a_n \cos nx + b_n \sin nx)$

Here, $a_0 = \frac{1}{\pi} \int_{-\pi}^{\pi} f(x)\, dx = \frac{1}{\pi} \int_{-\pi}^{\pi} \sinh ax\, dx = 0$

$\because$ the integrand $f(x)$ is odd.

$$a_n = \frac{1}{\pi} \int_{-\pi}^{\pi} \sinh ax \cos nx\, dx = 0 \quad \because \text{ the integrand is odd}$$

$$b_n = \frac{1}{\pi} \int_{-\pi}^{\pi} \sinh ax \sin nx\, dx$$

$$= \frac{2}{\pi} \int_{0}^{\pi} \sinh ax \sin nx\, dx \text{ since the integrand is even.}$$

$$= \frac{2}{\pi} \int_{0}^{\pi} \frac{e^{ax} - e^{-ax}}{2} \sin nx\, dx$$

$$= \frac{1}{\pi} \int_{0}^{\pi} e^{ax} \sin nx\, dx - \frac{1}{\pi} \int_{0}^{\pi} e^{-ax} \sin nx\, dx$$

$$= \frac{1}{\pi} \cdot \left[\frac{e^{ax}}{a^2 + n^2} \{a \sin nx - n \cos nx\} - \frac{e^{-ax}}{a^2 + n^2} \{-a \sin nx - n \cos nx\} \right]_0^{\pi}$$

$$= \frac{1}{\pi} \cdot \frac{1}{a^2 + n^2} \left[-ne^{a\pi} \; (-1)^n + n + ne^{-a\pi} \; (-1)^n - n \right]$$

$$= \frac{1}{\pi} \cdot \frac{1}{a^2 + n^2} \, (-1)^{n-1} \cdot n(e^{a\pi} - e^{-a\pi})$$

$$= \frac{2n}{\pi} \frac{(-1)^{n-1}}{a^2 + n^2} \cdot \sinh a\pi$$

Hence $\sinh ax = \frac{2}{\pi} \sinh a\pi \sum_{n=1}^{\infty} \frac{(-1)^{n-1} n}{a^2 + n^2} \sin nx.$

(*ii*) Let $\cosh ax = \frac{a_0}{2} + \sum_{n=1}^{\infty} (a_n \cos nx + b_n \sin nx)$

$$a_0 = \frac{1}{\pi} \int_{-\pi}^{\pi} f(x)\, dx = \frac{1}{\pi} \int_{-\pi}^{\pi} \cosh ax \, dx$$

$$= \frac{2}{\pi} \int_0^{\pi} \cosh ax \, dx$$

$$= \frac{2}{\pi} \left(\frac{\sinh ax}{a} \right)_0^{\pi}$$

$$= \frac{2}{\pi a} \sinh a\pi$$

$$a_n = \frac{1}{\pi} \int_{-\pi}^{\pi} \cosh ax \cos nx \, dx$$

$$= \frac{2}{\pi} \int_0^{\pi} \cosh ax \cos nx \, dx$$

$$= \frac{2}{\pi} \int_0^{\pi} \frac{e^{ax} + e^{-ax}}{2} \cdot \cos nx \, dx$$

$$= \frac{1}{\pi} \left[\int_0^{\pi} e^{ax} \cos nx \, dx + \int_0^{\pi} e^{-ax} \cos nx \, dx \right]$$

$$= \frac{1}{\pi}\left[\frac{e^{ax}}{a^2+n^2}\{a\cos nx + n\sin nx\} + \frac{e^{-ax}}{a^2+n^2}\begin{Bmatrix}-a\cos nx\\ +n\sin nx\end{Bmatrix}\right]_0^{\pi}$$

$$= \frac{1}{\pi}\cdot\frac{1}{a^2+n^2}\left[e^{a\pi}\cdot a(-1)^n - a - e^{-a\pi}\cdot a(-1)^n + a\right]$$

$$= \frac{1}{\pi}\cdot\frac{1}{a^2+n^2}\,a(-1)^n\,[e^{a\pi} - e^{-a\pi}]$$

$$= \frac{2}{\pi}\quad\frac{1}{a^2+n^2}\,a(-1)^n\cdot\sinh a\pi$$

$$b_n = \frac{1}{\pi}\int_{-\pi}^{\pi}\cosh ax\sin nx\,dx = 0\quad(\because \text{ integrand is odd})$$

$$f(x) = \cosh ax = \frac{1}{\pi a}\sinh a\pi + \sum_1^{\infty}\cdot\frac{2}{\pi}\cdot\frac{1}{a^2+n^2}\,a(-1)^n\,\sinh a\pi\cos nx.$$

$$= \frac{2a}{\pi}\sinh a\pi\left[\frac{1}{2a^2} + \sum_1^{\infty}\frac{1}{a^2+n^2}(-1)^n\cdot\cos nx\right]$$

Example 11. *If* $f(x) = \sin x$ *in* $0 \le x \le \pi$

$= 0$ *in* $\pi \le x \le 2\pi$

find a Fourier series of periodicity 2π *and hence evaluate* $\frac{1}{1\cdot 3} + \frac{1}{3\cdot 5} + \frac{1}{5\cdot 7} + \cdots$ *to* ∞.

Let $$f(x) = \frac{a_0}{2} + \sum_{n=1}^{\infty}(a_n\cos nx + b_n\sin nx) \qquad ...(1)$$

$$a_0 = \frac{1}{\pi}\int_0^{2\pi} f(x)\,dx$$

$$= \frac{1}{\pi}\left[\int_0^{\pi}\sin x\,dx + \int_{\pi}^{2\pi} 0\cdot dx\right]$$

$$= \frac{1}{\pi}(-\cos x)_0^{\pi} = \frac{2}{\pi} \qquad ....(2)$$

$$a_n = \frac{1}{\pi}\cdot\int_0^{\pi}\sin x\cos nx\,dx$$

$$= \frac{1}{2\pi} \int_0^{\pi} [\sin (n+1) x - \sin (n-1) x]\, dx$$

$$= \frac{1}{2\pi} \left[-\frac{\cos (n+1) x}{n+1} + \frac{\cos (n-1) x}{n-1} \right]_0^{\pi} \text{ if } n \neq 1$$

$$= \frac{1}{2\pi} \left[\frac{1}{n+1} \{(-1)^n + 1\} + \frac{1}{n-1} \{(-1)^{n-1} - 1\} \right]$$

$$= \frac{1}{2\pi} \frac{1}{n^2 - 1} [(n-1)\{1 + (-1)^n\} - (n+1)\{(-1)^n + 1\}]$$

$$= \frac{1}{2\pi (n^2 - 1)} \{(-1)^n (-2) - 2 \}$$

$$= \frac{-1}{\pi (n^2 - 1)} [1 + (-1)^n]$$

$$= 0 \text{ if } n \text{ is odd}$$

$$= \frac{-2}{\pi (n^2 - 1)} \text{ if } n \text{ is even}, \quad n \neq 1. \qquad \text{...(3)}$$

$$a_1 = \frac{1}{\pi} \int_0^{\pi} \sin x \cos x \, dx$$

$$= \frac{1}{2\pi} \int_0^{\pi} \sin 2x \, dx$$

$$= \frac{1}{2\pi} \left(-\frac{1}{2} \cos 2x \right)_0^{\pi}$$

$$= -\frac{1}{4\pi} (1 - 1) = 0$$

$$b_n = \frac{1}{\pi} \int_0^{\pi} \sin x \sin nx \, dx$$

$$= \frac{1}{2\pi} \int_0^{\pi} [\cos (n-1)x - \cos (n+1)x] \, dx$$

$$= \frac{1}{2\pi} \left[\frac{\sin (n-1) x}{n-1} - \frac{\sin (n+1) x}{n+1} \right]_0^{\pi} \text{ if } n \neq 1$$

$$= 0 \text{ if } n \neq 1$$

$$b_1 = \frac{1}{\pi} \int_0^{\pi} \sin x \sin x \, dx$$

$$= \frac{1}{\pi} \int_0^{\pi} \frac{1 - \cos 2x}{2} \, dx$$

$$= \frac{1}{2\pi} \left[x - \frac{1}{2} \sin 2x \right]_0^{\pi} = \frac{1}{2}$$

$$\therefore \quad f(x) = \frac{1}{\pi} + \frac{1}{2} \sin x - \frac{2}{\pi} \sum_{n=2,4,6}^{\infty} \frac{1}{(n^2 - 1)} \cos nx$$

$x = 0$ is an end point in the range

$\therefore$ The value of the Fourier series at $x = 0$ is equal to $\dfrac{f(0) + f(2\pi)}{2}$

Putting $x = 0$ in the Fourier series,

$$\frac{1}{\pi} - \frac{2}{\pi} \sum_{n=2,4,6}^{\infty} \frac{1}{n^2 - 1} = \frac{0 + 0}{2} = 0$$

$$\sum_{n=2,4,6}^{\infty} \frac{1}{(n-1)(n+1)} = \frac{1}{2}$$

i.e.
$$\frac{1}{1 \cdot 3} + \frac{1}{3 \cdot 5} + \frac{1}{5 \cdot 7} + \cdots \text{to } \infty = \frac{1}{2}$$

Example 12. *Express* $f(x) = (\pi - x)^2$ *as a Fourier series of periodicity* 2π *in* $0 < x < 2\pi$ *and hence deduce the sum* $\sum_{n=1}^{\infty} \frac{1}{n^2}$ **(*MS. 1989 April*)**

Let
$$f(x) = \frac{a_0}{2} + \sum_{n=1}^{\infty} (a_n \cos nx + b_n \sin nx)$$

$$a_0 = \frac{1}{\pi} \int_0^{2\pi} f(x) \, dx$$

$$= \frac{1}{\pi} \int_0^{2\pi} (\pi - x)^2 \, dx$$

$$= \frac{1}{\pi}\left[\frac{-(\pi - x)^3}{3}\right]_0^{2\pi}$$

$$= -\frac{1}{3\pi}\left[-\pi^3 - \pi^3\right] = \frac{2}{3}\pi^2$$

$$a_n = \frac{1}{\pi}\int_0^{2\pi} (\pi - x)^2 \cos nx \, dx$$

$$= \frac{1}{\pi}\left[(\pi - x)^2\left(\frac{\sin nx}{n}\right) + 2(\pi - x)\left(-\frac{\cos nx}{n^2}\right) + 2\left(-\frac{\sin nx}{n^3}\right)\right]_0^{2\pi}$$

$$= \frac{1}{\pi}\left[\frac{2\pi}{n^2} + \frac{2\pi}{n^2}\right] = \frac{4}{n^2}$$

$$b_n = \frac{1}{\pi}\int_0^{2\pi} (\pi - x)^2 \sin nx \, dx$$

$$= \frac{1}{\pi}\left[(\pi - x)^2\left(-\frac{\cos nx}{n}\right) + 2(\pi - x)\left(-\frac{\sin nx}{n^2}\right) + 2\left(\frac{\cos nx}{n^3}\right)\right]_0^{2\pi}$$

$$= \frac{1}{\pi}\left[-\frac{\pi^2}{n} + \frac{\pi^2}{n} + \frac{2}{n^3}(1 - 1)\right] = 0$$

$$\therefore \qquad f(x) = (\pi - x)^2 = \frac{\pi^2}{3} + 4\sum_{n=1}^{\infty}\frac{1}{n^2}\cos nx$$

$x = 0$ is an end point. The value of Fourier series at $x = 0$ is $\dfrac{f(0) + f(2\pi)}{2}$

$$\therefore \qquad \frac{\pi^2}{3} + 4\sum_{n=1}^{\infty}\frac{1}{n^2} = \frac{\pi^2 + \pi^2}{2} = \pi^2$$

$$\therefore \qquad \sum_{1}^{\infty}\frac{1}{n^2} = \frac{\pi^2}{6}.$$

Example 13. *Find the Fourier series of* $f(x) = e^x$ *in* $(-\pi, \pi)$, *of periodicity* 2π **(*MS. 1991 Nov.*)**

Let $$f(x) = \frac{a_0}{2} + \sum_{n=1}^{\infty} (a_n \cos nx + b_n \sin nx) \qquad ...(1)$$

where $$a_0 = \frac{1}{\pi} \int_{-\pi}^{\pi} e^x \, dx$$

$$= \frac{1}{\pi} (e^x)_{-\pi}^{\pi}$$

$$= \frac{1}{\pi} (e^{\pi} - e^{-\pi})$$

$$= \frac{2}{\pi} \sinh \pi$$

$$a_n = \frac{1}{\pi} \int_{-\pi}^{\pi} e^x \cos nx \, dx$$

$$= \frac{1}{\pi} \left[\frac{e^x}{1+n^2} (\cos nx + n \sin nx) \right]_{-\pi}^{\pi}$$

$$= \frac{1}{\pi (1+n^2)} [e^{\pi} (-1)^n - e^{-\pi} (-1)^n]$$

$$= \frac{2(-1)^n}{\pi (1+n^2)} \sinh \pi$$

$$b_n = \frac{1}{\pi} \int_{-\pi}^{\pi} e^x \sin nx \, dx$$

$$= \frac{1}{\pi} \left[\frac{e^x}{1+n^2} (\sin nx - n \cos nx) \right]_{-\pi}^{\pi}$$

$$= \frac{1}{\pi (1+n^2)} [e^{\pi} (-n) (-1)^n + e^{-\pi} n (-1)^n]$$

$$= \frac{-2(-1)^n \cdot n}{\pi (1+n^2)} \sinh \pi$$

$$\therefore \qquad e^x = \frac{\sinh \pi}{\pi}\left[1 + \sum_{n=1}^{\infty} \frac{2(-1)^n}{1+n^2}(\cos nx - n \sin nx)\right]$$

Example 14. *Find the Fourier series of periodicity* 2π

$$\textit{for} \quad f(x) = \begin{cases} x \ \textit{when} \ -\pi < x < 0 \\ 0 \ \textit{when} \ 0 < x < \dfrac{\pi}{2} \\ x - \dfrac{\pi}{2} \ \textit{when} \ \dfrac{\pi}{2} < x < \pi \end{cases}$$

(*Kerala University*)

$$\text{Let} \quad f(x) = \frac{a_0}{2} + \sum_{n=1}^{\infty} (a_n \cos nx + b_n \sin nx) \qquad \text{...(1)}$$

$$\text{where} \qquad a_0 = \frac{1}{\pi}\int_{-\pi}^{\pi} f(x)\, dx$$

$$= \frac{1}{\pi}\left[\int_{-\pi}^{0} x\, dx + \int_{0}^{\frac{\pi}{2}} 0\, dx - \int_{\frac{\pi}{2}}^{\pi}\left(x - \frac{\pi}{2}\right) dx\right]$$

$$= \frac{1}{\pi}\left[\left(\frac{x^2}{2}\right)_{-\pi}^{0} + \left\{\frac{1}{2}\left(x - \frac{\pi}{2}\right)^2\right\}_{\frac{\pi}{2}}^{\pi}\right]$$

$$= \frac{1}{\pi}\left[-\frac{\pi^2}{2} + \frac{\pi^2}{8}\right] = -\frac{3\pi}{8}$$

$$a_n = \frac{1}{\pi}\int_{-\pi}^{\pi} f(x)\cos nx\, dx$$

$$= \frac{1}{\pi}\left[\int_{-\pi}^{0} x \cos nx\, dx + \int_{0}^{\frac{\pi}{2}} 0 \cdot \cos nx\, dx + \int_{\frac{\pi}{2}}^{\pi}\left(x - \frac{\pi}{2}\right)\cos nx\, dx\right]$$

$$= \frac{1}{\pi}\left[\left\{(x)\left(\frac{\sin nx}{n}\right)-(1)\left(-\frac{\cos nx}{n^2}\right)\right\}_{-\pi}^{0}\right.$$

$$\left.+\left\{\left(x-\frac{\pi}{2}\right)\left(\frac{\sin nx}{n}\right)-(1)\left(-\frac{\cos nx}{n^2}\right)\right\}_{\frac{\pi}{2}}^{\pi}\right]$$

$$= \frac{1}{\pi}\left[\frac{1}{n^2}\{1-(-1)^n\}+\frac{1}{n^2}\left\{(-1)^n-\cos\frac{n\pi}{2}\right\}\right]$$

$$= \frac{1}{\pi}\left[\frac{1}{n^2}\left(1-\cos\frac{n\pi}{2}\right)\right]$$

$$b_n = \frac{1}{\pi}\left[\int_{-\pi}^{0} x\sin nx\,dx + \int_{\frac{\pi}{2}}^{\pi}\left(x-\frac{\pi}{2}\right)\sin nx\,dx\right]$$

$$= \frac{1}{\pi}\left[\left\{(x)\left(-\frac{\cos nx}{n}\right)-(1)\left(-\frac{\sin nx}{n^2}\right)\right\}_{-\pi}^{0}\right.$$

$$\left.+\left\{\left(x-\frac{\pi}{2}\right)\left(-\frac{\cos nx}{n}\right)-(1)\left(-\frac{\sin nx}{n^2}\right)\right\}_{\frac{\pi}{2}}^{\pi}\right]$$

$$= \frac{1}{\pi}\left[-\frac{\pi}{n}\cos n\pi-\frac{\pi}{2n}\cos n\pi-\frac{1}{n^2}\sin\frac{n\pi}{2}\right]$$

$$= -\frac{1}{\pi}\left[\frac{3\pi}{2n}(-1)^n+\frac{1}{n^2}\sin\frac{n\pi}{2}\right]$$

$$\therefore \quad f(x) = \frac{-3\pi}{16}+\frac{1}{\pi}\sum_{n=1}^{\infty}\frac{1}{n^2}\left(1-\cos\frac{n\pi}{2}\right)\cos nx-\frac{1}{\pi}\sum_{n=1}^{\infty}\left(\frac{3\pi}{2n}(-1)^n+\frac{1}{n^2}\sin\frac{n\pi}{2}\right)\sin nx$$

Example 15. *Find the Fourier series of periodicity* 2π

for $f(x) = \begin{cases} x & \text{in } (0, \pi) \\ 2\pi - x & \text{in } (\pi, 2\pi) \end{cases}$

and hence deduce $\frac{1}{1^2}+\frac{1}{3^2}+\frac{1}{5^2}+\ldots = \frac{\pi^2}{8}$

Let $f(x) = \frac{a_0}{2} + \sum_{n=1}^{\infty} (a_n \cos nx + b_n \sin nx)$

$$a_0 = \frac{1}{\pi}\left[\int_0^{\pi} x\,dx + \int_{\pi}^{2\pi} (2\pi - x)\,dx\right]$$

$$= \frac{1}{\pi}\left[\left(\frac{x^2}{2}\right)_0^{\pi} - \left\{\frac{1}{2}(2\pi - x)^2\right\}_{\pi}^{2\pi}\right]$$

$$= \frac{1}{\pi}\left[\frac{\pi^2}{2} + \frac{\pi^2}{2}\right] = \pi$$

$$a_n = \frac{1}{\pi}\left[\int_0^{\pi} x \cos nx\,dx + \int_{\pi}^{2\pi} (2\pi - x)\cos nx\,dx\right]$$

$$= \frac{1}{\pi}\left[\left\{(x)\left(\frac{\sin nx}{n}\right) - (1)\left(-\frac{\cos nx}{n^2}\right)\right\}_0^{\pi}\right.$$

$$\left. + \left\{(2\pi - x)\left(\frac{\sin nx}{n}\right) - (-1)\left(-\frac{\cos nx}{n^2}\right)\right\}_{\pi}^{2\pi}\right]$$

$$= \frac{1}{\pi}\left[\frac{1}{n^2}\{(-1)^n - 1\} - \frac{1}{n^2}\{1 - (-1)^n\}\right]$$

$$= \frac{2}{\pi n^2}[(-1)^n - 1]$$

$$\therefore b_n = \frac{1}{\pi}\left[\int_0^{\pi} x \sin nx\,dx + \int_{\pi}^{2\pi} (2\pi - x)\sin nx\,dx\right]$$

$$= \frac{1}{\pi}\left[\left\{(x)\left(-\frac{\cos nx}{n}\right) - (1)\left(-\frac{\sin nx}{n^2}\right)\right\}_0^{\pi}\right.$$

$$\left. + \left\{(2\pi - x)\left(\frac{-\cos nx}{n}\right) - (-1)\left(-\frac{\sin nx}{n^2}\right)\right\}_{\pi}^{2\pi}\right]$$

$$= \frac{1}{\pi}\left[\frac{-\pi}{n}(-1)^n + \frac{\pi}{n}(-1)^n\right] = 0$$

Hence $$f(x) = \frac{\pi}{2} + \frac{2}{\pi} \sum_{n=1}^{\infty} \frac{1}{n^2} [(-1)^n - 1] \cos nx$$

$$= \frac{\pi}{2} - \frac{4}{\pi} \left[\cos x + \frac{1}{3^2} \cos 3x + \frac{1}{5^2} \cos 5x + \cdots \text{to } \infty \right]$$

$x = 0$ is an end point of the range.

The value of Fourier series at an end is equal to the average of $[f(0) + f(2\pi)]$

$\therefore$ At $x = 0$,

$$\frac{\pi}{2} - \frac{4}{\pi} \left(\frac{1}{1^2} + \frac{1}{3^2} + \frac{1}{5^2} + \cdots \right) = \frac{1}{2} [f(0) + f(2\pi)]$$

$$= \frac{1}{2} [0 + 0] = 0$$

$$\frac{4}{\pi} \left(\frac{1}{1^2} + \frac{1}{3^2} + \frac{1}{5^2} + \cdots \right) = \frac{\pi}{2}$$

$$\frac{1}{1^2} + \frac{1}{3^2} + \frac{1}{5^2} + \cdots = \frac{\pi^2}{8}.$$

Example 16. *In $(-\pi, \pi)$, find the Fourier series of periodicity 2π for*

$$f(x) = \begin{cases} 1 + x & \text{in } 0 < x < \pi \\ -1 + x & \text{in } -\pi < x < 0 \end{cases}$$

The function $f(x)$ is odd evidently; for

Let $0 < \alpha < \pi$; Than $f(\alpha) = 1 + \alpha$

$$f(-\alpha) = -1 - \alpha = -(1 + \alpha) = -f(\alpha)$$

Let $$f(x) = \frac{a_0}{2} + \sum_{n=1}^{\infty} (a_n \cos nx + b_n \sin nx)$$

where $$a_0 = \frac{1}{\pi} \int_{-\pi}^{\pi} f(x)\, dx = 0$$

$$a_n = \frac{1}{\pi} \int_{-\pi}^{\pi} f(x) \cos nx\, dx = 0 \qquad (\because \text{ integrand is o})$$

$$b_n = \frac{1}{\pi} \int_{-\pi}^{\pi} f(x) \sin nx\, dx$$

$$= \frac{2}{\pi} \int_0^\pi f(x) \sin nx \, dx$$

$$= \frac{2}{\pi} \int_0^\pi (1 + x) \sin nx \, dx$$

$$= \frac{2}{\pi} \left[(1 + x) \left(- \frac{\cos nx}{n} \right) - (1) \left(- \frac{\sin nx}{n^2} \right) \right]_0^\pi$$

$$= \frac{2}{\pi} \left[(1 + \pi) \left(- \frac{\cos n\pi}{n} \right) + \frac{1}{n} \right]$$

$$= \frac{2}{n\pi} [1 - (1 + \pi)(-1)^n]$$

$$\therefore \quad f(x) = \frac{2}{\pi} \sum_{n=1}^{\infty} \frac{1}{n} [1 - (1 + \pi)(-1)^n] \sin nx.$$

Example 17. *Write down the analytic expressions of the following functions given in the graph. Hence find Fourier series of $f(x)$.*

(*a*)

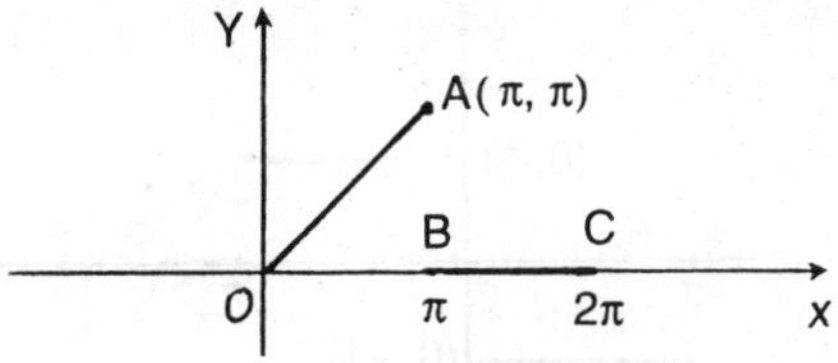

$$f(x) = x, \quad 0 < x < \pi$$
$$= 0 \quad \pi < x < 2\pi$$

(*b*) From the figure on page 36,

Slope of $BC = \frac{1-0}{0-\pi} = \frac{-1}{\pi}$

equation of BC is $y = -\frac{1}{\pi}(x - \pi)$

Slope of AC is $\frac{1-0}{0+\pi} = \frac{1}{\pi}$

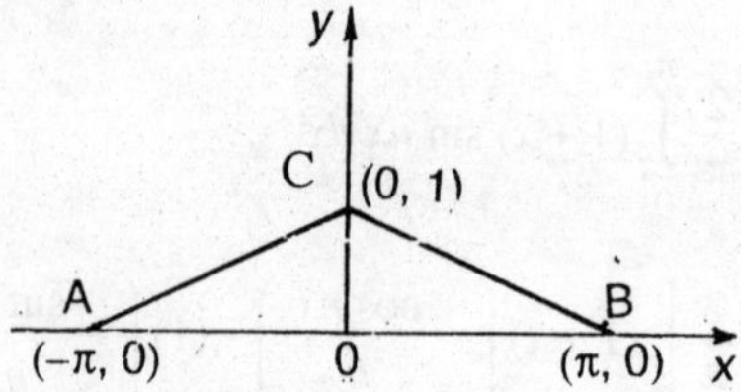

Equation of AC is $y = \frac{1}{\pi}\ (x + \pi)$

$\therefore\ f(x) = \frac{1}{\pi}\ (x + \pi), \qquad -\pi < x < 0$

$= -\frac{1}{\pi}\ (x - \pi),$ if $0 < x < \pi$

{$f(x)$ is an even function}

(*c*)

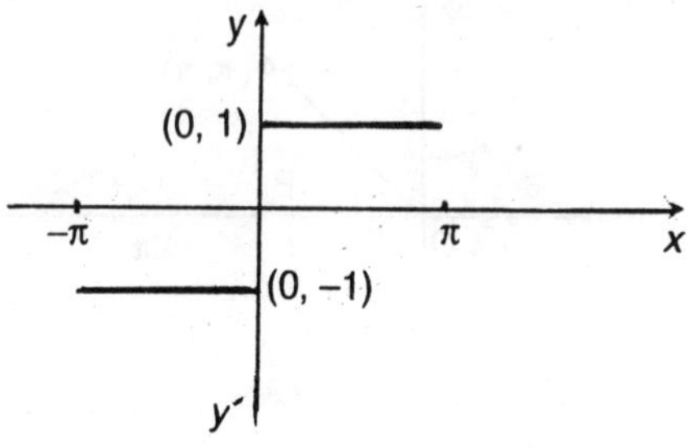

The two segments are parallel to x axis

$\therefore\ f(x) = -1, \quad -\pi < x < 0$

$= 1, \qquad 0 < x < \pi$

(*d*) Refer to figure on page 36

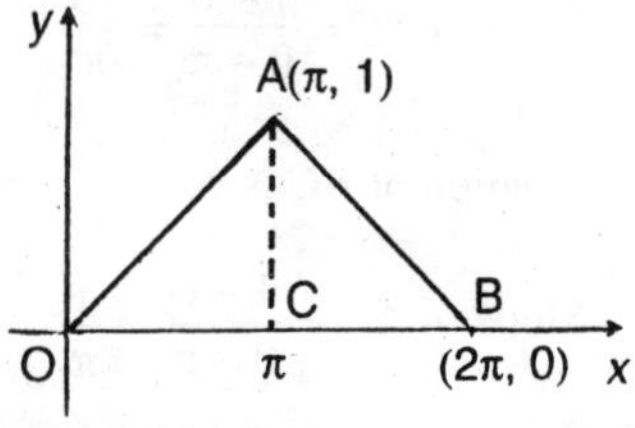

Slope of $OA = \frac{1}{\pi}$

Equation of OA is $y = \frac{1}{\pi}x$

$$\text{Slope of } AB = \frac{1-0}{\pi - 2\pi} = \frac{-1}{\pi}$$

$$\text{Equation of } AB \text{ is } y = -\frac{1}{\pi}(x - 2\pi)$$

$$\therefore \quad f(x) = \frac{x}{\pi}, \quad 0 < x < \pi$$

$$= \frac{1}{\pi}(2\pi - x), \pi < x < 2\pi$$

Fourier series of the above functions can be got in the usual way.

Example 18. *If $f(x)$ is defined in $(-\pi, \pi)$ and if*

$f(x) = x + 1$ in $(0, \pi)$ find $f(x)$ in $(-\pi, 0)$ if

(*i*) $f(x)$ is odd (*ii*) $f(x)$ is even.

Ans: If $f(x)$ is even then,

$$f(x) = -x + 1 \qquad \text{in } (-\pi, 0)$$

If $f(x)$ is odd, then

$$f(x) = x - 1 \qquad \text{in } (-\pi, 0)$$

Exercises 1(c)

Expand each of the functions in 1 to 17 in a Fourier series. It is given that each of the functions has the period 2π.

1. $f(x) = e^x, \ -\pi < x < \pi$. Derive a series for $\dfrac{\pi}{\sinh \pi}$.
2. $f_1(x) = x, \ -\pi < x < \pi$. Deduce the value of $\dfrac{\pi}{4}$.
3. $F(x) = x + e^x, \ -\pi < x < \pi$.
4. $f(x) = x^2, -\pi < x < \pi$. Deduce the relations

$$1 + \frac{1}{2^2} + \frac{1}{3^2} + \frac{1}{4^2} + \cdots \text{ to } \infty = \frac{\pi^2}{6};$$

$$1 - \frac{1}{2^2} + \frac{1}{3^2} - \frac{1}{4^2} + \cdots \text{ to } \infty = \frac{\pi^2}{12};$$

$$1 + \frac{1}{3^2} + \frac{1}{5^2} + \frac{1}{7^2} + \cdots \text{ to } \infty = \frac{\pi^2}{8}.$$

5. $f(x) = x^3, -\pi < x < \pi$.
6. $f(x) = \begin{cases} -\dfrac{\pi}{4}, & \text{when } -\pi < x < 0 \\ \dfrac{\pi}{4}, & \text{when } 0 < x < \pi, \end{cases}$

and $f(-\pi) = f(0) = f(\pi) = 0$, and $f(x + 2\pi) = f(x)$ for all x.

Deduce that $\frac{\pi}{4} = 1 - \frac{1}{3} + \frac{1}{5} - \frac{1}{7} + \text{....to } \infty$. **(*Madras, 72 Engg.*)**

7. $f(x) = 0$, if $-\pi \le x < 0$; and $f(x) = \pi$, if $0 \le x < \pi$. Show how this can be got from example 6.

8. $F(x) = -x$, if $-\pi \le x \le 0$; and $f(x) = \pi - x$, if $0 < x < \pi$. Show how this can be got from example 2 and example 7.

9. $f(x) = \pi$, if $-\pi < x \le \frac{1}{2}\pi$; and $f(x) = 0$, if $\frac{1}{2}\pi < x \le \pi$.

10. $f(x) = 1$, if $-\frac{1}{2}\pi < x < \frac{1}{2}\pi$; and $f(x) = -1$, if $\frac{1}{2}\pi < x < \frac{3}{2}\pi$.

11. $f(x) = 1$, if $0 < x < \frac{1}{2}\pi$; and $f(x) = 0$, if $\frac{1}{2}\pi < x < 2\pi$.

12. $f(x) = \begin{cases} 2, \text{ when } 0 < x < \frac{2}{3}\pi \\ 1, \text{ when } \frac{2}{3}\pi < x < \frac{4}{3}\pi \\ 0, \text{ when } \frac{4}{3}\pi < x < 2\pi \end{cases}$

13. $f(x) = 0$, when $-\pi < x < 0$ and $f(x) = x$, when $0 < x < \pi$.

14. $f(x) = x$, if $-\frac{1}{2}\pi < x < \frac{1}{2}\pi$ and $f(x) = 0$, if $\frac{1}{2}\pi < x < \frac{3}{2}\pi$.

15. $f(x) = 0$, when $-\pi < x < 0$ and $f(x) = x^2$, when $0 < x < \pi$.

16. $f(x) = \sin \frac{1}{2}x$, when $-\pi < x < \pi$.

17. $f(x) = 0$, when $-\pi < x < 0$; and $f(x) = \cos x$, when $0 \le x \le \pi$.

18. An alternating current of period 2π after passing through a rectifier is of the form

$$i = \begin{cases} 0, \text{ for } -\pi \le \theta < 0, \\ I \sin \theta, \text{ for } 0 \le \theta < \pi, \text{ where } I \text{ is the maximum current.} \end{cases}$$

Obtain the Fourier expansion for i.

19. Find the Fourier series of periodicity 2π for the following functions

(i) $f(x) = \begin{cases} 1, & 0 < x < \pi \\ -1 & \pi < x < 2\pi \end{cases}$ and $f(x + 2\pi) = f(x)$.

(ii) $f(x) = 2, \quad -\pi < x < 0$

$\quad = 1 \quad 0 \le x \le \pi$ with period 2π

(iii) $f(x) = -\frac{1}{2}$ if $-\pi < x < 0$

$\quad = \frac{1}{2}$ if $0 < x < \pi$ and $f(x + 2\pi) = f(x)$

(iv) $f(x) = x + 1, \quad 0 < x < \pi$

$\quad = x - 1, \quad -\pi < x < 0$

(v) $f(x) = \begin{cases} 0, & -\pi < x < 0 \\ 1, & 0 < x < \pi \end{cases}$

(vi) $f(x) = (\pi - x)^2,\quad 0 < x < 2\pi$ and hence deduce $1 + \frac{1}{2^2} + \frac{1}{3^2} + \cdots = \frac{\pi^2}{6}$

(MS. 1989 April)

Example 19. *Obtain the Fourier series of periodicity 2π for $f(x) = x$, in $-\pi < x < \pi$.*

$f(x)$ is an odd function of x. Hence by Th. 5(*a*), $a_n = 0$, and $b_n = \frac{2}{\pi}\int_0^{\pi} x \sin nx\, dx$. In worked example 3, this integral was calculated as $-\frac{\pi}{n}\cos n\pi$. Hence $b_n = -2\,\frac{\cos n\pi}{n}$. Using the values of the coefficients, in the given interval,

$$f(x) = 2\left(\sin x - \frac{\sin 2x}{2} + \frac{\sin 3x}{3} - \cdots\right).$$

Note. Putting $x = \frac{\pi}{2}$, we get $1 - \frac{1}{3} + \frac{1}{5} - \frac{1}{7} + \cdots = \frac{\pi}{4}$.

Example 20. *Obtain the Fourier series of periodicity 2π for*

(*i*) $f(x) = -x$, *when* $-\pi < x \le 0$, *and* $f(x) = x$, *when* $0 < x < \pi$,

(*ii*) $f(x) = |x|$, *when* $-\pi < x < \pi$.

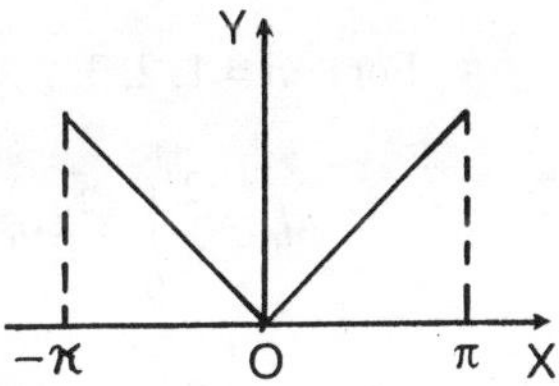

Note that both the questions are identical; they are merely different forms of wording the same question. The figure gives the graph of the function. The given function is even. Hence by Theorem 5, $b_n = 0$; and

$$a_0 = \frac{2}{\pi}\int_0^{\pi} f(x)\, dx = \frac{2}{\pi}\int_0^{\pi} x\, dx = \frac{2}{\pi}\left[\frac{x^2}{2}\right]_0^{\pi} = \frac{2}{\pi} \times \frac{\pi^2}{2} = \pi.$$

Again by Theorem 5, $a_n = \frac{2}{\pi}\int_0^{\pi} x \cos nx\, dx$. In worked example 3, this integral was calculated to be $\frac{\cos n\pi - 1}{n^2}$.

Hence $a_n = \frac{2}{\pi} \times \frac{\cos n\pi - 1}{n^2}$. Substituting the coefficients,

$$f(x) = \frac{\pi}{2} - \frac{4}{\pi}\left\{\cos x + \frac{\cos 3x}{3^2} + \frac{\cos 5x}{5^2} + \cdots\right\}$$

Note. Putting $x = 0$, we easily get $\frac{\pi^2}{8} = 1 + \frac{1}{3^2} + \frac{1}{5^2} + \cdots$ to ∞.

Example 21. *Expand* $f(x) = x^2$, *when* $-\pi < x < \pi$, *in a Fourier series of periodicity* 2π. *Hence deduce that*

(*i*) $\frac{1}{1^2} + \frac{1}{2^2} + \frac{1}{3^2} + \cdots \text{to } \infty = \frac{\pi^2}{6}$ ***(Calicut, 71 Engg.)***

(*ii*) $\frac{1}{1^2} - \frac{1}{2^2} + \frac{1}{3^2} - \cdots \text{to } \infty = \frac{\pi^2}{12}$

(*iii*) $\frac{1}{1^2} + \frac{1}{3^2} + \frac{1}{5^2} + \cdots \text{to } \infty = \frac{\pi^2}{8}$

$f(x)$ is an even function of x in $-\pi < x < \pi$. Hence by theorem 5, $b_n = 0$ and only cosine terms will be present. Therefore,

$$f(x) = \frac{a_0}{2} + \sum_{n=1}^{\infty} a_n \cos nx \qquad \text{....}(i)$$

$$a_0 = \frac{2}{\pi}\int_0^{\pi} x^2 dx = \frac{2\pi^2}{3}$$

For $n = 1, 2, 3, \ldots$

$$a_n = \frac{2}{\pi}\int_0^{\pi} x^2 \cos nx \, dx$$

$$= \frac{2}{\pi}\left[(x^2)\left(\frac{\sin nx}{n}\right) - (2x)\left(\frac{-\cos nx}{n^2}\right) + (2)\left(\frac{-\sin nx}{n^3}\right)\right]_0^{\pi}$$

$$= \frac{2}{\pi}\left[\frac{2\pi}{n^2}\cos n\pi\right] = \frac{4(-1)^n}{n^2}.$$

Substituting these values in (*i*),

$$f(x) = \frac{\pi^2}{3} + 4\sum_{n=1}^{\infty} \frac{(-1)^n}{n^2} \cos nx \qquad \text{....}(ii)$$

i.e., $x^2 = \frac{\pi^2}{3} - 4\left[\frac{\cos x}{1^2} - \frac{\cos 2x}{2^2} + \frac{\cos 3x}{3^2} - \cdots\right]$, in $-\pi < x < \pi$.

The function $f(x) = x^2$ is continuous at $x = 0$. Hence the sum of the Fourier series equals the value of the function at $x = 0$. Putting $x = 0$, in (*ii*),

$$0 = \frac{\pi^2}{3} - 4\left[\frac{1}{1^2} - \frac{1}{2^2} + \frac{1}{3^2} - \cdots\right]$$

$$\therefore \quad \frac{1}{1^2} - \frac{1}{2^2} + \frac{1}{3^2} - \cdots = \frac{\pi^2}{12} \qquad \ldots(iii)$$

$x = \pi$ is an end point. Hence the sum of the Fourier series at $x = \pi$ equals $\frac{1}{2}\{f(-\pi+0) + f(\pi-0)\}$

Putting $x = \pi$ in the series of (*ii*),

$$\frac{\pi^2}{3} + 4\sum_{n=1}^{\infty} \frac{(-1)^n \cos n\pi}{n^2} = \frac{1}{2}\,[f(-\pi+0) + f(\pi-0)]$$

$$= \frac{1}{2}\,[\pi^2 + \pi^2] = \pi^2$$

$$\therefore \quad 4\sum_{n=1}^{\infty} \frac{1}{n^2} = \pi^2 - \frac{\pi^2}{3} = \frac{2}{3}\pi^2$$

i.e.,

$$\sum_{n=1}^{\infty} \frac{1}{n^2} = \frac{1}{1^2} + \frac{1}{2^2} + \frac{1}{3^2} + \cdots = \frac{\pi^2}{6} \qquad \ldots(iv)$$

Adding (*iii*) and (*iv*),

$$2\left(\frac{1}{1^2} + \frac{1}{3^2} + \frac{1}{5^2} + \cdots\right) = \frac{\pi^2}{4}$$

i.e.,

$$\frac{1}{1^2} + \frac{1}{3^2} + \frac{1}{5^2} + \cdots \text{ to } \infty = \frac{\pi^2}{8}.$$

Example 22. *Obtain the Fourier expansion of* $f(x) = |\sin x|$, *in* $-\infty < x < \infty$.

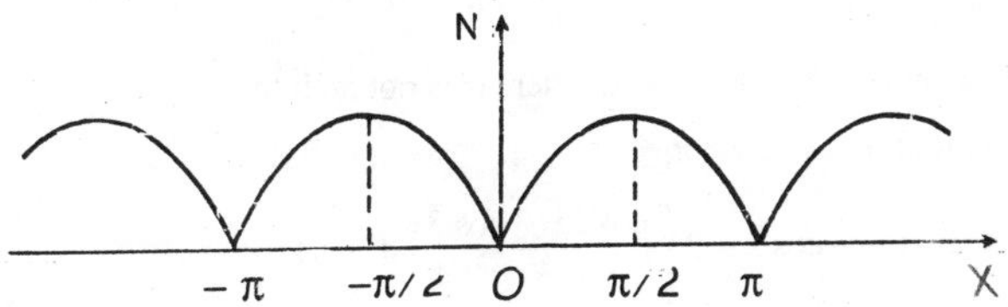

The graph of $y = |\sin x|$ is as shown above. It is periodic, with a period of π; without loss of generally, we can take a period from (0 to π). Again, the curve is evidently symmetrical about the y-axis. Hence $f(x)$ is an *even* function of x; therefore, its expansion will not contain terms in sine.

In the interval $-\pi$ to $\pi, f(x)$ can also be defined in the form $f(x) = -\sin x$, if $-\pi < x < 0$, and $f(x) = \sin x$, if $0 < x < \pi$.

Now, $a_1 = \frac{2}{\pi}\int_0^{\pi} \sin x \cos x \, dx = \frac{2}{\pi} \times 0 = 0$

For $n = 0, 2, 3, 4, \ldots$

$$a_n = \frac{2}{\pi}\int_0^{\pi} \sin x \cos nx \, dx = -\frac{2}{\pi} \cdot \frac{1 + \cos n\pi}{(n+1)(n-1)}$$

(For the evaluation of the integral, refer worked example 4,)

$$\text{Thus } f(x) = \frac{2}{\pi} - \frac{4}{\pi}\left\{\frac{\cos 2x}{3} + \frac{\cos 4x}{15} + \frac{\cos 6x}{35} + \frac{\cos 8x}{63} + \ldots \text{ to } \infty\right\}.$$

Example 23. *Expand $F(x) = x - x^2$, in $-\pi < x < \pi$ in a Fourier series.*

If $f(x) = x$ and $f_1(x) = x^2$, then evidently $F = f - f_1$. Hence the answer is written down using worked examples 19 and 21.

Exercises 1(d)

Obtain the Fourier expansion of the following:

1. $f(x) = \begin{cases} -x^2, & \text{if } -\pi < x < 0 \\ x^2, & \text{if } 0 < x < \pi \end{cases}$

2. $f(x) = x^3$, in $-\pi < x < \pi$.

3. $f(x) = -\cos x$, in $-\pi < x < 0$, and $f(x) = \cos x$, in $0 < x < \pi$.

4. $f(x) = \cos \alpha x$, in $-\pi < x < \pi$, where α is not an integer.

Deduce that $\cot \pi\alpha = \frac{1}{\pi}\left\{\frac{1}{\alpha} - \sum_{n=1}^{\infty} \frac{2\alpha}{n^2 - \alpha^2}\right\}$. **(*Calicut, 71 Engg.*)**

5. $f(x) = \sin \alpha x$, in $-\pi < x < \pi$, where α is not an integer.

6. Show that in $-\pi < x < \pi$,

$$x \sin x = 1 - \frac{1}{2}\cos x - 2\left(\frac{\cos 2x}{1\cdot 3} - \frac{\cos 3x}{2\cdot 4} + \ldots\right)$$

7. If $f(x) = 1 + \frac{2x}{\pi}$ in $-\pi \le x \le 0$ (*Anna Ap 2005*)

$= 1 - \frac{2x}{\pi}$ in $0 \le x \le \pi$

show that $f(x) = \frac{8}{\pi^2}\left(\cos x + \frac{1}{3^2}\cos 3x + \frac{1}{5^2}\cos 5x + \ldots\right)$

Hence show that $\sum_{n=1}^{\infty}(2n-1)^{-2}=\frac{\pi^2}{8}$.

Half-range series. In some problems, we are concerned with the interval 0 to π, (instead of the usual interval of length 2π.) Further, the conditions of the problem may require us to expand the given function in a series of sines alone or a series of consines alone.

Half-range sine series. This is achieved as follows: suppose $f(x)$ is given for $0 < x < \pi$, and we need a sine expansion for $f(x)$ in that interval. Then, we introduce a new function $F(x)$, which is defined as below.

$$F(x) = \begin{cases} f(x), \text{ when } 0 < x < \pi. \\ -f(-x), \text{ when } -\pi < x < 0. \end{cases}$$

From the definition of $F(x)$ made here, $F(x)$ is an odd function of x; and it is defined in the interval $-\pi$ to π. From theorem 5(a), the Fourier expansion of $F(x)$ will contain sine terms only. Since we are concerned with only the range $0 < x < \pi$, and since in this range $F(x) \equiv f(x)$, the sine series of $F(x)$ is the required series for $f(x)$ in the range $0 < x < \pi$. Further, by Theorem 5(a).

$$b_n = \frac{2}{\pi}\int_0^{\pi} F(x) \sin nx \, dx.$$

But as stated above, within the limits of integration, $F(x)$ coincides with $f(x)$.

Therefore, $f(x) = \sum_{1}^{\infty} b_n \sin nx$

where $$b_n = \frac{2}{\pi}\int_0^{\pi} f(x) \sin nx \, dx.$$

Since b_n depends only on $f(x)$, the actual creation of $F(x)$ is not necessary.

Half-range cosine series. On the other hand, if we need a cosine series for $f(x)$ defined in $0 < x < \pi$, define $F(x)$ by

$$F(x) = \begin{cases} f(x), \text{ when } 0 < x < \pi \\ f(-x), \text{ when } -\pi < x < 0. \end{cases}$$

Then $F(x)$ is an even function of x in the interval $(-\pi, \pi)$. By theorem 5, the Fourier series of $F(x)$ will have only the cosine terms, (including the constant). But, as before $F(x)$ coincides with $f(x)$ in the required interval $0 < x < \pi$. Hence the cosine series for $F(x)$ is the required cosine series for $f(x)$ in the given interval.

$$\text{Now, } a_n = \frac{2}{\pi}\int_0^{\pi} F(x)\cos nx\,dx = \frac{2}{\pi}\int_0^{\pi} f(x)\cos nx\,dx.$$

Therefore,

$$f(x) = \frac{a_0}{2} + \sum_{n=1}^{\infty} a_n \cos nx$$

where $$a_0 = \frac{2}{\pi}\int_0^{\pi} f(x)\,dx$$

and $$a_n = \frac{2}{\pi}\int_0^{\pi} f(x)\cos nx\,dx.$$

Thus a function $f(x)$ defined over the interval $0 < x < \pi$ is capable of expansion in two distinct types of series as above. Since these two series are valid over the half-range $0 < x < \pi$, we call them half-range series.

Since in the above discussions, we did not stipulate any condition on $f(x)$, the required type of series can be got whether $f(x)$ is odd, even or neither.

Example 24. *Expand the function $f(x) = \sin x$, $0 < x < \pi$ in Fourier cosine series.* ***(Calicut, '71 Engg.)***

Let $f(x) = \frac{a_0}{2} + \sum_{n=1}^{\infty} a_n \cos nx$ where

$$a_0 = \frac{2}{\pi}\int_0^{\pi} \sin x\,dx = \frac{2}{\pi}\Big[-\cos x\Big]_0^{\pi} = \frac{2}{\pi}\left(1+1\right) = \frac{2}{\pi}\times 2 = \frac{4}{\pi}$$

$$a_1 = \frac{2}{\pi}\int_0^{\pi} \sin x \cos x\,dx = \frac{1}{\pi}\Big[\sin^2 x\Big]_0^{\pi} = \frac{1}{\pi}\times 0 = 0$$

For $n = 2, 3, 4, \ldots$. (for the working, refer worked example 4)

$$a_n = \frac{2}{\pi}\int_0^{\pi} \sin x \cos nx\,dx = -\frac{2(1+\cos n\pi)}{\pi(n-1)(n+1)}$$

Hence in the given interval,

$$f(x) = \frac{2}{\pi} - \frac{4}{\pi}\left\{\frac{\cos 2x}{1\cdot 3} + \frac{\cos 4x}{3\cdot 5} + \frac{\cos 6x}{5\cdot 7} + \cdots \text{to } \infty\right\}$$

Example 25. *Find sine half-range series for the function* $f(x) = x$, *in* $0 < x \le \pi/2$, *and* $f(x) = \pi - x$ *in* $\frac{1}{2}\pi < x < \pi$.

Let $f(x) = \sum_{n=1}^{\infty} b_n \sin nx$, where

$$b_n = \frac{2}{\pi}\int_0^{\pi} f(x) \sin nx\, dx = \frac{2}{\pi}\left\{\int_0^{\pi/2} x \sin nx\, dx + \int_{\pi/2}^{\pi} (\pi - x) \sin nx\, dx\right\} \quad ....(i)$$

Using the method of integration by parts,

$$\int_0^{\pi/2} x \sin nx\, dx = \left[x\left(\frac{-\cos nx}{n}\right)\right]_0^{\pi/2} + \int_0^{\pi/2} \frac{\cos nx}{n}\, dx$$

$$= -\frac{\pi}{2n}\cos\frac{n\pi}{2} + \frac{1}{n^2}\sin\frac{n\pi}{2};$$

$$\int_{\pi/2}^{\pi} (\pi - x) \sin nx\, dx = \left[(\pi - x)\left\{-\frac{\cos nx}{n}\right\}\right]_{\pi/2}^{\pi} - \int_{\pi/2}^{\pi} \frac{\cos nx}{n}\, dx$$

$$= \frac{\pi}{2n}\cos\frac{n\pi}{2} + \frac{1}{n^2}\sin\frac{n\pi}{2}.$$

Using the values of the integrals in (*i*),

$$b_n = \frac{2}{\pi} \times \frac{2}{n^2}\sin\frac{n\pi}{2} = \frac{4}{\pi n^2}\sin\frac{n\pi}{2}.$$

Case 1. Let $n = 2m$; then $b_{2m} = \dfrac{4}{\pi (2m)^2}\sin m\pi = \dfrac{4}{4\pi m^2} \times 0 = 0;$

putting $m = 1, 2, 3, 4, ...$, we get $b_2 = b_4 = b_6 = \cdots = 0$.

Case 2. Let $n = 2r + 1$, where r is any positive integer. Then,

$$b_{2r+1} = \frac{4}{\pi (2r+1)^2}\sin\frac{(2r+1)\pi}{2} = \frac{4}{\pi (2r+1)^2}\sin\left(r\pi + \frac{\pi}{2}\right)$$

$$= \frac{4}{\pi (2r+1)^2}\cos r\pi = (-1)^r \frac{4}{\pi (2r+1)^2}.$$

Putting $r = 0, 1, 2$, we get

$$b_1 = +\frac{4}{\pi};\ b_3 = \frac{-4}{\pi \times 3^2};\ b_5 = \frac{4}{\pi \times 5^2};\ b_7 = \frac{-4}{\pi \times 7^2};\ .....$$

Substituting the values of the coefficients,

$$f(x) = \frac{4}{\pi}\left\{\sin x - \frac{\sin 3x}{3^2} + \frac{\sin 5x}{5^2} - \frac{\sin 7x}{7^2} + \cdots \text{ to } \infty\right\}$$

Example 26. *Find half-range Forurier cosine series and sine series. for* $f(x) = x$ *in* $0 < x < \pi$ **(*M.S. 1991 Ap.*)**

(*i*) **Half-range cosine series:**

Let $f(x) = \frac{a_0}{2} + \sum_{n=1}^{\infty} a_n \cos nx$

where $a_0 = \frac{2}{\pi} \int_0^{\pi} f(x)\, dx$

$$= \frac{2}{\pi} \int_0^{\pi} x dx$$

$$= \frac{2}{\pi} \left(\frac{x^2}{2} \right)_0^{\pi} = \pi$$

$$a_n = \frac{2}{\pi} \int_0^{\pi} f(x) \cos nx\, dx$$

$$= \frac{2}{\pi} \int_0^{\pi} (x) \cos nx\, dx$$

$$= \frac{2}{\pi} \left[(x) \left(\frac{\sin nx}{n} \right) - (1) \left(-\frac{\cos nx}{n^2} \right) \right]_0^{\pi}$$

$$= \frac{2}{\pi} \left[\frac{1}{n^2} \{(-1)^n - 1)\} \right]$$

$$= \frac{-4}{\pi n^2} \text{ if } n \text{ is odd}$$

$$= 0 \text{ if } n \text{ is even}$$

$$\therefore \quad x = \frac{\pi}{2} - \frac{4}{\pi} \sum_{n=1,3,5\cdots} \frac{1}{n^2} \cos nx$$

(*ii*) **Half-range-sine series:**

Let $x = \sum_{n=1}^{\infty} b_n \sin nx$

$$\text{where} \quad b_n = \frac{2}{\pi}\int_0^{\pi} x \sin nx \, dx$$

$$= \frac{2}{\pi}\left[(x)\left(-\frac{\cos nx}{n}\right) - (1)\left(-\frac{\sin nx}{n^2}\right)\right]_0^{\pi}$$

$$= \frac{2}{\pi}\left[-\frac{\pi}{n}(-1)^n\right]$$

$$= \frac{2}{n}(-1)^{n-1}$$

$$\therefore \quad x = 2\sum_{n=1}^{\infty} \frac{(-1)^{n-1}}{n} \sin nx$$

Example 27. *Express $f(x) = x(\pi - x)$, $0 < x < \pi$, as a Fourier series of periodicity 2π containing (i) Sine terms only (ii) cosine terms only.*

Hence deduce, $1 - \frac{1}{3^3} + \frac{1}{5^3} - \frac{1}{7^3} + \cdots = \frac{\pi^3}{32}$

and $\frac{1}{1^2} - \frac{1}{2^2} + \frac{1}{3^3} - \frac{1}{4^2} + \cdots = \frac{\pi^2}{12}$

(*i*) **Sine series**:

Let $f(x) = \sum_{1}^{\infty} b_n \sin nx$

$$\text{where} \quad b_n = \frac{2}{\pi}\int_0^{\pi} x(\pi - x) \sin nx \, dx$$

$$= \frac{2}{\pi}\left[\{\pi x - x^2\}\left(-\frac{\cos nx}{n}\right) - (\pi - 2x)\left(\frac{\sin nx}{n^2}\right) + (-2)\left(\frac{\cos nx}{n^3}\right)\right]_0^{\pi}$$

$$= \frac{2}{\pi}\left[-\frac{2}{n^3}\{(-1)^n - 1\}\right]$$

$$= \frac{4}{\pi n^3}\left[1 - (-1)^n\right]$$

$= 0$ if n is even

$= \frac{8}{\pi n^3}$ if n is odd

$$\therefore \quad f(x) = \frac{8}{\pi} \sum_{n=1}^{\infty} \frac{1}{(2n-1)^3} \sin (2n-1)\, x.$$

Setting $x = \pi/2$ which is a point of continuity we get first deduction.

(*ii*) **Cosine series:**

$$\text{Let } f(x) = \frac{a_0}{2} + \sum_{1}^{\infty} a_n \cos nx$$

where
$$a_0 = \frac{2}{\pi} \int_0^{\pi} x(\pi - x)\, dx$$

$$= \frac{2}{\pi} \left[\frac{\pi x^2}{2} - \frac{x^3}{3} \right]_0^{\pi} = \frac{\pi^2}{3}$$

$$a_n = \frac{2}{\pi} \int_0^{\pi} (\pi x - x^2) \cos nx\, dx$$

$$= \frac{2}{\pi} \left[(\pi x - x^2) \left(\frac{\sin nx}{n} \right) - (\pi - 2x) \left(-\frac{\cos nx}{n^2} \right) + (-2) \left(-\frac{\sin nx}{n^3} \right) \right]_0^{\pi}$$

$$= \frac{2}{\pi} \left[-\frac{\pi}{n^2} (-1)^n - \frac{\pi}{n^2} \right] = -\frac{2}{n^2} [1 + (-1)^n]$$

$$= 0 \text{ for } n \text{ odd}$$

$$= -\frac{4}{n^2} \text{ for } n \text{ even}$$

$$x(\pi - x) = \frac{\pi^2}{6} - 4 \sum_{n = 2, 4, 6 \ldots}^{\infty} \frac{1}{n^2} \cos nx$$

$$x(\pi - x) = \frac{\pi^2}{6} - \sum_{1}^{\infty} \frac{1}{n^2} \cos 2nx.$$

Setting $x = \pi/2$ which is a point of continuity,

$$\frac{\pi}{2} \left(\pi - \frac{\pi}{2}\right) = \frac{\pi^2}{6} - \sum \frac{1}{n^2} (-1)^n$$

$$\sum \frac{1}{n^2} (-1)^n = \frac{\pi^2}{6} - \frac{\pi^2}{4}$$

$$\frac{1}{1^2}-\frac{1}{2^2}+\frac{1}{3^2}\cdots\cdots=\frac{\pi^2}{12}$$

Example 28. *Expand $f(x) = x \sin x$ as a cosine series in $0 < x < \pi$ and show that*

$$1+\frac{2}{1\cdot 3}-\frac{2}{3\cdot 5}+\frac{2}{5\cdot 7}-\cdots=\pi/2$$

Let $f(x)=\dfrac{a_0}{2}+\displaystyle\sum_{n=1}^{\infty} a_n \cos nx$

$$a_0=\frac{2}{\pi}\int_0^{\pi} x \sin x \, dx$$

$$=\frac{2}{\pi}\left[(x)(-\cos x)-(1)(-\sin x)\right]_0^{\pi}$$

$$=\frac{2}{\pi}[\pi]=2$$

$$a_n=\frac{2}{\pi}\int_0^{\pi} x \sin x \cos nx \, dx$$

$$=\frac{1}{\pi}\int_0^{\pi} x\left[\sin(n+1)x-\sin(n-1)x\right]dx$$

$$=\frac{1}{\pi}\left[x\left\{-\frac{\cos(n+1)x}{n+1}\right\}-(1)\left\{-\frac{\sin(n+1)x}{(n+1)^2}\right\}\right.$$

$$\left.-x\left\{-\frac{\cos(n-1)x}{n-1}\right\}+(1)\left\{-\frac{\sin(n-1)x}{(n-1)^2}\right\}\right]_0^{\pi} \quad \text{if } n\neq 1$$

$$=\frac{1}{\pi}\left[\frac{-\pi(-1)^{n+1}}{n+1}+\frac{\pi}{n-1}(-1)^{n-1}\right]$$

$$=(-1)^n\left[\frac{1}{n+1}-\frac{1}{n-1}\right]=\frac{2(-1)^{n+1}}{n^2-1}, \text{ if } n\neq 1$$

$$a_1=\frac{2}{\pi}\int_0^{\pi} x \sin x \cos x \, dx$$

$$= \frac{1}{\pi} \int_0^{\pi} x \sin 2x \, dx$$

$$= \frac{1}{\pi} \left[(x) \left(-\frac{\cos 2x}{2} \right) - (1) \left(-\frac{\sin 2x}{4} \right) \right]_0^{\pi}$$

$$= \frac{1}{\pi} \left[-\frac{\pi}{2} \right] = -\frac{1}{2}$$

$$x \sin x = 1 - \frac{1}{2} \cos x - 2 \sum_{n=2}^{\infty} \frac{(-1)^n}{n^2 - 1} \cos nx$$

Deduction: Set $x = \pi/2$ which is a point of continuity.

$$\therefore \quad 1 - \frac{1}{2} \cos \frac{\pi}{2} - 2 \sum_{n=2}^{\infty} \frac{(-1)^n}{(n-1)(n+1)} \cos \frac{n\pi}{2} = \frac{\pi}{2} \sin \frac{\pi}{2}$$

$$1 - 2 \left[-\frac{1}{1 \cdot 3} + \frac{1}{3 \cdot 5} - \frac{1}{5 \cdot 7} + \cdots \right] = \pi/2$$

$$1 + 2 \left[\frac{1}{1 \cdot 3} - \frac{1}{3 \cdot 5} + \frac{1}{5 \cdot 7} \cdots \right] = \pi/2$$

Example 29. *If* $f(x) = \frac{\pi x}{4}, \ 0 < x < \pi/2$

$$= \frac{\pi}{4} (\pi - x), \ \pi/2 < x < \pi$$

Express f (x) in a series of cosines only (of periodicty 2π) **(M.S. 1990 A.)**

Let $f(x) = \frac{a_0}{2} + \sum_{n=1}^{\infty} a_n \cos nx$

where $a_0 = \frac{2}{\pi} \int_0^{\pi} f(x) \, dx$

$$= \frac{2}{\pi} \left[\int_0^{\pi/2} \frac{\pi x}{4} \, dx + \frac{\pi}{4} \int_{\pi/2}^{\pi} (\pi - x) \, dx \right]$$

$$= \frac{2}{\pi} \left[\frac{\pi}{4} \left(\frac{x^2}{2} \right)_0^{\pi/2} + \frac{\pi}{4} \left(\pi x - \frac{x^2}{2} \right)_{\pi/2}^{\pi} \right]$$

$$= \frac{1}{2}\left[\frac{\pi^2}{8} + \frac{\pi^2}{2} - \frac{3}{8}\pi^2\right] = \frac{\pi^2}{8}$$

$$a_n = \frac{2}{\pi}\left[\frac{\pi}{4}\int_0^{\pi/2} x \cos nx\, dx + \frac{\pi}{4}\int_{\pi/2}^{\pi} (\pi - x)\cos nx\, dx\right]$$

$$= \frac{1}{2}\left[(x)\left(+\frac{\sin nx}{n}\right) - (1)\left(-\frac{\cos nx}{n^2}\right)\right]_0^{\pi/2}$$

$$+ \frac{1}{2}\left[(\pi - x)\left(\frac{\sin nx}{n}\right) - (-1)\left(\frac{\cos nx}{n^2}\right)\right]_{\pi/2}^{\pi}$$

$$= \frac{1}{2}\left[\frac{\pi}{2n}\sin\frac{n\pi}{2} + \frac{1}{n^2}\cos\frac{n\pi}{2} - \frac{1}{n^2} - \frac{\pi}{2n}\sin\frac{n\pi}{2} - \frac{1}{n^2}\cos n\pi + \frac{1}{n^2}\cos\frac{n\pi}{2}\right]$$

$$= \frac{1}{2}\left[\frac{2}{n^2}\cos\frac{n\pi}{2} - \frac{1}{n^2} - \frac{1}{n^2}\cos n\pi\right]$$

Hence $f(x) = \dfrac{\pi^2}{16} + \dfrac{1}{2}\sum_{n=1}^{\infty}\left(\dfrac{2}{n^2}\cos\dfrac{n\pi}{2} - \dfrac{1}{n^2} - \dfrac{1}{n^2}\cos n\pi\right)\cos nx.$

Exercises 1(e)

1. Find the sine and cosine half range series in $0 < x < \pi$, for

   (*i*) $f(x) = x + 1$, (*ii*) $f(x) = e^x$, (*iii*) $f(x) = x^2$.

2. Find the cosine half-range series for the function

$$f(x) = \begin{cases} x, \text{ when } 0 < x < \pi/2, \\ \pi - x, \text{ when } \frac{1}{2}\pi < x < \pi. \end{cases}$$ **(*MS. 1991 A.*)**

   Deduce the sum of the series $\sum_{n=0}^{\infty} \dfrac{1}{(2n+1)^2}$

   Draw the graph of the function which the series represents in $(-2\pi, 2\pi)$.

   (*Madurai, 72 Engg.*)

3. Expand the function $f(x) = \sin x$, in $0 < x < \pi$ in a sine half-range series and half-range cosine series

4. Show that $\log\left(2\sin\dfrac{x}{2}\right) = -\sum_{n=1}^{\infty}\dfrac{\cos nx}{n}$, if $0 < x < \pi$.

5. If $f(x) = kx,\ 0 \le x \le \pi/2,$

$$= k(\pi - x),\ \frac{1}{2}\pi \le x \le \pi,$$

Show that $f(x) = \dfrac{4k}{\pi}\left(\dfrac{\sin x}{1^2} - \dfrac{\sin 3x}{3^2} + \dfrac{\sin 5x}{5^2} - \dots\right)$ **(S.V.U, '64)**

6. If c is a constant, then, in $0 < x < \pi$, show that

$$c = \frac{4c}{\pi}\left(\sin x + \frac{\sin 3x}{3} + \frac{1}{5}\sin 5x + \dots\right)$$

7. Show that in $(0, \pi)$

$$x^2 = \frac{2}{\pi}\left[\left(\frac{\pi^2}{1} - \frac{4}{1^3}\right)\sin x - \frac{\pi^2}{2}\sin 2x + \left(\frac{\pi^2}{3} - \frac{4}{3^3}\right)\sin 3x - \dots\right]$$

Deduce the sum of the series $\dfrac{1}{1^3} - \dfrac{1}{3^3} + \dfrac{1}{5^3} - \dfrac{1}{7^3} + \dots$ **(Madras, B.E. '63)**

8. Obtain cosine series for

$$f(x) = \begin{cases} \cos x \text{ in } 0 < x < \pi/2 \\ 0 \text{ in } \pi/2 < x < \pi \end{cases}$$ **(M.S. 1990 Ap.)**

[**Ans :** $a_0 = \dfrac{2}{\pi},\ a_n = \dfrac{-2\cos n\pi/2}{\pi(n^2 - 1)},\ a_1 = 1/2$]

9. Find half-range sine series for $f(x)$ in $(0, \pi)$ where $f(x) = \sin kx$ for k, not an integer.

$$\left[\textbf{Ans. } b_n = \frac{2n}{\pi}\frac{(-1)^{n+1}\sin k\pi}{\pi(n^2 - k^2)}\right]$$

10. Find half-range cosine series for $f(x) = x^2$ in $0 \le x \le \pi$ and hence deduce the sum $1 - \dfrac{1}{2^2} + \dfrac{1}{3^2} - \dfrac{1}{4^2} + \dots$ **(M.S. 1989 Nov.)**

$$\left[\textbf{Ans. } f(x) = \frac{\pi^2}{3} - 4\left(\cos x - \frac{\cos 2x}{2^2} + \frac{\cos 3x}{3^2} - \dots\right)\right]$$

11. Find the half-range sine series for $f(x) = x^3$ in $(0, \pi)$.

$$\left[\textbf{Ans. } x^3 = 2\sum_{1}^{\infty}(-1)^n\left[\frac{6}{n^3} - \frac{\pi^2}{n}\right]\sin nx\right]$$

12. Find Fourier cosine series in $(0, \pi)$ for

$f(x) = x$ in $0 < x < \pi/2$

$= \pi - x$ in $\pi/2 < x < \pi$

$$\left[\textbf{Ans. } f(x) = \frac{\pi}{4} - \frac{2}{\pi}\left(\frac{\cos 2x}{1^2} + \frac{\cos 6x}{3^2} + \frac{\cos 10x}{5^2} + \dots\right)\right]$$

13. Obtain cosine series for

$f(x) = \cos x,\ 0 < x < \pi/2$

$= 0,\ \pi/2 < x < \pi$

14. Obtain sine series for $f(x) = \pi x - x^2$ in $(0, \pi)$

15. In $(0, \pi)$, prove

$$\sin x = \frac{4}{\pi}\left(\frac{1}{2} - \frac{\cos 2x}{1 \cdot 3} - \frac{\cos 4x}{3 \cdot 5} - \frac{\cos 6x}{5 \cdot 7} - \dots\right)$$

and
$$\cos x = \frac{4}{\pi}\left(\frac{2 \sin 2x}{2^2 - 1} + \frac{4 \sin 4x}{4^2 - 1} + \frac{6 \sin 6x}{6^2 - 1} + \dots\right)$$

16. Find half range sine series and cosine series for

(*i*) $f(x) = x - x^2$ in $0 < x < \pi$

(*ii*) $f(x) = x$ in $0 < x < \pi/2$

$= 0$ in $\frac{\pi}{2} < x < \pi$

(*iii*) $f(x) = x \sin x$ in $0 < x < \pi$.

Change of interval. In very many engineering applications of Fourier series, we require an expansion of a given function over an interval of length other than π or 2π. It is therefore a matter of importance to determine how the foregoing theory can be applied to the representation of periodic functions of arbitrary period. The problem is not a difficult one, for all that is needed is a proportional change of scale.

Suppose we have to obtain the Fourier series of a function $f(x)$ defined for the interval $-L < x < L$, where L is *any* positive number. Introduce a new variable z given by $z = \frac{\pi}{L} x$ .... (*i*). When $x = -L$, $z = -\pi$; and when $x = L$, $z = \pi$. Thus, when x changes from $-L$ to L, z changes from $-\pi$ to π. From (*i*), $x = \frac{Lz}{\pi}$; hence in terms of the variable z, $f(x)$ becomes $f\left(\frac{Lz}{\pi}\right)$. For the sake of clarity, let us call $f\left(\frac{Lz}{\pi}\right)$ by $F(z)$. Now, $F(z)$ is a function of z; since $f(x)$ is defined in the interval $(-L, L)$ and since $F(z)$ is derived from $f(z)$ using (*i*), $F(z)$ is defined in the interval $(-\pi, \pi)$. This is precisely the situation that we have mastered. Expanding $F(z)$ into a Fourier series, we get

$$F(z) = \frac{a_0}{2} + \sum_{n=1}^{\infty} (a_n \cos nz + b_n \sin nz),$$

where, $$a_n = \frac{1}{\pi}\int_{-\pi}^{\pi} F(z) \cos nz \, dz, \text{ and } b_n = \frac{1}{\pi}\int_{-\pi}^{\pi} F(z) \sin nz \, dz.$$

We introduced the notation $F(z)$ only for the sake of familiarity. It was nothing but $f\left(\frac{Lz}{\pi}\right)$. Making that change throughout, we get

$$f\left(\frac{Lz}{\pi}\right) = \frac{a_0}{2} + \sum_{n=1}^{\infty} (a_n \cos nz + b_n \sin nz),$$

where $$a_n = \frac{1}{\pi}\int_{-\pi}^{\pi} f\left(\frac{Lz}{\pi}\right)\cos nz\, dz, \text{ and } b_n = \frac{1}{\pi}\int_{-\pi}^{\pi} f\left(\frac{Lz}{\pi}\right)\sin nz\, dz.$$

Now that we have managed to get an expansion in z, let us go back to x this is done with the help of (*i*). From (*i*), $dz = \frac{\pi}{L}dx$. In the above result, changing every z in favour of x, we have

$$f(x) = \frac{a_0}{2} + \sum_{n=1}^{\infty}\left\{a_n \cos\frac{n\pi x}{L} + b_n \sin\frac{n\pi x}{L}\right\},$$

where $$a_n = \frac{1}{\pi}\int_{-L}^{L} f(x)\left\{\cos\frac{n\pi x}{L}\right\}\frac{\pi}{L}dx, \text{ and } b_n = \frac{1}{\pi}\int_{-L}^{L} f(x)\left\{\sin\frac{n\pi x}{L}\right\}\frac{\pi}{L}dx.$$

i.e., $$a_n = \frac{1}{L}\int_{-L}^{L} f(x)\cos\frac{n\pi x}{L}dx, \text{ and } b_n = \frac{1}{L}\int_{-L}^{L} f(x)\sin\frac{n\pi x}{L}dx.$$

Note 1. The Fourier expansion of an even function $f(x)$ defined in the interval $-L$ to L will contain cosine terms only (i.e., $b_n = 0$, and $a_n = \frac{2}{L}\int_0^L f(x)\cos\frac{n\pi x}{L}dx$.

Note 2. The Fourier expansion of an odd function $f(x)$ defined in the interval $(-L, L)$ will contain sine terms only (*i.e.,* $a_n = 0$), and $b_n = \frac{2}{L}\int_0^L f(x)\sin\frac{n\pi x}{L}dx$.

Note 3. *Any* function $f(x)$ defined in the half-range 0 to L admits of

(*i*) a cosine expansion, where $a_n = \frac{2}{L}\int_0^L f(x)\cos\frac{n\pi x}{L}dx$

(*ii*) a sine expansion, where $b_n = \frac{2}{L}\int_0^L f(x)\sin\frac{n\pi x}{L}dx$re

Note 4. If n is any integer, then

$$\cos\frac{n\pi(x+2L)}{L} = \cos\left(\frac{n\pi x}{L} + 2n\pi\right) = \cos\frac{n\pi x}{L}$$

and similarly for sines. Hence each term of the expansion for $f(x)$ has period $2L$, and therefore the sum also has period $2L$. For this reason, the sum cannot represent an arbitrary function on $(-\infty, \infty)$; it represents periodic functions only.

Note 5. It is presumed that the function $f(x)$ introduced in the main discussion above or the notes, satisfies Dirichlet's conditions within the interval of its definition.

Note 6. If the function $f(x)$ is defined in the interval c to $c + 2L$, then the Fourier expansion is given by

$$f(x) = \frac{a_0}{2} + \sum_{n=1}^{\infty} \left(a_n \cos \frac{n\pi x}{L} + b_n \sin \frac{n\pi x}{L} \right),$$

where $a_n = \frac{1}{L} \int_c^{c+2L} f(x) \cos \frac{n\pi x}{L}\, dx$, and $b_n = \frac{1}{L} \int_c^{c+2L} f(x) \sin \frac{n\pi x}{L}\, dx.$

Example 30. *Obtain the Fourier series expansion of $f(x)$, given that*

$$f(x) = \begin{cases} 1 \text{ for } 0 < x < 1 \\ 2 \text{ for } 1 < x < 3 \end{cases}$$

and $f(x) = \frac{3}{2}$ when $x = 0, 1, 3$ and $f(x+3) = f(x)$ for all x.

(*Madras, '72 Engg.*)

$f(x)$ is defined in the interval 0 to 3. Further, we are informed that $f(x)$ has the period 3. Thus $2L = 3$, giving $L = \frac{3}{2}$. Hence

$$f(x) = \frac{a_0}{2} + \sum_{n=1}^{\infty} \left(a_n \cos \frac{n\pi x}{\frac{3}{2}} + b_n \sin \frac{n\pi x}{\frac{3}{2}} \right)$$

i.e., $$f(x) = \frac{a_0}{2} + \sum_{n=1}^{\infty} \left(a_n \cos \frac{2n\pi x}{3} + b_n \sin \frac{2n\pi x}{3} \right) \qquad \text{...(i)}$$

where, $$a_0 = \frac{1}{\frac{3}{2}} \int_0^3 f(x)\, dx = \frac{2}{3} \left\{ \int_0^1 1\, dx + \int_1^3 2\, dx \right\} = \frac{2}{3} \left\{ [x]_0^1 + [2x]_1^3 \right\}$$

$$= \frac{2}{3} \{ (1 - 0) + (6 - 2) \} = \frac{2}{3} \times 5 = \frac{10}{3}$$

For $n = 1, 2, 3, \cdots,$

$$a_n = \frac{1}{\frac{3}{2}} \int_0^3 f(x) \cos \frac{n\pi x}{\frac{3}{2}}\, dx = \frac{2}{3} \left\{ \int_0^1 1 \cos \frac{2n\pi x}{3}\, dx + \int_1^3 2 \cos \frac{2n\pi x}{3}\, dx \right\}.$$

The following adjustment about the integrals is available.

$$a_n = \frac{2}{3}\left\{\int_0^1 \cos\frac{2n\pi x}{3}\,dx + 2\int_1^3 \cos\frac{2n\pi x}{3}\,dx\right\}$$

$$= \frac{2}{3}\left\{\left[\frac{3}{2n\pi}\sin\frac{2n\pi x}{3}\right]_0^1 + \left[\frac{6}{2n\pi}\sin\frac{2n\pi x}{3}\right]_1^3\right\}$$

$$= -\frac{1}{n\pi}\sin\frac{2n\pi}{3}$$

When n is a multiple of 3, that is when $n = 3m$, where m is an integer, we get

$$a_n = -\frac{1}{3m\pi}\sin 2m\pi = -\frac{1}{3m\pi}\times 0 = 0.$$

Thus $a_3 = a_6 = a_9 = \cdots = 0.$

Now $a_1 = -\frac{1}{\pi}\sin\frac{2\pi}{3} = -\frac{1}{\pi}\times\frac{\sqrt{3}}{2} = -\frac{\sqrt{3}}{2\pi}$

$$a_2 = -\frac{1}{2\pi}\sin\frac{4\pi}{3} = -\frac{1}{2\pi}\left(-\frac{\sqrt{3}}{2}\right) = \frac{\sqrt{3}}{2\pi}\left(\frac{1}{2}\right).$$

Similarly, $a_4 = -\left(\frac{\sqrt{3}}{2\pi}\right)\frac{1}{4};\ a_5 = \frac{\sqrt{3}}{2\pi}\left(\frac{1}{5}\right);\ \ldots$

Again, $b_n = \frac{2}{3}\left\{\int_0^1 \sin\frac{2n\pi x}{3}\,dx + 2\int_1^3 \sin\frac{2n\pi x}{3}\,dx\right\}$

$$= \frac{2}{3}\left\{\left[-\frac{3}{2n\pi}\cos\frac{2n\pi x}{3}\right]_0^1 + \left[-\frac{6}{2n\pi}\cos\frac{2n\pi x}{3}\right]_1^3\right\}$$

$$= -\frac{1}{n\pi}\left(1 - \cos\frac{2n\pi}{3}\right)$$

When n is a multiple of 3, that is when $n = 3m$, where m is an integer,

$$b_n = -\frac{1}{3m\pi}(1 - \cos 2m\pi) = -\frac{1}{3m\pi}(1 - 1) = 0.$$

Also, $b_1 = -\frac{1}{\pi}\left(1 - \cos\frac{2\pi}{3}\right) = -\frac{1}{\pi}\left(1 + \frac{1}{2}\right) = -\frac{3}{2\pi}$

$$b_2 = -\frac{1}{2\pi}\left(1 - \cos\frac{4\pi}{3}\right) = -\frac{1}{2\pi}\left(1 + \frac{1}{2}\right) = -\frac{3}{2\pi}\left(\frac{1}{2}\right)$$

$$b_4 = -\frac{3}{2\pi}\left(\frac{1}{4}\right);\ b_5 = -\frac{3}{2\pi}\left(\frac{1}{5}\right);\ \cdots$$

Substituting the coefficients in (*i*), we have

$$f(x) = \frac{5}{3} - \frac{\sqrt{3}}{2\pi}\left\{\cos\frac{2\pi x}{3} - \frac{1}{2}\cos\frac{4\pi x}{3} + \frac{1}{4}\cos\frac{8\pi x}{3} - \frac{1}{5}\cos\frac{10\pi x}{3} + \cdots\right\}$$
$$-\frac{3}{2\pi}\left\{\sin\frac{2\pi x}{3} + \frac{1}{2}\sin\frac{4\pi x}{3} + \frac{1}{4}\sin\frac{8\pi x}{3} + \frac{1}{5}\sin\frac{10\pi x}{3} + \cdots\right\}$$

Note : Since the period of the function is 3, $f(0-) = f(3-) = 2$, by hypothesis. Again by hypothesis, $f(0+) = 1$. Hence $\frac{f(0-)+f(0+)}{2} = \frac{3}{2}$. According to Dirichlet's theorem, the sum of the series for $x = 0$ will be $\frac{3}{2}$; the hypothesis, defines $f(0) = \frac{3}{2}$. Hence the derived series represents the $f(x)$ at the discontinuity 0 also. This is true of the other discontinuities $x = 1$ and 3.

Example 31. *Find the Fourier series of the function*

$$f(t) = \begin{cases} 0, \text{ when } -2 < t < -1 \\ k, \text{ when } -1 < t < 1 \\ 0, \text{ when } 1 < t < 2 \end{cases} \quad \text{and } f(t+4) = f(t).$$

[Note that the name of the variable need not be necessarily *x*; here, it is *t*,]

The period is 4; the graph is symmetrical about the *y*-axis in the interval – 2 to 2. That is because $f(x)$ is even. Hence there will be no sine terms in its expansion. Since $L = 2$,

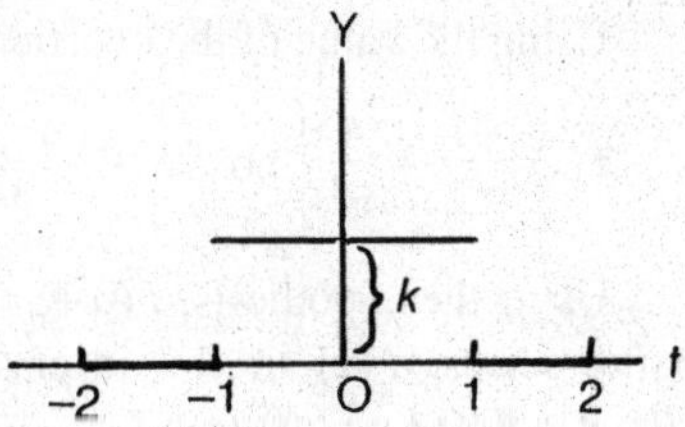

$$a_0 = \frac{2}{2}\int_0^2 f(t)\,dt = \int_0^1 k\,dt + \int_1^2 0\,dt = [\,kt\,]_0^1 = k$$

$$a_n = \frac{2}{2}\int_0^2 f(t)\cos\frac{n\pi t}{2}\,dt = \int_0^1 k\cos\frac{n\pi t}{2}\,dt = \frac{2k}{n\pi}\left[\sin\frac{n\pi t}{2}\right]_0^1$$

i.e., $a_n = \frac{2k}{n\pi}\sin\frac{n\pi}{2}$. Hence when *n* is even, $a_n = 0$.

Further, when $n = 1, 5, 9, \ldots, a_n = \frac{2k}{n\pi}$ and for $n = 3, 7, 11, \ldots,$

$$a_n = -\frac{2k}{n\pi}.$$

$$\therefore \ f(t) = \frac{k}{2} + \frac{2k}{\pi}\left\{ \cos\frac{\pi t}{2} - \frac{1}{3}\cos\frac{3\pi t}{2} + \frac{1}{5}\cos\frac{5\pi t}{2} - \cdots \right\}$$

Example 32. *Find the half-range cosine series for the function $f(x) = (x-1)^2$ in the interval $0 < x < 1$. Hence show that*

$$\pi^2 = 6\left\{ 1 + \frac{1}{2^2} + \frac{1}{3^2} + \cdots \right\}.$$ **(*Madras, '71 Engg.*)**

Here $L = 1$

$$a_0 = \frac{2}{1}\int_0^1 (x-1)^2\,dx = 2\left[\frac{(x-1)^3}{3}\right]_0^1 = 2\left\{0 - \left(-\frac{1}{3}\right)\right\} = \frac{2}{3}$$

$$a_n = \frac{2}{1}\int_0^1 (x-1)^2 \cos\frac{n\pi x}{1}\,dx = 2\int_0^1 (x-1)^2 \cos n\pi x\,dx$$

$$= 2\left[(x-1)^2\left(\frac{\sin n\pi x}{n\pi}\right) - 2(x-1)\left(-\frac{\cos n\pi x}{n^2\pi^2}\right) + (2)\left(-\frac{\sin n\pi x}{n^3\pi^3}\right)\right]_0^1$$

$$= 2\left[\frac{2}{n^2\pi^2}\right] = \frac{4}{n^2\pi^2}.$$

Using the value of the coefficients,

$$f(x) = \frac{1}{3} + \frac{4}{\pi^2}\left\{ \cos\pi x + \frac{\cos 2\pi x}{2^2} + \frac{\cos 3\pi x}{3^2} + \frac{\cos 4\pi x}{4^2} + \cdots \right\}$$

Using the hypothesis, $f(0+) = 1$; since we have developed the half range *cosine* series, $F(x)$ in the interval $(-1, 1)$ is an even function. Hence $F(0-) = F(0+) = f(0+) = 1$.

$\therefore \ \dfrac{F(0-) + F(0+)}{2} = \dfrac{1+1}{2} = 1$. Though the function is *not defined* at $x = 0$, by Dirichlet's theorem, the value of the series at $x = 0$, will be 1. Hence

$$1 = \frac{1}{3} + \frac{4}{\pi^2}\left\{ 1 + \frac{1}{2^2} + \frac{1}{3^2} + \cdots \right\} \text{ etc.}$$

Example 33. *If* $f(x) = \begin{cases} \sin x \text{ for } 0 \le x \le \pi/4 \\ \cos x \text{ for } \frac{1}{4}\pi \le x \le \pi/2 \end{cases}$

express $f(x)$ in a series of sines. **(*Madras, 71 Engg.*)**

In this question, the half range is 0 to $\pi/2$; hence $L = \pi/2$.

Now, $b_n = \frac{2}{\frac{\pi}{2}} \int_0^{\pi/2} f(x) \sin \frac{n\pi x}{\frac{\pi}{2}} \, dx = \frac{4}{\pi} \int_0^{\pi/2} f(x) \sin 2nx \, dx$

$$= \frac{4}{\pi} \left\{ \int_0^{\pi/4} \sin x \sin 2nx \, dx + \int_{\pi/4}^{\pi/2} \cos x \sin 2nx \, dx \right\}$$

$$= \frac{2}{\pi} \left\{ \int_0^{\pi/4} [\cos (2n-1)x - \cos (2n+1)x] \, dx + \int_{\pi/4}^{\pi/2} [\sin (2n+1)x + \sin (2n-1)x] \, dx \right\}$$

$$= \frac{2}{\pi} \left\{ \left[\frac{\sin (2n-1)x}{2n-1} - \frac{\sin (2n+1)x}{2n+1} \right]_0^{\pi/4} - \left[\frac{\cos (2n+1)x}{2n+1} + \frac{\cos (2n-1)x}{2n-1} \right]_{\pi/4}^{\pi/2} \right\}$$

$$= \frac{2}{\pi} \left\{ \frac{\sin\left(\frac{n\pi}{2} - \frac{\pi}{4}\right)}{2n-1} - \frac{\sin\left(\frac{n\pi}{2} + \frac{\pi}{4}\right)}{2n+1} - \frac{\cos\left(n\pi + \frac{\pi}{2}\right)}{2n+1} - \frac{\cos\left(n\pi - \frac{\pi}{2}\right)}{2n-1} + \frac{\cos\left(\frac{n\pi}{2} + \frac{\pi}{4}\right)}{2n+1} + \frac{\cos\left(\frac{n\pi}{2} - \frac{\pi}{4}\right)}{2n-1} \right\}$$

Expanding expressions like $\sin\left(\frac{n\pi}{2} - \frac{\pi}{4}\right)$, and using the value

$\sin \frac{\pi}{4} = \frac{1}{\sqrt{2}} = \cos \frac{\pi}{4}$, we get

$$b_n = \frac{2}{\pi} \left[\frac{1}{2n-1} \left\{ \left(\frac{1}{\sqrt{2}} \sin \frac{n\pi}{2} - \frac{1}{\sqrt{2}} \cos \frac{n\pi}{2} \right) + \left(\frac{1}{\sqrt{2}} \cos \frac{n\pi}{2} + \frac{1}{\sqrt{2}} \sin \frac{n\pi}{2} \right) \right\} - \frac{1}{2n+1} \left\{ \left(\frac{1}{\sqrt{2}} \sin \frac{n\pi}{2} + \frac{1}{\sqrt{2}} \cos \frac{n\pi}{2} \right) - \left(\frac{1}{\sqrt{2}} \cos \frac{n\pi}{2} - \frac{1}{\sqrt{2}} \sin \frac{n\pi}{2} \right) \right\} \right],$$

$$\text{since } \cos\left(n\pi + \frac{\pi}{2}\right) = 0 = \cos\left(n\pi - \frac{\pi}{2}\right)$$

$$= \frac{2}{\pi}\left\{\frac{\sqrt{2}}{2n-1}\sin\frac{n\pi}{2} - \frac{\sqrt{2}}{2n+1}\sin\frac{n\pi}{2}\right\}$$

$$= \frac{2\sqrt{2}}{\pi}\left(\sin\frac{n\pi}{2}\right)\frac{(2n+1)-(2n-1)}{(2n-1)(2n+1)} = \frac{4\sqrt{2}}{\pi\,(2n-1)(2n+1)}\sin\frac{n\pi}{2}$$

It is evident that when n is even, $b_n = 0$; further for $n = 1, 5, 9, \cdots$

$$b_n = \frac{4\sqrt{2}}{\pi\,(2n-1)\,(2n+1)} \quad \text{and for } n = 3, 7, 11, \ldots,$$

$$b_n = -\frac{4\sqrt{2}}{\pi\,(2n-1)(2n+1)}$$

$$\therefore\ f(x) = \frac{4\sqrt{2}}{\pi}\left\{\frac{\sin 2x}{1\cdot 3} - \frac{\sin 6x}{5\cdot 7} + \frac{\sin 10x}{9\cdot 11} - \frac{\sin 14x}{13\cdot 15} + \cdots\right\}$$

Exercises 1(f)

Develop the Fourier expansion of (1 to 11):

1. (*a*) $f(x) = 0$, in $-2 < x < 0$ and $f(x) = 1$, in $0 < x < 2$. ***(Madurai, 72 Engg.)***

(*b*) $f(x) = 0$, in $-2 < x < 0$ and $f(x) = k$, in $0 < x < 2$, k being $\neq 0$.

2. $f(x) = -1$, in $(-2, 0)$ and $f(x) = 1$, in $(0, 2)$.

3. $f(x) = 1$, in $-1 < x < 1$ and $f(x) = 0$, in $1 < x < 3$.

4. $f(x) = x$, in $-1 < x < 1$. **5.** $f(x) = |x|$, in $-1 < x < 1$.

6. $f(x) = \frac{1}{4} - x$, in $0 < x < \frac{1}{2}$, and $f(x) = x - \frac{3}{4}$, in $\frac{1}{2} < x < 1$.

7. $f(x) = \begin{cases} 0, \text{ in } -2 < x < -1 \\ 1 + x, \text{ in } -1 < x < 0 \\ 1 - x, \text{ in } 0 < x < 1 \\ 0, \text{ in } 1 < x < 2. \end{cases}$ **8.** $f(x) = x^2$, in $-1 < x < 1$.

9. $f(x) = \cos x$, when $-\frac{\pi}{2} < x < \frac{\pi}{2}$. **10.** $f(x) = 1 + \sin x$, in $-1 < x < 1$.

11. $f(x) = \begin{cases} 0, \text{ in } -3 < x < -1 \\ 1 + \cos \pi x, \text{ in } -1 < x < 1 \\ 0, \text{ in } 1 < x < 3. \end{cases}$

12. A sinusoidal voltage $E \sin \omega t$ is passed through a half wave rectifier which clips the negative portion of the wave. The resulting periodic function is given by.

$$u(t) = \begin{cases} 0, \text{ when } -\dfrac{\pi}{\omega} < t < 0 \\ E \sin \omega t, \text{ when } 0 < t < \pi/\omega. \end{cases}$$

Develop $u(t)$, in a Fourier series.

13. Find the Fourier series of $f(x)$ of period 4 given by

$$f(x) = \begin{cases} 1, \text{ when } -2 < x < 0 \\ e^{-x}, \text{ when } 0 < x < 2 \end{cases}$$

14. Show that the series $\frac{l}{\pi} \sum_{n=1}^{\infty} \frac{1}{n} \sin \frac{2n\pi x}{l}$ represents $\frac{1}{2} l - x$, when $0 < x < l$.

15. Find the expansion in the series of cosines, if $f(x) = 1,\ 0 < x < \pi$, and $f(x) = 0,\ \pi < x < 2\pi$.

16. Find the half-range expansions of the function $f(t) = t - t^2, 0 < t < 1$,

17. Find the half-range expansions of the function

$$f(t) = \begin{cases} \dfrac{2kt}{l}, \text{ when } 0 < t < \dfrac{l}{2}. \\ \dfrac{2k(l-t)}{l}, \text{ when } \dfrac{l}{2} < t < l. \end{cases}$$

Expand the following (18 to 24) :

18. $f(t) = \begin{cases} t^2, 0 < t < 1 \\ 2 - t, 1 < t < 2 \end{cases}$ in a cosine series.

19. $f(x) = x^2 - 2, 0 < x < 2$, in a sine series.

Deduce that $\pi^3 = 32\left(1 - \frac{1}{3^3} + \frac{1}{5^3} - \frac{1}{7^3} + \cdots\right)$

20. $f(x) = (x-1)^2, 0 < x < 1$, in a sine series.

21. $\cos t$ in a sine series for $0 < t < 1$.

22. $\cos \frac{1}{2} x$ in a sine series in $(0, \pi)$.

23. $\sin \frac{\pi x}{l}$ in a cosine series for $0 < x < l$.

24. e^x in a sine series for $0 < x < 1$.

25. If $f(x) = lx - x^2$ in $(0, l)$, show that the half range sine series for

$f(x)$ is $\frac{8l^2}{\pi^3} \sum_{n=1}^{\infty} \frac{1}{(2n+1)^3} \sin \frac{(2n+1)\pi x}{l}$.

26. For the function defined by the graph *OAB* in the figure below, show that the half range Fourier sine series of periodicity $2l$ is

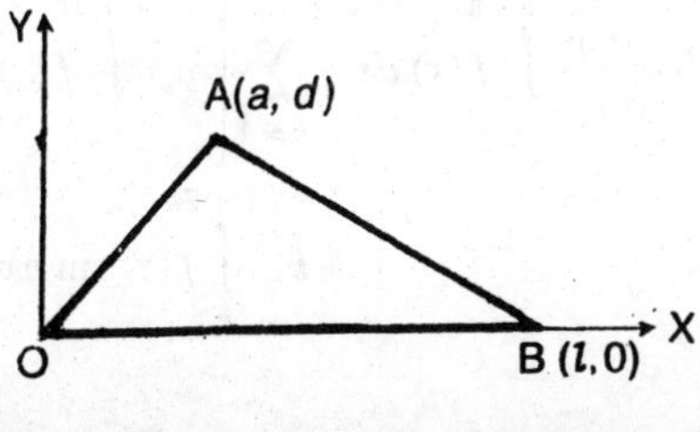

$$\frac{2dl^2}{a\pi^2 (l-a)} \sum_{n=1}^{\infty} \frac{l}{n^2} \sin \frac{n\pi a}{l} \sin \frac{n\pi x}{l}$$

Root-Mean square value of a function. PARSEVAL'S THEOREM.

In a number of physical applications, we have occasion to deal with the concept of root-mean-value (r.m.s), or effective value of a function.

The root-mean-square value of a function $y = f(x)$ over a given interval (a, b) is defined as

$$\bar{y} = \sqrt{\left\{ \frac{\int_a^b y^2\, dx}{b-a} \right\}} \qquad ...(i)$$

If the interval is taken as $(c, c + 2\pi)$, then

$$\bar{y}^2 = \frac{1}{2\pi} \int_c^{c+2\pi} y^2\, dx$$

Suppose that $y = f(x)$ is expressed as a Fourier-series of periodicity 2π in $(c, c + 2\pi)$. then,

$$y = f(x) = \frac{a_0}{2} + \sum_{n=1}^{\infty} (a_n \cos nx + b_n \sin nx) \qquad ...(ii)$$

$$\left.\begin{aligned} \text{where} \quad a_0 &= \frac{1}{\pi} \int_c^{c+2\pi} f(x)\, dx \\ a_n &= \frac{1}{\pi} \int_c^{c+2\pi} f(x) \cos nx\, dx \\ \text{and} \quad b_n &= \frac{1}{\pi} \int_c^{c+2\pi} f(x) \sin nx\, dx. \end{aligned}\right\} \qquad ...(iii)$$

Multiply (*ii*) by $f(x)$ and integrate term by term with respect to x over the given range. Thus,

$$\int_c^{c+2\pi} [f(x)]^2\, dx = \frac{a_0}{2} \int_c^{c+2\pi} f(x)\, dx + \sum_{n=1}^{\infty} \left[a_n \int_c^{c+2\pi} f(x) \cos nx\, dx + b_n \int_c^{c+2\pi} f(x) \sin nx\, dx \right]$$

$$= \frac{a_0}{2}(\pi a_0) + \sum_{n=1}^{\infty} [a_n(\pi a_n) + b_n(\pi b_n)] \text{ using } (iii)$$

$$\int_c^{c+2\pi} [f(x)]^2\, dx = 2\pi\left[\frac{a_0^{\,2}}{4} + \frac{1}{2}\sum_{n=1}^{\infty}(a_n^{\,2} + b_n^{\,2})\right]$$

$$= (\text{Range})\left[\frac{a_0^{\,2}}{4} + \frac{1}{2}\sum_{n=1}^{\infty}(a_n^{\,2} + b_n^{\,2})\right]$$

$$\bar{y}^2 = \frac{a_0^{\,2}}{4} + \frac{1}{2}\sum_{n=1}^{\infty}(a_n^{\,2} + b_n^{\,2}).$$

Ex. 34. *Find the Fourier series of periodicity 2π for $f(x) = x^2$, in $-\pi < x < \pi$. Hence show that*

$$\frac{1}{1^4} + \frac{1}{2^4} + \frac{1}{3^4} + \cdots + \text{to } \infty = \frac{\pi^4}{90}.$$ **(*Madras, M.Sc. Engg.*)**

In example 21, we have proved

$f(x) = \frac{\pi^2}{3} + 4\sum_{n=1}^{\infty} \frac{(-1)^n}{n^2} \cos nx$, which is the first part of this problem. The coefficients a_0, a_n, b_n were seen to be

$$a_0 = \frac{2\pi^2}{3},\quad a_n = \frac{4(-1)^n}{n^2},\quad b_n = 0.$$

Hence using the root-mean square value in series,

$$2\pi\left[\frac{a_0^{\,2}}{4} + \frac{1}{2}\sum_{n=1}^{\infty}(a_n^{\,2} + b_n^{\,2})\right] = \int_{-\pi}^{\pi} [f(x)^2]\, dx = \int_{-\pi}^{\pi} x^4 dx$$

$$2\pi\left[\frac{\pi^4}{9} + \frac{1}{2}\sum_{n=1}^{\infty}\frac{16}{n^4}\right] = \frac{2}{5}\pi^5$$

$$8\sum_{n=1}^{\infty}\frac{1}{n^4} = \frac{\pi^4}{5} - \frac{\pi^4}{9} = \frac{4\pi^4}{45}$$

$$\therefore \sum_{n=1}^{\infty}\frac{1}{n^4} = \frac{\pi^4}{90}.$$

Example 35. *The instantaneous current i at time t of an alternating-current wave is given by*

$i = I_1 \sin(\omega t + \alpha_1) + I_3 \sin(3\omega t + \alpha_3) + I_5 \sin(5\omega t + \alpha_5) + \ldots$ *Find the effective value of the current.*

Expanding expressions like $\sin(\omega t + \alpha_1)$ using the trigonometric addition formula we get

$$i = I_1 (\sin \omega t \cos \alpha_1 + \cos \omega t \sin \alpha_1) + I_3 (\sin 3\omega t \cos \alpha_3 + \cos 3\omega t \sin \alpha_3) + I_5 (\sin 5\omega t \cos \alpha_5 + \cos 5\omega t \sin \alpha_5) + \cdots$$

Let $A_n = I_n \sin \alpha_n$ and $B_n = I_n \cos \alpha_n$ ...(i)

Then, $i = A_1 \cos \omega t + A_3 \cos 3\omega t + A_5 \cos 5\omega t + \ldots$

$+ B_1 \sin \omega t + B_3 \sin 3\omega t + B_5 \sin 5\omega t + \ldots$

The R.H.S. is the Fourier expansion for i. If i denotes the required (r.m.s.) value of i, we have

$$i^2 = \tfrac{1}{2} \{(A_1^2 + B_1^2) + (A_3^2 + B_3^2) + (A_5^2 + B_5^2) + \ldots\}$$

But from (*i*), $A_n^2 + B_n^2 = I_n^2 (\sin^2\alpha_n + \cos^2 \alpha_n) = I_n^2 \times 1 = I_n^2$.

Hence $i^2 = \tfrac{1}{2}(I_1^2 + I_3^2 + I_5^2 + \ldots)$.

Example 36. *If* $f(x) = \dfrac{x}{l},\ 0 < x < l$

$$= \frac{2l - x}{l},\ l < x < 2l$$

express $f(x)$ as a Fourier series of periodicity $2l$.

$$\text{Let } f(x) = \frac{a_0}{2} + \sum_{n=1}^{\infty} \left(a_n \cos \frac{n\pi x}{l} + b_n \sin \frac{n\pi x}{l} \right)$$

where
$$a_0 = \frac{1}{l} \int_0^{2l} f(x)\, dx$$

$$= \frac{1}{l} \left[\int_0^l \frac{x}{l}\, dx + \int_l^{2l} \frac{2l - x}{l}\, dx \right]$$

$$= \frac{1}{l^2} \left[\left(\frac{x^2}{2} \right)_0^l + \left(2lx - \frac{x^2}{2} \right)_l^{2l} \right]$$

$$= \frac{1}{l^2}\left[\frac{l^2}{2} + 2l^2 - \frac{3}{2}l^2\right]$$

$$= 1$$

$$a_n = \frac{1}{l}\left[\int_0^l \frac{x}{l}\cos\frac{n\pi x}{l}\,dx + \int_l^{2l}\frac{(2l-x)}{l}\cos\frac{n\pi x}{l}\,dx\right]$$

$$= \frac{1}{l^2}\left[(x)\left(\frac{\sin\frac{n\pi x}{l}}{\frac{n\pi}{l}}\right) - (1)\left(-\frac{\cos\frac{n\pi x}{l}}{\frac{n^2\pi^2}{l^2}}\right)\right]_0^l$$

$$+ \frac{1}{l^2}\left[(2l-x)\left(\frac{\sin\frac{n\pi x}{l}}{\frac{n\pi}{l}}\right) - (-1)\left(-\frac{\cos\frac{n\pi x}{l}}{\frac{n^2\pi^2}{l^2}}\right)\right]_l^{2l}$$

$$= \frac{1}{l^2}\left[\frac{l^2}{n^2\pi^2}\{(-1)^n - 1\} - \frac{l^2}{n^2\pi^2}\{1 - (-1)^n\}\right]$$

$$= \frac{2}{n^2\pi^2}\{(-1)^n - 1\}$$

$= 0$ if n is even

$= -\dfrac{4}{n^2\pi^2}$ if n is odd

$$b_n = \frac{1}{l}\left[\int_0^l \frac{x}{l}\sin\frac{n\pi x}{l}\,dx + \int_l^{2l}\frac{2l-x}{l}\,.\,\sin\frac{n\pi x}{l}\,dx\right]$$

$$= \frac{1}{l^2}\left[(x)\left(-\frac{\cos\frac{n\pi x}{l}}{\frac{n\pi}{l}}\right) - (1)\left(-\frac{\sin\frac{n\pi x}{l}}{\frac{n^2\pi^2}{l^2}}\right)\right]_0^l$$

$$+ \frac{1}{l^2}\left[(2l-x)\left(-\frac{\cos\frac{n\pi x}{l}}{\frac{n\pi}{l}}\right) - (-1)\left(-\frac{\sin\frac{n\pi x}{l}}{\frac{n^2\pi^2}{l^2}}\right)\right]_l^{2l}$$

$$= \frac{1}{l^2}\left[-\frac{l^2}{n\pi}\cos n\pi + \frac{l^2}{n\pi}\cos n\pi\right] = 0$$

$$\therefore\ f(x) = \frac{1}{2} - \frac{4}{\pi^2} \sum_{n=1,3,5...}^{\infty} \left[\frac{1}{n^2} \cos \frac{n\pi x}{l} \right]$$

Example 37. *Find the Fourier series of periodicity 3 for $f(x) = 2x - x^2$, in $0 < x < 3$*

Here the range and the period are same (equal to 3)

It is a full range series.

$$\therefore\ 2l = 3\ ;\ l = \frac{3}{2}$$

$$\text{Let } f(x) = \frac{a_0}{2} + \sum_{n=1}^{\infty} \left(a_n \cos \frac{2n\pi x}{3} + b_n \sin \frac{2n\pi x}{3} \right)$$

where $$a_0 = \frac{1}{\frac{3}{2}} \int_0^3 (2x - x^2)\, dx = \frac{2}{3} \left[x^2 - \frac{x^3}{3} \right]_0^3 = 0$$

$$a_n = \frac{1}{\frac{3}{2}} \int_0^3 (2x - x^2) \cos \frac{2n\pi x}{3}\, dx$$

$$= \frac{2}{3} \left[(2x - x^2) \left(\frac{\sin \frac{2n\pi x}{3}}{\frac{2n\pi}{3}} \right) - (2 - 2x) \left(- \frac{\cos \frac{2n\pi x}{3}}{\frac{4n^2\pi^2}{9}} \right) + (-2) \left(- \frac{\sin \frac{2n\pi x}{3}}{\frac{8n^3\pi^3}{27}} \right) \right]_0^3$$

$$= \frac{2}{3} \left[\frac{-9}{n^2\pi^2} - \frac{9}{2n^2\pi^2} \right]$$

$$= - \frac{9}{n^2\pi^2}$$

$$b_n = \frac{1}{\frac{3}{2}} \int_0^3 (2x - x^2) \sin \frac{2n\pi x}{3}\, dx$$

$$= \frac{2}{3}\left[(2x - x^2)\left(-\frac{\cos\frac{2n\pi x}{3}}{\frac{2n\pi}{3}}\right) - (2-2x)\left(-\frac{\sin\frac{2n\pi x}{3}}{\frac{4n^2\pi^2}{9}}\right) + (-2)\left(\frac{\cos\frac{2n\pi x}{3}}{\frac{8n^3\pi^3}{27}}\right)\right]_0^3$$

$$= \frac{3}{n\pi}$$

$$\therefore \quad f(x) = -\frac{9}{\pi^2}\sum_{n=1}^{\infty}\frac{1}{n^2}\cos\left(\frac{2n\pi x}{3}\right) + \frac{3}{\pi}\sum_{n=1}^{\infty}\frac{1}{n}\sin\left(\frac{2n\pi x}{3}\right)$$

Example 38. *$f(x)$ is defined in (–2, 2) as follows. Express $f(x)$ in a Fourier series of periodicity 4.* **(*MS. 1987 Nov.*)**

$$f(x) = \begin{cases} 0 & -2 < x < -1 \\ 1 + x & -1 < x < 0 \\ 1 - x & 0 < x < 1 \\ 0 & 1 < x < 2 \end{cases}$$

Sol. Here $2l = 4 \quad \therefore \; l = 2$

$$\text{Hence } f(x) = \frac{a_0}{2} + \sum_{n=1}^{\infty}\left(a_n \cos\frac{n\pi x}{2} + b_n \sin\frac{n\pi x}{2}\right)$$

$$\text{where} \quad a_0 = \frac{1}{2}\int_{-2}^{2} f(x)\, dx$$

$$= \frac{1}{2}\left[\int_{-2}^{-1} a dx + \int_{-1}^{0}(1+x)dx + \int_{0}^{1}(1-x)\, dx + \int_{1}^{2} 0 \,.\, dx\right]$$

$$= \frac{1}{2}\left[\frac{(1+x)^2}{2}\right]_{-1}^{0} + \frac{1}{2}\left[\frac{(1-x)^2}{-2}\right]_0^1$$

$$= \frac{1}{4} + \frac{1}{4} = \frac{1}{2}$$

$$a_n = \frac{1}{2}\int_{-2}^{2} f(x)\cos\frac{n\pi x}{2}\, dx$$

$$= \frac{1}{2}\left[\int_{-1}^{0}(1+x)\cos\frac{n\pi x}{2}\,dx + \int_{0}^{1}(1-x)\cos\frac{n\pi x}{2}\,dx\right],$$

since the other integrals vanish

$$= \frac{1}{2}\left[(1+x)\left(\frac{\sin\frac{n\pi x}{2}}{\frac{n\pi}{2}}\right) - (1)\left(-\frac{\cos\frac{n\pi x}{2}}{\frac{n^2\pi^2}{4}}\right)\right]_{-1}^{0}$$

$$+ \frac{1}{2}\left[(1-x)\left(\frac{\sin\frac{n\pi x}{2}}{\frac{n\pi}{2}}\right) - (-1)\left(-\frac{\cos\frac{n\pi x}{2}}{\frac{n^2\pi^2}{4}}\right)\right]_{0}^{1}$$

$$= \frac{1}{2}\left[\frac{4}{n^2\pi^2} - \frac{4}{n^2\pi^2}\cos\frac{n\pi}{2} - \frac{4}{n^2\pi^2}\cos\frac{n\pi}{2} + \frac{4}{n^2\pi^2}\right]$$

$$= \frac{4}{n^2\pi^2}\left[1 - \cos\frac{n\pi}{2}\right]$$

$$b_n = \frac{1}{2}\int_{-2}^{2} f(x)\,.\,\sin\frac{n\pi x}{2}\,dx$$

$$= \frac{1}{2}\left[\int_{-1}^{0}(1+x)\sin\frac{n\pi x}{2}\,dx + \int_{0}^{1}(1-x)\sin\frac{n\pi x}{2}\,dx\right],$$

the other integrals being zero.

$$= \frac{1}{2}\left[(1+x)\left(-\frac{\cos\frac{n\pi x}{2}}{\frac{n\pi}{2}}\right) - (1)\left(-\frac{\sin\frac{n\pi x}{2}}{\frac{n^2\pi^2}{4}}\right)\right]_{-1}^{0}$$

$$+ \frac{1}{2}\left[(1-x)\left(-\frac{\cos\frac{n\pi x}{2}}{\frac{n\pi}{2}}\right) - (-1)\left(-\frac{\sin\frac{n\pi x}{2}}{\frac{n^2\pi^2}{4}}\right)\right]_{0}^{1}$$

$$= \frac{1}{2}\left[-\frac{2}{n\pi} + \frac{4}{n^2\pi^2}\sin\frac{n\pi}{2} - \frac{4}{n^2\pi^2}\sin\frac{n\pi}{2} + \frac{2}{n\pi}\right]$$

$$= 0.$$

Therefore, $f(x) = \frac{1}{4} + \frac{4}{\pi^2} \sum_{n=1}^{\infty} \frac{1}{n^2} \left(1 - \cos \frac{n\pi}{2} \right) \cos \frac{n\pi x}{2}$

$$= \frac{1}{4} + \frac{8}{\pi^2} \sum_{n=1}^{\infty} \frac{1}{n^2} \sin^2 \frac{n\pi}{4} \cos \frac{n\pi x}{2}.$$

Example 39. *Find Fourier series of periodicity 2 for $f(x)$*

given $f(x) = \begin{cases} 0 \text{ in } -1 < x < 0 \\ 1 \text{ in } 0 < x < 1 \end{cases}$ **(*MS. 1986 Ap.*)**

Here the period and the range are equal.

Hence, $2l = 2;\ l = 1$

Let $f(x) = \frac{a_0}{2} + \sum_{n=1}^{\infty} (a_n \cos n\pi x + b_n \sin n\pi x)$

where $a_0 = \frac{1}{1} \int_{-1}^{1} f(x) dx$

$$= \int_0^1 1\, dx$$

$$= (x)_0^1 = 1$$

$$a_n = \frac{1}{1} \int_{-1}^{1} f(x) \cos n\pi x\, dx$$

$$= \int_0^1 \cos n\pi x\, dx$$

$$= \left(\frac{\sin n\pi x}{n\pi} \right)_0^1$$

$$= 0$$

$$b_n = \frac{1}{1} \int_{-1}^{1} f(x) \sin n\pi x\, dx$$

$$= \int_0^1 \sin n\pi x\, dx$$

$$= -\left(\frac{\cos n\pi x}{n\pi}\right)_0^1$$

$$= -\frac{1}{n\pi}((-1)^n - 1)$$

$$= 0 \text{ if } n \text{ is even}$$

$$= \frac{2}{n\pi} \text{ if } n \text{ is odd}$$

$$\therefore \quad f(x) = \frac{1}{2} + \frac{2}{\pi} \sum_{n=1,3,5...}^{\infty} \frac{1}{n} \sin n\pi x.$$

$$= \frac{1}{2} + \frac{2}{\pi} \sum_{n=1}^{\infty} \frac{1}{(2n-1)} \sin (2n-1)\pi x.$$

Example 40. *Find Fourier series of periodicity 2 for*

$$f(x) = \begin{cases} x & in -1 < x \le 0 \\ x+2 & in \quad 0 < x \le 1 \end{cases}$$

and hence deduce the sum of $1 - \frac{1}{3} + \frac{1}{5} - \frac{1}{7} + \ldots$ *to* ∞.

The periodicity and range are equal to 2.

Hence $2l = 2$; $l = 1$

Let $f(x) = \frac{a_0}{2} + \sum_{n=1}^{\infty}\left(a_n \cos \frac{n\pi x}{1} + b_n \sin \frac{n\pi x}{1}\right)$

where $$a_0 = \frac{1}{1}\int_{-1}^{1} f(x)dx$$

$$= \int_{-1}^{0} x\, dx + \int_{0}^{1} (x+2)\, dx$$

$$= \left(\frac{x^2}{2}\right)_{-1}^{0} + \left[\frac{(x+2)^2}{2}\right]_0^1$$

$$= \frac{-1}{2} + \frac{9}{2} - \frac{4}{2}$$

$$= 2$$

$$a_n = \int_{-1}^{0} x \cos n\pi x\, dx + \int_{0}^{1} (x+2) \cos n\pi x\, dx$$

$$= \left[(x)\left(\frac{\sin n\pi x}{n\pi}\right) - (1)\left(-\frac{\cos n\pi x}{n^2\pi^2}\right)\right]_{-1}^{0}$$

$$+ \left[(x+2)\left(\frac{\sin n\pi x}{n\pi}\right) - (1)\left(-\frac{\cos n\pi x}{n^2\pi^2}\right)\right]_{0}^{1}$$

$$= \frac{1}{n^2\pi^2}[1-(-1)^n] + \frac{1}{n^2\pi^2}[(-1)^n - 1]$$

$$= 0$$

$$b_n = \int_{-1}^{0} x \sin n\pi x \, dx + \int_{0}^{1} (x+2) \sin n\pi x \, dx$$

$$= \left[(x)\left(-\frac{\cos n\pi x}{n\pi}\right) - \left(-\frac{\sin n\pi x}{n^2\pi^2}\right)\right]_{-1}^{0}$$

$$+ \left[(x+2)\left(-\frac{\cos n\pi x}{n\pi}\right) - \left(-\frac{\sin n\pi x}{n^2\pi^2}\right)\right]_{0}^{1}$$

$$= \left[-\frac{1}{n\pi}(-1)^n - \frac{3}{n\pi}(-1)^n + \frac{2}{n\pi}\right]$$

$$\therefore \quad f(x) = 1 + \frac{2}{\pi}\sum_{n=1}^{\infty}\frac{1}{n}\{1 - 2(-1)^n\} \sin n\pi x$$

$$= 1 + \frac{2}{\pi}\left[\frac{3}{1}\sin \pi x - \frac{1}{2}\sin 2\pi x + \frac{3}{3}\sin 3\pi x - \cdots\right]$$

Putting $x = \frac{1}{2}$ which is a point of continuity, the value of the Fourier series at $x = \frac{1}{2}$ is $f\left(\frac{1}{2}\right)$

$$\therefore \quad 1 + \frac{2}{\pi}\left[3 \cdot \frac{1}{1} - 3 \cdot \frac{1}{3} + 3 \cdot \frac{1}{5} - \cdots\right] = \frac{1}{2} + 2$$

$$\frac{6}{\pi}\left(1 - \frac{1}{3} + \frac{1}{5} - \cdots\right) = \frac{3}{2}$$

$$\therefore \quad 1 - \frac{1}{3} + \frac{1}{5} - \frac{1}{7} + \cdots = \frac{\pi}{4}$$

Example 41. *Express $f(x) = x$ in half range cosine series and sine series of periodicity $2l$ in the range $0 < x < l$ and deduce the value of $\frac{1}{1^4} + \frac{1}{3^4} + \frac{1}{5^4} + \cdots$ to ∞.*

Case 1. *Cosine series* :

Let $f(x) = \frac{a_0}{2} + \sum_{n=1}^{\infty} a_n \cos \frac{n\pi x}{l}$

where $a_0 = \frac{2}{l} \int_0^l f(x)\, dx$

$= \frac{2}{l} \int_0^l x\, dx$

$= \frac{2}{l} \left(\frac{x^2}{2} \right)_0^l$

$= l$

$a_n = \frac{2}{l} \int_0^l x \cos \frac{n\pi x}{l}\, dx$

$$= \frac{2}{l} \left[(x) \left(\frac{\sin \frac{n\pi x}{l}}{\frac{n\pi}{l}} \right) - (1) \left(- \frac{\cos \left(\frac{n\pi x}{l} \right)}{\frac{n^2\pi^2}{l^2}} \right) \right]_0^l$$

$= \frac{2}{l} \left[\frac{l^2}{n^2\pi^2} \{(-1)^n - 1\} \right]$

$= \frac{-4l}{n^2\pi^2}$ if n is odd

$= 0$ if n is even

Hence $x = \frac{l}{2} - \frac{4l}{\pi^2} \sum_{n=1}^{\infty} \frac{1}{(2n-1)^2} \cos \frac{(2n-1)\pi x}{l}$ ...(1)

Case 2. *Sine series* :

Let $f(x) = \sum_{n=1}^{\infty} b_n \sin \frac{n\pi x}{l}$

where $b_n = \frac{2}{l} \int_0^l x \sin \frac{n\pi x}{l}\, dx$

$$= \frac{2}{l}\left[(x)\left(-\frac{\cos\frac{n\pi x}{l}}{\frac{n\pi}{l}}\right) - (1)\left(-\frac{\sin\frac{n\pi x}{l}}{\frac{n^2\pi^2}{l^2}}\right)\right]_0^l$$

$$= \frac{2}{l}\left[-\frac{l^2}{n\pi}(-1)^n\right]$$

$$= \frac{2l}{n\pi}(-1)^{n+1}$$

$$\therefore \qquad x = \frac{2l}{\pi}\sum_{n=1}^{\infty}\frac{(-1)^{n+1}}{n}\sin\frac{n\pi x}{l} \qquad \ldots(2)$$

Deduction: By Parsevel's theorem,

$$(\text{Range})\left[\frac{a_0^{\ 2}}{4} + \frac{1}{2}\Sigma a_n^{\ 2}\right] = \int_0^l [f(x)]^2\,dx \quad (\text{for cosine series})$$

$$l\left[\frac{l^2}{4} + \frac{1}{2}\sum_{1,3,5,\ldots}\frac{16l^2}{n^4\pi^4}\right] = \int_0^l x^2 dx = \frac{l^3}{3}$$

$$8\sum_{1,3,5,\ldots}^{\infty}\frac{1}{n^4} = \frac{\pi^4}{12}$$

$$\frac{1}{1^4} + \frac{1}{3^4} + \frac{1}{5^4} + \cdots = \frac{\pi^4}{96}$$

Again

$$(\text{Range})\left[\frac{1}{2}\sum b_n^{\ 2}\right] = \int_0^l [f(x)]^2\,dx \quad (\text{for sine series})$$

$$l\left[\frac{1}{2}\sum_1^{\infty}\frac{4l^2}{n^2\pi^2}\right] = \int_0^l x^2\,dx = \frac{l^3}{3}$$

$$\sum_1^{\infty}\frac{1}{n^2} = \frac{\pi^2}{6}.$$

Example 42. *Expand $f(x) = x - x^2$ as a Fourier series in $-1 < x < 1$ and using this series find the R.M.S. value of $f(x)$ in the interval.* **(*BR. 1995 Ap.*)**

Here $2l = 2\,;\; l = 1$

Let $f(x) = \frac{a_0}{2} + \sum_{n=1}^{\infty} (a_n \cos n\pi x + b_n \sin n\pi x)$

where $a_0 = \frac{1}{1}\int_{-1}^{1} f(x)\, dx$

$$= \int_{-1}^{1} (x - x^2)\, dx$$

$$= \int_{-1}^{1} x dx - \int_{-1}^{1} x^2 dx$$

$$= 0 - 2\int_{0}^{1} x^2 dx$$

$$= -2\left(\frac{x^3}{3}\right)_0^1$$

$$= -\frac{2}{3}$$

$$a_n = \frac{1}{1}\int_{-1}^{1} (x - x^2) \cos n\pi x\, dx$$

$$= \int_{-1}^{1} x \cos n\pi x\, dx - \int_{-1}^{1} x^2 \cos n\pi x\, dx$$

$$= -2\int_{0}^{1} x^2 \cos n\pi x\, dx \quad (\because \text{ first integral vanishes})$$

$$= -2\left[(x^2)\left(\frac{\sin n\pi x}{n\pi}\right) - (2x)\left(-\frac{\cos n\pi x}{n^2\pi^2}\right) + (2)\left(-\frac{\sin n\pi x}{n^3\pi^3}\right)\right]_0^1$$

$$= -2\left[\frac{2}{n^2\pi^2}(-1)^n\right] = \frac{4(-1)^{n+1}}{n^2\pi^2}.$$

$$b_n = \int_{-1}^{1} (x - x^2) \sin n\pi x\, dx$$

$$= \int_{-1}^{1} x \sin n\pi x \, dx, \quad \text{(the other integral vanishes)}$$

$$= 2 \int_{0}^{1} x \sin n\pi x \, dx$$

$$= 2 \left[(x)\left(-\frac{\cos n\pi x}{n\pi} \right) - \left(-\frac{\sin n\pi x}{n^2\pi^2} \right) \right]_0^1$$

$$= -\frac{2}{n\pi} (-1)^n$$

$$\therefore \; x - x^2 = -\frac{1}{3} + \sum_{n=1}^{\infty} \left[\frac{4}{n^2\pi^2} (-1)^{n+1} \cos n\pi x + \frac{2}{n\pi} (-1)^{n+1} \sin n\pi x \right]$$

for $-1 < x < 1$

$$\text{R.M.S. value of } f(x) = \sqrt{\frac{\int_{-1}^{1} (x - x^2)^2 \, dx}{2}}$$

$$= \sqrt{\frac{1}{2} \int_{-1}^{1} (x^2 + x^4 - 2x^3) \, dx}$$

$$= \sqrt{\frac{1}{2} \times 2 \int_{0}^{1} (x^2 + x^4) \, dx}$$

$$= \sqrt{\left(\frac{x^3}{3} + \frac{x^5}{5} \right)_0^1}$$

$$= \sqrt{\frac{8}{15}}$$

Example 43. *Find Fourier cosine series of* $f(x) = \begin{cases} x^2, & 0 < x < 1 \\ 2 - x, & 1 < x < 2 \end{cases}$

Here half-range $l = 2$.

$$\text{Let} \quad f(x) = \frac{a_0}{2} + \sum_{n=1}^{\infty} a_n \cos \frac{n\pi x}{2}$$

$$\text{where} \quad a_0 = \frac{2}{2}\int_0^2 f(x)\,dx$$

$$= \int_0^1 x^2\,dx + \int_1^2 (2-x)\,dx$$

$$= \left(\frac{x^3}{3}\right)_0^1 + \left[\frac{(2-x)^2}{-2}\right]_1^2$$

$$= \frac{1}{3} + \frac{1}{2} = \frac{5}{6}$$

$$a_n = \int_0^1 x^2 \cos\frac{n\pi x}{2}\,dx + \int_1^2 (2-x)\cos\frac{n\pi x}{2}\,dx$$

$$= \left[(x)^2\left(\frac{\sin\frac{n\pi x}{2}}{\frac{n\pi}{2}}\right) - (2x)\left(-\frac{\cos\frac{n\pi x}{2}}{\frac{n^2\pi^2}{4}}\right) + (2)\left(-\frac{\sin\frac{n\pi x}{2}}{\frac{n^3\pi^3}{8}}\right)\right]_0^1$$

$$+ \left[(2-x)\left(\frac{\sin\frac{n\pi x}{2}}{\frac{n\pi}{2}}\right) - (-1)\left(-\frac{\cos\frac{n\pi x}{2}}{\frac{n^2\pi^2}{4}}\right)\right]_1^2$$

$$= \frac{2}{n\pi}\sin\frac{n\pi}{2} + \frac{8}{n^2\pi^2}\cos\frac{n\pi}{2} - \frac{16}{n^3\pi^3}\sin\frac{n\pi}{2}$$

$$+ \frac{4}{n^2\pi^2}\cos\frac{n\pi}{2} - \frac{4}{n^2\pi^2}\cos n\pi - \frac{2}{n\pi}\sin\frac{n\pi}{2}$$

$$= \frac{12}{n^2\pi^2}\cos\frac{n\pi}{2} - \frac{16}{n^3\pi^3}\sin\frac{n\pi}{2} - \frac{4}{n^2\pi^2}\cos n\pi$$

$$\therefore \quad f(x) = \frac{5}{12} + \sum_{n=1}^{\infty}\left(\frac{12}{n^2\pi^2}\cos\frac{n\pi}{2} - \frac{16}{n^3\pi^3}\sin\frac{n\pi}{2} - \frac{4}{n^2\pi^2}\cos n\pi\right)\cos\frac{n\pi x}{2}.$$

Example 44. *Expand $f(x) = (x-1)^2$, $0 < x < 1$ in a Fourier series of sines only.*

Here $l = 1$; we want half range sine series.

$$\text{Let } f(x) = \sum_{1}^{\infty} b_n \sin n\pi x$$

where $b_n = \frac{2}{1}\int_0^1 (x-1)^2 \sin n\pi x dx$

$$= 2\left[(x-1)^2\left(-\frac{\cos n\pi x}{n\pi}\right) - 2(x-1)\left(-\frac{\sin n\pi x}{n^2\pi^2}\right) + 2\left(\frac{\cos n\pi x}{n^3\pi^3}\right)\right]_0^1$$

$$= 2\left[\frac{2(-1)^n}{n^3\pi^3} + \frac{1}{n\pi} - \frac{2}{n^3\pi^3}\right]$$

$$= \frac{2}{\pi^3}\left[\frac{2(-1)^n - 2}{n^3} + \frac{\pi^2}{n}\right]$$

$$\therefore \quad f(x) = \frac{2}{\pi^3}\sum_{n=1}^{\infty}\left(\frac{2(-1)^n - 2}{n^3} + \frac{\pi^2}{n}\right)\sin n\pi x.$$

Example 45. *Find the Fourier series for $f(x) = |\cos x|$ in $(-\pi, \pi)$ of periodicity 2π.* ***(MS. 1992 Ap.)***

$f(x) = |\cos x|$ is even since $f(-x) = f(x)$.

Hence $f(x)$ will contain only cosine terms.

Let $\quad f(x) = \frac{a_0}{2} + \sum_{n=1}^{\infty} a_n \cos nx$

$$a_0 = \frac{1}{\pi}\int_{-\pi}^{\pi} |\cos x|\, dx$$

$$= \frac{2}{\pi}\int_0^{\pi} |\cos x|\, dx$$

$$= \frac{2}{\pi}\left[\int_0^{\pi/2} \cos x\, dx + \int_{\pi/2}^{\pi} (-\cos x)\, dx\right]$$

(since in $(0, \pi/2)$, $\cos x$ is positive and in $(\pi/2, \pi)$ $\cos x$ is negative.

$$= \frac{2}{\pi}\left[(\sin x)_0^{\pi/2} - (\sin x)_{\pi/2}^{\pi}\right] = \frac{4}{\pi}$$

$$a_n = \frac{1}{\pi}\int_{-\pi}^{\pi} |\cos x| \cos nx\, dx$$

$$= \frac{2}{\pi} \int_0^{\pi} |\cos x| \cos nx \, dx \quad \text{since integrand is even}$$

$$= \frac{2}{\pi} \left[\int_0^{\pi/2} \cos x \cos nx \, dx + \int_{\pi/2}^{\pi} (-\cos x \cos nx) \, dx \right]$$

$$= \frac{1}{\pi} \left[\int_0^{\pi/2} [\cos (n+1) x + \cos (n-1) x] \, dx \right.$$

$$\left. - \int_{\pi/2}^{\pi} [\cos (n+1)x + \cos (n-1) x] \, dx \right]$$

$$= \frac{1}{\pi} \left[\left\{ \frac{\sin (n+1)x}{n+1} + \frac{\sin (n-1)x}{n-1} \right\}_0^{\pi/2} \right.$$

$$\left. - \left\{ \frac{\sin (n+1)x}{n+1} + \frac{\sin (n-1)x}{n-1} \right\}_{\pi/2}^{\pi} \right]$$

$$= \frac{1}{\pi} \left[\frac{\sin (n+1)\pi/2}{n+1} + \frac{\sin (n-1)\pi/2}{n-1} + \frac{\sin (n+1)\pi/2}{n+1} \right.$$

$$\left. + \frac{\sin (n-1)\pi/2}{n-1} \right] \quad \text{if } n \neq 1$$

$$= \frac{2}{\pi} \left[\frac{1}{n+1} \left\{ \sin \frac{n\pi}{2} \cos \frac{\pi}{2} + \cos \frac{n\pi}{2} \sin \frac{\pi}{2} \right\} + \frac{1}{n-1} \right.$$

$$\left. \times \left\{ \sin \frac{n\pi}{2} \cos \frac{\pi}{2} - \cos \frac{n\pi}{2} \sin \frac{\pi}{2} \right\} \right] \quad \text{if } n \neq 1$$

$$= \frac{2}{\pi} \left[\frac{1}{n+1} - \frac{1}{n-1} \right] \cos \frac{n\pi}{2}$$

$$= -\frac{4}{\pi (n^2 - 1)} \cos \frac{n\pi}{2} \quad \text{if } n \neq 1$$

$$a_1 = \frac{1}{\pi} \int_{-\pi}^{\pi} |\cos x| \cos x \, dx$$

$$= \frac{2}{\pi} \left[\int_0^{\pi} |\cos x| \cos x \right]$$

$$= \frac{2}{\pi}\left[\int_0^{\pi/2} \cos^2 x\,dx - \int_{\pi/2}^{\pi} \cos^2 x\,dx\right]$$

$$= \frac{2}{\pi}\left[\frac{1}{2}\cdot\frac{\pi}{2} - \int_{\pi/2}^{\pi} \frac{1+\cos 2x}{2}\,dx\right]$$

$$= \frac{2}{\pi}\left[\frac{\pi}{4} - \frac{1}{2}\left(x + \frac{\sin 2x}{2}\right)_{\pi/2}^{\pi}\right]$$

$$= \frac{2}{\pi}\left[\frac{\pi}{4} - \frac{\pi}{4}\right]$$

$$= 0.$$

$$\therefore \quad |\cos x| = \frac{2}{\pi} - \frac{4}{\pi}\sum_{n=2}^{\infty} \frac{1}{n^2-1}\cos\frac{n\pi}{2}\cos nx$$

Example 46. *Find the Fourier series of $f(x) = |\sin x|$ in $(-\pi, \pi)$ of periodicity 2π.* **(MS. 1991 Ap., BR Ap. 1997)**

$f(x) = |\sin x|$ is an even function of x though $\sin x$ is odd.

$\therefore$ $f(x)$ will contain only cosine terms in its Fourier series.

Let $f(x) = \dfrac{a_0}{2} + \displaystyle\sum_1^{\infty} a_n \cos nx$

$$a_0 = \frac{1}{\pi}\int_{-\pi}^{\pi} |\sin x|\,dx$$

$$= \frac{2}{\pi}\int_0^{\pi} |\sin x|\,dx \qquad (\because \text{ even integrand})$$

$$= \frac{2}{\pi}\int_0^{\pi} \sin x\,dx$$

$$= \frac{2}{\pi}(-\cos x)_0^{\pi}$$

$$= \frac{2}{\pi}[1+1] = \frac{4}{\pi}$$

$$a_n = \frac{1}{\pi}\int_{-\pi}^{\pi} |\sin x| \cos nx\,dx$$

$$= \frac{2}{\pi}\int_0^{\pi} \sin x \cos nx\, dx$$

$$= \frac{1}{\pi}\int_0^{\pi} [\sin (n+1)x - \sin (n-1)x]\, dx$$

$$= \frac{1}{\pi}\left[-\frac{\cos (n+1)x}{n+1} + \frac{\cos (n-1)x}{n-1}\right]_0^{\pi} \text{ if } n \neq 1$$

$$= \frac{1}{\pi}\left[\frac{-1}{n+1}\{(-1)^{n+1} - 1\} + \frac{1}{n-1}\{(-1)^{n-1} - 1\}\right]$$

$$= \frac{1}{\pi}\left[\frac{1}{n+1}\{1 + (-1)^n\} - \frac{1}{n-1}\{1 + (-1)^n\}\right]$$

$$= \frac{1}{\pi}\left(\frac{-2}{n^2 - 1}\right)\{1 + (-1)^n\}$$

$= 0$ if n is odd

$= -\dfrac{4}{\pi(n^2 - 1)}$ if n is even and $n \neq 1$

$$a_1 = \frac{2}{\pi}\int_0^{\pi} \sin x \cos x\, dx = \frac{1}{\pi}\left(-\frac{\cos 2x}{2}\right)_0^{\pi} = 0$$

$$\therefore \quad f(x) = \frac{2}{\pi} - \frac{4}{\pi}\sum_{n=1}^{\infty} \frac{1}{(4n^2 - 1)} \cos 2nx.$$

Exercises 1(g)

1. If $f(x) = \dfrac{a_0}{2} + \displaystyle\sum_{n=1}^{\infty}\left(a_n \cos \frac{n\pi x}{l} + b_n \sin \frac{n\pi x}{l}\right)$

in $0 < x < 2l$, prove $\displaystyle\int_0^{2l} [f(x)]^2\, dx = 2l\left[\frac{a_0^2}{4} + \frac{1}{2}\ \Sigma(a_n^2 + b_n^2)\right]$.

2. Show that whenever

$f(x) = \displaystyle\sum_{n=1}^{\infty} b_n \sin \frac{n\pi x}{L}$ in the interval $0 < x < L$, the coefficients are given by

$b_n = \dfrac{2}{L}\displaystyle\int_0^{L} f(x) \sin \frac{n\pi x}{L}\, dx$ and then show that $\displaystyle\int_0^{L} [f(x)]^2 dx = \frac{1}{2}L \sum_{n=1}^{\infty} b_n^2$.

(*Madras, B.E. 64*)

3. Check that for $0 < x < L$,

$$1 = \frac{4}{\pi}\left(\sin\frac{\pi x}{L} + \frac{1}{3}\sin\frac{3\pi x}{L} + \frac{1}{5}\sin\frac{5\pi x}{L} + \cdots\right)$$

and deduce that

$$(i)\ 1 - \frac{1}{3} + \frac{1}{5} - \frac{1}{7} + \cdots = \frac{\pi}{4} \quad \text{and } (ii)\ \frac{1}{1^2} + \frac{1}{3^2} + \frac{1}{5^2} + \ldots = \frac{\pi^2}{8}.$$

4. If $f(x) = \frac{a_0}{2} + \sum_{n=1}^{\infty} a_0 \cos\frac{n\pi x}{l}$, in $0 < x < l$ show that

$$\int_0^l [f(x)]^2 dx = \tfrac{1}{2} l (\tfrac{1}{2} a_0^2 + a_1^2 + a_2^2 + \ldots)$$

5. The impressed voltage e of an alternating-current wave is given by

$$e = E_1 \sin(\omega t + \beta_1) + E_3 \sin(3\omega t + \beta_3) + E_5 \sin(5\omega t + \beta_5)$$

$+ \ldots$. Show that the effective voltage $\bar{e}$ is given by

$$\bar{e}^2 = \tfrac{1}{2}(E_1^2 + E_3^2 + E_5^2 + \cdots).$$

6. Prove $\sin\frac{\pi x}{l} = \frac{2}{\pi} - \frac{4}{\pi}\left[\frac{\cos\frac{2\pi x}{l}}{1\cdot 3} + \frac{\cos\frac{4\pi x}{l}}{3\cdot 5} + \frac{\cos\frac{6\pi x}{l}}{5\cdot 7} + \cdots\right]$, in $0 < x < l$

7. Prove $e^x = \pi \sum_{1}^{\infty} \frac{n[1 - e(-1)^n]}{1 + n^2\pi^2} \sin n\pi x$, in $0 < x < 1$.

8. Prove $x = -\frac{4}{\pi} \sum_{1}^{\infty} \frac{(-1)^n}{n} \sin\frac{n\pi x}{2}$, in $0 < x < 2$.

9. If $f(x) = x$, in $0 < x < \frac{l}{2}$

$= \frac{2a}{l}(l - x)$, in $\frac{l}{2} < x < l$, prove

$$f(x) = \frac{8a}{\pi^2}\left[\sin\frac{\pi x}{l} - \frac{1}{3^2}\sin\frac{3\pi x}{l} + \frac{1}{5^2}\sin\frac{5\pi x}{l} - \cdots\right]$$

10. If $f(x) = \frac{2kx}{l}$, $0 < x < \frac{l}{2}$

$= \frac{2k}{l}(l - x)$, $\frac{l}{2} < x < l$ prove

$$f(x) = \frac{k}{2} - \frac{16k}{\pi^2}\left(\frac{1}{2^2}\cos\frac{2\pi x}{l} + \frac{1}{6^2}\cos\frac{6\pi x}{l} + \cdots\right)$$

11. If $f(x) = \sin x, 0 < x < \pi/4$

$= \cos x, \pi/4 < x < \pi/2$

expand $f(x)$ as a sine series.

12. If $f(x) = \frac{\pi}{3}$, in $0 < x < \frac{\pi}{3}$

$= 0$, in $\frac{\pi}{3} < x < \frac{2\pi}{3}$

$= -\frac{\pi}{3}$, in $\frac{2\pi}{3} < x < \pi$

prove $f(x) = \sin 2x + \frac{1}{2} \sin 4x + \frac{1}{4} \sin 8x + \ldots$

13. In $0 < x < 4$, find cosine series and sine series for $f(x) = 3x - 2$.

14. Find half range sine series and cosine series of period 4 for $f(x)$ where

$f(x) = 2x$, in $0 < x < 1$

$= 4 - 2x$, in $1 < x < 2$.

15. Find half range sine series for

$f(x) = \frac{2x}{l}$, in $0 < x < \frac{l}{2}$

$= \frac{2}{l}(l - x)$, in $\frac{l}{2} < x < l$.

16. If $f(x)) = c - x$, in $0 < x < c$, prove

$$f(x) = \frac{c}{2} + \frac{2c}{\pi^2}\left[\frac{\cos \frac{\pi x}{c}}{1^2} + \frac{\cos \frac{3\pi x}{c}}{3^2} + \frac{\cos \frac{5\pi x}{c}}{5^2} + \cdots\right]$$

17. If $f(x) = \begin{cases} \pi x, & \text{in } 0 < x < 1 \\ 0 & \text{at } x = 1 \\ \pi(x-2), & \text{in } 1 < x < 2 \end{cases}$

Prove $f(x) = 2\left[\sin \pi x - \frac{1}{2} \sin 2\pi x + \frac{1}{3} \sin 3\pi x - \cdots\right]$

18. If $f(x) = \frac{1}{4} - x$, in $0 < x < \frac{1}{2}$

$= x - \frac{3}{4}$, in $\frac{1}{2} < x < 1$

Prove $f(x) = \left(\frac{1}{\pi} - \frac{4}{\pi^2}\right) \sin \pi x + \left(\frac{1}{3\pi} - \frac{4}{3^2\pi^2}\right) \sin 3\pi x + \cdots$

19. If $f(x) = -1$, in $-2 \le x < -1$

$= x$, in $-1 < x < 1$

$= 1$, in $1 < x \le 2$

find Fourier series of $f(x)$, in $(-2, 2)$ of periodicity 4.

Rectangular rule for integration. In engineering applications, we come across definite integrals, whose integrands are given by a table of numerical values which are obtained from a physical experiment. In such a situation, we resort to numerical integration. The simplest formula for numerical integration is the rectangular rule.

Suppose we require $\int_a^b f(x)\,dx$. Let us draw the graph $y = f(x)$, and the ordinates $x = a$, $x = b$. Then we know that the required integral is the area enclosed by the x-axis and the curves drawn above. Now divide the length AB into q equal parts. Let these point of division be $(a =)\ x_0, x_1, x_2, \ldots, x_q\ (= b)$. Then the length of each part is $h = \dfrac{b-a}{q}$. Let $x_1^{1} x_2^{1}, x_3^{1}, \ldots, x_q^{1}$ be the mid-points of these n intervals. The n rectangles of the above figure have areas given by $hf(x_1^{1})$, $hf(x_2^{1})$. ... $hf(x_q^{1})$. Since the sum of the areas of the rectangles is nearly equal to the area representing the definite integral, we have

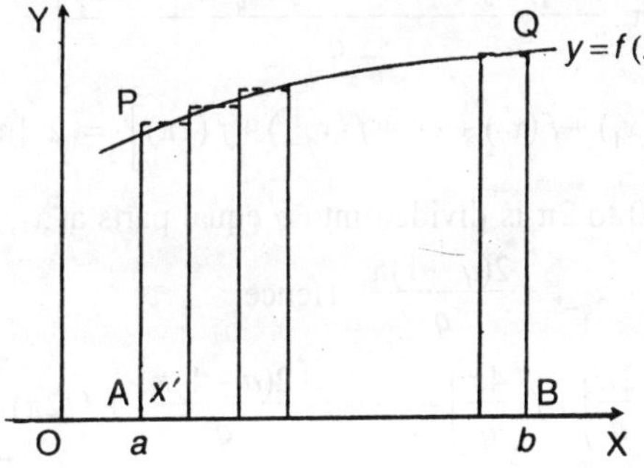

$$\int_a^b f(x)dx \approx h[f(x_1^{1}) + f(x_2^{1}) + \cdots + f(x_q^{1})].$$ Substituting for h, we get that

$$\int_a^b f(x)\,dx \approx (b-a)\frac{f(x_1^{1}) + f(x_2^{1}) + \cdots + f(x_q^{1})}{q}.$$

This is the rectangular rule. Since $\dfrac{f(x_1^{1}) + f(x_2^{1}) + \cdots + f(x_q^{1})}{q}$ is the mean of the mid-values (of the function), we have

$$\int_a^b f(x)\,dx \approx \text{interval} \times \text{mean of the mid-values.}$$

Note. An approximation to the above approximate formula is got by changing the mid-values into the end values. Thus, we get

$$\int_a^b f(x)\,dx \approx (b-a)\frac{f(x_1)+f(x_2)+\cdots+f(x_{q-1})+f(b)}{q}.$$

Harmonic analysis. When a function $f(x)$ is given by its numerical values at q equally spaced points, the coefficients in the Fourier series representing $f(x)$ can be obtained by numerical integration—for example, by the rectangular rule as given in the above note.

By theorem, $a_0 = \frac{1}{\pi}\int_0^{2\pi} f(x)\,dx.$

Hence remembering that in the present case $b - a = 2\pi$, and using the result in the above note, we get

$$a_0 = \frac{1}{\pi}\left[2\pi \,.\, \frac{f(x_1)+f(x_2)+\cdots+f(x_{q-1})+f(2\pi)}{q}\right].$$

i.e., $a_0 = \frac{2}{q}\left[f(x_1)+f(x_2)+\cdots+f(x_{q-1})+f(2\pi)\right] \cdot = 2$ [mean value of $f(x$

Since the interval 0 to 2π is divided into q equal parts at $x_1, x_2, \ldots$ we have $x_1 = \frac{2\pi}{q}, x_2 = \frac{4\pi}{q}, \ldots, x_{q-1} = \frac{2(q-1)\pi}{q}$. Hence

$$a_0 = \frac{2}{q}\left[f\left(\frac{2\pi}{q}\right)+f\left(\frac{4\pi}{q}\right)+\cdots+f\left\{\frac{2(q-1)\pi}{q}\right\}+f(2\pi)\right]$$

Working similarly,

$$a_n = \frac{2}{q}\left[f\left(\frac{2\pi}{q}\right)\cos\frac{2n\pi}{q}+f\left(\frac{4\pi}{q}\right)\cos\frac{4n\pi}{q}+\cdots+f\left\{\frac{2(q-1)\pi}{q}\right\}\right.$$

$$\left.\cos\frac{2(q-1)n\pi}{q}+f(2\pi)\cos 2n\pi\right] = 2\ [\text{mean value of } f(x)\cos nx]$$

and $$b_n = \frac{2}{q}\left[f\left(\frac{2\pi}{q}\right)\sin\frac{2n\pi}{q}+f\left(\frac{4\pi}{q}\right)\sin\frac{4n\pi}{q}+\cdots\right.$$

$$\left.+f\left\{\frac{2(q-1)\pi}{q}\right\}\sin\frac{2(q-1)n\pi}{q}+f(2\pi)\sin 2n\pi\right]\cdot$$

$$= 2\ [\text{mean value of } f(x)\sin nx]$$

The process of finding the Fourier series for a function given by numerical values is known as harmonic analysis.

Note 1. Since the values of sine and cosine are repeated in the four quadrants, it is advantageous to have q as a multiple of 4.

Note 2. In a Fourier expansion, the term $(a_1 \cos x + b_1 \sin x)$ is called the *fundamental* or *first harmonic*, the term $(a_2 \cos 2x + b_2 \sin 2x)$, the *second* *harmonic*, and so on. The harmonics of successive harmonics decrease rapidly; hence a few earlier harmonics will in general give a good approximation of the given function.

Example 1. *Compute the first three harmonics of the Fourier series for $f(x)$ from the following data:*

x	30°	60°	90°	120°	150°	180°	210°	240°	270°	300°	330°	360°
$f(x)$	2·34	3·01	3·68	4·15	3·69	2·20	0·83	0·51	0·88	1·09	1·19	1·64

(*Madurai, 72 Engg.*)

Let the Fourier series upto the third harmonic representing $f(x)$ in $(0, 2\pi)$ be

$$f(x) = \tfrac{1}{2} a_0 + a_1 \cos x + a_2 \cos 2x + a_3 \cos 3x$$
$$+ b_1 \sin x + b_2 \sin 2x + b_3 \sin 3x \qquad \text{...(i)}$$

To evaluate the coefficients, the following table is constructed.

x	$f(x)$	$\cos x$	$\sin x$	$\cos 2x$	$\sin 2x$	$\cos 3x$	$\sin 3x$
30°	2·34	0·866	0·5	0·5	0·866	0	1
60°	3·01	0·5	·866	–0·5	0·866	–1	0
90°	3·68	0	1	–1	0	0	–1
120°	4·15	–0·5	0·866	– 0·5	–0·866	1	0
150°	3·69	–0·866	0·5	0·5	–0·866	0	1
180°	2·20	–1	0	1	0	–1	0
210°	0·83	–0·866	–0·5	0·5	0·866	0	–1
240°	0·51	–0·5	–0·866	–0·5	0·866	1	0
270°	0·88	0	–1	–1	0	0	1
300°	1·09	0·5	–0·866	–0·5	–0·866	–1	0
330°	1·19	0·866	–0·5	0·5	–0·866	0	–1
360°	1·64	1	0	1	0	1	0

$$a_0 = \tfrac{2}{12} \Sigma f(x)$$
$$= \tfrac{1}{6} \times 25{\cdot}21 = 4{\cdot}202$$

$$a_1 = \tfrac{2}{12} \times \Sigma f(x) \cos x$$
$$= \tfrac{1}{6} [2{\cdot}34 \times 0{\cdot}866 + 3{\cdot}01 \times 0{\cdot}5 - 4{\cdot}15 \times 0{\cdot}5 - 3{\cdot}69 \times 0{\cdot}86 - 2{\cdot}20$$
$$\times 1 - 0{\cdot}83 \times 0{\cdot}866 - 0{\cdot}51 \times 0{\cdot}5 + 1{\cdot}09 \times 0{\cdot}5$$
$$+ 1{\cdot}19 \times 0{\cdot}866 + 1{\cdot}64 \times 1]$$
$$= \tfrac{1}{6} [(2{\cdot}34 - 3{\cdot}69 - 0{\cdot}83 + 1{\cdot}19) \times 0{\cdot}866 + (3{\cdot}01 - 4{\cdot}15 - 0{\cdot}51$$
$$+ 1{\cdot}09) \times 0{\cdot}5 + (-2{\cdot}20 + 1{\cdot}64)]$$

$$= -0{\cdot}280$$

$$b_1 = \frac{2}{12} \times \Sigma f(x) \sin x$$

$$= \frac{1}{6} [(2{\cdot}34 + 3{\cdot}69 - 0{\cdot}83 - 1{\cdot}19) \times 0{\cdot}5$$

$$+ (3{\cdot}01 + 4{\cdot}15 - 0{\cdot}51 - 1{\cdot}09) \times 0{\cdot}866 + (3{\cdot}68 - 0{\cdot}88)]$$

$$= 1{\cdot}618$$

$$a_2 = \frac{2}{12} \times \Sigma f(x) \cos 2x$$

$$= \frac{1}{6} [(2{\cdot}34 - 3{\cdot}01 - 4{\cdot}15 + 3{\cdot}69 + 0{\cdot}83 - 0{\cdot}51 - 1{\cdot}09 + 1{\cdot}19) \times 0{\cdot}5$$

$$+ (-3{\cdot}68 + 2{\cdot}20 - 0{\cdot}88 + 1{\cdot}64)]$$

$$= -0{\cdot}178.$$

$$b_2 = \frac{2}{12} \times \Sigma f(x) \sin 2x$$

$$= \frac{1}{6} [(2{\cdot}34 + 3{\cdot}01 - 4{\cdot}15 - 3{\cdot}69 + 0{\cdot}83 + 0{\cdot}51 - 1{\cdot}09 - 1{\cdot}19) \times 0{\cdot}866]$$

$$= -0{\cdot}495$$

$$a_3 = \frac{2}{12} \times \Sigma f(x) \cos 3x$$

$$= \frac{1}{6} [-3{\cdot}01 + 4{\cdot}15 - 2{\cdot}20 + 0{\cdot}51 - 1{\cdot}09 + 1{\cdot}64]$$

$$= 0$$

$$b_3 = \frac{2}{12} \times \Sigma f(x) \sin 3x$$

$$= \frac{1}{6} [2{\cdot}34 - 3{\cdot}68 + 3{\cdot}69 - 0{\cdot}83 + 0{\cdot}88 - 1{\cdot}19]$$

$$= 0{\cdot}202.$$

$$f(x) = \frac{1}{2} a_0 + a_1 \cos x + a_2 \cos 2x + a_3 \cos 3x + \cdots$$

$$+ b_1 \sin x + b_2 \sin 2x + b_3 \sin 3x + \cdots$$

$$= 2{\cdot}101 - 0{\cdot}280 \cos x - 0{\cdot}178 \cos 2x + 0(\cos 3x) + \cdots$$

$$+ 1{\cdot}618 \sin x - 0{\cdot}495 \sin 2x + 0{\cdot}202 \sin 3x + \cdots$$

Example 2. *Compute the first three harmonics of the Fourier series of $f(x)$ given by the following table.* ***(MS. 1991 Ap.)***

x	0	$\pi/3$	$2\pi/3$	π	$4\pi/3$	$5\pi/3$	2π
$f(x)$	1·0	1·4	1·9	1·7	1·5	1·2	1·0

We will form the table for the convenience of work.

We exclude the last point $x = 2\pi$.

x	$f(x)$	cos x	sin x	cos $2x$	sin $2x$	cos $3x$	sin $3x$
0	1·0	1	0	1	0	1	0
$\frac{\pi}{3}$	1·4	0·5	0·866	–0·5	0·866	–1	0
$\frac{2\pi}{3}$	1·9	–0·5	0·866	–0·5	–0·866	1	0
π	1·7	–1	0	1	0	–1	0
$\frac{4\pi}{3}$	1·5	–0·5	–0·866	–0·5	0·866	1	0
$\frac{5\pi}{3}$	1·2	0·5	–0·866	–0·5	–0·866	–1	0

$$a_0 = \frac{2}{6}\Sigma f(x) = \frac{1}{3}(1{\cdot}0 + 1{\cdot}4 + 1{\cdot}9 + 1{\cdot}7 + 1{\cdot}5 + 1{\cdot}2) = 2{\cdot}9$$

$$a_1 = \frac{2}{6}\Sigma f(x)\cos x = \frac{1}{3}(1 + 0{\cdot}7 - 0{\cdot}95 - 1{\cdot}7 - 0{\cdot}75 + 0{\cdot}6)$$

$$= -\,0{\cdot}37$$

$$a_2 = \frac{2}{6}\Sigma f(x)\cos 2x = -\,0{\cdot}1$$

$$a_3 = \frac{2}{6}\Sigma f(x)\cos 3x = 0{\cdot}03$$

$$b_1 = \frac{2}{6}\Sigma f(x)\sin x = 0{\cdot}17$$

$$b_2 = \frac{2}{6}\Sigma f(x)\sin 2x = -\,0{\cdot}06$$

$$b_3 = \frac{2}{6}\Sigma f(x)\sin 3x = 0$$

$$\therefore\ f(x) = 1{\cdot}45 - 0{\cdot}33\cos x - 0{\cdot}1\cos 2x + 0{\cdot}03\cos 3x + \ldots$$
$$+ 0{\cdot}17\sin x - 0{\cdot}06\sin 2x$$
$$+ 0\sin 3x + \ldots$$

Example 3. *Compute the first three harmonics of Fourier series of $f(x)$ given by the following table:*

x	0°	30°	60°	90°	120°	150°	180°	210°	240°	270°	300°	330°
$y = f(x)$	6·824	7·976	8·026	7·204	5·676	3·674	1·764	0·552	0·262	0·904	2·492	4·736

We will form the table to calculate a's and b's of the Fourier coefficients.

x	$y = f(x)$	$\cos x$	$y \cos x$	$\cos 2x$	$y \cos 2x$	$\cos 3x$	$y \cos 3x$	$\sin x$	$y \sin x$	$\sin 2x$	$y \sin 2x$	$\sin 3x$	$y \sin 3x$
0°	6·824	1·000	6·824	1	6·824	1	6·824	0	0	0	0	0	0
30°	7·976	0·866	6·907	0·5	3·988	0	0	·5	3·988	0·866	6·907	1	7·976
60°	8·026	0·5	4·013	–0·5	–4·013	–1	–8·026	0·866	6·950	0·866	6·950	0	0
90°	7·204	0	0	–1	–7·204	0	0	1	7·204	0	0	–1	–7·24
120°	5·676	–0·5	–2·838	–0·5	–2·838	1	5·676	0·866	4·916	–·866	–4·918	0	0
150°	3·674	–0·866	–3·182	0·5	1·837	0	0	0·5	1·187	–·866	–3·182	1	3·674
180°	1·764	–1	–1·764	1	1·764	–1	–1·764	0	0	0	0	0	0
210°	0·552	–0·866	–0·478	0·5	0·276	0	0	–·5	–0·276	0·866	0·478	–1	–0·552
240°	0·262	–0·5	0·131	–0·5	–0·131	1	0·262	–·866	–0·227	0·866	0·227	0	0
270°	0·904	0	0	–1	–·904	0	0	–1	–0·904	0	0	1	0·904
300°	2·492	0·5	1·246	–0·5	–1·246	–1	–2·492	–·866	–2·158	–·866	–2·158	0	0
330°	4·736	0·866	4·102	0·5	2·368	0	0	–0·5	–2·368	–·866	–4·102	–1	–4·736
	50·090		14·699		0·721		0·48		18·962		0·204		0·062

$$a_0 = \tfrac{2}{12}\,\Sigma f(x) = 8{\cdot}348 \qquad b_1 = \tfrac{2}{12}\,\Sigma y \sin x = 3{\cdot}160$$

$$a_1 = \tfrac{2}{12}\,\Sigma f(x) \cos x = 2{\cdot}450 \qquad b_2 = \tfrac{2}{12} \times 0{\cdot}204 = 0{\cdot}034$$

$$a_2 = \tfrac{2}{12}(0{\cdot}721) = 0{\cdot}120 \qquad b_3 = \tfrac{2}{12} \times 0{\cdot}062 = 0{\cdot}010$$

$$a_3 = \tfrac{2}{12}(0{\cdot}48) = 0{\cdot}080$$

$$\therefore \quad f(x) = 4{\cdot}174 + 2{\cdot}450 \cos x + 0{\cdot}120 \cos 2x$$
$$+ 0{\cdot}080 \cos 3x + 3{\cdot}160 \sin x + 0{\cdot}034 \sin 2x$$
$$+ 0{\cdot}010 \sin 3x.$$

Example 4. *The values of x and the corresponding values of $f(x)$ over a period T are given below. Show that $f(x) = 0{\cdot}75 + 0{\cdot}37 \cos \theta + 1{\cdot}004 \sin \theta$* where $\theta = \dfrac{2\pi x}{T}$.

x	0	$T/6$	$T/3$	$T/2$	$2T/3$	$5T/6$	T
$f(x)$	1·98	1·30	1·05	1·30	–0·88	–0·25	1·98

We omit the last value since $f(x)$ at $x = 0$ is known. $\theta = \dfrac{2\pi x}{T}$

When x varies from 0 to T, θ varies from 0 to 2π with an increase of $\dfrac{2\pi}{6}$

Let $f(x) = F(\theta) = \dfrac{a_0}{2} + a_1 \cos \theta + b_1 \sin \theta$

$$a_0 = \frac{2}{6}\,\Sigma f(x) = \frac{4{\cdot}6}{3} = 1{\cdot}5$$

θ	y	cos θ	sin θ	y cos θ	y sin θ
0	1·98	1·0	0	1·98	0
π/3	1·30	0·5	0·866	0·65	1·1258
2π/3	1·05	–0·5	0·866	–0·525	0·9093
π	1·30	–1	0	–1·3	0
4π/3	–0·88	–0·5	–0·866	0·44	0·762
5π/3	–0·25	0·5	–·866	–0·125	0·2165
Σ	4·6			1·12	3·013

$$a_1 = 2\,\frac{(1\cdot 12)}{6} = 0\cdot 37$$

$$b_1 = \frac{2}{6}(3\cdot 013) = 1\cdot 005 \quad \therefore\ f(x) = 0\cdot 75 + 0\cdot 37 \cos\theta + 1\cdot 005 \sin\theta$$

Example 5. *Find the first three harmonics of Fourier series of $y = f(x)$ from the following data.*

x	0°	30°	60°	90°	120°	150°	180°	210°	240°	270°	300°	330°
y	298	356	373	337	254	155	80	51	60	93	147	221

Let $\quad y = \dfrac{a_0}{2} + \displaystyle\sum_{n=1}^{\infty}\left(a_n \cos nx + b_n \sin nx\right)$

Here $a_0 = \dfrac{2}{12}\Sigma y = \dfrac{1}{6}(2425) \approx 404$

$$a_1 = \frac{2}{12}\Sigma y \cos x = 107\cdot 048 \approx 107$$

$$a_2 = \frac{2}{12}\Sigma y \cos 2x \approx -13$$

$$a_3 = \frac{2}{12}\Sigma y \cos 3x \approx 2\cdot 0$$

$$b_1 = \frac{2}{12}\Sigma y \sin x \approx 121$$

$$b_2 = \frac{2}{12}\Sigma y \sin 2x \approx 9$$

$$b_3 = \frac{2}{12}\Sigma y \sin 3x \approx -1$$

The table can be formulated by the reader in the usual way.

$$\therefore\ y \approx 202 + 107 \cos x - 13 \cos 2x + 2 \cos 3x + 121 \sin x + 9 \sin 2x - \sin 3x.$$

Exercises 1 (h)

1. The following values of x and y are taken from a periodic curve which has a period 2π. Express y as a Fourier series up to the third harmonic.

x	0°	30°	60°	90°	120°	150°	180°	210°	240°	270°	300°	330°
y	0·50	0·60	0·83	1	0·80	0·42	0·00	– 0·34	–0·50	–0·20	0·67	0·70

[**Note.** In this question, take the approximation to the rectangular rule in the form:

$$\int_a^b f(x)\,dx = (b-a) \times \frac{f(a) + f(x_1) + f(x_2) + \cdots + f(x_{q-1})}{q}\Bigg]$$

2. The table of values of the function $y = f(x)$ is given below:

x	0	$\frac{\pi}{3}$	$\frac{2\pi}{3}$	π	$\frac{4\pi}{3}$	$\frac{5\pi}{3}$	2π
y	1·0	1·4	1·9	1·7	1·5	1·2	1·0

Find a Fourier series upto the third harmonic to represent $f(x)$ in terms of x.

(Calicut, 72 Engg.)

[**Note.** For the application of the formula, *one* of the end points 0 or 2π can be ignored.]

3. Analyse the current i given by the table below into its constituent harmonics as far as the third one:

θ (deg.)	0	30	60	90	120	150	180	210	240	270	300	330
i (amp.)	0	24·0	32·5	27·5	18·2	13·0	0	–24	–33·5	–27·5	–18·2	–13

[**Note.** The table shows that $f(180° + x) = -f(x)$; hence, the even harmonics will be zero.]

4. The displacement y of a part of a mechanism are tabulated with the corresponding angular movement of the crank. Express y as a Fourier series neglecting the harmonics above the third:

$x°$	0	30	60	90	120	180	210	240	270	300	330	150
y	1·8	1·1	0·30	0·16	0·50	2·16	1·25	1·30	1·52	1·76	2·00	1·3

5. The turning moment T lb-ft. on the crank-shaft of a steam engine is given for a series of values of the crank-angle θ degrees:

θ	0	30	60	90	120	150	180
T	0	5224	8097	7850	5499	2626	0

Obtain the first four terms in a series of sines to represent T; also calculate T when $\theta = 75°$.

6. The following table gives the variations of a periodic current over a period:–

t sec.	0	T/6	T/3	T/2	2T/3	5T/6	T
A amps.	1·98	1·30	1·05	1·30	–0·88	0·25	1·98

By numerical analysis, show that there is a direct current part of 0·75 amp. in the variable current; obtain the amplitude of the first harmonic.

7. Using six ordinates analyse harmonically the following data to two harmonics. **(*MS. 1988 Nov.*)**

x	0	$\pi/3$	$2\pi/3$	π	$4\pi/3$	$5\pi/3$	π
y	10	12	15	20	17	11	10

8. Complete the first two harmonics of the Fourier series of $f(x)$ from the data given below. **(*MS 1989 Nov.*)**

x	30°	60°	90°	120°	150°	180°	210°	240°	270°	300°	330°	360°
y	2·34	3·01	3·68	4·15	3·69	2·2	0·83	0·51	0·88	1·09	1·19	1·64

9. From the following table find the first two harmonics of the Fourier series of $f(x)$.

x	0°	30°	60°	90°	120°	150°	180°	210°	240°	270°	300°	330°
y	137	164	265	325	156	–54	–137	–164	–265	–325	–156	54

10. Determine the first two harmonics of the Fourier series for $f(x)$ from the table below.

x	30°	60°	90°	120°	150°	180°	210°	240°	270°	300°	330°	360°
y	3·5	6·09	7·82	8·58	8·43	7·73	6·98	6·19	6·04	5·55	5·01	5·35

Fourier series in complex form. The series for $f(x)$ defined in the interval c to $c + 2\pi$, and satisfying Dirichlet's conditions can be given in the form

$$f(x) = \sum_{n=-\infty}^{\infty} c_n e^{inx} \qquad \text{...(i)}$$

where $$c_n = \frac{1}{2\pi}\int_c^{c+2\pi} f(x)\, e^{-inx}\, dx.$$

Proof. Multiply both the sides of (i) by e^{-imx}, where m is an integer; then, integrate between the limits c and $c + 2\pi$. Thus we have

$$\int_c^{c+2\pi} f(x) e^{-imx}\, dx = \sum_{n=-\infty}^{\infty} c_n \int_c^{c+2\pi} e^{i(n-m)x}\, dx \qquad \text{...(ii),}$$

on the assumption that term by term integration is permissible.

When $n = m$, $$\int_c^{c+2\pi} e^{i(n-m)x}\, dx = \int_c^{c+2\pi} e^0\, dx = \int_c^{c+2\pi} dx = \Big[x\Big]_c^{c+2\pi} = 2\pi \qquad \text{...(iii)}$$

For all other values of n, that is if $n \neq m$,

$$\int_c^{c+2\pi} e^{i(n-m)x}\, dx = \left[\frac{e^{i(n-m)x}}{i(n-m)}\right]_c^{c+2\pi} = \frac{e^{i(n-m)(c+2\pi)} - e^{i(n-m)c}}{i(n-m)}$$

$$= \frac{e^{i(n-m)c}}{i(n-m)}\left[e^{2i(n-m)\pi}\right] \qquad \text{...(iv)}$$

Since n and m are integers, $n - m$ of an integer; let us denote it by p. Then,

$$e^{2i(n-m)\pi} - 1 = e^{2ip\pi} - 1 = \cos 2p\pi + i \sin 2p\pi - 1 = 1 + i \times 0 - 1 = 0.$$

Using this in (iv), we get that when $n \neq m$,

$$\int_c^{c+2\pi} e^{i(n \neq m)x}\, dx = 0 \qquad \text{...(v)}$$

Using the results (iii) and (v) in (ii), we get

$$\int_c^{c+2\pi} f(x)\, e^{imx}\, dx = c_m \times 2\pi.$$

$$\text{Hence } c_m = \frac{1}{2\pi} \cdot \int_c^{c+2\pi} f(x) e^{-imx}\, dx \text{ ; or } c_n = \frac{1}{2\pi} \int_c^{c+2\pi} f(x)\, e^{-inx}\, dx.$$

This proves the statement (i).

Note 1. When the R.H.S. of (i) is written out in full, (i) becomes

$$f(x) = c_0 + c_1 e^{ix} + c_2 e^{2ix} + c_3 e^{3ix} + \cdots + c_n e^{inx} + \cdots \text{ to } \infty$$

$$+ c_{-1}\, e^{-ix} + c_{-2}\, e^{-2ix} + c_{-3} e^{-3ix} + \cdots + c_{-n} e^{-inx} + \cdots \text{ to } \infty.$$

Thus, the index n runs through all positive and negative values including 0.

The term independent of the exponential factor is c_0 and *not* $\frac{1}{2}c_0$: remember that in the trigonometric form of the series, the independent term was given by $\frac{1}{2}a_0$. Putting $n = 0$, $c_0 = \frac{1}{2\pi} \int_c^{c+2\pi} f(x)\, dx.$

Note 2. In very many cases, the lower limit c will be either 0 or $-\pi$. Thus, for example when $f(x)$ satisfies Dirichlet's conditions in the interval $-\pi$ to π,

$$f(x) = \sum_{n=-\infty}^{\infty} c_n e^{inx}, \text{ where } c_n = \frac{1}{2\pi} \int_{-\pi}^{\pi} f(x)\, e^{-inx}\, dx.$$

Note 3. We can show that if $f(x)$ is defined in $(-L, L)$, then

$$f(x) = \sum_{n=-\infty}^{\infty} c_n\, e^{in\pi x/L}, \text{ where } c_n = \frac{1}{2L} \int_{-L}^{L} f(x)\, e^{-in\pi x/L}\, dx.$$

Theorem. Derive the trigonometric form of Fourier series from the complex form.

We have

$$f(x) = c_0 + c_1 e^{ix} + c_2 e^{2ix} + \cdots + c_n e^{inx} + \cdots \text{ to } \infty$$
$$+ c_{-1} e^{-ix} + c_{-2} e^{-2ix} + \cdots + c_{-n} e^{-inx} + \cdots \text{ to } \infty.$$

This can be written in the form

$$f(x) = c_0 + \sum_{n=1}^{\infty} (c_n e^{inx} + c_{-n} e^{-inx})$$

$$= c_0 + \sum_{n=1}^{\infty} \{c_n (\cos nx + i \sin nx) + c_{-n} (\cos nx - i \sin nx)\}$$

i.e. $$f(x) = c_0 + \sum_{n=1}^{\infty} [(\cos nx)(c_n + c_{-n}) + (\sin nx)\{i(c_n - c_{-n})\}] \qquad ...(i)$$

Let us denote c_0 by $\frac{1}{2}a_0$, $c_n + c_{-n}$ by a_n and $i(c_n - c_{-n})$ by b_n.

Then $a_0 = 2c_0 = 2 \times \dfrac{1}{2\pi} \displaystyle\int_c^{c+2\pi} f(x)\, dx = \dfrac{1}{\pi} \int_c^{c+2\pi} f(x)\, dx$

$$a_n = c_n + c_{-n} = \frac{1}{2\pi} \left\{ \int_c^{c+2\pi} f(x)\, e^{-inx}\, dx + \int_c^{c+2\pi} f(x)\, e^{inx}\, dx \right\}$$

$$= \frac{1}{2\pi} \int_c^{c+2\pi} f(x)\,(e^{-inx} + e^{inx})\, dx = \frac{1}{2\pi} \int_c^{c+2\pi} \{f(x)\}\, 2\cos nx\, dx$$

i.e., $$a_n = \frac{1}{\pi} \int_c^{c+2\pi} f(x) \cos nx\, dx.$$

Again, $b_n = i(c_n - c_{-n}) = i \left[\dfrac{1}{2\pi} \displaystyle\int_c^{c+2\pi} f(x)\,(e^{-inx} - e^{inx})\, dx \right]$

$$= \frac{i}{2\pi} \int_c^{c+2\pi} \{f(x)\}\,(-2i \sin nx)\, dx$$

$$= \frac{-i^2}{\pi} \int_c^{c+2\pi} f(x) \sin nx\, dx.$$

Thus $$b_n = \frac{1}{\pi} \int_c^{c+2\pi} f(x) \sin nx\, dx.$$

Changing the coefficients c_n and c_{-n} in (*i*) in favour of a_n and b_n, we get,

$$f(x) = \frac{a_0}{2} + \sum_{n=1}^{\infty} (a_n \cos nx + b_n \sin nx),$$

where we have calculated that

$$a_n = \frac{1}{\pi} \int_c^{c+2\pi} f(x) \cos nx \, dx, \text{ and } b_n = \frac{1}{\pi} \int_c^{c+2\pi} f(x) \sin nx \, dx.$$

Example 1. *Derive the complex form of the Fourier series for $f(x) = e^{ax}$, $-\pi < x < \pi$, given that a is a real constant. Deduce that (i) when α is a constant other than an integer,*

$$\cos \alpha x = \frac{\sin \pi \alpha}{\pi} \sum_{n=-\infty}^{\infty} (-1)^n \frac{\alpha}{\alpha^2 - n^2} e^{inx}, -\pi < x < \pi,$$

$$(ii) \sum_{n=-\infty}^{\infty} \frac{(-1)^n}{n^2 + a^2} = \frac{\pi}{a \sinh a\pi}.$$

By theorem, $f(x) = \sum_{n=-\infty}^{\infty} c_n e^{inx}$ ...(*i*),

where $c_n = \frac{1}{2\pi} \int_{-\pi}^{\pi} e^{ax} e^{-inx} dx = \frac{1}{2\pi} \int_{-\infty}^{\infty} e^{(a-in)x} dx$

$$= \frac{1}{2\pi} \left[\frac{e^{(a-in)x}}{a - in} \right]_{-\pi}^{\pi} = \frac{1}{2\pi (a - in)} \left\{ e^{(a-in)\pi} - e^{-(a-in)\pi} \right\}$$

i.e., $$c_n = \frac{1}{2\pi (a - in)} \left\{ e^{a\pi} . e^{-in\pi} - e^{-a\pi} . e^{in\pi} \right\}$$

But $e^{\pm in\pi} = \cos (\pm n\pi) + i \sin (\pm n\pi) = (-1)^n + i \times 0 = (-1)^n$.

Hence $$c_n = \frac{1}{2\pi (a - in)} (-1)^n (e^{a\pi} - e^{-a\pi})$$

$$= (-1)^n \frac{1}{\pi (a - in)} \times \frac{e^{a\pi} - e^{-a\pi}}{2}$$

$$= (-1)^n \frac{a + in}{\pi (a - in)(a + in)} \sinh a\pi$$

i.e. $$c_n = \frac{\sinh a\pi}{\pi} \left\{ (-1)^n \frac{a + in}{a^2 + n^2} \right\}.$$

Substituting the value of c_n in (*i*) we get

$$f(x) = \sum_{n=-\infty}^{\infty} \frac{\sinh a\pi}{\pi} (-1)^n \frac{a+in}{a^2+n^2} e^{inx}$$

Since $f(x) = e^{ax}, -\pi < x < \pi$, we get

$$e^{ax} = \frac{\sinh a\pi}{\pi} \sum_{n=-\infty}^{\infty} (-1)^n \frac{a+in}{a^2+n^2} e^{inx}, \text{ in } -\pi < x < \pi \qquad \ldots(ii)$$

(*i*) Now $\cos \alpha x = \frac{1}{2}(e^{i\alpha x} + e^{-i\alpha x})$ ...(*iii*)

The Fourier series for $e^{i\alpha x}$ is got from (*ii*) by putting $a = i\alpha$

Thus

$$e^{i\alpha x} = \frac{\sinh i\,\alpha\,\pi}{\pi} \sum_{n=-\infty}^{\infty} (-1)^n \frac{i\alpha + in}{(i\alpha)^2+n^2} e^{inx}$$

$$= \frac{\sinh i\,\alpha\,\pi}{\pi} \sum_{n=-\infty}^{\infty} (-1)^n \frac{i(\alpha+n)}{-\alpha^2+n^2} e^{inx}$$

i.e., $$e^{i\alpha x} = \frac{-i\sinh i\,\alpha\,\pi}{\pi} \sum_{n=-\infty}^{\infty} (-1)^n \frac{\alpha+n}{\alpha^2-n^2} e^{inx}$$

But we know that $\sin i\theta = i \sinh \theta$; putting $\theta = i\alpha\pi$, $\sin i^2 \alpha\pi = i \sinh i\alpha\pi$ *i.e*, $-\sin \alpha\pi = i \sinh i\alpha\pi$ giving $\sin \alpha\pi = -i \sinh i\,\alpha\,\pi$. Hence

$$e^{i\alpha x} = \frac{\sin\alpha\,\pi}{\pi} \sum_{n=-\infty}^{\infty} (-1)^n \frac{\alpha+n}{\alpha^2-n^2} e^{inx}$$

Changing α into $-\alpha$, we have

$$-e^{-i\alpha x} = \frac{\sin\alpha\,\pi}{\pi} \sum_{n=-\infty}^{\infty} (-1)^n \frac{\alpha-n}{\alpha^2-n^2} e^{inx}$$

Adding, $$e^{i\alpha x} + e^{-i\alpha x} = \frac{2\sin\alpha\pi}{\pi} \sum_{n=-\infty}^{\infty} (-1)^n \frac{\alpha}{\alpha^2-n^2} e^{inx}$$

Hence $\cos \alpha x = \frac{1}{2}(e^{i\alpha x} + e^{-i\alpha x})$

$$= \frac{\sin\alpha\,\pi}{\pi} \sum_{n=-\infty}^{\infty} (-1)^n \frac{\alpha}{\alpha^2-n^2} e^{inx}$$

Note. Subtracting the value of $e^{-i\alpha x}$ from that of $e^{i\alpha x}$, we get

$$\sin \alpha x = \frac{\sin \alpha\pi}{\pi i} \sum_{n=-\infty}^{\infty} (-1)^n \frac{n}{\alpha^2 - n^2} e^{inx}$$

(*ii*) In (*ii*), putting $x = 0$, we get

$$1 = \frac{\sinh a\pi}{\pi} \sum_{n=-\infty}^{\infty} (-1)^n \frac{a + in}{a^2 + n^2}$$

Equating the real parts we have

$$1 = \frac{\sinh a\pi}{\pi} \sum_{n=-\infty}^{\infty} (-1)^n \frac{a}{a^2 + n^2}$$

i.e.,
$$1 = \frac{a \sinh a\pi}{\pi} \sum_{n=-\infty}^{\infty} \frac{(-1)^n}{n^2 + a^2}$$

$$\therefore \quad \frac{\pi}{a \sinh a\pi} = \sum_{n=-\infty}^{\infty} \frac{(-1)^n}{n^2 + a^2}$$

Note. If the range is $(c, c + 2l)$ the complex form is

$$f(x) = \sum_{n=-\infty}^{\infty} c_n e^{\frac{in\pi x}{l}} \qquad \text{...(1)}$$

where
$$c_n = \frac{1}{2l} \int_c^{c+2l} f(x)\, e^{\frac{-in\pi x}{l}}\, dx \qquad \text{...(2)}$$

Example 2. *Find the complex form of the Fourier series of* $f(x) = e^{-x}$ *in* $-1 < x < 1$ **(*M.S. 1997*)**

Here $2l = 2$; $l = 1$

Let
$$f(x) = \sum_{n=-\infty}^{\infty} c_n\, e^{in\pi x}$$

where
$$c_n = \frac{1}{2} \int_{-1}^{1} e^{-x}\, e^{-in\pi x}\, dx$$

$$= \frac{1}{2} \int_{-1}^{1} e^{-(1 + in\pi)x}\, dx$$

$$= \frac{1}{2} \left[\frac{e^{-(1 + in\pi)x}}{-(1 + in\pi)} \right]_{-1}^{1}$$

$$= -\frac{1}{2(1+in\pi)}\left[e^{-(1+in\pi)} - e^{(1+in\pi)}\right]$$

$$= \frac{-(1-in\pi)}{2(1+n^2\pi^2)}\left[e^{-1}(\cos n\pi - i\sin n\pi) - e(\cos n\pi + i\sin n\pi)\right]$$

$$= \frac{(1-in\pi)}{2(1+n^2\pi^2)}\cos n\pi\,(e - e^{-1})$$

$$= \frac{(1-in\pi)}{1+n^2\pi^2}(-1)^n \sinh 1,$$

$$f(x) = \sum_{n=-\infty}^{\infty} \frac{(1-in\pi)}{1+n^2\pi^2}(-1)^n \sinh 1\, e^{in\pi x}$$

Exercises 1(i)

1. Starting with the complex form of the Fourier series for e^{ax} obtained in the worked example 1, derive the real Fourier series for e^{ax}.
2. Using the worked example 1, deduce the real Fourier series for (*i*) $\cos \alpha x$, (*ii*) $\sin \alpha x$.
3. Find the complex form of the Fourier series of the function whose definition in one period is $f(x) = e^{-x}, -1 \le x \le 1$.
   Deduce the real Fourier series for e^{-x}, in the given interval. ***(Calicut, 72 Engg.)***
4. Find the complex form of the Fourier series of the periodic function whose definition in one period is given by $f(x) = \sin x$, in $-\pi < x < \pi$.
5. Show the complex form of the Fourier series of $f(x) = e^x$, when $-\pi < x < \pi$ and $f(x+2\pi) = f(x)$ is

$$f(x) = \frac{\text{Sin}\,h\,\pi}{\pi}\sum_{n=-\infty}^{\infty}(-1)^n \frac{1+in}{1+n^2}e^{inx}$$

6. Find the complex form of the Fourier series of the periodic function $f(x) = \text{Sin}\,x, 0 < x < \pi$ ***(B.E.M.U. 1976)***
7. Find the complex form of the Fourier series of e^{ax}, in $-l < x < l$. ***(B.E. 1980, Madurai Uni.)***
8. Find the complex Fourier series for $f(x)$
   where
   (*i*) $f(x) = \sin ax$, a not an integer in $(-\pi, \pi)$
   (*ii*) $f(x) = e^x$, in $-\pi < x < \pi$
   (*iii*) $f(x) = e^{-x}$, in $-1 < x < 1$

CHAPTER II

PARTIAL DIFFERENTIAL EQUATIONS

2·1. Introduction

You have already studied ordinary differential equations in the previous semesters. Now we will proceed to the study of partial differential equations. A partial differential equation is one which involves partial derivatives. The *order* of a partial differential equation is the order of the highest derivative occuring in it.

In what follows, z will be taken as the dependent variable and x, y the independent variables so that $z = f(x, y)$. We will use the following notations hereafter:

$$\frac{\partial z}{\partial x} = p, \frac{\partial z}{\partial y} = q, \frac{\partial^2 z}{\partial x^2} = r, \frac{\partial^2 z}{\partial x\, \partial y} = s \text{ and } \frac{\partial^2 z}{\partial y^2} = t$$

2·2. Formation of differential equations

Partial differential equations can be formed either by the elimination of arbitrary constants or by the elimination of arbitrary functions from a relation involving three or more variables.

2·3. By elimination of arbitrary constants

Let us take the function

$$f(x, y, z, a, b) = 0 \qquad ...(1)$$

where a and b are arbitrary constants.

Now we have to eliminate a and b while forming the differential equation.

Differentiating the equation (1) partially with respect to the independent variables x, y we get,

$$\frac{\partial f}{\partial x} + \frac{\partial f}{\partial z} \cdot \frac{\partial z}{\partial x} + 0$$

i.e.,

$$\frac{\partial f}{\partial x} + p\frac{\partial f}{\partial z} = 0 \qquad ...(2)$$

Similarly,

$$\frac{\partial f}{\partial y} + q\frac{\partial f}{\partial z} = 0 \qquad ...(3)$$

Now, eliminating the two arbitrary constants a and b from (1), (2) and (3), we get a partial differential equation of the first order of the form

$$\phi(x, y, z, p, q) = 0. \quad ...(4)$$

Note. Equation (1) is said to be the *primitive* or *complete solution* of the first order differential equation (4). If the number of constants to be eliminated is equal to the number of independent variables, the differential equation got after elimination of arbitrary constants will be of the first order. On the other hand, if the number of constants to be eliminated is more than the number of independent variables, the resulting partial differential equation will be of the second or higher orders. This is an important difference between the partial differential equation and the ordinary differential equation where the order of the differential equation equals the number of arbitrary constants eliminated.

Example 1. *Form the partial differential equation by eliminating the arbitrary constants a and b from* $z = (x^2 + a)(y^2 + b)$. **(A.U. '65 B.E.)**

Now, $z = (x^2 + a)(y^2 + b)$ ...(1)

Differentiating partially w.r.t. x and y in turn,

$$\frac{\partial z}{\partial x} = 2x(y^2 + b),$$

and

$$\frac{\partial z}{\partial y} = 2y(x^2 + a).$$

Therefore, $x^2 + a = \dfrac{q}{2y}$

and

$$y^2 + b = \frac{p}{2x}$$

Substituting these in (1), we get,

$$z = \frac{p}{2y} \cdot \frac{q}{2x}$$

i.e.

$$4xyz = pq.$$

Example 2. *Find the partial differential equation of all planes cutting equal intercepts from the x and y axes.*

Let a, c be the intercepts on x and z axes respectively.

Hence, the equation of the plane is

$$\frac{x}{a} + \frac{y}{a} + \frac{z}{c} = 1 \quad ...(1)$$

Differentiating partially w.r.t. x and y in turn,

$$\frac{1}{a} + \frac{1}{c}\frac{\partial z}{\partial x} = 0 \quad ...(2)$$

and $$\frac{1}{a} + \frac{1}{c}\frac{\partial z}{\partial y} = 0 \qquad ...(3)$$

From (2) and (3), we get $\frac{1}{c}(p - q) = 0$

Therefore, the equation is $p - q = 0$.

Example 3. *Form the partial differential equation by elimination a and b from* $\log(az - 1) = x + ay + b$ ***(MS. 1991. Ap.)***

Differentiating the given function partially w.r.t. x,

$$\frac{1}{az - 1}\, ap = 1 \qquad ...(1)$$

Differentiating w.r.t y partially,

$$\frac{1}{az - 1} \cdot aq = a \qquad ...(2)$$

Dividing (2) by (1), we get

$$a = \frac{q}{p}\text{; i.e. } ap = q$$

Putting in (1)

$$q = az - 1$$

$$q = \frac{q}{p} z - 1$$

$p(q + 1) = qz$ is the required equation.

Example 4. *Obtain the partial differential equation of all spheres whose centres lie on Z = 0 and whose radius is constant and equal to r.*

(MS. 1991. Ap.)

The centre of sphere is $(a, b, 0)$

Hence its equation is $(x - a)^2 + (y - b)^2 + z^2 = r^2 \qquad ...(1)$

Differentiate w.r.t. x and y partially in order

$$2(x - a) + 2z\frac{\partial z}{\partial x} = 0$$

$$2(y - b) + 2z\frac{\partial z}{\partial y} = 0$$

$$\therefore \quad x - a = -pz$$

$$y - b = -qz$$

Substituting in (1),

$$p^2z^2 + q^2z^2 + z^2 = r^2$$

i.e., $z^2(p^2 + q^2 + 1) = r^2$ is the required equation.

Example 5. *Obtain the partial differential equation by eliminating a, b, c from* $\frac{x^2}{a^2}+\frac{y^2}{b^2}+\frac{z^2}{c^2}=1$

Since the number of arbitrary constants is more than the number of independent variables, we will get the partial differential equation of order greater than 1.

$$\frac{x^2}{a^2}+\frac{y^2}{b^2}+\frac{z^2}{c^2}=1$$

Differentiate partially w.r.t. x and y. (z is dependent)

$$\frac{2x}{a^2}+\frac{2z}{c^2}p=0 \qquad (i)$$

$$\frac{2y}{b^2}+\frac{2z}{c^2}q=0 \qquad (ii)$$

Differentiate (*i*) w.r.t. x again.

$$\frac{2}{a^2}+\frac{2}{c^2}\left[z\,.\,\frac{\partial^2 z}{\partial x^2}+\left(\frac{\partial z}{\partial x}\right)^2\right]=0 \qquad (iii)$$

Differentiate (*ii*) w.r.t. y again.

$$\frac{2}{b^2}+\frac{2}{c^2}\left[z\,.\,\frac{\partial^2 z}{\partial y^2}+\left(\frac{\partial z}{\partial y}\right)^2\right]=0 \qquad (iv)$$

Differentiate (*i*) w.r.t. y

$$0+\frac{2}{c^2}\left[z\frac{\partial^2 z}{\partial y\partial x}+\frac{\partial z}{\partial x}\cdot\frac{\partial z}{\partial y}\right]=0 \qquad (v)$$

From (*v*), we get

$$z\frac{\partial^2 z}{\partial y\partial x}+\frac{\partial z}{\partial x}\cdot\frac{\partial z}{\partial y}=0 \quad \text{(required equation)}$$

[**Note.** We may also get different partial differential equations. The answer is not unique].

2·4. By elimination of arbitrary. functions

Let u and v be any two given functions of x, y and z. Let u and v be connected by an arbitrary function φ by the relation

$$\varphi(u, v)=0. \qquad (1)$$

Now, we want to eliminate φ.

Differentiating partially, w.r.t. x and y, we get,

$$\frac{\partial \varphi}{\partial u}\left(\frac{\partial u}{\partial x}+\frac{\partial u}{\partial z}\cdot p\right)+\frac{\partial \varphi}{\partial v}\left(\frac{\partial v}{\partial x}+\frac{\partial v}{\partial z}\cdot p\right)=0 \quad ...(2)$$

and
$$\frac{\partial \varphi}{\partial u}\left(\frac{\partial u}{\partial y}+\frac{\partial u}{\partial z}\cdot q\right)+\frac{\partial \phi}{\partial v}\left(\frac{\partial v}{\partial y}+\frac{\partial v}{\partial z}\cdot q\right)=0 \quad ...(3)$$

Eliminating $\frac{\partial \varphi}{\partial u}$ and $\frac{\partial \varphi}{\partial v}$ from (2) and (3), we obtain,

$$\begin{vmatrix} \frac{\partial u}{\partial x}+p\frac{\partial u}{\partial z} & \frac{\partial v}{\partial x}+p\frac{\partial v}{\partial z} \\ \frac{\partial u}{\partial y}+q\frac{\partial u}{\partial z} & \frac{\partial v}{\partial y}+q\frac{\partial v}{\partial z} \end{vmatrix}=0$$

which simplifies to

$$Pp+Qq=R \quad ...(4),$$

where $P=\frac{\partial u}{\partial y}\cdot\frac{\partial v}{\partial z}-\frac{\partial u}{\partial z}\cdot\frac{\partial v}{\partial y}\equiv\frac{\partial (u, v)}{\partial (y, z)}$

$$Q=\frac{\partial u}{\partial z}\cdot\frac{\partial v}{\partial x}-\frac{\partial u}{\partial x}\cdot\frac{\partial v}{\partial z}\equiv\frac{\partial (u, v)}{\partial (z, x)}$$

$$R=\frac{\partial u}{\partial x}\cdot\frac{\partial v}{\partial y}-\frac{\partial u}{\partial y}\cdot\frac{\partial v}{\partial x}\equiv\frac{\partial (u, v)}{\partial (x, y)}$$

Equation (4) is the required one which is called Lagrange's linear equation.

The relation $\varphi (u, v) = 0$ is a solution of (4), whatever may the arbitrary function φ be.

Example 6. *Form the partial differential equation by eliminating the arbitrary functions from (i)* $z=f(x^2+y^2)$ and

(ii) $z = f(x+ct)+\phi (x-ct)$

(i) Now $z=f(x^2+y^2)$

Differentiating w.r.t x and y in turn,

$$\frac{\partial z}{\partial x}=p=f'(x^2+y^2)\,.\,2x$$

$$\frac{\partial z}{\partial y}=q=f'(x^2+y^2)\,.\,2y$$

Dividing, $\frac{p}{q}=\frac{x}{y}$

$\therefore$ $py-qx=0$ is the required one.

(ii) $z=f(x+ct)+\phi (x-ct)$

$$\therefore \quad \frac{\partial z}{\partial x}=p=f'(x+ct)+\phi'(x-ct)$$

$$\frac{\partial^2 z}{\partial x^2} = f''(x+ct) + \phi''(x-ct) \quad ...(i)$$

$$\frac{\partial z}{\partial t} = cf'(x+ct) - c\phi'(x-ct)$$

$$\frac{\partial^2 z}{\partial t^2} = c^2 f''(x+ct) + c^2\phi''(x-ct)$$

i.e., $$\frac{\partial^2 z}{\partial t^2} = c^2 \{f''(x+ct) + \phi''(x-ct)\}$$

Now using (*i*) , $\frac{\partial^2 z}{\partial t^2} = c^2 \frac{\partial^2 z}{\partial x^2}$.

Therefore, the required differential equation is

$$\frac{\partial^2 z}{\partial t^2} = c^2 \frac{\partial^2 z}{\partial x^2}$$

Example 7. *Form the partial differential equation by eliminating the arbitrary function from the relation* $\phi(x^2+y^2+z^2, lx+my+nz) = 0$.

Now the given relation is of the form $\phi(u, v) = 0$,

where $u = x^2 + y^2 + z^2$,

and $v = lx + my + nz$.

Hence the partial differential equation is

$Pq + Qq = R$, where

$$P = \frac{\partial u}{\partial y} \cdot \frac{\partial v}{\partial z} - \frac{\partial u}{\partial z} \cdot \frac{\partial v}{\partial y} = 2y \cdot n - 2z \cdot m,$$

$$Q = \frac{\partial u}{\partial z} \cdot \frac{\partial v}{\partial x} - \frac{\partial u}{\partial x} \cdot \frac{\partial u}{\partial z} = 2z \cdot l - 2x \cdot n,$$

and $$R = \frac{\partial u}{\partial x}\frac{\partial v}{\partial y} - \frac{\partial u}{\partial y}\frac{\partial v}{\partial x} = 2x \cdot m - 2y \cdot l$$

Therefore the required equation is,

$$2(ny - mz)p + 2(lz - nx)q = 2(mx - ly).$$

i.e., $$(ny - mz)p + (lz - nx)q = (mx - ly).$$

Aliter: The equation $\phi(x^2+y^2+z^2, lx+my+nz) = 0$ may be written as $x^2+y^2+z^2 = f(lx+my+nz)$ where f is an arbitrary function. Now differentiate this equation w.r.t x and y partially, treating z as dependent variable.

$$2x + 2zp = f'(lx+my+nz) \times (l+np) \quad ...(1)$$

$$2y + 2zq = f'(lx+my+nz) \times (m+nq) \quad ...(2)$$

Divide (1) and (2) to eliminate f'

$$\frac{x+pz}{y+qz}=\frac{l+np}{m+nq}$$

$$\therefore \quad mx+nqx+pmz+pqnz=ly+lqz+npy+pqnz.$$

i.e. $\quad (ny-mz)\,p+(lz-nx)\,q=mx-ly.$

Example 8. *Form the partial differential equation by eliminating f from*

$$z=x^2+2f\left(\frac{1}{y}+\log x\right)$$ **(MS. 1988. Ap.)**

Let $\quad z=x^2+2f\left(\frac{1}{y}+\log x\right)$ ...(1)

Differentiate w.r.t. x and y.

$$\frac{\partial z}{\partial x}=p=2x+2f'\left(\frac{1}{y}+\log x\right)\times\left(\frac{1}{x}\right) \quad ...(2)$$

$$\frac{\partial z}{\partial y}=q=2f'\left(\frac{1}{y}+\log x\right)\times\left(-\frac{1}{y^2}\right) \quad ...(3)$$

Eliminate f' from (2) and (3).

$$\therefore \quad \frac{p-2x}{q}=\frac{-1}{x}\times y^2$$

$$px-2x^2=-qy^2$$

$$px+qy^2=2x^2$$

Example 9. *Form the partial differential equation by eliminating f and g from*

$$z=f(ax+by)+g(\alpha x+\beta y).$$

Differentiate w.r.t. x and y partially,

$$\frac{\partial z}{\partial x}=p=f'(ax+by)\,.\,a+g'(\alpha x+\beta y)\,.\,\alpha$$

$$\frac{\partial^2 z}{\partial x^2}=f''(ax+by)\,a^2+g''(\alpha x+\beta y)\,\alpha^2 \quad ...(i)$$

$$\frac{\partial^2 z}{\partial y\partial x}=f''(ax+by)\,ab+g''(\alpha x+\beta y)\,\alpha\beta \quad ...(ii)$$

$$\frac{\partial z}{\partial y}=f'(ax+by)\,.\,b+g'(\alpha x+\beta y)\,.\,\beta$$

$$\frac{\partial^2 z}{\partial y^2}=f''(ax+by)\,b^2+g''(\alpha x+\beta y)\,\beta^2 \quad (iii)$$

$(i)\times b\beta-(a\beta+b\alpha)\quad (ii)+a\alpha\quad (iii)$ gives

$$b\beta \frac{\partial^2 z}{\partial x^2} - (a\beta + b\alpha) \frac{\partial^2 z}{\partial y \partial x} + a\,\alpha \frac{\partial^2 y}{\partial y^2}$$

$$= \left[a^2 b\beta - (a\beta + b\alpha)\, ab + a\alpha \,.\, b^2 \right] f''\,(ax + by)$$

$$+ \left[\alpha^2\, b\beta - (a\beta + b\alpha)\, \alpha\beta + a\alpha\beta^2 \right] g''\,(\alpha x + \beta y)$$

$$= [0]\, f'' + [0]\, g''$$

$$= 0$$

Hence, $b\beta \frac{\partial^2 z}{\partial x^2} - (a\,\beta + b\alpha) \frac{\partial^2 z}{\partial y \partial x} + a\alpha \frac{\partial^2 z}{\partial y^2} = 0.$

Example 10. *Form the partial differential equation by eliminating f from* $z = xy + f\,(x^2 + y^2 + z^2)$

Let $z = xy + f\,(x^2 + y^2 + z^2)$.

We will keep f on RHS and all other terms in the LHS.

i.e., $$z - xy = f\,(x^2 + y^2 + z^2) \quad \text{...(1)}$$

Differentiate w.r.t x and y.

$$p - y = f'\,(x^2 + y^2 + z^2) \times (2x + 2zp) \quad \text{...(2)}$$

$$q - x = f'\,(x^2 + y^2 + z^2) \times (2y + 2zq) \quad \text{...(3)}$$

Divide (2) and (3)

$$\frac{p - y}{q - x} = \frac{x + pz}{y + qz}$$

$$pqz + py - y^2 - qyz = qx + pqz - x^2 - pxz$$

$$(y + xz)\, p - (yz + x)\, q = y^2 - x^2$$

Example 11. *Form the partial differential equation by eliminating f from*

$$f\,(x^2 + y^2 + z^2, x + y + z) = 0$$

Rewriting the given equation as

$$x^2 + y^2 + z^2 = f\,(x + y + z) \quad \text{...(1)}$$

differentiate w.r.t. x and y partially.

$$2x + 2zp = f'\,(x + y + z) \times (1 + p) \quad \text{...(2)}$$

$$2y + 2zq = f'\,(x + y + z) \times (1 + q) \quad \text{...(3)}$$

Divide (2) and (3)

$$\frac{x + pz}{y + qz} = \frac{1 + p}{1 + q}$$

$$x + qx + pz + pqz = y + py + qz + pqz$$

$$(z - y)\, p + (x - z)\, q = y - x.$$

Note. In all problems, where $f(u, v) = 0$ is given, write as $u = \phi(v)$ or $v = \psi(u)$ and proceed so that the elimination of arbitrary function is very easy.

Example 12. *Form the partial differential equation by eliminating f and ϕ from*

$$z = xf\left(\frac{y}{x}\right) + y\,\phi(x) \qquad \textbf{(MS. 1987)}$$

Since there are two arbitrary functions f and ϕ, we will get a second order partial differential equation.

Differentiate w.r.t. x and y.

$$p = xf'\left(\frac{y}{x}\right)\left(-\frac{y}{x^2}\right) + f\left(\frac{y}{x}\right) + y\,\phi'(x)$$

$$p = -\frac{y}{x}f'\left(\frac{y}{x}\right) + f\left(\frac{y}{x}\right) + y\,\phi'(x) \qquad ...(1)$$

$$q = xf'\left(\frac{y}{x}\right) \times \frac{1}{x} + \phi(x)$$

$$q = f'\left(\frac{y}{x}\right) + \phi(x) \qquad ...(2)$$

Differentiate once again w.r.t. x and y.

$$r = \frac{\partial^2 z}{\partial x^2} = \cdots\cdots$$

Differentiate (2) w.r.t. x

$$s = \frac{\partial^2 z}{\partial x\,\partial y} = f''\left(\frac{y}{x}\right) \times \left(-\frac{y}{x^2}\right) + \phi'(x) \qquad ...(3)$$

$$t = \frac{\partial^2 z}{\partial y^2} = f''\left(\frac{y}{x}\right) \times \frac{1}{x} \qquad ...(4)$$

(1) $\times x$ + (2) $\times y$ gives

$$px + qy = -yf'\left(\frac{y}{x}\right) + xf\left(\frac{y}{x}\right) + xy\,\phi'(x) + yf'\left(\frac{y}{x}\right) + y\,\varphi(x)$$

$$= xy\,\phi'(x) + xf\left(\frac{y}{x}\right) + y\,\phi(x)$$

$$px + qy = xy\,\phi'(x) + z \qquad ...(5)$$

use (4) in (3)

$$s = -\frac{y}{x} \times t + \phi'(x)$$

$$\frac{xs + yt}{x} = \varphi'(x). \text{ Use this } \varphi'(x) \text{ in (5)}$$

$$px + qy = z + xy\left[\frac{xs + yt}{x}\right]$$

$$= z + xys + y^2t$$

$\therefore$ $z = px + qy - xys - y^2t$ is the required equation.

Exercises 2(a)

1. Form the partial differential equations by eliminating the arbitrary constants a and b from the following equations.

(i) $z = ax^3 + by^3$.

(ii) $z = ax + by + ab$.

(iii) $z = a(x + y) + b$.

(iv) $(x - a)^2 + (y - b)^2 + z^2 = 1$.

(v) $ax^2 + by^2 + z^2 = 1$.

(vi) $2z = (ax + y)^2 + b$.

(vii) $z = (x + a)(y + b)$.

2. Form the partial differential equation by eliminating the arbitrary constants.

(i) $z = ax^n + by^n$ *(Anna Ap. 2005)*

(ii) $(x - a)^2 + (y - b)^2 + z^2 = a^2 + b^2$

(iii) $z = a(x + \log y) - \frac{x^2}{2} - b$

(iv) $z = a \log\left[\frac{b(y - 1)}{1 - x}\right]$

(v) $z = \frac{1}{2}[\sqrt{x + a} + \sqrt{y - a}] + b$.

(vi) $(x - a)^2 + (y - b)^2 = z^2 \cot^2 \alpha$, α known constant

(vii) $z = ax + by + \frac{a}{b} - b$

3. Obtain partial differential equations by eliminating the arbitrary functions:

(i) $z = F(x^2 + y^2)$. **(M.U. 64 B.E.)**

(ii) $z = f(x^2 + y^2 + z^2)$. **(M.U. 64 B.E.)**

(iii) $xyz = \phi(x + y + z)$. **(M.U. 64 B.E.)**

(iv) $z = yf(x) + xg(y)$. **(S.V.U. 66 B.E.)**

(v) $z = f(2x + y) + g(3x - y)$. **(M.U. 63 B.E.)**

(vi) $z = xy + f(x^2 + y^2)$. **(M.U. 65 B.E.)**

(vii) $z = f\left(\frac{y}{x}\right)$. (M.U. 64 B.E.)

(viii) $z = f\left(\frac{xy}{z}\right)$. (M.U. 65 B.E.)

(ix) $z = x + y + f(xy)$. (M.U. 65 B.E.)

(x) $z = xf(ax + by) + g(ax + by)$. (M.U. 64 B.E.)

(xi) $z = f(x + iy) + (x + iy)\, g(x - iy)$. (M.U. 66 B.E.)

(xii) $xy + yz + zx = f\left(\frac{z}{x + y}\right)$. (M.U. 65 B.E.)

(xiii) $z = e^y f(x + y)$.

(xiv) $f(x^2 + y^2 + z^2, z^2 - 2xy) = 0$.

(xv) $f(x^2 + y^2 + z^2, x + y + z) = 0$.

(xvi) $z = xf\left(\frac{y}{x}\right) + y\,\phi(x)$.

(xvii) $z = f_1(x + y) + x f_2(x + y) + f_3(x - y) + x f_4(x - y)$. (S.V.U. 67 B.E.)

(xviii) $z = y^2 + 2f\left(\frac{1}{x} + \log y\right)$. (M.U. 72 B.E.)

(xix) $f(x + y + z, xy + z^2) = 0$ (Madurai 79 B.E.)

(xx) $z = f(2r + 3y) + \phi(y + 2x)$.

4. Find the partial differential equation of all spheres whose centres lie on the z-axis.

5. Find the partial differential equation of all spheres of radius k units having their centres on the xy-plane.

6. Show that the partial differential equation

$$\frac{\partial^2 u}{\partial x^2} - \frac{\partial^2 u}{\partial y^2} = 2u/x \text{ is satisfied by } u = \frac{1}{x}\phi(y - x) + \phi'(y - x),$$

where ϕ is an arbitrary function.

7. If $z = f(x + iy) + F(x - iy)$, prove that

$$\frac{\partial^2 z}{\partial x^2} + \frac{\partial^2 z}{\partial y^2} = 0 \text{ where } f, F \text{ are arbitrary.}$$

8. If $u = f(x^2 + y) + F(x^2 - y)$, show that

$$\frac{\partial^2 u}{\partial x^2} - \frac{1}{x}\cdot\frac{\partial u}{\partial x} - 4x^2\frac{\partial^2 u}{\partial y^2} = 0.$$ (M.U. 64 B.E.)

9. Find the differential equation of all planes which are at constant distance k from the origin.

2·5. Solution of Partial differential equations.

A solution or integral of a partial differential equation is a relation between the independent and the dependent variables which satisfies the given partial differential equation. There are two distinct types of solution for partial differential equations, one type of solution containing arbitrary constants and the other type of solution containing an arbitrary function. Both these types of solutions may be given as solutions of the same partial differential equation.

Consider the equations

$$z = ax + by \qquad \ldots(1)$$

and $$z = xf\left(\frac{x}{y}\right) \qquad \ldots(2)$$

where a and b are arbitrary constants and f an arbitrary function. By eliminating the arbitrary constants a and b from (1) and the arbitrary function from (2), we obtain the same partial differential equation $px + qy = z$. Therefore, equations (1) and (2) are both solutions of $px + qy = z$.

A solution which contains as many arbitrary constants as there are independent variables is called a *complete integral.*

A solution got by giving particular values to the arbitrary constants in a complete integral is called a *particular integral.*

2·6. To find the singular integral

Suppose that $f(x, y, z, p, q) = 0 \qquad \ldots(1)$

is the partial differential equation whose complete integral is

$$\phi(x, y, z, a, b) = 0 \qquad \ldots(2)$$

where a and b are arbitrary constants.

Differentiating (2) partially w.r.t. a and b, we obtain,

$$\frac{\partial \phi}{\partial a} = 0 \qquad \ldots(3)$$

and $$\frac{\partial \phi}{\partial b} = 0 \qquad \ldots(4)$$

The eliminant of a and b from the equations (2), (3) and (4), when it exists, is called the singular integral of (1). Geometrically, the singular integral represents the envelope of the surfaces represented by the complete integral (2). The singular integral is not contained in the complete integral whereas the particular integral is obtained from the complete integral.

2·7. To find the general integral

In the complete integral (2), assume that one of the constants is a function of the other. That is, $b = f(a)$

Then (2) becomes,

$$\varphi\,[x, y, z, a, f(a)\,] = 0 \qquad \ldots(5)$$

Differentiating (2), partially w.r.t. a,

$$\frac{\partial \varphi}{\partial a} + \frac{\partial \varphi}{\partial b} \cdot f'(a) = 0 \qquad ...(6)$$

The eliminant of a between the two equations (5) and (6), when it exists, is called the *general integral* of (1).

2·8. Solution of partial differential equations by direct integration

A partial differential equation can be solved by successive integration in all cases where the dependent variable occurs only in the partial derivatives. We shall study a few examples.

Example 13. *Solve* $\dfrac{\partial^2 z}{\partial x\, \partial y} = \sin x$

Integrating w.r.t. x,

$$\frac{\partial z}{\partial y} = -\cos x + f(y), \text{ where } f \text{ is arbitrary.}$$

Again integration w.r.t. y,

we get, $$z = -y\cos x + F(y) + \varphi(x).$$

Example 14. *Solve* $\dfrac{\partial^2 z}{\partial x^2} = xy.$

Integrating w.r.t. x,

$$\frac{\partial z}{\partial x} = \frac{x^2}{2} y + f(y). \text{ Integrating again w.r.t. } x$$

$$z = \frac{x^3}{6} y + xf(y) + \varphi(y).$$

Example 15. *Solve* $\dfrac{\partial^2 u}{\partial x \partial t} = e^{-t}\cos x$, *given that* $u = 0$ *when* $t = 0$ *and* $\dfrac{\partial u}{\partial t} = 0$ *when* $x = 0$. *Show also that as* $t \to \infty$, $u \to \sin x$. **(MS. 1989 Nov.)**

$$\frac{\partial^2 u}{\partial x \partial t} = e^{-t}\cos x$$

Integrating w.r.t. x,

$$\frac{\partial u}{\partial t} = e^{-t}\sin x + f(t).$$

When $x = 0$, $\dfrac{\partial u}{\partial t} = 0.$

$$\therefore \qquad 0 = f(t).$$

Hence $\dfrac{\partial u}{\partial t} = e^{-t}\sin x.$

Integrating this equation w.r.t. t,

$$u(x, t) = -e^{-t} \sin x + \phi(x)$$

When $t = 0$, $u = 0$.

$\therefore \quad 0 = -\sin x + \phi(x)$

Hence $\varphi(x) = \sin x$

$\therefore \quad u(x, t) = \sin x (1 - e^{-t}).$

Example 16. *Solve* $\dfrac{\partial^2 z}{\partial x^2} = a^2 z$, *given that when* $x = 0$, $\dfrac{\partial z}{\partial x} = a \sin y$ *and* $\dfrac{\partial z}{\partial y} = 0.$ **(*MS. 1991 N.*)**

If z were a function of x alone, solving $\dfrac{\partial^2 z}{\partial x^2} = a^2 z$, we get,

$$z = Ae^{ax} + Be^{-ax}.$$

Since z is a function of x and y, A and B will be functions of y alone.

Hence $z = f(y)e^{ax} + \varphi(y)e^{-ax}$ ...(1)

where $f(y)$ and $\phi(y)$ are functions of y alone.

$\therefore \quad \dfrac{\partial z}{\partial x} = a f(y) e^{ax} - a\varphi(y)e^{-ax}$

By hyp., when $x = 0$, $\dfrac{\partial z}{\partial x} = a \sin y$.

$\therefore \quad a \sin y = a [f(y) - \varphi(y)]$

i.e., $\quad f(y) - \varphi(y) = \sin y$ ...(2)

From (1), $\dfrac{\partial z}{\partial y} = e^{ax} f'(y) + e^{-ax} \varphi'(y).$

By hyp., when $x = 0$, $\dfrac{\partial z}{\partial y} = 0.$

$\therefore \quad 0 = f'(y) + \varphi'(y)$. Intergrating $f(y) + \phi(y) = k$ (constant) ...(3)

From (2) and (3)

$$f(y) = \tfrac{1}{2}(\sin y + k), \text{ and}$$

$$\varphi(y) = \tfrac{1}{2}(k - \sin y)$$

$\therefore \quad z = \tfrac{1}{2}(\sin y + k)\, e^{ax} + \tfrac{1}{2}(k - \sin y)e^{-ax}$

$z = \sin y \sinh ax + k \cosh ax$, where k is any constant.

Example 17. *By changing the independent variables by the relations*

$r = x + at$, $s = x - at$, show that the equation $\dfrac{\partial^2 y}{\partial t^2} = a^2 \dfrac{\partial^2 y}{\partial x^2}$ gets transformed into the equation $\dfrac{\partial^2 y}{\partial r \partial s} = 0$. Hence find a general solution of the partial differential equation.

Here y is a function of x and t.

Using $\quad r = x + at,$

and $\quad s = x - at,$

y can be expressed as a function of r and s.

Hence y is a function of r and s, where r and s are functions of t and x.

$$\frac{\partial y}{\partial t} = \frac{\partial y}{\partial r} \cdot \frac{\partial r}{\partial t} + \frac{\partial y}{\partial s} \cdot \frac{\partial s}{\partial t}$$

$$= \frac{\partial y}{\partial r} \cdot a + \frac{\partial y}{\partial s}(-a).$$

$$= a\left(\frac{\partial y}{\partial r} - \frac{\partial y}{\partial s}\right), \text{ since } \frac{\partial r}{\partial t} = a \text{ and } \frac{\partial s}{\partial t} = -a.$$

$$= a\left(\frac{\partial}{\partial r} - \frac{\partial}{\partial s}\right) y.$$

Thus
$$\frac{\partial}{\partial t} \equiv a\left(\frac{\partial}{\partial r} - \frac{\partial}{\partial s}\right)$$

$$\frac{\partial^2 y}{\partial t^2} = \frac{\partial}{\partial t}\, a\left(\frac{\partial}{\partial r} - \frac{\partial}{\partial}\right) y$$

$$= a\left(\frac{\partial}{\partial r} - \frac{\partial}{\partial s}\right) . \, a\left(\frac{\partial}{\partial r} - \frac{\partial}{\partial s}\right) y$$

$$= a^2 \left(\frac{\partial}{\partial r} - \frac{\partial}{\partial s}\right)^2 y$$

$$\frac{\partial^2 y}{\partial t^2} = a^2 \left[\frac{\partial^2 y}{\partial r^2} - \frac{2\partial^2 y}{\partial r \partial s} + \frac{\partial^2 y}{\partial s^2}\right] \qquad ...(1)$$

$$\frac{\partial y}{\partial x} = \frac{\partial y}{\partial r} \cdot \frac{\partial r}{\partial x} + \frac{\partial y}{\partial s} \cdot \frac{\partial s}{\partial x}$$

Now
$$\frac{\partial r}{\partial x} = 1, \text{ and } \frac{\partial s}{\partial x} = 1$$

Hence
$$\frac{\partial y}{\partial x} = \frac{\partial y}{\partial r}(1) + \frac{\partial y}{\partial s}(1)$$

$$= \left(\frac{\partial}{\partial r} + \frac{\partial}{\partial s}\right) y.$$

$$\therefore \qquad \frac{\partial}{\partial x} \equiv \left(\frac{\partial}{\partial r}+\frac{\partial}{\partial s}\right)$$

Hence $$\frac{\partial^2 y}{\partial x^2} = \left(\frac{\partial}{\partial r}+\frac{\partial}{\partial s}\right)^2 y$$

i.e., $$\frac{\partial^2 y}{\partial x^2} = \frac{\partial^2 y}{\partial r^2}+\frac{\partial^2 y}{\partial s^2}+2\frac{\partial^2 y}{\partial r \partial s} \qquad ...(2)$$

$$\therefore \quad \frac{\partial^2 y}{\partial t^2} - a^2 \frac{\partial^2 y}{\partial x^2} = 0,$$

gives $-4a^2 \dfrac{\partial^2 y}{\partial r \partial s} = 0.$

Hence $\dfrac{\partial^2 y}{\partial r \partial s} = 0.$

Integrating w.r.t. r,

$$\frac{\partial y}{\partial s} = \varphi(s).$$

Integrating this equation w.r.t. s,

$$y = \int \phi(s)\, ds$$

$$= F(s) + f(r)$$

$$\therefore \quad y = F(x - at) + f(x + at),$$

where F and f are arbitrary functions.

Example 18. *Solve* $\dfrac{\partial z}{\partial x} = 6x + 3y$; $\dfrac{\partial z}{\partial y} = 3x - 4y.$

From $\dfrac{\partial z}{\partial x} = 6x + 3y,$

we get $z = 3x^2 + 3xy + \phi(y)$

$$\therefore \qquad \frac{\partial z}{\partial y} = 3x + \varphi'(y)$$

Hence $3x + \varphi'(y) = 3x - 4y$, using the hyp.

i.e., $\varphi'(y) = -4y$, giving

$$\varphi(y) = -2y^2 + k$$

$\therefore$ $z = 3x^2 + 3xy - 2y^2 + k$, where k is *a* constant.

Exercises 2(b)

Solve the following equations (1–12):

1. $\frac{\partial z}{\partial x} = 0.$ **2.** $\frac{\partial^2 z}{\partial y^2} = 0$ **3.** $\frac{\partial^2 z}{\partial x^2} = \cos x.$

4. $\frac{\partial^2 z}{\partial x \partial y} + \frac{x}{y} = 6.$ **5.** $\frac{\partial^3 z}{\partial x^2 \partial y} = \cos(ax + by).$

6. $\frac{\partial^3 z}{\partial x^2 \partial y} + 18xy^2 + \sin(2x - y) = 0.$ **7.** $x\frac{\partial z}{\partial x} = 2x + y + 3z.$

8. $\frac{\partial^2 z}{\partial x \partial y} = \sin x \sin y$, for which $\frac{\partial z}{\partial y} = -2 \sin y$ when $x = 0$ and $z = 0$, when y is an odd multiple of $\pi/2$.

9. $\frac{\partial^2 z}{\partial y^2} - 5\frac{\partial z}{\partial y} + 6z = 12y.$ **10.** $\frac{\partial^2 z}{\partial x \partial y} = x^2 + y^2.$

11. $\frac{\partial^2 z}{\partial y^2} = z$ given that $z = e^x$ and $\frac{\partial z}{\partial y} = e^{-x}$ when $y = 0$.

12. $\frac{\partial z}{\partial x} = 3x - y$ and $\frac{\partial z}{\partial y} = -x + \cos y.$

13. By interchanging the independent variables by the relations $z = x + iy$, $\bar{z} = x - iy$, show that the equation $\frac{\partial^2 u}{\partial x^2} + \frac{\partial^2 u}{\partial y^2} = 0$ transforms into $\frac{\partial^2 u}{\partial z \partial \bar{z}} = 0$. Hence obtain a general solution of the equation. **(*M.U. 67 B.E.*)**

14. With the help of the substitution $u = x + \alpha y$, $v = x + \beta y$ where α, β are suitable constants, transform the equation

$$\frac{\partial^2 z}{\partial x^2} - 5\frac{\partial^2 z}{\partial x \partial y} + 6\frac{\partial^2 z}{\partial y^2} = 0$$ to the form $\frac{\partial^2 z}{\partial u \partial v} = 0$ and hence obtain its general solution. **(*O.U. 66 B.E.*)**

15. Solve $\frac{\partial^2 z}{\partial x^2} = 0, \frac{\partial^2 z}{\partial y^2} = 0.$

2·9. Methods to solve the first order partial differential equations

The partial differential equation of the first order can be written as $F(x, y, z, p, q) = 0$, where $p = \frac{\partial z}{\partial x}$ and $q = \frac{\partial z}{\partial y}$. We shall see some standard forms of such equations and solve them by special methods.

2·10. Type I. F (p, q) = 0. *i.e., the equations contain p and q only*. Suppose that $z = ax + by + c$ is a solution of the equation $F(p, q) = 0$.

Then $p = \dfrac{\partial z}{\partial x} = a$ and $q = \dfrac{\partial z}{\partial y} = b$.

Substituting these in the given equation, we get $F(a, b) = 0$.

Hence the complete solution of the given equation is

$$z = ax + by + c, \text{ where } F(a, b) = 0.$$

Solving for b from $F(a, b) = 0$ we get $b = \varphi(a)$, say

Then $z = ax + \varphi(a)\, y + c$ ...(1),

is the complete integral of the given equation since it contains two arbitrary constants.

Singular integral is got by eliminating a and c from

$z = ax + \varphi(a)\, y + c,$

$0 = x + \varphi'(a)\, y,$

and $0 = 1.$

The last equation being absurd, there is no singular integral for the given partial differential equation.

To find the general integral, put $c = f(a)$, f being arbitrary.

Then $z = ax + y\varphi(a) + f(a)$ ...(2)

Differentiating partially w.r.t. a,

we get $0 = x + y\varphi'(a) + f'(a)$ ...(3)

Eliminating a between (2) and (3), we get the general solution.

Example 19. *Solve* $\sqrt{p} + \sqrt{q} = 1$.

This is of the form $F(p, q) = 0$.

Hence the complete integral is $z = ax + by + c$, where $\sqrt{a} + \sqrt{b} = 1$

i.e., $b = (1 - \sqrt{a})^2$

$\therefore$ the complete solution is

$$z = ax + (1 - \sqrt{a})^2\, y + c \qquad \text{...(1)}$$

Differentiating partially w.r.t. c, we find that there is no singular solution.

Taking $c = f(a)$ where f is arbitrary,

$$z = ax + (1 - \sqrt{a})^2\, y + f(a) \qquad \text{...(2)}$$

Differentiating (2) partially w.r.t. a, we get

$$x + 2y(1 - \sqrt{a})\left(-\frac{1}{2\sqrt{a}}\right) + f'(a) = 0 \qquad \text{...(3)}$$

Eliminating a between (2) and (3), we get the general integral.

Example 20. *Solve* $p^2 + q^2 = npq$

The solution of this equation is $z = ax + by + c$

Subject to $a^2 + b^2 = abn$

Solving for b, we get $b = \dfrac{na \pm \sqrt{a^2n^2 - 4a^2}}{2}$

i.e.
$$b = \frac{a}{2}\left[n \pm \sqrt{n^2 - 4} \right]$$

Hence the complete solution is

$$z = ax + \frac{a}{2}[n \pm \sqrt{n^2 - 4}] y + c \qquad ...(1)$$

Differentiating partially w.r.t. c, we get $0 = 1$ which is absurd.

Hence, there is no singular integral.

To find general solution, put $c = \phi(a)$

$$Z = ax + \frac{a}{2}\left[n \pm \sqrt{n^2 - 4} \right] y + \phi(a) \qquad ...(2)$$

Differentiate partially w.r.t. 'a'.

$$0 = x + \frac{1}{2}\left[n \pm \sqrt{n^2 - 4} \right] y + \phi'(a) \qquad ...(3)$$

Eliminating 'a' between (2) and (3), we get the general solution of the given equation.

2·11. Type II. Clairaut's form . z = px + qy + f (p, q).

Suppose that the given equation is of the form

$$z = px + qy + f(p, q) \qquad ...(1),$$

We can easily prove that

$$z = ax + by + f(a, b) \qquad ...(2),$$

is the complete solution of (1), where a, b are arbitrary constants. Differentiating (2) partially w.r.t a and b, we get

$$x + \frac{\partial f}{\partial a} = 0 \qquad ...(3)$$

and
$$y + \frac{\partial f}{\partial b} = 0 \qquad ...(4)$$

By eliminating a and b from (2), (3), and (4), we get the singular integral of (1).

Taking $b = \varphi(a)$, (2) becomes,

$$z = ax + \varphi(a)\, y + f[a, \varphi(a)] \qquad ...(5)$$

Differentiating partially w.r.t. a, we get

$$0 = x + y\varphi'(a) + f'(a) \quad \text{...(6)}$$

Eliminating a between (5) and (6), we get the general integral of (1).

Example 21. *Solve* $z = px + qy + \sqrt{1 + p^2 + q^2}$. **(*M.S. 1991 Nov.***

This is of the form $z = px + qy + f(p, q)$.

Hence the complete integral is $z = ax + by + \sqrt{1 + a^2 + b^2}$, where a and b are arbitrary constants.

Singular solution is found as follows.

Differentiate $z = ax + by + \sqrt{1 + a^2 + b^2}$ *wrt* a and b

We get
$$0 = x + \frac{a}{\sqrt{1 + a^2 + b^2}}$$

$$0 = y + \frac{b}{\sqrt{1 + b^2 + a^2}}$$

$$x = -\frac{a}{\sqrt{1 + a^2 + b^2}} \quad \text{...(1)}$$

$$y = -\frac{b}{\sqrt{1 + a^2 + b^2}} \quad \text{...(2)}$$

$$x^2 + y^2 = \frac{a^2 + b^2}{1 + a^2 + b^2}$$

$$1 - x^2 - y^2 = 1 - \frac{a^2 + b^2}{1 + a^2 + b^2} = \frac{1}{1 + a^2 + b^2}$$

$$\therefore \quad \sqrt{1 + a^2 + b^2} = \frac{1}{\sqrt{1 - x^2 - y^2}}$$

$\therefore$ (1) and (2) become,

$$x = -a\sqrt{1 - x^2 - y^2}$$

$$y = -b\sqrt{1 - x^2 - y^2}$$

$$\therefore \quad a = \frac{-x}{\sqrt{1 - x^2 - y^2}}$$

$$b = \frac{-y}{\sqrt{1 - x^2 - y^2}}$$

Substituting in the given equation,

$$z = \frac{-x^2}{\sqrt{1-x^2-y^2}} - \frac{y^2}{\sqrt{1-x^2-y^2}} + \frac{1}{\sqrt{1-x^2-y^2}}$$

$$= \frac{1-x^2-y^2}{\sqrt{1-x^2-y^2}} = \sqrt{1-x^2-y^2}$$

$z^2 = 1 - x^2 - y^2$ *i.e.,* $x^2 + y^2 + z^2 = 1$ is the singular integral.

Example 22. *Solve* $z = px + qy + p^2q^2$

This is Clairaut's form

The complete solution is $z = ax + by + a^2 b^2$ ...(1)

Differentiate w.r.t. a and b,

$$0 = x + 2ab^2$$

and $$0 = y + 2ba^2$$

$\therefore$ $$x = -2ab^2 \qquad ...(2)$$

$$y = -2ba^2 \qquad ...(3)$$

$$\frac{x}{b} = \frac{y}{a} = -2ab = \frac{1}{k} \text{ (say)}$$

$$a = ky \text{ and } b = kx$$

Put in (2),

$$x = -2k^3 yx^2$$

$$k^3 = -\frac{1}{2xy}$$

Put a and b in (1),

$$z = kxy + kxy + k^4 x^2 y^2$$

$$= 2kxy + k x^2 y^2 \left(-\frac{1}{2xy}\right)$$

$$= 2k xy - \frac{k}{2} xy$$

$$= \frac{3}{2} k xy$$

$$z^3 = \frac{27}{8} k^3 x^3 y^3$$

$\therefore$ $$z^3 = \frac{27}{8} x^3 y^3 \left(-\frac{1}{2xy}\right)$$

$$z^3 = -\frac{27}{16} x^2 y^2$$

$16z^3 + 27x^2y^2 = 0$ is the singular solution

Put $b = \phi(a)$ in (1).

$$z = ax + \phi(a)\,y + a^2\,[\phi(a)]^2 \qquad ...(4)$$

Differentiate (4) w.r.t. a and eliminate a to get the general solution.

Example 23. *Solve:* $z = px + qy + p^2 - q^2$

This is Clairaut's form. Hence complete solution is

$$z = ax + by + a^2 - b^2. \qquad ...(1)$$

To get singular solution:

Differentiate (1) w.r.t. a and b.

$$0 = x + 2a \qquad ...(2)$$

$$0 = y - 2b \qquad ...(3)$$

$$\therefore \qquad a = -\frac{x}{2} \text{ and } b = \frac{y}{2}$$

Putting in (1),

$$z = -\frac{x^2}{2} + \frac{y^2}{2} + \frac{x^2 - y^2}{4}$$

$4z = y^2 - x^2$ is the singular solution.

Put $b = \phi(a)$ in (1).

$$z = ax + \varphi(a)\,y + a^2 - [\varphi(a)]^2 \qquad ...(4)$$

Differentiate (4) w.r.t. a.

$$0 = x + y\cdot\varphi'(a) + 2a - 2\varphi(a).\,\varphi'(a) \qquad ...(5)$$

Eliminating 'a' between (4) and (5) we get the general solution.

2·12. Type III. (a) **F (*z, p, q*) = 0.** *i.e., equations not containing x and y explicitly.*

As a trial solution, assume that z is a function of $u = x + ay$, where a is an arbitrary constant.

Now $\qquad z = f(u) = f(x + ay).$

$$\therefore \qquad p = \frac{\partial z}{\partial x} = \frac{dz}{du}\cdot\frac{\partial u}{\partial x} = \frac{dz}{du}\cdot 1 = \frac{dz}{du}$$

$$q = \frac{\partial z}{\partial y} = \frac{dz}{du}\cdot\frac{\partial u}{\partial y} = \frac{dz}{du}\cdot a = a\frac{dz}{du}.$$

Substituting these values of p and q in $F(z, p, q) = 0$, we get $F\left(z, \frac{dz}{du}, a\frac{dz}{du}\right) = 0$, which is an ordinary differential equation of the first order. Solving for $\frac{dz}{du}$, we obtain $\frac{dz}{du} = \varphi(z, a)$ (say) *i.e.*, $$\frac{dz}{\varphi(z, a)} = du$$

Integrating, $$\int \frac{dz}{\varphi(z, a)} = u + c$$

i.e., $$f(z, a) = u + c$$

i.e., $$f(z, a) = x + ay + c \quad ...(1).$$

This is the complete integral.

Singular and general integrals are found out as usual.

(*b*) Suppose that the given equation is of the form

$$\boldsymbol{F(x, p, q) = 0} \quad ...(1)$$

Since z is a function of x and y

$$dz = \frac{\partial z}{\partial x}\, dx + \frac{\partial z}{\partial y}\, dy = p\, dx + q\, dy$$

Assume that $q = a$.

Then the equation becomes $F(x, p, a) = 0$.

Solving for p, we obtain $p = \phi(x, a)$.

$\therefore$ $$dz = \varphi(x, a)\, dx + a\, dy$$

Hence $$z = \int \varphi(x, a)\, dx + ay + c$$

$$z = f(x, a) + ay + c \quad ...(2).$$

(2) is the complete integral of (1) since it contains two arbitrary constants a and c.

(*c*) If the given equation is of the form

$\boldsymbol{F(y, p, q) = 0}$, assume $p = a$ and proceed as before.

The complete integral will be of the form

$$z = ax + \int f(y, a)\, dy + c.$$

Example 24. *Solve* $p(1 + q) = qz$. ***(M.S. 1987 Ap.)***

Assume $$u = x + ay$$

Then $$p = \frac{dz}{du} \text{ and } q = \frac{dz}{du} \cdot a$$

Substituting these values in the given equation, we get

$$\frac{dz}{du}\left(1 + a\frac{dz}{du}\right) = az\frac{dz}{du}$$

$\therefore$ $$a\frac{dz}{du} = az - 1$$

$$\therefore \qquad \frac{a\,dz}{az-1} = du$$

Integrating, $\quad a\int \frac{dz}{az-1} = u + c$

$$\therefore \qquad \log(az-1) = u + c$$

i.e., $\log(az-1) = x + ay + c$, which is the required complete integral.

Singular and general integrals are found out as usual.

Example 25. *Solve* $p = 2qx$.

Let $\quad q = a$.

Then $\quad p = 2ax$.

But $\quad dz = p\,dx + q\,dy$

$$= 2ax\,dx + a\,dy$$

$$\therefore \qquad z = ax^2 + ay + c \qquad ...(1)$$

(1) Is the complete integral of the given equation. Differentiating partially w.r.t. c, we get $1 = 0$. Hence there is no singular integral. General integral can be found out in the usual way.

Example 26. *Solve:* $q = px + p^2$ **(*MS. ' 86 Ap.*)**

This is of the form $\phi(x, p, q) = 0$.

$\therefore$ Assume $q = a$ = constant

Then $p^2 + px - a = 0$.

$$\therefore \quad p = \frac{-x \pm \sqrt{x^2+4a}}{2}$$

Since, $\quad dz = pdx + qdy$

$$= \left(\frac{-x \pm \sqrt{x^2+4a}}{2}\right) dx + a\,dy$$

Integrating,

$$z = -\frac{x^2}{4} \pm \frac{1}{2}\int \sqrt{x^2+4a}\,dx + ay + b$$

$$z = -\frac{x^2}{4} \pm \frac{1}{2}\left\{2a\sinh^{-1}\left(\frac{x}{2\sqrt{a}}\right) + \frac{x}{2}\sqrt{x^2+4a}\right\} + ay + b$$

is the complete solution.

Singular integral does not exist; find the general solution as usual.

Example 27. *Solve* $pq = y$.

This is of the form $f(p, q, y) = 0$.

Assume $p = a$ = constent.

Then $aq = y \therefore q = \frac{y}{a}$.

$$dz = pdx + qdy$$

$$= adx + \frac{y}{a} dy$$

Integrating, $z = ax + \frac{y^2}{2a} + b$ is the complete solution.

There is no singular integral since $\frac{\partial \varphi}{\partial b} = 0$ gives $1 = 0$ which is absurd.

Put $b = \varphi(a)$

$$z = ax + \frac{y^2}{2a} + \varphi(a) \qquad ...(3)$$

Differentiate (3) w.r.t. a

$$0 = x - \frac{y^2}{2a^2} + \varphi'(a) \qquad ...(4)$$

Eliminate a between (3) and (4) to get general solution.

Example 28. *Solve* $9(p^2 z + q^2) = 4$.

This is of the form $f(p, q, z) = 0$

Assume $z = f(x + ay)$ and $u = x + ay$.

$$p = \frac{\partial z}{\partial x} = \frac{dz}{du}.$$

$$q = \frac{\partial z}{\partial y} = a\frac{dz}{du}$$

Substituting in the given equation,

$$9\left[z\left(\frac{dz}{du}\right)^2 + a^2\left(\frac{dz}{du}\right)^2\right] = 4$$

$$\left(\frac{dz}{du}\right)^2 = \frac{4}{9(z + a^2)}$$

$$\frac{dz}{du} = \frac{2}{3}\frac{1}{\sqrt{z + a^2}}$$

$$3\sqrt{z + a^2}\, dz = 2du$$

Integrating, $3\frac{(z + a^2)^{3/2}}{3/2} = 2u + 2b$

$(z+a^2)^{3/2} = (x+ay)+b$

$(z+a^2)^3 = (x+ay+b)^2$. This is the complete solution.

Singular solution and general solution can be found in the usual manner (exercise to the reader).

Example 29. *Solve:* $z = p^2 + q^2$ (*MS. 1986 A*)

Let $z = f(x+ay)$ where $u = x+ay$.

$\therefore p = \dfrac{dz}{du}, q = a\dfrac{dz}{du}$. Substituting in the given equation,

$$z = (1+a^2)\left(\frac{dz}{du}\right)^2$$

$$\frac{dz}{du} = \frac{\sqrt{z}}{\sqrt{1+a^2}}$$

$$\frac{dz}{\sqrt{z}} = \frac{1}{\sqrt{1+a^2}}\,du$$

Integrating, $2\sqrt{z} = \dfrac{1}{\sqrt{1+a^2}}(x+ay)+b$ is the complete solution.

Example 30. *Solve:* $Ap + Bq + cz = 0$ (*MS. 1986 Nov.*)

Let $z = f(x+ay)$ where $u = x+ay$

Hence $p = \dfrac{dz}{du},\quad q = a\dfrac{dz}{du}$.

The equation becomes, $A\dfrac{dz}{du} + Ba\dfrac{dz}{du} + cz = 0$

$$\therefore \quad \frac{dz}{du} = \frac{-cz}{a+Ba}$$

$$\frac{dz}{z} = -\frac{c}{A+Ba}\,du$$

Integrating, $\log z = -\dfrac{c}{A+Ba}(x+ay)+b$ is the complete solution.

2·12. Type IV. Separable equations

We say that a first order partial differential equation is *separable* if it can be written as $f(x, p) = \phi(y, q)$.

We first put each of these equal expressions equal to an arbitrary constant a, say.

Hence $f(x, p) = \phi(y, q) = a$.

Solving for p and q, we get $p = f_1(x, a)$ and $q = \varphi_1(y, a)$

But $dz = \frac{\partial z}{\partial x} dx + \frac{\partial z}{\partial y} dy.$

Hence $dz = p\,dx + q\,dy = f_1(x, a)\,dx + \varphi_1(y, a)\,dy.$

$$\therefore \quad z = \int f_1(x, a)\,dx + \int \varphi_1(y, a)\,dy + b \qquad ...(3)$$

Now (3) contains two arbitrary constants and hence it is the complete integral. The singular and general integrals are found out as usual.

Example 31. *Solve* $p^2 y(1 + x^2) = qx^2.$

The equation is separable.

$\therefore \quad p^2 \frac{(1 + x^2)}{x^2} = \frac{q}{y} = a$, where a is an arbitrary constant.

Thus $\quad p^2 \frac{1 + x^2}{x^2} = a$

Hence $\quad p = \frac{x\sqrt{a}}{\sqrt{1 + x^2}};$

Again $\quad q = ay$

But $\quad dz = pdx + qdy$

$$= \frac{x\sqrt{a}}{\sqrt{1 + x^2}} dx + aydy$$

$$\therefore \quad z = \sqrt{a} \int \frac{x}{\sqrt{1 + x^2}} dx + a \int ydy$$

$$z = \sqrt{a(1 + x^2)} + \tfrac{1}{2} ay^2 + b$$

This is the complete integral where a and b are arbitrary constants.

Differentiating partially w.r.t. b, we find that there is no singular integral.

Example 32. *Solve* $p^2 + q^2 = x + y.$ ...(1)

The equation is separable.

$\therefore \quad p^2 - x = y - q^2 = a$, say.

Thus $\quad p^2 = x + a$, and hence $p = \sqrt{(x + a)}$,

Again $y - q^2 = a$, and hence $q = \sqrt{(y - a)}$

But $\quad dz = pdx + qdy$

$$= \sqrt{(x + a)}\,dx + \sqrt{(y - a)}\,dy$$

$$\therefore \quad z = \tfrac{2}{3}(x + a)^{\frac{3}{2}} + \tfrac{2}{3}(y - a)^{\frac{3}{2}} + b \qquad ...(2)$$

This is the complete integral of the given equation. Differentiating (2) partially w.r.t. b, we find that there is no singular integral since we get $1 = 0$ which is absurd.

Example 33. *Solve* $p^2 + q^2 = x^2 + y^2$.

This is a separable equation.

$$\therefore \quad p^2 - x^2 = y^2 - q^2 = a^2 \text{ (say)}$$

$$p^2 - x^2 = a^2 \text{ and } y^2 - q^2 = a^2$$

Hence $\quad p = \sqrt{x^2 + a^2}$ and $q = \sqrt{y^2 - a^2}$

$$dz = pdx + qdy$$

$$= \sqrt{x^2 + a^2}\, dx + \sqrt{y^2 - a^2}\, dy$$

Integrating, $\int dz = \int \sqrt{x^2 + a^2}\, dx + \int \sqrt{(y^2 - a^2)}\, dy$

$$z = \frac{a^2}{2} \sinh^{-1}\frac{x}{a} + \frac{x}{2}\sqrt{a^2 + x^2} + \frac{y}{2}\sqrt{y^2 - a^2} - \frac{a^2}{2}\cosh^{-1}\frac{y}{a} + b$$

is the complete solution.

There is no singular integral.

Example 34. *Solve:* $p - x^2 = q + y^2$

Let $\quad p - x^2 = q + y^2 = a = \text{constants}$

$$p = a + x^2 \text{ and } q = a - y^2$$

$$dz = p\, dx + q\, dy$$

$$= (a + x^2)\, dx + (a - y^2)\, dy$$

Integrating, $z = ax + \frac{x^3}{3} + ay - \frac{y^3}{3} + b$ is the complete integral.

There is no singular integral.

put $\quad b = \varphi(a)$

$$z = ax + \frac{x^3}{3} + ay - \frac{y^3}{3} + \phi(a) \qquad ...(1)$$

Differentiate w.r.t. a.

$$0 = x + y + \varphi'(a) \qquad ...(2)$$

Eliminate a between (1) and (2) to get the general solution.

2·14. Equations reducible to standard forms

Many non-linear partial differential equations of the first order do not fall under any of the four standard types discussed so far. However, in some cases, it is possible to transform the given partial differential equation into one of the standard types by change of variables. We will see below a few types of equation, reducible in each case to one of the standard types.

Case 1. An equation of the form $F(x^m p, y^n q) = 0$, where m and n are constants can always be transformed into an equation of the *First Type.*

By putting $x^{1-m} = X$ and $y^{1-n} = Y$ where $m \neq 1$ and $n \neq 1$, we get

$$p = \frac{\partial z}{\partial x} = \frac{\partial z}{\partial X} \cdot \frac{dX}{dx} = (1-m)\, x^{-m} \frac{\partial z}{\partial X} = (1-m)\, x^{-m} P,$$

and $$q = \frac{\partial z}{\partial y} = \frac{\partial z}{\partial Y} \cdot \frac{dY}{dy} = (1-n)\, y^{-n} \frac{\partial z}{\partial Y} = (1-n)\, y^{-n} Q,$$

where $$P = \frac{\partial z}{\partial X} \text{ and } Q = \frac{\partial z}{\partial Y}.$$

Hence the equation reduces to

$F[(1-m)P, (1-n)Q] = 0$, which is of the form $f(P, Q) = 0$.

Case 2. An equation of the form $F(x^m p, y^n q, z) = 0$ can also be transformed to the standard type $f(P, Q, z) = 0$ by the substitutions $x^{1-m} = X$ and $y^{1-n} = Y$ if $m \neq 1$; $n \neq 1$.

Case 3. In the above two cases, if $m = 1$, put $X = \log x$; and if $n = 1$, put $Y = \log y$; whence we get

$$p = \frac{\partial z}{\partial x} = \frac{\partial z}{\partial X} \cdot \frac{dX}{dx} = P \cdot \frac{1}{x} \quad \text{i.e., } xp = P,$$

and $$q = \frac{\partial z}{\partial y} = \frac{\partial z}{\partial Y} \cdot \frac{dY}{dy} = Q \cdot \frac{1}{y} \quad \text{i.e., } yq = Q.$$

Case 4. An equation of the form $F(z^k p, z^k q) = 0$, where k is any constant, can be transformed into the *First Type* by proper substitution.

If $k \neq -1$, put $Z = z^{k+1}$.

Then, $$\frac{\partial Z}{\partial x} = (k+1)\, z^k p,$$

and $$\frac{\partial Z}{\partial y} = (k+1)\, z^k q.$$

Hence the given equation reduces to the form

$$f(P, Q) = 0, \text{ where } \frac{\partial Z}{\partial x} \text{ and } Q = \frac{\partial Z}{\partial y}.$$

If $k = -1$, set $Z = \log z$.

$$\therefore \quad \frac{\partial Z}{\partial x} = \frac{1}{z} p$$

and $$\frac{\partial Z}{\partial y} = \frac{1}{z} q$$

Hence the given equation again reduces to the form $f(P, Q) = 0$.

Case 5. An equation of the form $F(x^m z^k p, y^n z^k q) = 0$ may be transformed into the standard type $f(P, Q) = 0$ by putting $X = x^{1-m}$, $Y = y^{1-n}$ and $Z = z^{k+1}$ if $m \neq 1, n \neq 1$ and $k \neq -1$ or by putting $X = \log x$, $y = \log y$, $Z = \log z$ if $m = 1$, $n = 1$ and $k = -1$.

Example 35. *Solve:* $x^2 p^2 + y^2 q^2 = z^2$. **(*MS. 1988 April*)**

This equation is not in any of the four standard types. But this is reducible to one of the standard types by proper substitution of the variables. Rewriting the equation, we get,

$$\left(\frac{xp}{z}\right)^2 + \left(\frac{yq}{z}\right)^2 = 1.$$

This is of the form explained in case (5), where $m = 1$, $n = 1$ and $k = -1$. Hence put

$X = \log x$, $Y = \log y$ and $Z = \log z$.

Then $$P = \frac{\partial Z}{\partial X} = \frac{\partial Z}{\partial z} \cdot \frac{\partial z}{\partial x} \cdot \frac{\partial x}{\partial X}$$

$$= \frac{1}{z} \cdot p \cdot x = \frac{px}{z}$$

and $$Q = \frac{\partial Z}{\partial Y} = \frac{qy}{z}$$

$\therefore$ The equation reduces to $P^2 + Q^2 = 1$.

$\therefore$ the complete solution is

$$Z = aX + bY + c, \text{ where } a^2 + b^2 = 1.$$

i.e., $\log z = a \log x + \sqrt{1 - a^2} \log y + c$.

The other solutions can be got in the usual manner.

Example 36. *Solve:* $2x^4 p^2 - yzq - 3z^2 = 0$. **(*MS. 1987 Nov.*)**

Rewriting this equation, we get

$$2\left(\frac{x^2 p}{z}\right)^2 - \frac{yq}{z} - 3 = 0.$$

This is of the form explained in case 5 where $m = 2$, $n = 1$, $k = -1$

Hence put $X = x^{-1}$, $y = \log y$ and $Z = \log z$.

$$P = \frac{\partial Z}{\partial X} = \frac{\partial Z}{\partial z} \cdot \frac{\partial z}{\partial x} \cdot \frac{\partial x}{\partial X} = \frac{1}{z} p(-x^2) = -\frac{px^2}{z};$$

$$Q = \frac{\partial Z}{\partial y} = \frac{\partial Z}{\partial z} \cdot \frac{\partial z}{\partial y} \cdot \frac{\partial y}{\partial y} = \frac{1}{z} \cdot q \cdot y = \frac{qy}{z}$$

$\therefore$ The equation becomes,

$$2P^2 - Q - 3 = 0.$$

This is of the standard type.

$\therefore$ The complete integral is

$$Z = aX + bY + c, \text{ where } 2a^2 - b - 3 = 0.$$

i.e., $\log z = \frac{a}{x} + (2a^2 - 3)\log y + c.$

Example 37. *Solve:* $z^2(p^2 + q^2) = x^2 + y^2$. **(*MS. 1988 April*)**

Rewriting this equation, we get,

$$(zp)^2 + (zq)^2 = x^2 + y^2$$

Put $Z = z^2$, Then $P = \frac{\partial Z}{\partial x} = 2zp$, $Q = \frac{\partial Z}{\partial y} = 2zq$

$\therefore$ the equation reduces to

$$P^2 + Q^2 = 4(x^2 + y^2)$$

i.e., $P^2 - 4x^2 = 4y^2 - Q^2 = 4a$ (say)

$\therefore$ $P = \sqrt{(4a + 4x^2)}$, and $Q = \sqrt{(4y^2 - 4a)}$.

$$dZ = 2\sqrt{(a + x^2)}\, dx + 2\sqrt{(y^2 - a)}\, dy$$

$\therefore$

$$Z = 2\left[\frac{x}{2}\sqrt{x^2 + a} + \frac{a}{2}\sinh^{-1}\frac{x}{\sqrt{a}} + \frac{y}{2}\sqrt{(y^2 - a)} - \frac{a}{2}\cosh^{-1}\frac{x}{\sqrt{a}}\right] + t$$

i.e., $z^2 = x\sqrt{(x^2 + a)} + a\sinh^{-1}\frac{x}{\sqrt{a}} + y\sqrt{(y^2 - a)} - a\cosh^{-1}\frac{y}{\sqrt{a}} + b.$

Example 38. *Solve:* $p^2x^4 + y^2zq = 2z^2$

This can be written as

$$(px^2)^2 + (qy^2)z = 2z^2$$

which is of the form $F(x^m p, y^n q, z) = 0$ (Case 2)

where $m = 2, n = 2$

Put $X = x^{1-m} = \frac{1}{x}$; $Y = y^{1-n} = \frac{1}{y}$

$$P = \frac{\partial Z}{\partial X} = \frac{\partial Z}{\partial x} \cdot \frac{\partial x}{\partial X} = p \cdot (-x^2) = -px^2$$

$$Q = \frac{\partial Z}{\partial Y} = \frac{\partial z}{\partial y} \cdot \frac{\partial y}{\partial Y} = q(-y^2) = -qy^2$$

Substituting in the given equation,

$$P^2 - Qz = 2z^2$$

This of the form $f(p, q, z) = 0$

$\therefore$ Let $Z = f(X + aY)$ where $u = X + aY$

$$P = \frac{dz}{du},\ Q = a\frac{dz}{du}.$$

Equation becomes,

$$\left(\frac{dz}{du}\right)^2 - az\frac{dz}{du} - 2z^2 = 0$$

Solving for $\frac{dz}{du}$, we get

$$\frac{dz}{du} = \frac{az \pm \sqrt{a^2z^2 + 8z^2}}{2}$$

$$\frac{dz}{z} = \frac{a \pm \sqrt{a^2 + 8}}{2}\ du$$

$$\log z = \frac{a \pm \sqrt{a^2 + 8}}{2}\ (X + aY) + b$$

$\therefore \quad \log z = \left(\frac{a \pm \sqrt{a^2 + 8}}{2}\right)\left(\frac{1}{x} + \frac{a}{y}\right) + b$ is the complete solution.

Example 39. *Solve :* $z^2(p^2x^2 + q^2) = 1$ **(*BR. 1995 April*)**

This can be written as $(pxz)^2 + (qy^0z)^2 = 1$.

This is of the form $F(x^mz^kp, y^nz^kq) = 0$

where $m = 1, n = 0, k = 1$.

$\therefore$ Put $\log x = X$, (refer to Case 5)

$$p = \frac{\partial z}{\partial x} = \frac{\partial z}{\partial X} \cdot \frac{\partial X}{\partial x} = P \cdot \frac{1}{x}.$$

$$px = P \text{ where } P = \frac{\partial z}{\partial X}$$

Hence the given equation reduces to

$$P^2 + q^2 = \frac{1}{z^2}$$

This is of the form $f(P, q, z) = 0$

$\therefore$ Let $Z = f(X + ay)$ where $u = X + ay$

$$P = \frac{dz}{du}, \quad q = a\frac{dz}{du}.$$ Hence, the equation becomes,

$$(a^2 + 1)\left(\frac{dz}{du}\right)^2 = \frac{1}{z^2}$$

$$\sqrt{a^2 + 1}\; z\frac{dz}{du} = 1$$

$$\int \sqrt{a^2 + 1}\; z\,dz = \int du$$

$$\sqrt{a^2 + 1} \cdot \frac{z^2}{2} = u + b$$

$$\sqrt{a^2 + 1} \cdot \frac{z^2}{2} = X + ay + b$$

$$\sqrt{a^2 + 1}\; \frac{z^2}{2} = \log x + ay + b$$ is the complete solution.

Example 40. *Solve:* $p^2 + x^2y^2q^2 = x^2z^2$.

Dividing by x^2, $(px^{-1})^2 + (qy)^2 = z^2$

This is of the form $F(px^m, qy^n, z) = 0$.

where $m = -1, n = 1$

put $X = x^{1-m}$ and $Y = \log y$ (Case 2)

i.e., $X = x^2$ and $Y = \log y$.

$$p = \frac{\partial z}{\partial x} = \frac{\partial z}{\partial X} \cdot \frac{\partial X}{\partial x} = P \cdot 2x$$

$$q = \frac{\partial z}{\partial y} = \frac{\partial z}{\partial Y} \cdot \frac{\partial Y}{\partial y} = Q \cdot \frac{1}{y}$$

$$\therefore \quad p = 2x\,P, \quad q = \frac{Q}{y}$$

Hence the given equation reduces to

$$4P^2 + Q^2 = z^2 \qquad ...(1)$$

This is of the form $F(p, q, z) = 0$

$\therefore$ Let $z = f(u)$ where $u = X + aY$

$$\therefore P = \frac{dz}{du}, \quad Q = a\frac{dz}{du}.$$

Hence, (1) becomes, $(a^2+4)\left(\dfrac{dz}{du}\right)^2 = z^2$

$$\sqrt{a^2+4}\ \frac{dz}{z} = du$$

Integrating, $\sqrt{a^2+4}\ \log z = X + aY + b$

$\sqrt{a^2+4}\ \log z = x^2 + a\log y + b$ is the complete solution.

Example 41. *Solve:* $p^2 + q^2 = z^2(x^2+y^2)$

This can be rewritten as $\left(\dfrac{p}{z}\right)^2 + \left(\dfrac{q}{z}\right)^2 = x^2 + y^2.$

Hence putting $Z = \log z$ *i.e.*, $z = e^Z$ (case 5)

$$p = \frac{\partial z}{\partial x} = \frac{\partial}{\partial x} e^Z = e^Z \cdot \frac{\partial Z}{\partial x}$$

$$= zP \text{ where } P = \frac{\partial Z}{\partial x}$$

$$q = \frac{\partial z}{\partial y} = \frac{\partial}{\partial y} e^Z = e^Z \frac{\partial Z}{\partial y} = z\,Q$$

where $\quad Q = \dfrac{\partial Z}{\partial y}$

$\therefore \quad \dfrac{p}{z} = P$ and $\dfrac{q}{z} = Q$

Hence equation reduces to,

$$P^2 + Q^2 = x^2 + y^2$$

$$P^2 - x^2 = y^2 - Q^2 = a^2$$

$\therefore \qquad P = \sqrt{x^2+a^2}$ and $Q = \sqrt{y^2-a^2}$

$$dZ = P\,dx + Q\,dy$$

$$= \sqrt{x^2+a^2}\,dx + \sqrt{y^2-a^2}\,dy$$

Integrating,

$$\log z = Z = \frac{a^2}{2}\sinh^{-1}\frac{x}{a} + \frac{x}{2}\sqrt{a^2+x^2}$$

$$+\frac{y}{2}\sqrt{y^2-a^2} - \frac{a^2}{2}\cosh^{-1}\frac{y}{a} + b$$

is the complete solution

Example 42. *Solve:* $(x+pz)^2 + (y+qz)^2 = 1$

Put $\quad z^2 = Z$

$$P=\frac{\partial Z}{\partial x}=\frac{\partial Z}{\partial z}\cdot\frac{\partial z}{\partial x}=2zp\ ;\ pz=\frac{P}{2}$$

$$Q=\frac{\partial Z}{\partial y}=\frac{\partial Z}{\partial z}\cdot\frac{\partial z}{\partial y}=2zq\ ;\ qz=\frac{Q}{2}$$

Substituting in the given equation

$$\left(x+\frac{P}{2}\right)^2+\left(y+\frac{Q}{2}\right)^2=1$$

This is separable equation.

$$\left(x+\frac{P}{2}\right)^2=1-\left(y+\frac{Q}{2}\right)^2=a^2$$

$$\therefore\ x+\frac{P}{2}=a \qquad \left(y+\frac{Q}{2}\right)^2=1-a^2$$

$$P=2\,(a-x) \qquad Q=2\,[\sqrt{1-a^2}-y]$$

$$\therefore\ dZ=P\,dx+Q\,dy$$

$$dZ=2\,(a-x)dx+2\,[\sqrt{1-a^2}-y]\,dy$$

Integrating,

$$Z=-\,(a-x)^2+2\left[\sqrt{1-a^2}\cdot y-\frac{y^2}{2}\right]+b$$

$$\therefore\ z^2=-\,(a-x)^2+2\sqrt{1-a^2}\,y-y^2+b$$

is the complete solution.

Exercises 2(c)

Solve the following equations (1 to 10):

1. $p+q=pq.$ **2.** $pq=1.$ **3.** $p^2+q^2=m^2.$

4. $p=q^2.$ **5.** $2p+3q=1.$ **6.** $p^2+q^2=npq.$

7. $q^2-3q+p=2.$ **8.** $q^2-p^2=9.$ **9.** $q+\sin p=0.$

10. $p^2-2pq+3q=5.$

Obtain the complete solutions of the following equations:

11. $z=xp+yq+p^2-q^2.$ **12.** $z=px+qy+3pq.$ **13.** $z=px+qy-4p^2q^2.$

14. $(p+q)\ (z-px-qy)=1.$ **15.** $z=px+qy+\sqrt{(p^2+q^2)}.$ **16.** $z^2=pq.$

17. $z=p^2+q^2.$ **18.** $p^2z^2+q^2=1.$ **19.** $p(1+q^2)=q(z-a).$

20. $z^4q^2-z^2p=1.$ **21.** $z^2(p^2+q^2+1)=1.$ **22.** $q^2=(1-p^2)z^2p^2.$

23. $p(1+q)=qz.$ **24.** $p^2+q^2=x-y.$ **25.** $\sqrt{p}+\sqrt{q}=\sqrt{x}.$

26. $pq = xy.$ **27.** $p^2 + q^2 = x + y.$ **28.** $p + q = x + y.$

29. $p + q = \sin x + \sin y.$ **30.** $(1 - x^2)yp^2 + x^2q = 0.$ **31.** $q(p - \sin x) = \cos y.$

32. $px \tan y = q + 1.$ **33.** $(p^2 - q^2)\, z = x - y.$

34. $(x + y)\,(p + q)^2 + (x - y)\,(p - q)^2 = 1.$ **35.** $(x^2 + y^2)(p^2 + q^2) = 1.$

36. $p^2 + x^2y^2q^2 = x^2z^2.$ **37.** $p^2 + q^2 = z^2\,(x^2 + y^2).$ **38.** $q = xp + p^2.$

39. $p^2x + q^2y = z.$ **40.** $(p^2x^2 + q^2)z^2 = 1$ (Put $X = \log x$)

41. $z\,(p^2 - q^2) = x - y$ $\left(\text{Put } z = \frac{2}{3} z^{3/2}\right)$ **42.** $pqz = p^2\,(qx + p^2) + q^2\,(py + q^2)$

43. $\dfrac{x^2}{p} + \dfrac{y^2}{q} = z$ (**Hint:** put $x^3 = X, y^3 = Y$ $\left(\textbf{Ans: } \dfrac{3z^2}{2} = \dfrac{x^3}{a} + \dfrac{y^3}{1 - a} + b\right)$

44. $z^4\, q^2 - z^2p = 1.$ (**Hint:** put $z^3 = Z$ (**Ans:** $z^3 = ax \pm \sqrt{3a + 9}\; y + b$)

45. $q^2y^2 = z(z - px)$ (put $\log x = X$, $\log y = Y$

$$\left(\textbf{Ans: } \log z = \frac{-1 \pm \sqrt{1 + 4a^2}}{2a^2}\,(X + aY) + b\right)$$

46. $\dfrac{p}{x^2} + \dfrac{q}{y^2} = z$ $\left(\text{put } x^3 = X, y^3 = Y;\ \log z = \dfrac{1}{3(1 + a)}\,(x^3 + ay^3) + b\right)$

47. $px^2 + qy^2 = z^2$ $\left(\textbf{Ans: } \dfrac{(1 + a)}{z} = \dfrac{1}{x} + \dfrac{a}{y} + b\right)$

48. $z^2\,(p^2 - q^2) = 1$ (**Ans:** $z^2 = ax \pm \sqrt{a^2 - 4}\; y + b$)

49. $z^2\left(\dfrac{p^2}{x^2} + \dfrac{q^2}{y^2}\right) = 1$ (**Ans:** $z^4\,(1 + a^2) = (x^2 + ay^2)^2$

50. $z^2\,(p^2 + q^2) = x + y$ $\left(\textbf{Ans: } \dfrac{3z^2}{2} = (4x + a)^{3/2} + (4y - a)^{3/2} + b\right)$

51. $4\,z^2q^2 = y - x + 2zp.$ (put $Z = z^2$;

$$\left(\textbf{Ans: } z^2 = \frac{2}{3}\,(y + a)^{3/2} + \frac{(x + a)^2}{2} + b\right)$$

52. $p^2z^2 \sin^2 x + q^2z^2 \cos^2 y = 1.$

(**Ans:** $z^2 = 2\sqrt{a}\, \log \tan x/2 + 2\sqrt{1-a}\, \log (\sec y + \tan y) + c$)

2·15. Lagrange's linear equation. A linear partial differential equation of the first order known as Lagrange's linear equation is of the form

$$Pp + Qq = R \qquad ...(1),$$

where P, Q and R are functions of x, y, z. We have already seen, under article 2·4, that by eliminating the arbitrary function ϕ from the relation

$$\phi\,(u, v) = 0 \qquad ...(2),$$

where u, v are functions of x, y, z, we get a partial differential equation of the form

$$Pp + Qq = R, \text{ where}$$

$$P = \frac{\partial u}{\partial y} \cdot \frac{\partial v}{\partial z} - \frac{\partial u}{\partial z} \cdot \frac{\partial v}{\partial y}$$

$$Q = \frac{\partial u}{\partial z} \cdot \frac{\partial v}{\partial x} - \frac{\partial u}{\partial x} \cdot \frac{\partial v}{\partial z}$$

and $$R = \frac{\partial u}{\partial x} \cdot \frac{\partial v}{\partial y} - \frac{\partial u}{\partial y} \cdot \frac{\partial v}{\partial x}$$

Hence $\phi(u, v) = 0$ is the general solution of (1), ϕ being any arbitrary function.

Now suppose $u = a$ and $v = b$ where a, b are constants.

$$\therefore \quad du = \frac{\partial u}{\partial x} dx + \frac{\partial u}{\partial y} dy + \frac{\partial u}{\partial z} \cdot dz = 0,$$

and $$dv = \frac{\partial v}{\partial x} dx + \frac{\partial v}{\partial y} dy + \frac{\partial v}{\partial z} dz = 0$$

From these equations, we get

$$\frac{dx}{\frac{\partial u}{\partial y} \cdot \frac{\partial v}{\partial z} - \frac{\partial u}{\partial z} \cdot \frac{\partial v}{\partial y}} = \frac{dy}{\frac{\partial u}{\partial z} \cdot \frac{\partial v}{\partial x} - \frac{\partial u}{\partial x} \cdot \frac{\partial v}{\partial z}} = \frac{dz}{\frac{\partial u}{\partial x} \cdot \frac{\partial v}{\partial y} - \frac{\partial u}{\partial y} \cdot \frac{\partial v}{\partial x}}$$

or

$$\frac{dx}{P} = \frac{dy}{Q} = \frac{dz}{R} \qquad ...(3)$$

The solutions of equations (3) are $u = a$ and $v = b$.

$\therefore$ $\phi(u, v) = 0$ is the general solution of (1), where $u = a$ and $v = b$ are the solutions of (3).

Thus to solve the equation $Pp + Qq = R$,

(*i*) *form the auxiliary simultaneous equations*

$$\frac{dx}{P} = \frac{dy}{Q} = \frac{dz}{R};$$

(*ii*) *solve these auxiliary simultaneous equations giving two independent solutions* $u = a$ and $v = b$;

(*iii*) *then write down the solution as* $\phi(u, v) = 0$ *or* $u = f(v)$ *or* $v = F(u)$, *where the function is arbitrary.*

2·16. Solution of the subsidiary equation by the method of multipliers

The subsidiary equations

$\frac{dx}{P} = \frac{dy}{Q} = \frac{dz}{R}$ can be solved as follows.

By algebra, we know,

$$\frac{dx}{P}=\frac{dy}{Q}=\frac{dz}{R}=\frac{l\,dx+m\,dy+n\,dz}{lP+mQ+nR}=\frac{l'dx+m'dy+n'dz}{l'P+m'Q+n'R}$$

where the two sets of multipliers $l, m, n; l', m', n'$ may be constants or variables in x, y, z. Choosing l, m, n such that $lP + mQ + nR = 0$, we have

$$ldx + mdy + ndz = 0 \qquad \text{....(1)}$$

If $ldx + mdy + ndz$ is a perfect differential of some function, say, $u(x, y, z)$ then $du = 0$, by (1). Hence integrating (1), we get

$u = a$, as one solution.

Similarly, the other set of multipliers l', m', n' can be found out so that $l'P + m'Q + n'R = 0$.

Hence $l'\,dx + m'\,dy + n'\,dz = 0$.

This yields another solution $v = b$.

Therefore the general solution is

$$\phi(u, v) = 0, \text{ or } u = f(v).$$

Here, the set of multipliers l, m, n and l', m', n' are called Lagrangian multipliers.

Example 43. *Find the general integral of* $px + qy = z$.

Comparing the equation with $Pp + Qq = R$, we get

$$P = x, Q = y \text{ and } R = z.$$

The subsidiary equations are

$$\frac{dx}{x}=\frac{dy}{y}=\frac{dz}{z}.$$

In these, the variables are separated.

From $\frac{dx}{x}=\frac{dy}{y}$, we get $\log x = \log y + \log a$.

i.e., $\frac{x}{y} = a$.

Similarly from $\frac{dy}{y}=\frac{dz}{z}$, we get $\frac{y}{z} = b$.

Hence the general integral is $\phi\left(\frac{x}{y}, \frac{y}{z}\right) = 0$.

Example 44. *Solve:* $(mz - ny)\,p + (nx - lz)\,q = ly - mx$.

Lagrange's subsidiary equations are

$$\frac{dx}{mz-ny}=\frac{dy}{nx-lz}=\frac{dz}{ly-mx} \qquad \text{...(1)}$$

Using the two sets of multipliers $x, y, z; l, m, n$ each of the ratio in (1)

$$= \frac{x\,dx + y\,dy + z\,dz}{x\,(mz - ny) + y\,(nx - lz) + z\,(ly - mx)}$$

$$= \frac{l\,dx + m\,dy + n\,dz}{l\,(mz - ny) + m\,(nx - lz) + n\,(ly - mx)}$$

$$= \frac{x\,dx + y\,dy + z\,dz}{0} = \frac{l\,dx + m\,dy + n\,dz}{0}$$

Hence $x\,dx + y\,dy + z\,dz = 0$ and $l\,dx + m\,dy + n\,dz = 0$.

Integrating we get,

$$x^2 + y^2 + z^2 = a \text{ and } lx + my + nz = b.$$

Hence the general integral is

$$\phi\,(x^2 + y^2 + z^2,\, lx + my + nz) = 0.$$

Example 45. *Find the general solution of*

$x(z^2 - y^2)p + y(x^2 - z^2)q = z(y^2 - x^2)$ **(*B.E. 1963, 78; 1976 Madurai*)**

The subsidiary equations are

$$\frac{dx}{x(z^2 - y^2)} = \frac{dy}{y(x^2 - z^2)} = \frac{dz}{z(y^2 - x^2)} \qquad \text{....(1)}$$

Taking the two sets of multipliers as x, y, z and

$\frac{1}{x}, \frac{1}{y}, \frac{1}{z}$, each of ratio in (1)

$$= \frac{x\,dx + y\,dy + z\,dz}{\Sigma x^2(z^2 - y^2)} = \frac{\frac{1}{x}\,dx + \frac{1}{y}\,dy + \frac{1}{z}\,dz}{\Sigma\,(z^2 - y^2)}$$

$$= \frac{x\,dx + y\,dy + z\,dz}{0} = \frac{\frac{1}{x}\,dx + \frac{1}{y}\,dy + \frac{1}{z}\,dz}{0}$$

Hence $x\,dx + y\,dy + z\,dz = 0$ and $\frac{1}{x}\,dx + \frac{1}{y}\,dy + \frac{1}{z}\,dz = 0$.

Integrating, $x^2 + y^2 + z^2 = a$ and $\log x + \log y + \log z = k$.

i.e., $x^2 + y^2 + z^2 = a$ and $x\,yz = b$.

Hence the general integral is $\phi\,(x^2 + y^2 + z^2, xyz) = 0$.

Example 46. *Solve:* $\frac{y^2 z}{x}\,p + xzq = y^2$. **(*BR. 1995 April*)**

The subsidiary equations are

$$\frac{xdx}{y^2z} = \frac{dy}{xz}, = \frac{dz}{y^2}$$

From the first equality, $x^2dx = y^2dy$, giving $x^3 - y^3 = c$.

From the first and the last ratios,

$$x\,dx = z\,dz, \text{ giving } x^2 - z^2 = k.$$

Hence the general integral is $F(x^3 - y^3, x^2 - z^2) = 0$.

Example 47. *Find the general solution of* $(y + z)p + (z + x)q = x + y$.

The auxiliary equation are

$$\frac{dx}{y+z} = \frac{dy}{z+x} = \frac{dz}{x+y}$$

each is equal to $$\frac{dx+dy+dz}{2(x+y+z)} = \frac{dx-dy}{y-x} = \frac{dy-dz}{z-y} \quad ....(1)$$

Taking the first two ratios,

$$\frac{d\,(x+y+z)}{2\,(x+y+z)} = \frac{-d\,(x-y)}{(x-y)}$$

Integrating,

$$\frac{1}{2}\log\,(x+y+z) = -\log\,(x-y) + \log c.$$

$$\therefore\ \log\,(x+y+z) = \log\,(x-y)^{-2} + \log k.$$

$$\therefore\ (x+y+z) = k\,(x-y)^{-2}$$

i.e., $$(x+y+z)\,(x-y)^2 = k \quad ....(2)$$

Taking the last two ratios of equations (1),

$$\frac{d\,(x-y)}{x-y} = \frac{d\,(y-z)}{y-z}$$

Integrating, $\log\,(x - y) = \log\,(y - z) + \log b$

$$\therefore\ \frac{x-y}{y-z} = b \quad ....(3)$$

Solutions given by (2) and (3) are independent.

Hence, the general solution is

$$(x+y+z)(x-y)^2 = f\left(\frac{x-y}{y-z}\right)$$

Example 48. *Find the general solution of* $p\tan x + q\tan y = \tan z$.

This is Lagrange's equation.

The auxiliary equations are

$$\frac{dx}{\tan x} = \frac{dy}{\tan y} = \frac{dz}{\tan z}$$

$$\int \cot x \, dx = \int \cot y \, dy = \int \cot z \, dz$$

Taking the first two ratios,

$\log \sin x = \log \sin y + \log a$

$$\frac{\sin x}{\sin y} = a \qquad \text{...(1)}$$

Similarly, taking the last two ratios,

we get $$\frac{\sin y}{\sin z} = b \qquad \text{...(2)}$$

Hence, the general solution is

$$f\left(\frac{\sin x}{\sin y}, \frac{\sin y}{\sin z}\right) = 0, f \text{ being arbitrary}$$

Example 49. *Solve :* $(y - z)\, p + (z - x)\, q = x - y$

The auxiliary equations are

$$\frac{dx}{y - z} = \frac{dy}{z - x} = \frac{dz}{x - y}.$$

Each is equal to $$\frac{d\,(x + y + z)}{y - z + z - x + x - y} = \frac{d\,(x + y + z)}{0}$$

Since the denominator is zero, $d\,(x + y + z) = 0$

$$\therefore \quad x + y + z = a \qquad \text{....(1)}$$

Also, taking Lagrange's multipliers x, y, z.

each radio is $$= \frac{x\,dx + y\,dy + z\,dz}{x\,(y - z) + y\,(z - x) + z\,(x - y)} = \frac{\frac{1}{2}\, d\,(x^2 + y^2 + z^2)}{0}$$

Hence, $\quad d\,(x^2 + y^2 + z^2) = 0$

i.e., $\quad x^2 + y^2 + z^2 = b \qquad$ ....(2)

$\therefore$ The general solution is $\phi\,(x + y + z, x^2 + y^2 + z^2) = 0$

Example 50. *Solve*: $x\,(y - z)\, p + y\,(z - x)\, q = z\,(x - y)$

The auxiliary eqations are

$$\frac{dx}{x\,(y - z)} = \frac{dy}{y\,(z - x)} = \frac{dz}{z\,(x - y)}.$$

Each is equal to $$\frac{dx + dy + dz}{\Sigma\, x\,(y - z)} = \frac{d\,(x + y + z)}{0}$$

Hence, $\quad d\,(x + y + z) = 0$

i.e., $x+y+z=a$ ...(1)

Taking the Lagrangian multipliers $\frac{1}{x}, \frac{1}{y}, \frac{1}{z}$, we have

$$\frac{\frac{1}{x}dx}{y-z}=\frac{\frac{1}{y}dy}{z-x}=\frac{\frac{1}{z}dz}{x-y}=\frac{\frac{dx}{x}+\frac{dy}{y}+\frac{dz}{z}}{\Sigma(y-z)}$$

$$=\frac{\frac{dx}{x}+\frac{dy}{y}+\frac{dz}{z}}{0}$$

Hence, $\frac{dx}{x}+\frac{dy}{y}+\frac{dz}{z}=0$

Integrating, $\log x+\log y+\log z=\log b$

$\therefore \quad xyz=b$ ...(2)

Hence, the general solution is $\phi(xyz, x+y+z)=0$.

Example 51. *Solve* : $(y-xz)p+(yz-x)q=(x+y)(x-y)$

(Anna 2002 Nov.)

Solution. The auxiliary equations are

$$\frac{dx}{y-xz}=\frac{dy}{yz-x}=\frac{dz}{x^2-y^2}$$

Each is equal to $\dfrac{y\,dx+x\,dy+dz}{y(y-xz)+x(yz-x)+(x^2-y^2)}$

$$=\frac{d(xy+z)}{0}$$

$\therefore \quad d(xy+z)=0$, Hence $xy+z=a$ ....(1)

Again each ratio is equal to $\dfrac{x\,dx+y\,dy}{xy-x^2z+y^2z-xy}$

Therefore, $\dfrac{\frac{1}{2}d(x^2+y^2)}{z(y^2-x^2)}=\dfrac{dz}{x^2-y^2}$

i.e. $\quad d(x^2+y^2)=-2z\,dz$

i.e. $\quad d(x^2+y^2+z^2)=0$

$\therefore \quad x^2+y^2+z^2=b$ ...(2)

Using (1) and (2), the general solution is

$$\phi(x^2+y^2+z^2, xy+z)=0$$

Example 52. *Solve:* $pz+qy=x$.

The auxiliary equations are

$$\frac{dx}{z} = \frac{dy}{y} = \frac{dz}{x}$$

Taking the first and the last ratios, we get,

$$x\,dx = zdz$$

Integrating $\frac{x^2}{2} = \frac{z^2}{2} + c$

$\therefore \quad x^2 - z^2 = a$ ....(2)

each ratio of (1) is equal to $= \dfrac{dx + dy + dz}{x + y + z}$

Hence $\dfrac{dy}{y} = \dfrac{d\,(x + y + z)}{x + y + z}$

Integrating, $\log y = \log\,(x + y + z) + \log b$

$\therefore \quad \dfrac{y}{x + y + z} = b$ ....(2)

Hence, the general solution is $\phi\left(x^2 - z^2, \dfrac{y}{x + y + z}\right) = 0$

Example 53. *Solve:* $z\,(x - y) = px^2 - qy^2$ ***(MS. 1988 Nov.)***

This is Lagrange's equation.

The auxiliary equations are

$$\frac{dx}{x^2} = \frac{dy}{-y^2} = \frac{dz}{z\,(x - y)}$$

Each equal to $\dfrac{dx + dy}{x^2 - y^2}$

Hence $\dfrac{dx}{x^2} = -\dfrac{dy}{y^2}$. Integrating,

$$-\frac{1}{x} = \frac{1}{y} + k \quad \therefore \quad \frac{1}{x} + \frac{1}{y} = a \qquad \text{....(1)}$$

Also $\dfrac{dz}{z\,(x - y)} = \dfrac{d\,(x + y)}{x^2 - y^2}$

i.e., $\dfrac{dz}{z} = \dfrac{d\,(x + y)}{x + y}$

Integrating, $\log z = \log\,(x + y) + \log b$

$$\frac{z}{x + y} = b \qquad \text{....(2)}$$

Hence the general solution is $\phi\left(\frac{1}{x}+\frac{1}{y}, \frac{z}{x+y}\right)=0$

Example 54. *Solve* $p-q=\log(x+y)$

The auxiliary equations are

$$\frac{dx}{1}=\frac{dy}{-1}=\frac{dz}{\log(x+y)}$$

Hence, $\frac{dx}{1}=\frac{dy}{-1}$ gives, on integration.

$$x=-y+a \quad \therefore \quad x+y=a \qquad ....(1)$$

$$\frac{dx}{1}=\frac{dz}{\log(x+y)}=\frac{dz}{\log a}$$

Hence $x=\frac{1}{\log a}z+b$, on integration

$$x-\frac{z}{\log(x+y)}=b \qquad ....(2)$$

The general solutions is $\phi\left(x+y,\ x-\frac{z}{\log(x+y)}\right)=0$

Example 55. $(2z-y)\,p+(x+z)\,q+2x+y=0.$

This is Lagrange's equation. Hence, the auxiliary *equations are*

$$\frac{dx}{2z-y}=\frac{dy}{x+z}=\frac{dz}{-2x-y}$$

Each equal to $\frac{dz+2dy-dx}{-2x-y+2x+2z-2z+y}$

$$=\frac{d\,(z+2y-x)}{0}$$

Hence $d\,(z+2y-x)=0 \quad \therefore \quad z+2y-x=a \qquad ....(1)$

Also, each ratio is equal to

$$=\frac{x\,dx+y\,dy+z\,dz}{2\,xz-xy+xy+yz-2\,xz-yz}$$

$$=\frac{\frac{1}{2}d\,(x^2+y^2+z^2)}{0}$$

$$\therefore \quad d\,(x^2+y^2+z^2)=0$$

$$x^2+y^2+z^2=b \qquad ....(2)$$

Therefore, the general solution is

$$\phi\,(x^2+y^2+z^2, z+2y-x)=0$$

Example 56. *Solve:* $(y^2 + z^2 - x^2)\, p - 2\, xyq + 2\, xz = 0.$

(*MS. 1988 Nov.*)

The auxiliary equations are

$$\frac{dx}{y^2 + z^2 - x^2} = \frac{dy}{-2\, xy} = \frac{dz}{-2\, xz}$$

From the last two ratios,

$$\frac{dy}{y} = \frac{dz}{z}$$

Integrating, $\log y = \log z + \log a$

$$\frac{y}{z} = a \qquad \text{....(1)}$$

Taking Lagrangian multipliers x, y, z we get

each ratio is equal to $\dfrac{x\, dx + y\, dy + z\, dz}{-x\,(x^2 + y^2 + z^2)}$

$$\therefore \quad \frac{dy}{-2\, xy} = \frac{x\, dx + y\, dy + z\, dz}{-x\,(x^2 + y^2 + z^2)}$$

$$\therefore \quad \frac{dy}{y} = \frac{d\,(x^2 + y^2 + z^2)}{x^2 + y^2 + z^2}$$

$$\log y = \log\,(x^2 + y^2 + z^2) + \log b.$$

$$\therefore \quad \frac{y}{x^2 + y^2 + z^2} = b \qquad \text{....(2)}$$

Hence the general solution is

$$\phi\left(\frac{y}{z}, \frac{y}{x^2 + y^2 + z^2}\right) = 0$$

Example 57. *Solve:* $(3z - 4y)\, p + (4x - 2z)\, q = 2y - 3x.$

(*MS. 1991 April*)

The auxiliary equations are

$$\frac{dx}{3z - 4y} = \frac{dy}{4x - 2z} = \frac{dz}{2y - 3x}. \qquad \text{...(1)}$$

use Lagrangian multipliers $x, y, z.$ we get

each ratio is $\dfrac{x\, dx + y\, dy + z\, dz}{3\, xz - 4\, xy + 4\, xy - 2\, yz + 2\, yz - 3\, xz}$

$$= \frac{\frac{1}{2}\, d\,(x^2 + y^2 + z^2)}{0}$$

Hence $d\,(x^2 + y^2 + z^2) = 0 \quad \therefore \quad x^2 + y^2 + z^2 = a$ ...(2)

using multipliers 2, 3, 4,

each of equation (1) is

$$= \frac{2\,dx + 3\,dy + 4\,dz}{6z - 8y + 12x - 6z + 8y - 12x}$$

$$= \frac{2dx + 3\,dy + 4dz}{0}$$

$\therefore \quad 2dx + 3dy + 4dz = 0$

Hence $\quad 2x + 3y + 4z = b$ ....(3)

General solution is $\phi\,(x^2 + y^2 + z^2,\ 2x + 3y + 4z) = 0$

Example 58. *Solve:* $(x^2 - yz)\,p + (y^2 - zx)\,q = z^2 - xy.$ *(Anna Ap. 2005)*

The subsidiary equations are

$$\frac{dx}{x^2 - yz} = \frac{dy}{y^2 - zx} = \frac{dz}{z^2 - xy} \qquad \text{....(1)}$$

$$\text{each} = \frac{dx - dy}{(x^2 - yz) - (y^2 - zx)} = \frac{dy - dz}{y^2 - zx - z^2 + xy} = \frac{dx - dz}{x^2 - yz - z^2 + xy}$$

$$\textit{i.e.,} \quad \frac{d\,(x - y)}{(x^2 - y^2) + z\,(x - y)} = \frac{d\,(y - z)}{y^2 - z^2 + x\,(y - z)} = \frac{d\,(x - z)}{x^2 - z^2 + y\,(x - z)}$$

$$\textit{i.e.,} \quad \frac{d\,(x - y)}{(x - y)\,(x + y + z)} = \frac{d\,(y - z)}{(y - z)\,(x + y + z)} = \frac{d\,(x - z)}{(x - z)\,(x + y + z)} \qquad \text{...(2)}$$

$$\textit{i.e.,} \quad \frac{d\,(x - y)}{(x - y)} = \frac{d\,(y - z)}{y - z} = \frac{d\,(x - z)}{x - z} \qquad \text{...(3)}$$

Taking the first two ratios, and integrating

$\log\,(x - y) = \log\,(y - z) + \log a$

$$\therefore \quad \frac{x - y}{y - z} = a \qquad \text{...(4)}$$

Similarly taking the last two ratios of (3)

$$\text{we get} \quad \frac{y - z}{x - z} = b \qquad \text{...(5)}$$

But $\dfrac{x - y}{y - z}$, and $\dfrac{y - z}{x - z}$ *are not independent* solutions for $\frac{x-y}{y-z}+1$ gives $\frac{x-z}{y-z}$ which is the reciprocal of the second solution.

Therefore solution given by (4) and (5) are not independent. Hence we have to search for another independent solution.

using multipliers x, y, z in ...(1)

each ratio is $= \frac{x\,dx + y\,dy + z\,dz}{x^3+y^3+z^3-3xyz}$

using multipliers 1, 1, 1.

each ratio is $= \frac{dx+dy+dz}{x^2+y^2+z^2-xy-yz-zx}$

$$\therefore \quad \frac{x\,dx+y\,dy+dz}{x^3+y^3+z^3-3xyz} = \frac{dx+dy+dz}{x^2+y^2+z^2-xy-yz-zx}$$

$$\frac{\frac{1}{2}d(x^2+y^2+z^2)}{(x+y+z)(x^2+y^2+z^2-xy-yz-zx)} = \frac{d(x+y+z)}{x^2+y^2+z^2-xy-yz-zx}$$

Hence $\quad \frac{1}{2}d(x^2+y^2+z^2) = (x+y+z)\,d(x+y+z)$

Integrating $\frac{1}{2}(x^2+y^2+z^2) = \frac{(x+y+z)^2}{2} + k$

$$\therefore \quad (x^2+y^2+z^2) = (x+y+z)^2 + 2k$$

$\therefore \quad xy+yz+zx = b$ on simplification.

$\therefore \quad$ The general solution

$$\phi\left(xy+yz+zx, \frac{x-y}{y-z}\right) = 0.$$

Example 59. *Find the equation of the curve satisfying $px + qy = z$ and passing through the circle*

$$x^2+y^2+z^2 = 4, \; x+y+z = 2.$$

The auxiliary equations are

$$\frac{dx}{x} = \frac{dy}{y} = \frac{dz}{z}$$

Taking the first two ratios, $\log x = \log y + \log a$

$$\therefore \quad \frac{x}{y} = a \qquad ....(1)$$

Similarly, $\frac{y}{z} = b$ ....(2)

Both solutions are independent

The general solution is $\phi\left(\frac{x}{y}, \frac{y}{z}\right) = 0$ ...I

We have to find that function ϕ satisfying (I)

and also $x^2 + y^2 + z^2 = 4$ ....(3)

and $x + y + z = 2$ ....(4)

Hence, we will eliminate x, y, z from (1), (2), (3), (4).

From (2), $y = bz$ From (1), $x = ay = abz$

using these values in (3) and (4)

$$a^2 b^2 z^2 + b^2 z^2 + z^2 = 4$$

i.e., $z^2 (1 + b^2 + a^2 b^2) = 4$ ....(5)

Also, $abz + bz + z = 2$

$(1 + b + ab)\, z = 2.$ ....(6)

Eliminate z between (5) and (6)

Squaring (6),

$(1 + b + ab)^2 z^2 = 4$ ....(7)

From (5) and (7), $1 + b^2 + a^2b^2 = (1 + b + ab)^2$

Simplifying, $b + ab^2 + ab = 0.$

i.e., $1 + ab + a = 0$ ....(8)

using $a = \frac{x}{y}, b = \frac{y}{z}$ in (8)

$$1 + \frac{xy}{yz} + \frac{x}{y} = 0 \quad \therefore \quad 1 + \frac{x}{z} + \frac{x}{y} = 0$$

i.e., $yz + xy + xz = 0$ is the required surface.

Exercises 2(d)

Find the general solution of the following partial differential equations (1 to 18):

1. $px^2 + qy^2 = (x + y)\, z.$ **2.** $px - qy = xz.$ **3.** $(p - q)\, z = z^2 + (x + y)^2.$

4. $p\, zx + q\, zy = xy.$ **5.** $px^2 + qy^2 = z^2.$ **6.** $p + q = 1.$

7. $(a - x)\, p + (b - y)\, q = c - z.$ **8.** $p\sqrt{x} + q\sqrt{y} = \sqrt{z}.$ ***(M.U. 64 B.E.)***

9. $px + qy = x.$ **10.** $px + qy = nz.$ ***(M.U. 86 Ap.)***

11. $py^2 z + qx^2 z = xy^2.$ ***(M.U. 71 B.E.)***

12. $px (y^2 + z) - qy (x^2 + z) = z (x^2 - y^2).$ ***(MS. 1991 Ap.)***

13. $px^2 - qy^2 = z^2.$ **14.** $py^2 - xyq = x (z - 2y)$

15. $p \cot x + q \cot y = \cot z$. **(Os. 64 B.E.)**

16. $p\sqrt{x} + q\sqrt{y} = \sqrt{z}$. **(Os. U. 65 B.E.)**

17. $(1+y)\,p + (1+x)\,q = z$ **18.** $(x+y)\,zp + (x-y)\,zq = x^2 + y^2$.

19. Show that the surface which satisfies the differential equation $(x^2 - a^2)\,p + (xy - az \tan\alpha)\,q = xz - ay \cot\alpha$ and passes through the curve $x^2 + y^2 = a^2$, $z = 0$ is $x^2 + y^2 - a^2 = z^2 \tan^2\alpha$.

20. Show that the integral surface of the linear partial differential equation $x\,(y^2 + z)\,p - y\,(x^2 + z)\,q = (x^2 - y^2)\,z$ which contains the straight line $x + y = 0$, $z = 1$ is $x^2 + y^2 + 2xyz - 2z + 2 = 0$.

21. Show that the integral surface of the equation $2y\,(z-3)\,p + (2x - z)\,q = y\,(2x - 3)$ which passes through the circle $x^2 + y^2 = 2x$, $z = 0$ is $x^2 + y^2 - z^2 - 2x + 4z = 0$. **(M.U. 70 B.E.)**

22. If the expression $(y^2 + z)\,dx + (x^2 + z)\,dy$ is an exact differential in x and y, show that $z = 2xy + f(x+y)$, where f is arbitrary. Find f if $z = 2y + 1$ when $x = 0$. **(M.U. 65 B.E.)**

23. $pz + qy = x$ **24.** $x\,(z^2 - y^2)\,p + y\,(x^2 - z^2)\,q = z\,(y^2 - x^2)$

25. $pz - qz = z^2 + (x+y)^2$

2·17. Partial Differential Equation of Higher order

A thorough treatment of the subject of partial differential equations of order higher than the first is too vast to study here. We shall study only linear partial differential equations of higher order with *constant coefficients*. We can divide this study into two groups, *viz.* (*i*) homogeneous linear and (*ii*) non-homogeneous linear equations.

For example, the equation,

$$2\frac{\partial^3 z}{\partial x^3} + 3\frac{\partial^3 z}{\partial x^2 \partial y} + 4\frac{\partial^3 z}{\partial x \partial y^2} + 5\frac{\partial^3 z}{\partial y^3} = x^2 + y \qquad ...(i)$$

is an equation in which the partial derivatives occurring are all of the same order and the coefficients are constants whereas, the equation

$$\frac{\partial^3 z}{\partial x^3} + 2\frac{\partial^2 z}{\partial y^2} - 4\frac{\partial z}{\partial x} + z = x^2 + y^2 \qquad ...(ii)$$

possesses derivatives which are not all of the same order but with constant coefficients. (*i*) is called a homogeneous linear equation with constant coefficients whereas (*ii*) is called a non-homogenous linear equation with constant coefficients.

We shall use the differential operators D and D' to denote $D = \dfrac{\partial}{\partial x}$ and $D' = \dfrac{\partial}{\partial y}$ and hence (*i*) can be written symbolically as

$$(2D^3 + 3D^2D' + 4DD'^2 + 5D'^3)\, z = x^2 + y.$$

(*ii*) becomes $(D^3 + 2D'^2 - 4D + 1)\, z = x^2 + y^2$.

2·18. Homogeneous linear equation

A homogeneous linear partial differential equation of nth order with constant coefficients is of the form

$$a_0 \frac{\partial^n z}{\partial x^n} + a_1 \frac{\partial^n z}{\partial x^{n-1}\,\partial y} + a_2 \frac{\partial^n z}{\partial x^{n-2}\,\partial y^2} + \cdots\cdots + a_n \frac{\partial^n z}{\partial y^n} = F(x, y) \qquad ...(i)$$

where a's are constants and F is a known function of x, y. Writing symbolically, (*i*) can be written as

$$(a_0 D^n + a_1 D^{n-1} D' + a_2 D^{n-2} D'^2 + \cdots\cdots + a_n D'^n)\, z = F(x, y) \qquad ...(ii)$$

or $$f(D, D')\, z = F(x, y) \qquad ...(iii)$$

where $f(D, D')$ stands for the polynomial expression

$$a_0 D^n + a_1 D^{n-1} D' + a_2 D^{n-2} D'^2 + \cdots\cdots + a_n D'^n.$$

The method of solving (*iii*) is anologous to that of solving ordinary linear differential equation with constant coefficients. The complete solution of (*iii*) consists of two parts namely the *complementary function* and the *particular integral.* The complementary function of (*iii*) is the solution of

$$f(D, D')\, z = 0 \qquad ...(iv)$$

and the particular integral of (*iii*) is a particular solution of (*iii*) given symbolically by

$$\frac{1}{f(D, D')} F(x, y).$$

Hence the complete solution = complementary function + particular integral

= C.F. + P.I.

Note. If $z = z_1, z = z_2, \ldots\ldots, z = z_n$ are solutions of (*iv*), then $z = \lambda_1 z_1 + \lambda_2 z_2 + \cdots\cdots + \lambda_n z_n$ is also a solution of (*iv*), where λ's' are constants arbitrarily chosen.

2·19. Complementary function of homogeneous linear equations, with constant coefficients

Let $$(a_0 D^n + a_1 D^{n-1} D' + a_2 D^{n-2} D'^2 + \ldots\ldots + a_n D'^n)\, z = F(x, y) \qquad ...(1)$$

or $$f(D, D')\, z = F(x, y) \qquad ...(2)$$

be the given homogeneous linear equation with constant coefficients. Now the complementary function of (2) is the solution of

$$f(D, D')\, z = 0 \qquad ...(3)$$

Since $f(D, D')$ is a polynomial which is homogeneous of degree n in D and

D', we can factorise $f(D, D')$ into linear factors, and hence (3) can be written as

$$(D - m_1 D')(D - m_2 D')......(D - m_n D')\, z = 0 \quad ...(4)$$

where $m_1, m_2,......m_n$ are the roots of

$$(m - m_1)(m - m_2) \,......\, (m - m_n) = 0 \quad ...(5)$$

i.e., of $f(m, 1) = 0$.

Equation (3) will be satisfied by the solution of each of the component differential equations,

$$(D - m_1 D')\, z = 0, \ (D - m_2 D')\, z = 0, ... \ (D - m_n D') z = 0 \quad ...(6)$$

$(D - m_r D')\, z = 0$ is Lagrange's equation.

Hence the subsidiary equations are

$$\frac{dx}{1} = \frac{dy}{-m_r} = \frac{dz}{0}.$$

Therefore, $y + m_r x = c$ and $z = k$.

Hence the general solution of $(D - m_r D')\, z = 0$ is

$$z = \phi_r\, (y + m_r\, x).$$

Putting $r = 1, 2,.....,n$ we get the general solutions of the component equations (6). Hence the most general solution of (3) is $z = \phi_1\, (y + m_1 x) + \phi_2\, (y + m_2 x) + \cdots + \phi_n\, (y + m_n x)$, where $\phi_1, \phi_2,.....\phi_u$ are arbitrary functions.

That is, the complementary function of (1) is

$$z = \phi_1\, (y + m_1 x) + \phi_2\, (y + m_2 x) + \cdots + \phi_n\, (y + m_n x),$$

where $m_1, m_2,......., m_n$ *are the roots of the auxiliary* equation $f(m, 1) = 0$ which is got by replacing D by m and D' by 1 in $f(D, D')$ and equating it to zero.

Note. The above argument is valid only if $m_1 \neq m_2 \neq .... \neq m_n$.

2·20. Auxiliary equation with repeated roots

Suppose the auxiliary equation $f(m, 1) = 0$ possesses two equal roots, say $m_1 = m_2$. Then the above method will give a complementary function with $(n–1)$ arbitrary functions only which will not be the complete solution of the given equation.

If $m_1 = m_2$, we come across the component equation

$$(D - m_1 D')^2\, z = 0 \quad ...(7)$$

Let $(D - m_1 D')\, z = u$ ...(8)

Then (7) becomes

$$(D - m_1 D')\, u = 0.$$

$\therefore \quad u = \phi_1 (y + m_1 x)$

Substituting this value of u in (8), we get

$$(D - m_1 D') z = \phi_1 (y + m_1 x) \qquad \text{...(9)}$$

This is Lagrange's equation whose subsidiary equations are

$$\frac{dx}{1} = \frac{dy}{-m_1} = \frac{dz}{\phi_1 (y + m_1 x)}$$

Solving these, we get,

$$y + m_1 x = c \text{ and } z - x\phi_1 (y + m_1 x) = k.$$

Hence the solution of (7) is

$$z - x\phi_1 (y + m_1 x) = \phi_2 (y + m_1 x)$$

i.e.,

$$z = x\phi_1 (y + m_1 x) + \phi_2 (y + m_1 x) \qquad \text{...(10)}$$

(10) is the solution of (7).

Note. If the auxiliary equation $f(m, 1) = 0$ has r equal roots $m_1 = m_2 = m_3 = \cdots\cdots = m_r$, then the corresponding part in the complementary function is

$$z = \phi_1 (y + m_1 x) + x\phi_2 (y + m_1 x) + \cdots\cdots + x^{r-1} \phi_r (y + m_1 x).$$

2·21. The Particular integral

Evaluation of the particular integral in P.D.E., is analogous to that of the P.I. in an ordinary linear diferential equation. There are short methods to evaluate the particular integrals of the homogeneous linear equation with constant coefficients. The proofs are simple and the students can prove them by themselves. The methods are given below.

Type 1. $\dfrac{1}{f(D, D')} e^{ax+by} = \dfrac{1}{f(a, b)} e^{ax+by}$, if $f(a, b) \neq 0$.

Type 2. $\dfrac{1}{f(D, D')} x^r y^s = [f(D, D')]^{-1} x^r y^s$, where

$[f(D, D')]^{-1}$ is to be expanded in powers of D, D'.

Type 3. $\dfrac{1}{f(D^2, DD', D'^2)} \begin{matrix}\sin\\ \cos\end{matrix} \left(ax + by\right)$

$$= \frac{1}{f(-a^2, -ab, -b^2)} \begin{matrix}\sin\\ \cos\end{matrix} \left(ax + by\right)$$

Type 4. $\dfrac{1}{f(D, D')} e^{ax+by} \phi(x, y) = e^{ax+by} \dfrac{1}{f(D + a, D' + b)} \phi(x, y)$.

Type 5. $\dfrac{\sin ax \sin by}{f(D^2, D'^2)} = \dfrac{\sin ax \sin by}{f(-a^2, -b^2)}$ if denominator $\neq 0$

Type 6. $\dfrac{\cos ax \cos by}{f(D^2, D'^2)} = \dfrac{\cos ax \cos by}{f(-a^2 - b^2)}$ if denominator $\neq 0$

General rule to find $\dfrac{1}{D - mD'} F(x, y)$.

First change y to $y - mx$ in $F(x, y)$, integrate it w.r.t. x treating y as a constant and then in the resulting integral change y to $y + mx$. The result thus got is the value of $\dfrac{1}{D - mD'} F(x, y)$.

OR

Integrate $F(x, a - mx)$ w.r.t. x and after integration replace 'a' by $y + mx$

Example 60. $(D^2 - 4DD' + 4D'^2) z = 0.$

The auxiliary equation is

$m^2 - 4m + 4 = 0$ {Replace D by m and D' by 1 in $f(D, D')$ and equate to zero}

Solving $m = 2, 2$ (equal roots)

Since R.H.S. is zero, there is no Particular integral.

Hence $z = C.F.$ alone

i.e. $z = f(y + 2x) + x\,\phi\,(y + 2x)$

Example 61. *Solve* $(D^3 - 3D^2 D' + 2DD'^2) Z = 0$

The auxiliary equations $m^3 - 3m^2 + 2m = 0$

i.e. $m(m^2 - 3m + 2) = 0$

i.e. $m(m - 1)(m - 2) = 0$

$\therefore\ m = 0, 1, 2$

General solution is

$\therefore\ Z = f_1(y + 0 \cdot x) + f_2(y + x) + f_3(y + 2x)$

$Z = f_1(y) + f_2(y + x) + f_3(y + 2x)$

Example 62. $(D^3 + DD'^2 - D^2 D' - D'^3) Z = 0$

Auxiliary equation is

$$m^3 - m^2 + m - 1 = 0$$

$$m^2(m - 1) + (m - 1) = 0$$

$$(m - 1)(m^2 + 1) = 0$$

$$\therefore \quad m = 1, i, -i$$

General Solution is

$$Z = \phi(y + x) + f(y + ix) + F(y - ix)$$

Example 63. Solve $\dfrac{\partial^2 z}{\partial x^2} + \dfrac{\partial^2 z}{\partial x \partial y} - 2\dfrac{\partial^2 z}{\partial y^2} = 0.$ ***(M.U. 63 B.E.)***

Writing this equation symbolically, we get,

$$(D^2 + DD' - 2D'^2)\,z = 0 \qquad \text{...(1)}$$

The auxiliary equation is $m^2 + m - 2 = 0$,

giving $m = 1, -2$.

Hence the C.F. of (1) given by

$$z = \phi_1(y + x) + \phi_2(y - 2x). \qquad \text{...(2)}$$

R.H.S. of (1) is zero. Hence the complete solution of (1) is the C.F. itself.

Thus the complete solution is (2).

Example 64. *Solve* $(D^3 - 7DD'^2 - 6D'^3)\,z = x^2y + \sin(x + 2y)$.

(MS. 1988 Nov.)

The auxiliary equation is $m^3 - 7m - 6 = 0$

i.e., $(m + 1)(m + 2)(m - 3) = 0$

$\therefore\ m = -1, -2, 3$

The C.F. of the given P.D.E. is

$$z = \phi_1(y - x) + \phi_2(y - 2x) + \phi_3(y + 3x)$$

$$(\text{P.I.})_1 = \frac{1}{D^3 - 7DD'^2 - 6D'^3}(x^2 y)$$

$$= \frac{1}{D^3}\left[1 - \left(\frac{7D'^2}{D^2} + \frac{6D'^3}{D^3}\right)\right]^{-1}(x^2 y)$$

$$= \frac{1}{D^3}\left[1 + \frac{7D'^2}{D^2} + \frac{6D'^3}{D^3} + \cdots\right](x^2 y)$$

$$= \frac{1}{D^3}(x^2 y)$$

$$= \frac{1}{60}x^5 y.$$

$$(\text{P.I.})_2 = \frac{1}{D^3 - 7DD'^2 - 6D'^3}\sin(x + 2y)$$

$$= \frac{1}{-D + 28D + 24D'}\sin(x + 2y)$$

$$= \tfrac{1}{3} \cdot \frac{1}{9D + 8D'} \sin(x + 2y)$$

$$= \tfrac{1}{3} \cdot \frac{9D - 8D'}{81D^2 - 64D'^2} \sin(x + 2y)$$

$$= \tfrac{1}{3} \frac{9D - 8D'}{-81 + 256} \sin(x + 2y)$$

$$= \tfrac{1}{525} [(-7 \cos(x + 2y)]$$

$$= -\tfrac{1}{75} \cos(x + 2y)$$

Hence the complete solution of the given equation is

$$z = \phi_1(y - x) + \phi_2(y - 2x) + \phi_3(y + 3x) + \frac{x^5 y}{60} - \tfrac{1}{75} \cos(x + 2y).$$

Example 65. *Solve* $\dfrac{\partial^3 z}{\partial x^3} - 3\dfrac{\partial^3 z}{\partial x^2 \partial y} + 4\dfrac{\partial^3 z}{\partial y^3} = e^{x+2y}$.

Writing in the symbolic form, the equation is

$$(D^3 - 3D^2D' + 4D'^3)\, z = e^{x+2y}$$

The auxiliary equation is $m^3 - 3m^2 + 4 = 0$

i.e., $\quad (m + 1)(m - 2)^2 = 0$

i.e., $\quad m = -1, 2, 2$

Hence the C.F. is $\phi_1(y - x) + \phi_2(y + 2x) + x\phi_3(y + 2x)$

$$\text{P.I.} = \frac{1}{D^3 - 3D^2D' + 4D'^3} \cdot e^{x+2y}$$

$$= \frac{e^{x+2y}}{1 - 6 + 32}$$

$$= \frac{e^{x+2y}}{27}$$

The complete solution is

$$z = \phi_1(y - x) + \phi_2(y + 2x) + x\phi_3(y + 2x) + \tfrac{1}{27} e^{x+2y}.$$

Example 66. *Solve* $(D^3 + D^2D' - DD'^2 - D'^3)\, z = e^x \cos 2y$.

The auxiliary equation is

$$m^3 + m^2 - m - 1 = 0$$

i.e., $\quad (m + 1)^2 (m - 1) = 0$

$\therefore \quad m = 1, -1, -1$

$\therefore$ C.F. is $\phi_1(y+x)+\phi_2(y-x)+x\,\phi_3(y-x)$

$$\text{P.I.} = \frac{1}{D^3 + D^2D' - DD'^2 - D'^3}\, e^x \cos 2y$$

$$= e^x \frac{\cos 2y}{(D+1)^3 + (D+1)^2 D' - (D+1)\, D'^2 - D'^3}$$

$= e^x$. Real part of

$$\frac{e^{i2y}}{(D+1)^3 + (D+1)^2 D' - (D+1)\, D'^2 - D'^3}$$

$$= e^x \text{ R.P. of } \frac{e^{i2y}}{1 + 2i + 4 + 8i}$$

$$= \frac{e^x}{5} \text{. R.P. of } \frac{1-2i}{(1+2i)(1-2i)} (\cos 2y + i \sin 2y)$$

$$= \tfrac{1}{5} e^x \,.\, \tfrac{1}{5} (\cos 2y + 2 \sin 2y)$$

$$= \frac{e^x}{25} (\cos 2y + 2 \sin 2y)$$

The complete solution is

$$Z = \phi_1(y+x) + \phi_2(y-x) + x\phi_3(y-x) + \frac{e^x}{25} (\cos 2y + 2 \sin 2y)\,.$$

Example 67. *Solve:* $(D^3 - 7DD'^2 - 6D'^3)\, z = e^{2x+y}$ **(*BR. 1995 Ap.*)**

The auxiliary equation is

$$m^3 - 7m - 6 = 0$$

Evidently $m = -1$ is a root.

Factorizing,

$$(m+1)(m^2 - m - 6) = 0$$

$$(m+1)(m-3)(m+2) = 0$$

$\therefore\ m = -1, -2, 3$

$\therefore$ C.F. is $\phi_1(y-x)+\phi_2(y-2x)+\phi_3(y+3x)$.

$$\text{P.I.} = \frac{e^{2x+y}}{D^3 - 7DD'^2 - 6D'^3}$$

$$= \frac{e^{2x+y}}{8 - 7\,(2)(1) - 6\,(1)}$$

$$= -\frac{1}{12} e^{2x+y}$$

The complete solution is

$$z = \text{C.F.} + \text{P.I.}$$

$$z = \phi_1 (y - x) + \phi_2 (y - 2x) + \phi_3 (y + 3x) - \frac{1}{12} e^{2x + y}.$$

Example 68. $(D^2 - 2DD' + D'^2)\, z = e^{x + 2y}$

Auxiliary equation is

$$m^2 - 2m + 1 = 0$$

$$\therefore \quad m = 1, 1$$

C.F. is $\phi_1 (y + x) + x\, \phi_2 (y + x)$

$$\text{P.I.} = \frac{e^{x + 2y}}{(D - D')^2}$$

$$= \frac{e^x + 2y}{(1 - 2)^2} = e^{x + 2y}.$$

Hence, the complete solution is $z = \phi_1 (y + x) + x\, \phi_2 (y + x) + e^{x + 2y}$

Example 69. *Solve* $\frac{\partial^3 z}{\partial x^3} - 2\,\frac{\partial^3 z}{\partial x^2 \partial y} = e^{x + 2y}$

Writing in the symbolic form, we have

$$(D^3 - 2\, D^2\, D')Z = e^{x + 2y}$$

Auxiliary equation is $m^3 - 2m^2 = 0$

$$m^2 (m - 2) = 0$$

$$\therefore \quad m = 0, 0, 2.$$

C.F. is $\phi_1 (y) + x\, \phi_2 (y) + \phi_3 (y + 2x)$

P.I. $\quad \dfrac{e^{x + 2y}}{D^3 - 2D^2\, D'}$

$$= \frac{e^{x + 2y}}{1 - 2\,(2)}$$

$$= \frac{-1}{3}\, e^{x + 2y}$$

The complete solution is

$$z = \phi_1 (y) + x\, \phi_2 (y) + \phi_3 (y + 2x) - \frac{1}{3}\, e^{x + 2y}.$$

Example 70. *Solve* $(D^4 - D'^4)\, z = e^{x+y}$

The auxiliary equation is

$$m^4 - 1 = 0$$

$$(m^2 - 1)(m^2 + 1) = 0$$

$$\therefore \quad m = 1, -1, i, -i$$

C.F. is $\phi_1(y + x) + \phi_2(y - x) + \phi_3(y + ix) + \phi_4(y - ix)$

$$\text{P.I.} = \frac{e^{x+y}}{D^4 - D'^4}$$

(Now replacing D by 1, D' by 1, the denominator becomes zero.

$$= \frac{e^{x+y}}{(D - D')(D + D')(D^2 + D'^2)}$$

$$= \frac{e^{x+y}}{(D - D')(1 + 1)(1 + 1)}$$ (Replace D by 1, D' by 1 in these factors which do not vanish.)

$$= \frac{1}{4} \cdot \frac{e^{x+y}}{D - D'} \qquad \text{...(1)}$$

To evaluate $\dfrac{e^{x+y}}{D - D'}$ we remember the rule $\dfrac{F(x, y)}{D - mD'}$.

Change y as y-x and integrate w.r.t. x treating y as a constant and then change y into $y + x$.

Here, $F(x, y) = e^{x+y}$

Changing y as $y - x$, it becomes e^y

Integrating w.r.t x, we get xe^y

Changing y as $y + x$, we get xe^{x+y}

$$\therefore \quad \text{P.I.} = \frac{1}{4} x \,.\, e^{x+y}.$$

$\therefore$ The complete solution is

$$z = \phi_1(y + x) + \phi_2(y - x) + \phi_3(y + ix) + \phi_4(y - ix) + \frac{1}{4} x\, e^{x+y}.$$

Note 1. $\dfrac{e^{ax+by}}{D - \dfrac{a}{b} D'} = x\, e^{ax+by}$ (where Denominator vanishes when D is replaced by a and D' by b)

Note 2. $\dfrac{e^{ax+by}}{\left(D-\dfrac{a}{b}D'\right)^2}$

$= \dfrac{x^2}{2\,!}\, e^{ax+by}$

Note 3. using the result of Note 1 in the above Problem, $\text{P.I.} = \dfrac{1}{4}\, x e^{x+y}$.

Example 71. *Solve* $(D^2 - 4DD' + 4D'^2)\, z = e^{2x+y}$.

Auxiliary equation is $m^2 - 4m + 4 = 0$

Solving, $\therefore\ m = 2, 2.$

O.F. is $\phi_1\,(y+2x) + x\,\phi_2\,(y+2x)$.

$$\text{P.I.} = \frac{e^{2x+y}}{(D-2D')^2}$$

Replacing D by 2, D' by 1, the denominator is zero.

using the *Note 2* under the previous problem, $\text{P.I.} = \dfrac{x^2}{2}\, e^{2x+y}$

The complete solution is

$$z = \phi_1\,(y+2x) + x\,\phi_2\,(y+2x) + \frac{x^2}{2}\, e^{2x+y}.$$

Example 72. *Solve* $(D^2 + 2DD' + D'^2)\, z = \text{Sinh}\,(x+y) + e^{x+2y}$

The auxiliary equation is

$m^2 + 2m + 1 = 0$

$\therefore\ m = -1, -1$

C.F. is $\phi_1\,(y-x) + x\,\phi_2\,(y-x)$

$$(\text{P.I.})_1 = \frac{Sinh\,(x+y)}{(D+D')^2} = \frac{\frac{1}{2}\left(e^{x+y} - e^{-(x+y)}\right)}{(D+D')^2}$$

$$= \frac{1}{2}\,\frac{e^{x+y}}{(D+D')^2} - \frac{1}{2}\,\frac{e^{-x-y}}{(D+D')^2}$$

$$= \frac{1}{2}\,\frac{e^{x+y}}{4} - \frac{1}{2}\,\frac{e^{-x-y}}{4}$$

$$= \frac{1}{4}\,\text{Sinh}\,(x+y)$$

$$(PI)_2 = \frac{e^{x+2y}}{(D+D^1)^2}$$

$$= \frac{e^{x+2y}}{9}$$

$\therefore$ The complete solution is

$$z = C.F. + P.I.$$

$$z = \phi_1 (y-x) + x\,\phi_2 (y-x) + \frac{1}{4} \text{Sinh}\,(x+y) + \frac{1}{9} e^{x+2y}$$

Example 73. *Solve:* $(D^3 - 2\,D^2\,D')\,z = \sin\,(x+2y) + 3x^2y.$

The auxiliary equation is $m^3 - 2m^2 = 0$

$$\therefore\ \phi\, m = 0, 0, 2$$

C.F. is $\phi_1\,(y) + x\,\phi_2\,(y) + \phi_3\,(y+2x)$

$$(PI)_1 = \frac{\sin\,(x+2y)}{D^3 - 2\,D^2D'}$$

$$= \frac{\sin\,(x+2y)}{D\,(-1) - 2D\,(-2)}, \text{ replace } D^2 \text{ by } -1 \text{ and } DD' \text{ by } -2$$

$$= \frac{\sin\,(x+2y)}{3D}$$

$$= \frac{-1}{3} \cos\,(x+2y)$$

$$(PI)_2 = \frac{3x^2y}{D^3 - 2\,D^2D'}$$

$$= \frac{1}{D^3}\left(1 - \frac{2D'}{D}\right)^{-1} (3x^2\,y)$$

$$= \frac{1}{D^3}\left[1 + \frac{2D'}{D} + \frac{4D'^2}{D^2} + \cdots\right](3x^2\,y)$$

$$= \frac{1}{D^3}\left[3x^2y + \frac{6x^2}{D}\right]$$

$$= 3y \,.\, \frac{x^5}{3\times4\times5} + 6 \,.\, \frac{x^6}{3\times4\times5\times6} = \frac{x^5y}{20} + \frac{x^6}{60}$$

Complete solution is $Z = C.F. + P.I.$

i.e., $$z = \phi_1\,(y) + x\,\phi_2\,(y) + \phi_3\,(y+2x) - \frac{1}{3}\cos\,(x+2y) + \frac{x^5y}{20} + \frac{x^6}{60}.$$

Example 74. *Solve:* $(D^2 - 2DD')\,z = x^3y + e^{2x}$ **(MS. 1989 Nov.)**

The auxiliary equation is $m^2 - 2m = 0$

$\therefore \quad m = 0, 2$

O.F. is $\phi_1(y) + \phi_2(y + 2x)$

$$(\text{P.I.}) = \frac{x^3 y}{D^2 - 2DD'}$$

$$= \frac{1}{D^2}\left(1 - \frac{2D'}{D}\right)^{-1}(x^3 y)$$

$$= \frac{1}{D^2}\left[1 + \frac{2D'}{D} + \ldots\right](x^3 y)$$

$$= \frac{1}{D^2}\left[x^3 y + \frac{2x^3}{D}\right]$$

$$= \frac{x^5 y}{20} + 2 \cdot \frac{x^6}{4 \times 5 \times 6}$$

$$= \frac{x^5 y}{20} + \frac{x^6}{60}$$

$$(\text{PI})_2 = \frac{e^{2x}}{D^2 - 2DD'}$$

$$= \frac{e^{2x}}{4} \text{ replace } D \text{ by 2 and } D' \text{ by 0.}$$

$\therefore$ The complete solution is

$$Z = \phi(y) + \phi_2(y + 2x) + \frac{e^{2x}}{4} + \frac{x^5 y}{20} + \frac{x^6}{60}$$

Example 75. *Solve:* $(D^2 + 3DD' + 2D'^2)\,z = x + y.$

The auxiliary equation is

$$m^2 + 3m + 2 = 0$$

$$m = -1, -2$$

C.F. is $\phi_1(y - x) + \phi_2(y - 2x)$.

$$\text{P.I.} = \frac{x + y}{D^2 + 3DD' + 2D'^2}$$

$$= \frac{1}{D^2}\left(1 + \frac{3D'}{D} + \frac{2D'^2}{D^2}\right)^{-1}(x + y)$$

$$= \frac{1}{D^2}\left[1 - \left(\frac{3D'}{D} + \frac{2D'^2}{D^2}\right) + \left(\frac{3D'}{D} + \frac{2D'^2}{D^2}\right)^2 - \ldots\right](x+y)$$

$$= \frac{1}{D^2}\left[x + y - \frac{3}{D}\right] = \frac{1}{D^2}x + \frac{1}{D^2}y - \frac{3}{D^3}$$

$$= \frac{x^3}{6} + \frac{x^2y}{2} - \frac{x^3}{2}$$

$$= \frac{x^2y}{2} - \frac{1}{3}x^3$$

Hence, the complete solution of the equation is

$$z = \phi_1(y - x) + \phi_2(y - 2x) + \frac{x^2y}{2} - \frac{x^3}{3}.$$

Example 76. *Solve:* $(D^2 - DD')\,z = \sin x \sin 2y$

The auxiliary equation is $m^2 - m = 0$

Solving, $m = 0, 1$

C.F. is $\phi_1(y) + \phi_2(y + x)$.

$$\text{P.I.} = \frac{\sin x \sin 2y}{D^2 - DD'}$$

$$= \frac{D(D + D')\sin x \sin 2y}{D(D - D')\,.\,D(D + D')}$$

$$= D(D + D')\frac{\sin x \sin 2y}{D^2(D^2 - D'^2)}$$

$$= D(D + D')\frac{\sin x \sin 2y}{(-1)[(-1) - (-4)]}$$

$$= -\frac{1}{3}\,.\,D(D + D')\sin x \sin 2y$$

$$= -\frac{1}{3}(D^2 + DD')\sin x \sin 2y$$

$$= -\frac{1}{3}[-\sin x \sin 2y + 2\cos x \cos 2y]$$

Hence, the general solution is

$$z = \phi_1(y) + \phi_2(y + x) - \frac{1}{3}[2\cos x \cos 2y - \sin x \sin 2y]$$

Example 77. *Solve* $(D^2 - 4D'^2)\,z = \cos 2x \cos 3y$

The auxiliary equation is

$m^2 - 4 = 0$

$m = \pm 2$

C.F. is $\phi_1 (y + 2x) + \phi_2 (y - 2x)$

$$\text{P.I.} = \frac{\cos 2x \cos 3y}{D^2 - 4D'^2}$$

$$= \frac{\cos 2x \cos 3y}{-4 - 4(-9)}$$

$$= \frac{1}{32} \cos 2x \cos 3y$$

The general solution is

$$Z = \phi_1 (y + 2x) + \phi_2 (y - 2x) + \frac{1}{32} \cos 2x \cos 3y.$$

Example 78. *Solve* $(D^2 - 2DD' + D'^2) Z = \cos (x - 3y)$

(MS. 1990 Ap.)

The auxiliary equation is

$m^2 - 2m + 1 = 0$

$m = 1, 1$

C.F. is $\phi_1 (y + x) + x \phi_2 (y + x)$

$$\text{P.I.} = \frac{\cos (x - 3y)}{D^2 - 2DD' + D'^2}$$

$$= \frac{\cos (x - 3y)}{-1 - 2(3) - 9}$$

$$= -\frac{1}{16} \cos (x - 3y)$$

$\therefore$ The complete solution is,

$$z = \phi_1 (y + x) + x \phi_2 (y + x) - \frac{1}{16} \cos (x - 3y)$$

Example 79. *Solve:* $(D^2 - 3DD' + 2D'^2) z = \sin x \cos y$

(MS. 1989 Nov.)

The auxiliary equation is

$m^2 - 3m + 2 = 0$

$(m - 1)(m - 2) = 0$

$\therefore m = 1, 2$

C.F. is $\phi_1 (y + x) + \phi_2 (y + 2x)$

$$\text{P.I.} = \frac{\sin x \cos y}{D^2 - 3DD' + 2D'^2}$$

$$= \frac{1}{2} \cdot \frac{\sin (x+y) + \sin (x-y)}{D^2 - 3DD' + 2D'^2}$$

$$= \frac{1}{2} \cdot \frac{\sin (x+y)}{D^2 - 3DD' + 2D'^2} + \frac{1}{2} \frac{\sin (x-y)}{(-1-3-2)}$$

$$= \frac{1}{2} I - \frac{1}{12} \sin (x-y)$$

$$I = \frac{\sin (x+y)}{D^2 - 3DD' + 2D'^2}$$

$$= \frac{\sin (x+y)}{(D-2D')(D-D')}$$

$$= \frac{1}{D-2D'} \cdot \frac{\sin (x+y)}{D-D'}$$

$$= \frac{1}{D-2D'} \cdot x \cdot \sin (x+y) \quad \text{use general rule}$$

$$= \int x \sin (a-x)\, dx \quad \text{where } a = y + 2x$$

$$= x [+ \cos (a-x)] - \sin (a-x) \quad \text{where } a = y + 2x$$

$$= x \cos (x+y) - \sin (x+y)$$

Hence $z = \phi_1 (x+y) + \phi_2 (y+2x)$

$$+ \frac{1}{2} x \cos (x+y) - \tfrac{1}{2} \sin (x+y) - \frac{1}{12} \sin (x-y)$$

Example 80. *Solve* $(D^2 + DD' - 6D'^2)\, z = y \cos x$ ***(MS. 1990 Ap.)***

Auxiliary equation is $m^2 + m - 6 = 0$

$$m = 2, -3$$

C.F. is $\phi_1 (y+2x) + \phi_2 (y-3x)$

$$\text{PI} = \frac{y \cos x}{D^2 + DD' - 6D'^2}$$

$$= \frac{1}{(D-2D')} \frac{y \cos x}{(D+3D')}$$

$$= \frac{1}{(D-2D^1)} \left[\int (a+3x) \cos x\, dx \right] \quad \text{where } a + 3x = y$$

$$= \frac{1}{D-2D'} \left[(a+3x) \sin x - 3 \int \sin x\, dx \right] \text{where } a + 3x = y$$

$$= \frac{1}{D - 2D'} [\, y \sin x + 3 \cos x \,]$$

$$= \int [(a - 2x) \sin x + 3 \cos x]\, dx \text{ where } a - 2x = y$$

$$= [\, (a - 2x)(-\cos x) - (-2)(-\sin x) + 3 \sin x \,] \text{ where } a - 2x = y$$

$$= -y \cos x + \sin x$$

$$\therefore \quad z = \phi_1 (y + 2x) + \phi_2 (y - 3x) + \sin x - y \cos x.$$

Example 81. *Solve* $(D^2 + 2DD' + D'^2)\, z = 2 \cos y - x \sin y$.

Auxiliary equation is $m^2 + 2m + 1 = 0$

$\therefore \; m = -1, -1$

C.F. is $\phi_1 (y - x) + x\, \phi_2 (y - x)$

$$\text{PI} = \frac{2 \cos y - x \sin y}{(D + D')(D + D')}$$

$$= \frac{1}{D + D^1} \int [2 \cos (a + x) - x \sin (a + x)\,]\, dx \text{ where } y = x + a$$

$$= \frac{1}{D + D'} [2 \sin (a + x) + x \cos (a + x) - \sin (x + a)\,]$$

where $y = x + a$

$$= \frac{1}{D + D'} [\, x \cos (a + x) + \sin (a + x)\,]$$

$$= \frac{1}{D + D'} [\, x \cos y + \sin y\,]$$

$$= \int [x \cos (a + x) + \sin (a + x)]\, dx \text{ where } y = a + x$$

$$= x \sin (a + x) + \cos (a + x) - \cos (a + x) \text{ where } y = a + x$$

$$= x \sin y.$$

The general solution is

$$z = \phi_1 (y - x) + x\, \phi_2 (y - x) + x \sin y.$$

Example 82. *Solve* $(D^2 + 4DD' - 5D'^2)\, z = x + y^2 + \pi$

The auxiliary equation is

$$m^2 + 4m - 5 = 0$$

$$(m + 5)(m - 1) = 0$$

$$\therefore \qquad m = 1 \text{ or } -5$$

C.F. is $\phi_1 (y + x) + \phi_2 (y - 5x)$.

$$\text{P.I.} = \frac{x + y^2 + \pi}{D^2 + 4DD' - 5D'^2}$$

$$= \frac{1}{D^2}\left[1+\left(\frac{4D'}{D}-\frac{5D'^2}{D^2}\right)\right]^{-1}(x+y^2+\pi)$$

$$= \frac{1}{D^2}\left[1-\left(\frac{4D'}{D}-\frac{5D'^2}{D^2}\right)+\left(\frac{4D'}{D}-\frac{5D'^2}{D^2}\right)^2-\cdots\right](x+y^2+\pi)$$

$$= \frac{1}{D^2}\left[1-\frac{4D'}{D}+\frac{5D'^2}{D^2}+\frac{16D'^2}{D^2}+\cdots\right](x+y^2+\pi)$$

(collect only terms upto D'^2 in the numerator as we have y^2 only in the operand.)

$$= \frac{1}{D^2}\left[x+y^2+\pi-\frac{4}{D}(2y)+\frac{21}{D^2}(2)\right]$$

$$= \frac{x^3}{6}+(y^2+\pi)\frac{x^2}{2}-8\left(\frac{yx^3}{6}\right)+42\,.\,\frac{x^4}{24}$$

$$= \frac{x^3}{6}+\frac{x^2}{2}(y^2+\pi)-\frac{4}{3}x^3y+\frac{7}{4}x^4.$$

The general solution is

$$z = \phi_1(x+y)+\phi_2(y-5x)+\frac{x^3}{6}+\frac{x^2}{2}(y^2+\pi)$$

$$\frac{-4}{3}x^3y+\frac{7}{4}x^4$$

Example 83. *Solve* $(D^2-3DD'+2D'^2)\,z=(2+4x)\,e^{x+2y}$

The auxiliary equation is $m^2-3m+2=0$

$\therefore\ m = 1, 2$

C.F. is $\phi_1(y+x)+\phi_2(y+2x)$

$$\text{P.I.} = \frac{e^{x+2y}\times(2+4x)}{D^2-3DD'+2D'^2}$$

$$= \frac{e^{x+2y}\times(2+4x)}{(D-2D')(D-D')}$$

$$= e^{x+2y}\times\frac{(2+4x)}{[(D+1)-2(D'+2)]\,[(D+1)-(D'+2)]}$$

$$= e^{x+2y}\,\frac{2+4x}{(D-2D'-3)(D-D'-1)}$$

$$= e^{x+2y} \frac{2+4x}{3\left[1-\frac{D-2D'}{3}\right][1-(D-D']}$$

$$= \frac{1}{3} e^{x+2y} \cdot \left[1-\frac{D-2D'}{3}\right]^{-1} [1-(D-D')]^{-1} (2+4x)$$

$$= \frac{1}{3} e^{x+2y} \left[1+\frac{D-2D'}{3}+\cdots\right] [1+(D-D')+....] (2+4x)$$

$$= \frac{1}{3} e^{x+2y} \left[1+\frac{4D-5D'}{3}+\cdots\right] (2+4x)$$

$$= \frac{1}{3} e^{x+2y} \left[2+4x+\frac{1}{3}(16)\right]$$

$$= \frac{1}{3} e^{x+2y} \left[\frac{22}{3}+4x\right]$$

Hence the general solution is

$$Z = \phi_1 (y+x) + \phi_2 (y+2x) + \frac{2}{9} e^{x+2y} \times (11+6x)$$

Exercises 2 (e)

Solve the equations given below where

$$\frac{\partial^2 z}{\partial x^2} = r, \frac{\partial^2 z}{\partial x \partial y} = s \text{ and } \frac{\partial^2 z}{\partial y^2} = t$$

1. $(2D^2 + 5DD' + 2D'^2)\, z = 0.$

2. $(D^2 + DD' - 2D'^2)\, z = 0.$

3. $(D^2 + 6DD' + 9D'^2)\, z = 0.$

4. $\frac{\partial^3 z}{\partial x^3} + 3\frac{\partial^3 z}{\partial x^2 \partial y} + 3\frac{\partial^3 z}{\partial x \partial y^2} - \frac{\partial^3 z}{\partial y^3} = 0.$

5. $\frac{\partial^3 z}{\partial x^3} - \frac{\partial^3 z}{\partial x^2 \partial y} - 8\frac{\partial^3 z}{\partial x \partial y^2} + 12\frac{\partial^3 z}{\partial y^3} = 0.$

6. $\frac{\partial^2 y}{\partial t^2} = a^2 \frac{\partial^2 y}{\partial x^2}.$

7. $(2D^2 + DD' - D'^2)\, z = 1.$

8. $(D^3 - 4D^2 D' + 5DD'^2 - 2D'^3)\, z = e^{x+y} + e^{y-2x} + e^{y+2x}.$

9. $4r + 12s + 9t = e^{3x-2y}.$

10. $(D^2 - D'^2)z = e^{x+2y}.$ **(*S.V.V. 65 B.E.*)**

11. $(D^2 - 2DD' + D'^2)\, z = e^{x+2y}.$

12. $(D^2 - 2DD')\, z = e^{2x} + x^2 y.$

13. $(D^3 - 2D^2 D')\, z = 2e^{2x} + 3x^2 y.$

14. $(D^2 + 4DD' - 5D'^2)\, z = x + y^2.$ **(*S.V.U. 65 B.E.*)**

15. $(D^2 + 2DD' + D'^2)z = x^2 y.$ **(*S.V.U. 65 B.E.*)**

16. $(D^2 - DD' - 6D'^2)\, z = xy.$

17. $(D^2 + 3DD' + 2D'^2)z = x + y.$ **(*M.U. 64, 76 B.E.*)**

18. $(D^3 - 7DD'^2 - 6D'^3)\, z = x^2 + xy^2 + y^3 + \cos(x - y).$

19. $(D^3 + D^2D' - DD'^2 - D'^3)\, z = e^{2x+y} + \cos(x + y).$ ***(Anna Ap. 2005)***

20. $(D^2 - DD')z = \cos x \cos 2y.$ **21.** $(D^2 - 4DD' + 4D'^2)\, z = e^{2x+y}$

22. $(D^2 - 6DD' + 9D'^2)\, z = 6x + 2y$ **23.** $r + s - 6t = \cos(2x + y).$

24. $r - 2s + t = \sin x.$ **25.** $(D^2 - DD')\, z = \sin x \cos 2y.$

26. $(D^3 - 3D^2D' - 4DD'^2 + 12D'^3)\, z = \sin(2x + y).$

27. $(D^2 - DD' - 2D'^2)\, z = (y - 1)\, e^x$ **28.** $(D^2 + 4DD' - 5D'^2)z = x + y.$

29. $(D^2 - 4D'^2)z = \sin(2x + y)$ **30.** $(D^2 - D'^2)\, z = e^{x+y}$

31. $\dfrac{\partial^2 u}{\partial x^2} + \dfrac{\partial^2 u}{\partial y^2} = \dfrac{y}{x^2 + y^2}$ ***(B.E. 1966)***

32. $(D^3 + D^2D' - DD'^2 - D'^3)\, z = 3 \sin(x + y)$ ***(MS. 1986 Nov.)***

33. $(D^3 - 7DD'^2 - 6D'^3)\, z = \cos(x - y) + x^2 + xy^2 + y^3$

34. $(D - D')^2 z = 2e^{x+y} \cos^2\left(\dfrac{x+y}{2}\right)$

2·22. Non-homogenous linear equation

In the P.D. Equation,

$$f(D, D')\, z = F(x, y) \quad ...(1)$$

if the polynomial expression $f(D, D')$ in D, D' is not homogenous in D, D', then the equation (1) is called non-homogenous linear equation. Here also, the complete solution = C.F. + P.I. The particular integral is found by the same methods as in the case of homogeneous linear equations. To obtain the C.F., we have to find the solution of

$$f(D, D')\, z = 0 \quad ...(2)$$

Assume a trial solution

$$z = ce^{hx+ky} \quad ...(3)$$

where c, h, k are all arbitrary constants.

Substituting (3) in (2), we get

$$cf(h, k)\, e^{hx+ky} = 0. \quad ...(4)$$

Hence $\qquad f(h, k) = 0$

If $f(D, D')$ is of degree r in D', then $f(h, k) = 0$ will be of rth degree in k. Solving for k from (4) in terms of h, we get

$$K_s = f_s(h) \text{ where } s = 1, 2, ... r.$$

Hence $z = C_s e^{hx + f_s(h)y}$ where $s = 1, 2, ...r$ are separate solutions of (2).

By giving all possible arbitrary values to C_s and h we get

$z = \Sigma C_1\, e^{hx+f_1(h)y},\ \Sigma C_2 e^{hx+f_2(h)y}, \cdots$ to be the solutions of (2).

Hence the most general solution of (2) is

$$z = \Sigma C_1 e^{hx + f_1(h)\,y} + \Sigma C_2 e^{hx + f_2(h)\,y} + \ldots.. + \Sigma C_r e^{hx + f_r(h)\,y}.$$

Note. Suppose k is linear in h, say $k_1 = f_1(h) = \alpha h + \beta$ where α, β are constants.

The corresponding part of the C.F. is

$$\Sigma C_1 e^{hx + (\alpha h + \beta)\,y}$$

$$= \Sigma C_1 e^{h\,(x + \alpha y)} \,.\, e^{\beta y}$$

$$= e^{\beta y}\, \Sigma C_1\, e^{h\,(x + ay)}$$

$$= e^{\beta y}\, \phi\,(x + \alpha\, y), \quad \text{where } \phi \text{ is arbitrary function.}$$

Example 84. *Solve* $(D^2 + DD' + D' - 1)\, z = 5e^x$.

Assume $z = Ce^{hx + ky}$ to be a trial solution.

of $(D^2 + DD' + D' - 1)\, z = 0$.

Then we get,

$$h^2 + kh + k - 1 = 0$$

i.e., $(h + 1)\,(h + k - 1) = 0$

i.e., $h = -1$ or $h = 1 - k$.

Then the C.F. of the given equation is

$$\Sigma\, C_1 e^{-x + ky} + \Sigma\, C_2 e^{(1-k)\,x + ky}$$

$$= e^{-x}\, \Sigma C_1 e^{ky} + e^x\, \Sigma\, C_2 e^{k\,(y - x)}$$

$$= e^{-x}\, \phi_1\,(y) + e^x\, \phi_2\,(y - x).$$

P.I. $$= \frac{5e^x}{D^2 + DD' + D' - 1}$$

$$= \frac{5e^x}{D^2 - 1}$$

$$= \frac{5e^x}{(D+1)(D+1)}$$

$$= \frac{5}{2}\, \frac{1}{D-1}\, e^x \cdot = \frac{5}{2}\, xe^x.$$

The complete solution is $z = e^{-x}\, \phi_1\,(y) + e^x\, \phi_2\,(y - x) + \frac{5}{2}\, xe^x$.

Example 85. *Solve.* $(\,2D^4 - 3D^2\, D' + D'^2)\, z = e^{2x + y}$

$f\,(D, D')$ on the L.H.S is not homogeneous.

The equation can be written as

$$(2D^2 - D')(D^2 - D')\, z = e^{2x+y}$$

Let $z = ce^{hx+ky}$ be a solution of $f(D, D')\, z =$

Then $(2h^2 - k)(h^2 - k) = 0$

$\therefore\ k = 2h^2$ or $k = h^2$

Hence, C.F. is $\sum C_1\, e^{hx + 2h^2 y} + \sum C_2\, e^{h_1 x + h_1^2 y}$

$$\text{P.I.} = \frac{e^{2x+y}}{2D^4 - 3D^2D' + D'^2}$$

$$= \frac{e^{2x+y}}{2\,(2)^4 - 3\,(2)^2\,(1) + (1)^2} \text{ replacing } D \text{ by 2 and } D' \text{ by 1}$$

$$= \frac{e^{2x+y}}{21}$$

The complete solution is

$$z = \sum C_1\, e^{hx + 2h^2 y} + \sum C_2\, e^{h_1 x + h_1^2 y} + \frac{1}{21}\, e^{2x+y}.$$

2·23. None homogeneous linear equation $f(D, D')\, z = F(x, y)$ where $f(D, D')$ is factorisable into linear factors

Case 1. Consider $(D - m\,D' - c)\, z = 0$

Rewriting the equation as $p - mq = cz$

This is Lagrange's equation.

The auxiliary equations are

$$\frac{dx}{1} = \frac{dy}{-m} = \frac{dz}{cz} \qquad \text{...(1)}$$

$dy = -\, mdx$ gives

$$y + mx = k_1 = \text{constant} \qquad \text{...(2)}$$

Taking $dx = \dfrac{dz}{cz}$, and integrating,

$$\log z = cx + k_2$$

i.e., $z = e^{cx + k_2} = e^{cx} \cdot e^{k_2} = e^{cx} \cdot k_3$

$$\therefore \qquad \frac{z}{e^{cx}} = k_3 = \text{constant} \qquad \text{...(3)}$$

From (2) and (3), the complete solution is

$$\frac{z}{e^{cx}} = f(y + mx)$$

$\therefore$ $z = e^{cx} f(y + mx)$, where f is arbitrary ...(4)

Case 2. If $(D - m_1 D' - c_1)(D - m_2 D' - c_2) \cdots (D - m_n D' - C_n) z = 0$ is the partial differential equation, then the general solution of this equation is

$$z = e^{c_1 x} \phi_1 (y + m_1 x) + e^{c_2 x} \phi_2 (y + m_2 x) + \cdots e^{c_n x} \phi_n (y + m_n x) \quad ...(5)$$

In case of repeated factors, namely,

Case 3. $(D - mD' - c)^r z = 0$ has the complete solution.

$$z = e^{cx} \phi_1 (y + mx) + x e^{cx} \phi_2 (y + mx) + x^2 e^{cx} \phi_3 (y + mx) + \cdots + x^{r-1} e^{cx} \phi_r (y + mx) \quad ...(6)$$

Example 86. *Solve*: $(D - D' - 1)(D - D' - 2) z = e^{2x - y}$

Referring to case (2), $m_1 = 1, c_1 = 1, m_2 = 1, c_2 = 2$,

Hence, the C.F. is $z = e^x \phi_1 (y + x) + e^{2x} \phi_2 (y + x)$

$$\text{P.I.} = \frac{e^{2x-y}}{(D - D' - 1)(D - D' - 2)}$$

$$= \frac{e^{2x-y}}{(2 + 1 - 1)(2 + 1 - 2)} \text{ replacing } D \text{ by 2 and } D' \text{ by } -1$$

$$= \frac{1}{2} e^{2x - y}$$

Hence, the complete solution is

$$z = e^x \phi_1 (y + x) + e^{2x} \phi_2 (y + x) + \frac{1}{2} e^{2x - y}$$

Example 87. *Solve* $(D + D' - 1)(D + 2D' - 3) z = 4 + 3x + 6y.$

(*MI. 1969*)

Comparing with case 2, $m_1 = -1, c_1 = 1, m_2 = -2, c_2 = 3$

Hence the C.F. is $e^x \phi_1 (y - x) + e^{3x} \phi_2 (y - 2x)$ ...(1)

$$\text{P.I.} = \frac{4 + 3x + 6y}{(D + D' - 1)(D + 2D' - 3)}$$

$$= \frac{4 + 3x + 6y}{3[1 - (D + D')]\left[1 - \dfrac{D + D'}{3}\right]}$$

$$= \frac{1}{3} [1 - (D + D')]^{-1} \left[1 - \frac{D + D'}{3}\right]^{-1} (4 + 3x + 6y)$$

$$= \frac{1}{3}\,[\,1 + (D + D') + (D + D')^2 + \cdots\,]$$

$$\times \left[\,1 + \frac{1}{3}(D + 2D') + \frac{1}{9}(D + 2D')^2 + \cdots\,\right](4 + 3x + 6y)$$

$$= \frac{1}{3}\left[\,1 + \frac{4}{3}D + \frac{5}{3}D' + \cdots\,\right](4 + 3x + 6y)$$

$$= \frac{1}{3}\left[\,4 + 3x + 6y + \frac{4}{3}(3) + \frac{5}{3}(6)\,\right]$$

$$= x + 2y + 6.$$

The general solution is

$$z = e^x \phi_1(y - x) + e^{3x}\phi_2(y - 2x) + x + 2y + 6.$$

Example 88. *Solve:* $(D^2 - D'^2 - 3D + 3D')z = xy + 7$ *(Anna. Ap 2005)*

The equation can be rewritten as

$$(D - D')(D + D' - 3)\,z = xy + 7$$

Hence, $m_1 = 1, c_1 = 0, m_2 = -1, c_2 = 3$

$\therefore$ C.F. is $e^{ox}\,\phi(y + x) + e^{3x} f(y - x)$

i.e., $\phi(y + x) + e^{3x} f(y - x)$

$$\text{P.I.} = \frac{xy + 7}{(D - D')(D + D' - 3)}$$

$$= \frac{1}{D}\,\frac{xy + 7}{\left(1 - \frac{D'}{D}\right)(-3)\left(1 - \frac{D + D'}{3}\right)}$$

$$= -\frac{1}{3D}\left(1 - \frac{D'}{D}\right)^{-1}\left(1 - \frac{D + D'}{3}\right)^{-1}(xy + 7)$$

$$= -\frac{1}{3D}\left[\,1 + \frac{D'}{D} + \frac{D'^2}{D^2} + \cdots\right]$$

$$\times \left[\,1 + \frac{D + D'}{3} + \frac{1}{9}(D + D')^2 + \cdots\,\right](xy + 7)$$

$$= -\frac{1}{3D}\left[\,1 + \frac{D}{3} + \frac{D'}{3} + \frac{D^2}{9} + \frac{2DD'}{9} + \frac{DD'}{3D} + \cdots\,\right](xy + 7)$$

$$= -\frac{1}{3}\left[\frac{1}{D} + \frac{1}{3} + \frac{2D'}{3D} + \frac{D}{9} + \frac{D'}{3} + \frac{D'}{D^2} + \frac{4DD'}{27}\right](xy + 7)$$

$$= -\frac{1}{3}\left[\frac{x^2}{2}y + 7x + \frac{xy}{3} + \frac{67}{27} + \frac{x^2}{3} + \frac{x}{3} + \frac{x^3}{6} + \frac{y}{9}\right]$$

$$= -\frac{1}{3}\left[\frac{x^2y}{2} + \frac{xy}{3} + 7x + \frac{x^2}{3} + \frac{x}{3} + \frac{y}{9} + \frac{x^3}{6} + y + \frac{67}{27}\right]$$

Hence, the general solution is

$$z = \phi(y+x) + e^{3x} f(y-x) - \frac{1}{3}\left[\frac{x^2y}{2} + \frac{xy}{3} + 7x + \frac{x^2}{3} + \frac{x}{3} + \frac{y}{9} + \frac{x^3}{6} + \frac{67}{27}\right]$$

Example 89. *Solve:* $(D - D' - 1)(D - D' - 2)\, z = e^{2x+y}$

Here $m_1 = 1, c_1 = 1, c_2 = 2, m_2 = 1$

C.F. is $e^x\phi_1(y+x) + e^{2x}\phi_2(y+x)$

$$\text{P.I.} = \frac{e^{2x+y}}{(D - D' - 1)(D - D' - 2)}$$

$$= \frac{1}{(D - D' - 1)} \cdot \frac{e^{2x+y}}{(2 - 1 - 2)}$$

$$= -\frac{e^{2x+y}}{D - D' - 1}$$ replacing D by 2 and D' by 1 in second factor since other factor vanishes.

$$= -x\, e^{2x+y}$$

$\therefore$ The general solutions is

$$z = e^x\phi_1(y+x) + e^{2x}\phi_2(y+x) - x\, e^{2x+y}$$

Exercises 2(f)

Solve the equations (1 to 7):

1. $(D^2 + DD' + D' - 1)\, z = e^{-x}$.
2. $(D^2 + 2DD' + D'^2 - 2D - 2D')\, z = e^{x-y} + x^2y$. **(*S.V.U. 64 B.E.*)**
3. $D(D + D' - 1)(D + 3D' - 2)\, z = xy + e^{2x+3y}$. **(*S.V.U. 64 B.E.*)**
4. $(2DD' + D'^2 - 3D')\, z = 3\cos(3x - 2y)$.
5. $(2D^2 - DD' - D'^2 + 6D + 3D')\, z = 0$.
6. $(D^2 - DD' - 2D'^2 + 2D + 2D')\, z = e^{2x+3y} + \sin(2x + y) + xy$.
7. $(D^2 - DD' + D')\, z = z + e^y + \cos(x + 2y)$.
8. Substituting $u = \log x$, $v = \log y$, solve $(x^2D^2 - y^2D'^2 - yD' + xD)\, z = 0$.
9. $(r - s + p) = x^2 + y^2$ **10.** $r - s + q = z + e^v + \cos(x + 2y)$
11. $(D - D' - 1)(D - D' - 2)\, z = e^{-2x+y}$
12. $(D^2 - DD' + D' - 1)\, z = e^{y-x} + \cos(x + 2y)$
13. $(D^2 - DD' + D' - 1)\, z = \cos(x + 2y)$ **(*B.R. 1995A*)**

CHAPTER III

APPLICATIONS OF PARTIAL DIFFERENTIAL EQUATIONS –BOUNDARY VALUE PROBLEMS

3·1. Boundary value problems. In applied mathematics, the partial differential equations generally arise from the mathematical formulation of physical problems. Subject to certain given conditions, called *boundary conditions,* solving such an equation is known as solving a *boundary value problem.* The method of solution of such equations differs from that used in the case of ordinary differential equations. Here, from the start, we try to find particular solutions of the partial differential equations which satisfy all the boundary conditions. For imposing the boundary conditions of the physical problem on the solution of the partial differential equation, we solve the differential equation using a method called, *method of separation of variables.* In most of the applied partial differential equations, we employ this method of separation of variables to solve the equation. This method is best explained in the following example.

Example 1. *Solve* $2x\frac{\partial z}{\partial x} - 3y\frac{\partial z}{\partial y} = 0$ *by method of separation of variables.* **(*M.S. 1986 Nov.*)**

Assume a solution of the form

$z = X(x)\,Y(y)$, where X is a function of x alone and Y is a function of y only.

$$\frac{\partial z}{\partial x} = X'Y, \text{ and } \frac{\partial z}{\partial y} = X\,Y'.$$

Hence the given equation becomes

$$2x\,X'\,Y - 3yX\,Y' = 0.$$

Separating the variables, $\dfrac{2x\,X'}{X} = \dfrac{3y\,Y'}{Y}$ ...(1)

The L.H.S. is a function x only and the R.H.S. is a function of y only. Also, x and y are independent variables. Whey y varies, keeping x fixed, the L.H.S. is constant and hence the R.H.S. must also be the same constant. Therefore,

$$\frac{2x\,X'}{X} = \frac{3y\,Y'}{Y} = k, \text{ say where } k \text{ is a constant.}$$

$$\frac{X'}{X} = \frac{k}{2x}.$$

Integrating w.r.t.x., $\log X = \frac{k}{2} \log x + \log c_1$

i.e., $$\log X = \log (c_1 x^{k/2})$$

$\therefore$ $$X = c_1 x^{k/2}$$

Similarly, $$\frac{Y'}{Y} = \frac{k}{3y}.$$

Integrating w.r.t. y, we get $Y = c_2 y^{k/3}$

Hence $z = XY = c_1 c_2 x^{k/2} y^{k/3} = cx^{k/2} y^{k/3}$, where c and k are arbitrary constants.

Example 2. *Solve* $\frac{\partial^2 z}{\partial x^2} - \frac{2\partial z}{\partial x} + \frac{\partial z}{\partial y} = 0$ *by the method of separation of variables.*

Assume a solution of the form $z = X(x) \cdot Y(y)$ ...(1)

where X is a function of x alone and Y is a function of y alone. (It is possible such a solution may not exist).

$\frac{\partial z}{\partial x} = X'Y, \frac{\partial^2 z}{\partial x^2} = X''Y, \frac{\partial z}{\partial y} = XY'$ where primes (′) denote differentiation w.r.t. the independent variable of the function.

Hence the given equation reduces to

$$X''Y - 2X'Y + XY' = 0$$

i.e., $$(X'' - 2X')\,Y = -XY'$$

Dividing by XY, $\frac{X'' - 2X'}{X} = \frac{-Y'}{Y}$ ...(2)

As in the previous problem, L.H.S. is a function of x alone while the R.H.S. is a function of y alone and also x, y are independent.

$\therefore$ $$\frac{X'' - 2X'}{X} = -\frac{Y'}{Y} = k \text{ (constant)}$$

This reduces to two ordinary differential equations,

$$X'' - 2X' - kX = 0 \qquad \text{...(3)}$$

and $$Y' + kY = 0 \qquad \text{...(4)}$$

If $D = \frac{d}{dx}$, then the first of these two becomes, $(D^2 - 2D - k)X = 0$.

Auxiliary equation is $m^2 - 2m - k = 0$

$$m = \frac{2 \pm \sqrt{4+4k}}{2} = 1 \pm \sqrt{1+k}$$

$$\therefore \quad X = Ae^{(1+\sqrt{1+k})x} + Be^{(1-\sqrt{1+k})x}$$

$$= e^x \left[Ae^{\sqrt{1+k}\,x} + Be^{-\sqrt{1+k}\,x} \right]$$

Solving $Y' + kY = 0$ we have $\frac{Y'}{Y} = -k$

$$\log Y = -ky + c$$

$$\therefore \quad Y = e^{-ky+c} = e^c \cdot e^{-ky} = De^{-ky}$$

Hence $z = XY = e^x \left[Ae^{\sqrt{1+k}\,x} + Be^{-\sqrt{1+k}\,y} \right] \cdot De^{-ky}$

i.e., $$z = e^x \cdot e^{-ky} \left[ae^{\sqrt{1+k}\,x} + be^{-\sqrt{1+k}\,y} \right]$$

where a, b, k are arbitrary constants to be determined by boundary and initial conditions.

Example 3. *By the method of separation of variables, solve* $4\frac{\partial z}{\partial x} + \frac{\partial z}{\partial y} = 3z$ *subject to* $z = e^{-5y}$ *when* $x = 0$.

Let $z = X(x)\,Y(y) = XY$

The equation (given) reduces to,

$$4X'Y + XY' = 3XY$$

i.e., $$\frac{4X'}{X} + \frac{Y'}{Y} = 3, \text{ dividing by } XY.$$

$$\frac{4X'}{X} = 3 - \frac{Y'}{Y} = k = \text{ constant.}$$

$$\therefore \quad \frac{4X'}{X} = k \text{ and } 3 - \frac{Y'}{Y} = k$$

$$\frac{X'}{X} = \frac{k}{4} \text{ and } \frac{Y'}{Y} = 3 - k$$

Integrating,

$$\log X = \frac{k}{4}x + A \text{ and } \log Y = (3-k)y + B$$

$$\therefore \quad X = e^{A + \frac{k}{4}x} \text{ and } Y = e^{B + (3-k)y}$$

i.e., $X = \alpha e^{\frac{kx}{4}}$ and $Y = \beta e^{(3-k)y}$

Hence $z = XY = ce^{\frac{kx}{4}} e^{(3-k)y}$ where $c = \alpha\beta$

$$z = c\,e^{\frac{k}{4}x + (3-k)y}$$

when $x = 0, z = e^{-5y}$

At $x = 0,\ z = c\,e^{(3-k)y} = e^{-5y}$

$\therefore$ $c = 1, 3 - k = -5$

i.e., $c = 1$ and $k = 8$

Hence $z = e^{2x-5y}$ is the solution.

Example 4. *Solve, by the method of separation of variables* $\dfrac{\partial^2 u}{\partial x^2} - \dfrac{\partial u}{\partial t} = u$ *subject to* $u(0, t) = 0$, $u(\pi, t) = 0$

Evidently $u = u(x, t)$ where x, t are independent.

Let $u = X(x)\,T(t) = XT.$

$$\frac{\partial^2 u}{\partial x^2} = X''\,T,\ \frac{\partial u}{\partial t} = XT'$$

Hence, the given equation becomes,

$$X''\,T - XT' = XT$$

$$\frac{X''}{X} - \frac{T'}{T} = 1$$

i.e.,
$$\frac{X''}{X} = 1 + \frac{T'}{T} = k \text{ (constant)}$$

$$\frac{X''}{X} = k$$

$$X'' - kX = 0$$

$$(D^2 - k)X = 0 \qquad \text{...(1)}$$

$$m = \pm\sqrt{k}$$

Hence $X = Ae^{\sqrt{k}\,x} + Be^{-k\,x}$ ...(2) if $k = +ve$

$$1 + \frac{T'}{T} = k \therefore \frac{T'}{T} = k - 1$$

Integrating, $\log T = (k-1)\,t + c$

$\therefore$ $T = e^{c + (k-1)t} = De^{(k-1)t}$

Hence, $u = XT = \left(Ae^{\sqrt{k}\,x} + Be^{-\sqrt{k}\,x}\right) De^{(k-1)t}$

$$\therefore \quad u = \left(ae^{\sqrt{k}\,x} + be^{-\sqrt{k}\,x} \right) e^{(k-1)t} \qquad \text{...(3)}$$

Using, $u(0, t) = 0$ we get (put $x = 0$ in (3) $(a + b)\, e^{(k-1)t} = 0$ for all t.

$\therefore \quad a + b = 0$ *i.e.*, $b = -a$ ...(4)

Using $u(\pi, t) = 0$ we get $\left(ae^{\sqrt{k}\,\pi} + be^{-\sqrt{k}\,\pi} \right) e^{(k-1)t} = 0$

Using $b = -a, \quad a\left(e^{\sqrt{k}\,\pi} - e^{-\sqrt{k}\,\pi} \right) e^{(k-1)t} = 0$ for all $t > 0$.

$e^{\sqrt{k}\,\pi} - e^{-\sqrt{k}\,\pi} \neq 0$ if k positive

$\therefore \quad a = 0$ and hence $b = 0$

Equation (3) becomes $u(x, t) \equiv 0$ which is not the spirit of the problem.

So, we will assume $k = -ve$, in which case

$$m = \pm\sqrt{-ve} = \pm\sqrt{-\alpha^2} = \pm i\,\alpha \text{ where } k = -\alpha^2$$

Hence $X = A\cos\alpha x + B\sin\alpha x$

As before $T = D\, e^{(-\alpha^2 - 1)t}$

Hence from (1), we get,

$$u = XT = (a\cos\alpha x + b\sin\alpha x)\, e^{-(\alpha^2+1)t} \qquad \text{...(4)}$$

Now use $u(0, t) = 0$ for all t.

$\therefore \quad a \cdot e^{-(\alpha^2+1)t} = 0$ which implies $a = 0$

$u(\pi, t) = 0$ for all t.

$\therefore \quad b\sin\alpha\pi \cdot e^{-(\alpha^2+1)t} = 0$

$\therefore \quad b \neq 0\ \alpha =$ any integer n.

(if $b = 0$, then $u \equiv 0$)

Therefore, using values of a and α in (4),

$$u(x, t) = b\sin nx\, e^{-(n^2+1)t}$$

where n is any integer and b arbitrary constant.

Exercises 3(a)

Solve the following equations by the method of separation of variables.

1. $(D^2 - 2D + D')\,z = 0$

2. Find the solution of the equation $\dfrac{\partial^2 z}{\partial x^2} = \dfrac{\partial z}{\partial y} + 2z$ in the form $z = X(x) \cdot Y(y)$. Solve the equation subject to the conditions $z = 0$ and $\dfrac{\partial z}{\partial x} = 1 + e^{-3y}$ when $x = 0$ for all values of y.

3. $x^2 q + y^3 p = 0$.

4. $(2D^2 - D')\,z = 0$.

5. Show that the equation $(D^2 - D'^2 + 4D - 6D')\, z = 0$ has five types of solutions, according as $k < -9$, $k = -9$, $-9 < k < -4$, $k = -4$ and $k > -4$. (Refer worked ex. 1 above for k).

6. $\dfrac{\partial z}{\partial x} = 2\dfrac{\partial z}{\partial y} + z$ given $z(x, 0) = 6e^{-3x}$

7. $\dfrac{\partial u}{\partial x} + 4 = \dfrac{\partial u}{\partial t}$ given $u(x, 0) = 4e^{-3x}$

8. $3\dfrac{\partial u}{\partial x} + 2\dfrac{\partial u}{\partial y} = 0$ given $u(x, 0) = 4e^{-x}$.

Classification of Partial Differential Equations of the Second Order

The most general linear partial differential equation of second order can be written as

$$A\frac{\partial^2 u}{\partial x^2} + B\frac{\partial^2 u}{\partial x \partial y} + C\frac{\partial^2 u}{\partial y^2} + D\frac{\partial u}{\partial x} + E\frac{\partial u}{\partial y} + Fu = 0$$

i.e., $$A\, u_{xx} + Bu_{xy} + Cu_{yy} + Du_x + Eu_y + Fu = 0 \qquad \text{... (1)}$$

where A, B, C, D, E, F are in general functions of x and y.

The above equation of second order (linear) (1) is said to

(*i*) elliptic at a point (x, y) in the plane if $B^2 - 4AC < 0$

(*ii*) parabolic if $B^2 - 4AC = 0$

(*iii*) hyperbolic if $B^2 - 4AC > 0$.

Note: The same differential equation may be elliptic in one region, parabolic in another and hyperbolic in some other region. For example, $xu_{xx} + u_{yy} = 0$ is elliptic if $x > 0$, hyperbolic if $x < 0$ and parabolic if $x = 0$.

Examples

Elliptic Type	**Parabolic Type**	**Hyperbolic Type**
1. $\dfrac{\partial^2 u}{\partial x^2} + \dfrac{\partial^2 u}{\partial y^2} = 0$	$\dfrac{\partial^2 u}{\partial x^2} = \dfrac{1}{\alpha^2}\dfrac{\partial u}{\partial t}$	$\dfrac{\partial^2 u}{\partial x^2} = \dfrac{1}{\alpha^2}\dfrac{\partial^2 u}{\partial t^2}$
(Laplace Equation in two dimension)	(one dimensional heat equation)	(one dimensional wave equation)
2. $\dfrac{\partial^2 u}{\partial x^2} + \dfrac{\partial^2 u}{\partial y^2} = f(x, y)$		
(Poisson's equation)		

Example 1. *Classify the following equations:*

(*i*) $\dfrac{\partial^2 u}{\partial x^2} + 2\dfrac{\partial^2 u}{\partial x \partial y} + \dfrac{\partial^2 u}{\partial y^2} = 0$

(*ii*) $x^2 f_{xx} + (1-y^2) f_{yy} = 0$ (B.E. 1974)

(*i*) Here $A = 1, B = 2, C = 1$

$B^2 - 4AC = 4 - 4 = 0$, for all x, y.

Hence, the equation is parabolic at all points.

(*ii*) $A = x^2,\ B = 0,\ C = 1 - y^2$

$$B^2 - 4AC = -4x^2 (1-y^2)$$
$$= 4x^2 (y^2 - 1)$$

For all x except $x = 0$, x^2 is + ve.

If $-1 < y < 1$, $y^2 - 1$ is negative.

$\therefore$ $B^2 - 4AC$ is – ve if $-1 < y < 1, x \neq 0$

$\therefore$ For $-\infty < x < \infty\ (x \neq 0),\ -1 < y < 1$, the equation is elliptic.

For $-\infty < x < \infty,\ x \neq 0, y < -1$ or $y > 1$, the equation is hyperbolic.

For $x = 0$ for all y or for all, $x,\ y = \pm 1$ the equation is parabolic.

Example 2. *Classify the following partial differential equations:*

(*i*) $u_{xx} + 4u_{xy} + (x^2 + 4y^2)\, u_{yy} = \sin(x+y)$ (*MS. Nov. 91*)

(*ii*) $(x+1)\, u_{xx} - 2(x+2)\, u_{xy} + (x+3)\, u_{yy} = 0$ (*MS. Nov. 92*)

(*iii*) $x f_{xx} + y f_{yy} = 0,\ x > 0, y > 0$.

Solution. (*i*) Here, $A = 1, B = 4, C = (x^2 + 4y^2)$

$$B^2 - 4AC = 16 - 4(x^2 + 4y^2)$$
$$= 4[4 - x^2 - 4y^2]$$

The equation is elliptic if $4 - x^2 - 4y^2 < 0$

i.e., $$x^2 + 4y^2 > 4$$

i.e., $$\frac{x^2}{4} + \frac{y^2}{1} > 1$$

$\therefore$ It is elliptic in the region outside the ellipse

$$\frac{x^2}{4} + \frac{y^2}{1} = 1.$$

It is hyperbolic inside, the ellipse $\frac{x^2}{4} + \frac{y^2}{1} = 1$.

It is parabolic on the ellipse $\frac{x^2}{4} + \frac{y^2}{1} = 1$.

(*ii*) Here, $A = x + 1, B = -2(x+2), C = x + 3$

$$B^2 - 4AC = 4(x+2)^2 - 4(x+1)(x+3)$$

$$= 4\,[1] = 4 > 0$$

$\therefore$ The equation is hyperbolic at all points of the region.

(*iii*) $A = x, B = 0, C = y$

$$B^2 - 4AC = -4xy, (x > 0, y > 0 \text{ given})$$
$$= -\text{ve}$$

$\therefore$ It is elliptic for all $x > 0, y > 0$.

Example 3. *Classify the equations given below*

(*i*) $\dfrac{\partial^2 u}{\partial x^2} + 4\dfrac{\partial^2 u}{\partial x \partial y} + 4\dfrac{\partial^2 u}{\partial y^2} - 12\dfrac{\partial u}{\partial x} + \dfrac{\partial u}{\partial y} + 7u = x^2 + y^2$

(*ii*) $(1 + x^2) f_{xx} + (5 + 2x^2) f_{xy} + (4 + x^2) f_{yy} = 2\sin(x + y)$

Solution. (*i*) Here $A = 1, B = 4, C = 4$

$\therefore B^2 - 4AC = 16 - 4\,(1)\,(4) = 0$

Hence, the given equation is parabolic for all x, y.

(*ii*) Here, $A = 1 + x^2, B = 5 + 2x^2, C = 4 + x^2$

$\therefore B^2 - 4AC = (5 + 2x^2)^2 - 4\,(1 + x^2)\,(4 + x^2) = 9 > 0$

Hence, the given equation is hyperbolic for all x, y.

Example 4. *Classify the equations:*

(*i*) The Laplace equation $\dfrac{\partial^2 u}{\partial x^2} + \dfrac{\partial^2 u}{\partial y^2} = 0$

(*ii*) The Poisson equation $\dfrac{\partial^2 u}{\partial x^2} + \dfrac{\partial^2 u}{\partial y^2} = f(x, y)$

(*iii*) One dimensional heat equation $\alpha^2 \dfrac{\partial^2 u}{\partial x^2} = \dfrac{\partial u}{\partial t}$

(*iv*) One dimensional wave equation $\alpha^2 \dfrac{\partial^2 u}{\partial x^2} = \dfrac{\partial^2 u}{\partial t^2}$

Solution. (*i*) Here $A = 1, B = 0, C = 1$

$\therefore\ B^2 - 4AC = -4 < 0$ Hence the equation is elliptic

(*ii*) Here also $B^2 - 4AC = -4 < 0$ Hence it is elliptic

(*iii*) Here $A = \alpha^2,\ B = 0,\ C = 0\ \therefore\ B^2 - 4AC = 0$

Hence it is parabolic

(*iv*) Here $A = \alpha^2,\ B = 0,\ C = -1$

Hence, $B^2 - 4AC = 4\alpha^2 > 0$

Therefore, the equation is hyperbolic.

Exercises 3 (b)

1. Prove $f_{xx} + 2f_{xy} + 4f_{yy} = 0$ is elliptic.
2. Prove $f_{xx} = 2f_{xy}$ and $f_{xy} = f_x$ are hyperbolic
3. Prove $f_{xx} - 2f_{xy} + f_{yy} = 0$ and $u_{xx} = u_t$ are parabolic.
4. Classify : $(1 - x^2) f_{xx} - 2xy\, f_{xy} + (1 - y^2) f_{yy} = 0$

3·2. Transverse Vibrations of a stretched elastic string

Let us consider small transverse vibrations of an elastic string of length l, which is stretched and then fixed at its two ends. Now we will study the transverse vibration of the string when no external forces act on it. Take an end of the string as the origin and the string in the equilibrium position as the x-axis and the line through the origin and perpendicular to the x-axis as the y-axis. We make the following assumptions:

(1) The motion takes place entirely in one plane. This plane is chosen as the *xy plane.*

(2) In this plane, each particle of the string moves in a direction perpendicular to the equilibrium position of the string.

(3) The tension T caused by stretching the string before fixing it at the end points is constant at all times at all points of the deflected string.

(4) The tension T is very large compared with the weight of the string and hence the gravitational force may be neglected.

(5) The effect of friction is negligible.

(6) The string is perfectly flexible. It can transmit only tension but not bending or shearing forces.

(7) The slope of the deflection curve is small at all points and at all times.

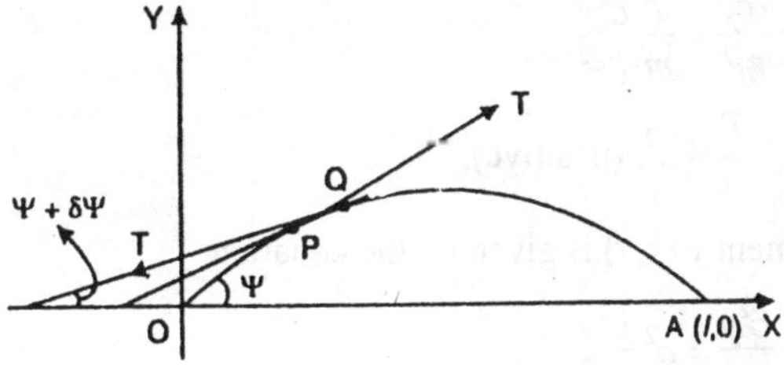

When the string is in motion in the xy plane, the displacement y of any point of the string is a function of x and time t. Let $P(x, y)$ and $Q(x + \delta x, y + \delta y)$ be two neighbouring points on the string. Let ψ and $\psi + \delta\psi$ be the inclinations made by the tangents at P and Q respectively with the x-axis. Let m be the mass per unit length of the string which is homogeneous. Consider the motion of the infinitesimal element PQ of the string. The vertical component of the force to which this element is subjected to is

$$T \sin(\psi + \delta\psi) - T \sin \psi$$

$= T[(\psi + \delta\psi) - \psi]$, since $\sin\psi = \psi$ (approximately).

$= T\,\delta\psi$, (approximately).

The acceleration of the element in the *oy* direction is $\frac{\partial^2 y}{\partial t^2}$. If the length of *PQ is* δs, then the mass of *PQ is* $m \cdot \delta s$.

Hence by the second law of Newton, the equation of motion becomes,

$$\cdot\, m \cdot \delta s \cdot \frac{\partial^2 y}{\partial t^2} = T \cdot \delta\psi$$

i.e.,
$$\frac{\partial^2 y}{\partial t^2} = \frac{T}{m}\frac{\delta\psi}{\partial s}$$

As $Q \to P$, $\delta s \to 0$. Therefore, taking limit as $\delta s \to 0$, the above equation becomes,

$$\frac{\partial^2 y}{\partial t^2} = \frac{T}{m}\frac{d\psi}{ds}, \text{ where}$$

$$\frac{d\psi}{ds} = \text{curvature at } P \text{ of the deflection curve} = \frac{\frac{\partial^2 y}{\partial x^2}}{\left[1 + \left(\frac{\partial y}{\partial x}\right)^2\right]^{3/2}}$$

$$= \frac{\partial^2 y}{\partial x^2}, \text{ approximately since } \left(\frac{\partial y}{\partial x}\right)^2 \text{ is negligible.}$$

Hence $\frac{\partial^2 y}{\partial t^2} = \frac{T}{m}\frac{\partial^2 y}{\partial x^2}$.

Putting $\frac{T}{m} = a^2$, (positive),

the displacement $y(x, t)$ is given by the equation

$$\frac{\partial^2 y}{\partial t^2} = a^2 \frac{\partial^2 y}{\partial x^2}.$$

Note. The partial differential equation is known as one-dimensional *wave equation.*

3·3. Solution of the wave equation $\frac{\partial^2 y}{\partial t^2} = a^2 \frac{\partial^2 y}{\partial x^2}$.

Method 1. (By the method of separation of variables).

$$\frac{\partial^2 y}{\partial t^2} = a^2 \frac{\partial^2 y}{\partial x^2}. \qquad ...(1)$$

Let $y = X(x) \,.\, T(t)$ be a solution of (1), where $X(x)$ is a function of x only and $T(t)$ is a function of t only.

$$\frac{\partial^2 y}{\partial t^2} = XT'' \text{ and } \frac{\partial^2 y}{\partial x^2} = X''T,$$

where $\quad X'' = \dfrac{d^2X}{dx^2}$ and $T'' = \dfrac{d^2T}{dt^2}$.

Hence (1) becomes, $XT'' = a^2 X'' T$

i.e.,
$$\frac{X''}{X} = \frac{T''}{a^2T} \qquad ...(2)$$

The L.H.S. of (2) is a function of x only whereas the R.H.S. is a function of time t only. But x and t are independent variables. Hence (2) is true only if each is equal to a constant.

$\therefore \quad \dfrac{X''}{X} = \dfrac{T''}{a^2T} = k$ (say) where k is any constant.

Hence $X'' - kX = 0$ and $T'' - a^2 kT = 0 \qquad ...(3)$

Solutions of these equations depend upon the nature of the value of k.

Case 1. Let $k = \lambda^2$, a positive value .

Now the equation (3) are $X'' - \lambda^2 X = 0$ and $T'' - a^2\lambda^2 T = 0$.

Solving the ordinary differential equations we get,

$$X = A_1 e^{\lambda x} + B_1 e^{-\lambda x}$$

and
$$T = C_1 e^{\lambda at} + D_1 e^{-\lambda at}$$

Case 2. Let $k = -\lambda^2$, a negative number.

Then the equations (3) are $X'' + \lambda^2 X = 0$ and $T'' + a^2\lambda^2 T = 0$.

Solving, we get,

$$X = A_2 \cos \lambda x + B_2 \sin \lambda x$$

and
$$T = C_2 \cos \lambda\, at + D_2 \sin \lambda\, at.$$

Case 3. Let $k = 0$.

Now the equations (3) are $X'' = 0$ and $T'' = 0$.

Then integrating, $X = A_3 x + B_3$

and
$$T = C_3 t + D_3$$

Thus the various possible solutions of the wave equation are

$$y = (A_1 e^{\lambda x} + B_1 e^{-\lambda x})(C_1 e^{\lambda at} + D_1 e^{-\lambda at}) \qquad ...(I)$$

$$y = (A_2 \cos \lambda x + B_2 \sin \lambda x)\,(C_2 \cos \lambda at + D_2 \sin \lambda at) \qquad ...(II)$$

$$y = (A_3 x + B_3)(C_3 t + D_3) \qquad ...(III)$$

Out of these solutions, we have to select that particular solution which suits the physical nature of the problem and the given boundary conditions. In the case of vibration of string, it is evident that y must be a periodic function of x and t. Hence we select the solution II as the probable solution of the wave equation. The constants are determined by using the boundary conditions in the problem. In doing problems, we shall select the solution II directly.

3·4. D' Alembert's solution of the wave equation

(Method 2)

The one-dimensional wave equation is $\dfrac{\partial^2 y}{\partial t^2} = a^2 \dfrac{\partial^2 y}{\partial x^2}$. ...(1)

Writing $\dfrac{\partial}{\partial t} = D$ and $\dfrac{\partial}{\partial x} = D'$, the equation becomes, $(D^2 - a^2 D'^2)\, y = 0$

The auxiliary equation is $m^2 - a^2 = 0.$ *i.e.,* $m = \pm a$

Hence the general solution of the wave equation is

$$y = f(x + at) + g(x - at) \quad \text{...(2),}$$

where f, g are arbitrary functions.

Suppose we assume that the displacement and the velocity of the string at $t = 0$ are

$$y = \eta(x) \text{ and } \frac{\partial y}{\partial t} = v(x) \quad \text{...(3)}$$

Then substituting $t = 0$ in (2), we get

$$\eta(x) = f(x) + g(x) \quad \text{...(4)}$$

and

$$v(x) = af'(x) - ag'(x) \quad \text{...(5)}$$

Integrating the last equation, we get,

$$f(x) - g(x) = \frac{1}{a}\int_c^x v(\theta)\, d\theta, \text{ where } c \text{ is arbitrary} \quad \text{...(6)}$$

From (4) and (6), $f(x) = \frac{1}{2}\eta(x) + \dfrac{1}{2a}\displaystyle\int_c^x v(\theta)\, d\theta$, and

$$g(x) = \tfrac{1}{2}\eta(x) - \frac{1}{2a}\int_c^x v(\theta)\, d\theta$$

Hence,

$$y = \tfrac{1}{2}[\eta(x + at) + \eta(x - at)] + \frac{1}{2a}\int_{x-at}^{x+at} v(\theta)\, d\theta.$$

This solution is called as D'Alembert's solution of the one-dimensional wave equation. If the string is at rest at $t = 0$, *i.e.*, $v = 0$, then $y = \frac{1}{2}[\eta(x + at) + \eta(x - at)]$. ..(7)

Example 1. *A tightly stretched string with fixed end points $x = 0$ and $x = l$ is initially in the position $y = f(x)$. It is set vibrating by giving to each of its points a velocity $\frac{\partial y}{\partial t} = g(x)$ at $t = 0$. Find $y(x, t)$ in the form of Fourier series.*

The displacement $y(x, t)$ is governed by

$$\frac{\partial^2 y}{\partial t^2} = a^2 \frac{\partial^2 y}{\partial x^2} \quad ...(1)$$

The boundary conditions under which (1) is to be solved are

(*i*) $y(0, t) = 0$ for $t \geq 0$

(*ii*) $y(l, t) = 0$ for $t \geq 0$

(*iii*) $y(x, 0) = f(x)$, for $0 < x < l$

(*iv*) $\left(\frac{\partial y}{\partial t}\right)_{t=0} = g(x)$, for $0 < x < l$.

Solving (1) by the method of separation of variables, we get,

$$y(x, t) = (A_1 e^{\lambda x} + B_1 e^{-\lambda x})(C_1 e^{\lambda at} + D_1 e^{-\lambda at}) \quad ...(I)$$

$$y = (A_2 \cos \lambda x + B_2 \sin \lambda x)(C_2 \cos \lambda at + D_2 \sin \lambda at) \quad ...(II)$$

$$y = (A_3 x + B_3)(C_3 t + D_3). \quad ...(III)$$

Since the solution should be periodic in t, we reject solutions (I) and (III) and select (II) to suit the boundary conditions (*i*), (*ii*), (*iii*) and (*iv*).

$$\therefore \; y(x, t) = (A \cos \lambda x + B \sin \lambda x)(C \cos \lambda at + D \sin \lambda at) \quad ...(2),$$

where A, B, C, D are arbitrary constants.

Using boundary condition (*i*) in (*ii*),

$A(C \cos \lambda at + D \sin \lambda at) = 0$ for all $t \geq 0$.

$\therefore A = 0$.

Applying the boundary condition (*ii*) in (2),

$B \sin \lambda l (C \cos \lambda at + D \sin \lambda at) = 0$, for all $t \geq 0$.

If $B = 0$, the solution becomes $y = 0$ which is not true.

$\sin \lambda l = 0, B \neq 0$.

i.e., $\lambda l = n\pi$, where n is any integer.

$\therefore \quad \lambda = \frac{n\pi}{l}$

$$\therefore\quad y(x, t) = B \sin\frac{n\pi x}{l}\left(C \cos\frac{n\pi at}{l} + D \sin\frac{n\pi at}{l}\right)$$

i.e., $$y(x, t) = \sin\frac{n\pi x}{l}\left(C_n \cos\frac{n\pi at}{l} + D_n \sin\frac{n\pi at}{l}\right) \qquad ...(3),$$

where $BC = C_n$ and $BD = D_n$.

Since the wave equation is linear and homogeneous, the most general solution of it is

$$y(x, t) = \sum_{n=1}^{\infty}\left(C_n \cos\frac{n\pi at}{l} + D_n \sin\frac{n\pi at}{l}\right)\sin\frac{n\pi x}{l} \qquad ...(4).$$

This satisfies boundary condition (*i*) and (*ii*). To find C_n and D_n we make use of the initial conditions (*iii*) and (*iv*).

$$y(x, 0) = \sum_{n=1}^{\infty} C_n \sin\frac{n\pi x}{l} = f(x) \qquad ...(5)$$

and $$\left(\frac{\partial y}{\partial t}\right)_{t=0} = \sum \frac{n\pi a}{l} D_n \sin\frac{n\pi x}{l} = g(x) \qquad ...(6)$$

The left hand sides of (5) and (6) are Fourier sine series of the right hand side functions.

Hence $$C_n = \frac{2}{l}\int_0^l f(x) \sin\frac{n\pi x}{l}\,dx \qquad ...(7)$$

and $$\frac{n\pi a}{l} D_n = \frac{2}{l}\int_0^l g(x)\sin\frac{n\pi x}{l}\,dx \qquad ...(8)$$

Substituting the values of C_n and D_n in (4), we get the solution of the wave equation satisfying the given boundary conditions.

Example 2. *A tightly stretched string with fixed end points $x = 0$ and $x = l$ is initially in a position given by $y(x, 0) = y_0 \sin^3\left(\frac{\pi x}{l}\right)$. If it is released from rest from this position, find the displacement y at any time and at any distance from the end $x = 0$.* **(BR. 1995 Ap.)**

The displacement y of the particle at a distance x from the end $x = 0$ at time t is governed by

$$\frac{\partial^2 y}{\partial t^2} = a^2 \frac{\partial^2 y}{\partial x^2} \qquad ...(1)$$

The boundary, conditions are

$$y(0, t) = 0, \quad \text{for all } t \geq 0 \quad (i)$$

$$y(l, t) = 0, \quad \text{for all } t \geq 0, \quad (ii)$$

$$\left(\frac{\partial y}{\partial t}\right)_{t=0} = 0, \quad \text{for } 0 \leq x \leq l \quad (iii)$$

$$y(x, 0) = y_0 \sin^3\left(\frac{\pi x}{l}\right), \quad \text{for } 0 \leq x \leq l \quad (iv)$$

Now solving (1) and selecting the proper solution to suit the physical nature of the problem and making use of the boundary conditions (*i*) and (*ii*) as in the previous problem, we get

$$y(x, t) = B \sin\frac{n\pi x}{l}\left(C \cos\frac{n\pi at}{l} + D \sin\frac{n\pi at}{l}\right) \qquad ...(2)$$

Again using the boundary condition (*iii*),

$$\left(\frac{\partial y}{\partial t}\right)_{t=0} = 0 = B \sin\frac{n\pi x}{l}\left(D \cdot \frac{n\pi a}{l}\right).$$

If $B = 0$, (2) takes the form $y(x, t) = 0$. Hence B cannot be zero.

$\therefore \; D = 0.$

Hence (2) becomes,

$y(x, t) = B_n \frac{n\pi x}{l} \cos\frac{n\pi at}{l}$, where n is any integer and B_n is any constant.

The most general solution satisfying (1) and the boundary conditions (*i*), (*ii*) and (*iii*) is

$$y(x, t) = \sum_{n=1}^{\infty} B_n \sin\frac{n\pi x}{l} \cos\frac{n\pi at}{l} \qquad ...(3).$$

To find B_n use the boundary condition (*iv*).

$$y(x, 0) = \sum_{n=1}^{\infty} B_n \sin\frac{n\pi x}{l} = y_0 \sin^3\left(\frac{\pi x}{l}\right)$$

$$= \frac{y_0}{4}\left(3 \sin\frac{\pi x}{l} - \sin\frac{3\pi x}{l}\right)$$

This is true only if $B_1 = \frac{3y_0}{4}$, $B_3 = -\frac{y_0}{4}$ and $B_n = 0$, for $n \neq 1, 3$.

Using these values in (3), the solution of the equation is

$$y(x, t) = \frac{3y_0}{4}\sin\frac{\pi x}{l}\cos\frac{\pi at}{l} - \frac{y_0}{4}\sin\frac{3\pi x}{l}\cos\frac{3\pi at}{l}$$

Example 3. *A tightly stretched string with fixed end points $x = 0$ and $x = l$ is initially at rest in its equilibrium position. If it is set vibrating giving each point a velocity $3x(l - x)$, find the displacement.* **(*Madras, 1991 Ap.*)**

The equation to be solved is $\dfrac{\partial^2 y}{\partial t^2} = a^2\dfrac{\partial^2 y}{\partial x^2}$ ...(1)

The ends are fixed at $x = 0$ and $x = l$. The boundary conditions are

$y(0, t) = 0,$ for $t \geq 0$ (*i*);

$y(l, t) = 0,$ for $t \geq 0$ (*ii*);

$y(x, 0) = 0,$ for $0 \leq x \leq l$ (*iii*);

$\left(\dfrac{\partial y}{\partial t}\right)_{t=0} = 3x(l - x),$ for $0 \leq x \leq l$ (*iv*).

Solving (1) and selecting the suitable solution and making use of the boundary conditions (*i*) and (*ii*) as before,

$$y(x, t) = B\sin\frac{n\pi x}{l}\left(C\cos\frac{n\pi at}{l} + D\sin\frac{n\pi at}{l}\right) \qquad ...(2)$$

Again using (*iii*),

$$y(x, 0) = BC\sin\frac{n\pi x}{l} = 0, \text{ for all } 0 \leq x < l.$$

Since $B \neq 0$, $C = 0$.

Therefore, $y(x, t) = B_n\sin\dfrac{n\pi x}{l}\sin\dfrac{n\pi at}{l}$,

where B_n is an arbitrary constant, and n is any integer.

Since (1) is homogeneous and linear, the most general solution of (1) is.

$$y(x, t) = \sum_{n=1}^{\infty} B_n\sin\frac{n\pi x}{l}\sin\frac{n\pi at}{l} \qquad ...(3)$$

Making use of (*iv*),

$$\left(\frac{\partial y}{\partial t}\right)_{t=0} = \sum_{n=1}^{\infty} B_n\frac{n\pi a}{l}\sin\frac{n\pi x}{l} = 3x(l - x) \text{ for } 0 \leq x \leq l.$$

If Fourier sine series of $3x(l - x)$ in $(0, l)$ is

$$3x(l-x) = \sum_{n=1}^{\infty} b_n \sin\frac{n\pi x}{l}, \text{ then}$$

$$B_n \frac{n\pi a}{l} = b_n = \frac{2}{l}\int_0^l 3x(l-x)\sin\frac{n\pi x}{l}\,dx$$

$$= \frac{6}{l}\left[(lx - x^2)\left(-\frac{\cos\frac{n\pi x}{l}}{\frac{n\pi}{l}}\right) - (l-2x)\left(-\frac{\sin\frac{n\pi x}{l}}{\frac{n^2\pi^2}{l^2}}\right) + (-2)\left(\frac{\cos\frac{n\pi x}{l}}{\frac{n^3\pi^3}{l^3}}\right)\right]_0^l$$

$$= \frac{6}{l}\left[\frac{-2l^3}{n^3\pi^3}(\cos n\pi - 1)\right]$$

$$= \frac{12l^2}{n^3\pi^3}[1 - (-1)^n]$$

$= 0$, if n is even, and

$= \dfrac{24l^2}{n^3\pi^3}$, if n is odd.

$\therefore \quad B_n = \dfrac{24l^3}{an^4\pi^4}$, if n is odd

$= 0$, if n is even.

Substituting the value of B_n in (3), we get,

$$y(x,t) = \frac{24l^3}{\pi^4 a}\sum_{n=1,3,5,\cdots}^{\infty} \frac{1}{n^4}\sin\frac{n\pi x}{l}\sin\frac{n\pi at}{l}.$$

$$= \frac{24l^3}{\pi^4 a}\sum_{r=1}^{\infty}\frac{1}{(2r-1)^4}\sin\frac{(2r-1)\pi x}{l}\sin\frac{(2r-1)\pi at}{l}.$$

Example 4. *The points of trisection of a tightly stretched string of length l with fixed ends are pulled aside through a distance d on opposite sides of the position of equilibrium, and the string is released from rest. Obtain an expression for the displacement of the string at any subsequent time and show that the mid point of the string always remains at rest.*

(*Madras, 65 B.E.*)

Let B and C be the points of trisection of the string OA. The initial position of the string is shown by lines $ODEA$, where

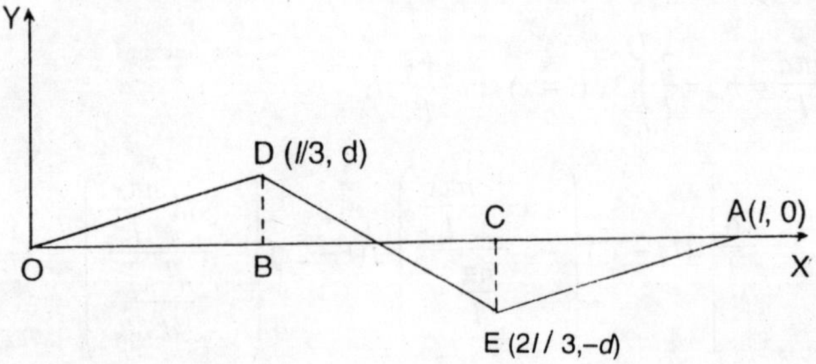

$BD = CE = d.$

The displacement $y\,(x, t)$ is governed by

$$\frac{\partial^2 y}{\partial t^2} = a^2 \frac{\partial^2 y}{\partial x^2} \qquad ...(1)$$

The boundary conditions here are

$$y\,(0, t) = 0 \qquad \text{for } t \geq 0 \qquad ...(i);$$

$$y\,(l, t) = 0 \qquad \text{for } t \geq 0 \qquad ...(ii);$$

and $$\left(\frac{\partial y}{\partial t}\right)_{t=0} = 0, \qquad \text{for } 0 \leq x \leq l \qquad ...(iii).$$

To find the initial position of the string, we require the equation of $ODEA$.

The equation of OD is $\quad y = \frac{d}{l/3}\,x = \frac{3dx}{l}\,.$

The equation of DE is $y - d = -\frac{d}{(l/6)}\,(x - l/3)$

i.e., $$y = \frac{3d}{l}\,(l - 2x).$$

The equation of EA is $\quad y = \frac{3d}{l}\,(x - l).$

The fourth initial condition is

$$y\,(x, 0) = \begin{cases} \dfrac{3dx}{l} & \text{for } 0 \leq x \leq l/3 \\ \dfrac{3d}{l}\,(l - 2x) & \text{for } \dfrac{l}{3} \leq x \leq \dfrac{2l}{3} \\ \dfrac{3d}{l}\,(x - l) & \text{for } \dfrac{2l}{3} \leq x \leq l \end{cases} \qquad ...(iv)$$

Solving (1) and selecting the suitable solution and using the boundary conditions (*i*), (*ii*) and (*iii*) as in example 2, we get

$$y(x, t) = \sum_{n=1}^{\infty} B_n \sin \frac{n\pi x}{l} \cos \frac{n\pi at}{l} \qquad ...(2)$$

Using the initial condition (*iv*), we get,

$$\sum_{n=1}^{\infty} B_n \sin \frac{n\pi x}{l} = y(x, 0) = \frac{3dx}{l} \text{ for } 0 \le x \le l/3$$

$$= \frac{3d}{l}(l - 2x), \text{ for } \frac{l}{3} \le x \le \frac{2l}{3}$$

$$= \frac{3d}{l}(x - l), \text{ for } \frac{2l}{3} \le x \le l.$$

Finding Fourier sine series of $y(x, 0)$ in $(0, l)$ we get in the usual

way $y(x, 0) = \sum_{n=1}^{\infty} b_n \sin \frac{n\pi x}{l}$.

$$\therefore \qquad B_n = b_n = \frac{2}{l} \int_0^1 y(x, 0) \sin \frac{n\pi x}{l} \, dx$$

$$\therefore \qquad B_n = \frac{2}{l}\left[\int_0^{l/3} \frac{3dx}{l} \sin \frac{n\pi x}{l} dx + \int_{l/3}^{2l/3} \frac{3d}{l}(l - 2x) \sin \frac{n\pi x}{l} dx + \int_{2l/3}^{l} \frac{3d}{l}(x - l) \sin \frac{n\pi x}{l} dx\right]$$

$$= \frac{6d}{l^2}\left[x\left(-\frac{\cos \frac{n\pi x}{l}}{\frac{n\pi}{l}}\right) - (1)\left(-\frac{\sin \frac{n\pi x}{l}}{\frac{n^2\pi^2}{l^2}}\right)\right]_0^{l/3}$$

$$+ \frac{6d}{l^2}\left[(l - 2x)\left(-\frac{\cos \frac{n\pi x}{l}}{\frac{n\pi}{l}}\right) - (-2)\left(-\frac{\sin \frac{n\pi x}{l}}{\frac{n^2\pi^2}{l^2}}\right)\right]_{l/3}^{2l/3}$$

$$+\frac{6d}{l^2}\left[(x-l)\left(-\frac{\cos\frac{n\pi x}{l}}{\frac{n\pi}{l}}\right)-(1)\left(-\frac{\sin\frac{n\pi x}{l}}{\frac{n^2\pi^2}{l^2}}\right)\right]_{2l/3}^{l}$$

$$=\frac{18d}{n^2\pi^2}\left[\sin\frac{n\pi}{3}-\sin\frac{2n\pi}{3}\right]$$

$$=\frac{18d}{n^2\pi^2}\left[\sin\frac{n\pi}{3}-\sin\left(n\pi-\frac{n\pi}{3}\right)\right]$$

$$=\frac{18d}{n^2\pi^2}\left[\sin\frac{n\pi}{3}+\cos n\pi\cdot\sin\frac{n\pi}{3}\right]$$

$$=\frac{18d}{n^2\pi^2}\sin\frac{n\pi}{3}\left[1+(-1)^n\right]$$

$= 0$ if n is odd.

$$=\frac{36d}{n^2\pi^2}\sin\frac{n\pi}{3} \text{ if } n \text{ is even.}$$

Hence,

$$y(x,t)=\frac{36d}{\pi^2}\sum_{n=2,4,6,\cdots}^{\infty}\frac{1}{n^2}\sin\frac{n\pi}{3}\sin\frac{n\pi x}{l}\cos\frac{n\pi at}{l}$$

i.e., $$y(x,t)=\frac{9d}{\pi^2}\sum_{n=1}^{\infty}\frac{1}{n^2}\sin\frac{2n\pi}{3}\sin\frac{2n\pi x}{l}\cdot\cos\frac{2n\pi at}{l}.$$

By putting $x = l/2$, we get the displacement of the midpoint.

$\therefore\ y\left(\frac{l}{2},t\right)=0$, since $\sin\frac{2n\pi x}{l}$ becomes $\sin n\pi = 0$ when $x = l/2$.

Example 5. *A string is stretched between two fixed points at a distance 2l apart and the points of the string are given initial velocities v where*

$$v=\frac{cx}{l}, \text{ in } 0<x<l$$

$$=\frac{c}{l}(2l-x), \text{ in } l<x<2l,$$

x being the distance from an end point. Find the displacement of the string at any subsequent time. **(Madras, 1988 N.)**

Take the origin at the end referred to. Let y be the displacement of any point at a distance x from the origin.

Then $$\frac{\partial^2 y}{\partial t^2} = a^2 \frac{\partial^2 y}{\partial x^2} \qquad ...(1)$$

The boundary conditions are

$$y(0, t) = 0, \text{ for } t \geq 0 \qquad ...(i)$$

$$y(2l, t) = 0, \text{ for } t \geq 0 \qquad ...(ii)$$

$$y(x, 0) = 0, \text{ for } 0 \leq x \leq 2l \qquad ...(iii)$$

$$\left(\frac{\partial y}{\partial t}\right)_{t=0} = \frac{cx}{l}, \text{ in } 0 < x < l$$

$$= \frac{c}{l}(2l - x), \text{ in } l < x < 2l \qquad ...(iv).$$

As in the previous examples, using boundary conditions (*i*) and (*ii*), we get

$$y(x, t) = \sin\frac{n\pi x}{2l}\left[C_n \cos\frac{n\pi at}{2l} + D_n \sin\frac{n\pi at}{2l}\right]$$

Using (*iii*), $C_n = 0$.

$$\therefore \quad y(x, t) = D_n \sin\frac{n\pi x}{2l} \sin\frac{n\pi at}{2l}$$

The most general solution of the equation (1) is

$$y(x, t) = \sum_{n=1}^{\infty} D_n \sin\frac{n\pi x}{2l} \sin\frac{n\pi at}{2l} \qquad ...(2)$$

$$\frac{\partial y}{\partial t}(x, t) = \sum_{n=1}^{\infty} D_n \left(\frac{n\pi a}{2l}\right) \sin\frac{n\pi x}{2l} \cos\frac{n\pi at}{2l}.$$

Using (*iv*),

$$\sum_{n=1}^{\infty} D_n \left(\frac{n\pi a}{2l}\right) \sin\frac{n\pi x}{2l} = v = \frac{cx}{l}, \text{ in } 0 < x < 1$$

$$= \frac{c}{l}(2l - x), \text{ in } l < x < 2l.$$

Expanding v in Fourier sine series, we get

$$D_n \cdot \frac{n\pi a}{2l} = \frac{2}{2l}\left[\frac{c}{l}\int_0^l x \sin\frac{n\pi x}{2l}\, dx + \frac{c}{l}\int_l^{2l} (2l - x) \sin\frac{n\pi x}{2l}\, dx\right]$$

$$\therefore\ D_n = \frac{2c}{n\pi al}\left[\left\{x\left(-\frac{\cos\frac{n\pi x}{2l}}{\frac{n\pi}{2l}}\right) - (1)\left(-\frac{\sin\frac{n\pi x}{2l}}{\frac{n^2\pi^2}{4l^2}}\right)\right\}_0^l\right.$$

$$\left. + \left\{(2l - x)\left(-\frac{\cos\frac{n\pi x}{2l}}{\frac{n\pi}{2l}}\right) - (-1)\left(-\frac{\sin\frac{n\pi x}{2l}}{\frac{n^2\pi^2}{4l^2}}\right)\right\}_l^{2l}\right]$$

$$= \frac{2c}{n\pi al}\left[\frac{-2l^2}{n\pi}\cos\frac{n\pi}{2} + \frac{4l^2}{n^2\pi^2}\sin\frac{n\pi}{2} + \frac{2l^2}{n\pi}\cos\frac{n\pi}{2}\right.$$

$$\left. + \frac{4l^2}{n^2\pi^2}\sin\frac{n\pi}{2}\right]$$

$$= \frac{2c}{n\pi al}\cdot\frac{8l^2}{n^2\pi^2}\sin\frac{n\pi}{2}$$

$$= \frac{16lc}{n^3\pi^3 a}\sin\frac{n\pi}{2}$$

Substituting this value of D_n in (2),

$$y(x, t) = \frac{16cl}{a\pi^3}\sum_{n=1}^{\infty}\frac{1}{n^3}\sin\frac{n\pi}{2}\sin\frac{n\pi x}{2l}\sin\frac{n\pi at}{2l}$$

Example 6. *If a string of length l is initially at rest in equilibrium position and each point of it is given the velocity*

$\left(\frac{\partial y}{\partial t}\right)_{t=0} = v_0 \sin^3\frac{\pi x}{l}$, $0 < x < l$, *determine the transverse displacement* $y(x, t)$.

(*MS. 1990 Ap.*)

The displacement $y(x, t)$ is governed by

$$\frac{\partial^2 y}{\partial t^2} = a^2\frac{\partial^2 y}{\partial x^2} \quad ...(1)$$

Solving this, by the method of separation of variables,

we get, $y(x, t) = (A_1 e^{\lambda x} + B_1 e^{-\lambda x})(C_1 e^{\lambda at} + D_1 e^{-\lambda at})$ ...I

$= (A_2 \cos\lambda x + B_2 \sin\lambda x)(C_2 \cos\lambda\, at + D_2 \sin\lambda at)$ ...II

$= (A_3 x + B_3)(C_3 t + D)$ ...III

The boundary conditions are

$$y(0, t) = 0, \text{ for } t \geq 0 \quad \text{...(i)}$$

$$y(l, t) = 0, \text{ for } t \geq 0 \quad \text{...(ii)}$$

$$y(x, 0) = 0, \text{ for } 0 \leq x \leq l \quad \text{...(iii)}$$

$$\left(\frac{\partial y}{\partial t}\right)_{t=0} = v_0 \sin^3 \frac{\pi x}{l} \text{ for } 0 \leq x \leq l \quad \text{...(iv)}$$

Selecting the solution II, and using boundary conditions (*i*) and (*ii*)

we get $y(x, t) = B \sin \dfrac{n\pi x}{l}\left(C \cos \dfrac{n\pi at}{l} + D \sin \dfrac{n\pi at}{l}\right)$

using (*iii*), $C = 0$

Therefore $y(x, t) = B_n \sin \dfrac{n\pi x}{l} \sin \dfrac{n\pi at}{l}$, n any integer

The most general solution is

$$y(x, t) = \sum_{n=1}^{\infty} B_n \sin \frac{n\pi x}{l} \sin \frac{n\pi at}{l} \quad \text{...(3)}$$

$$\frac{\partial y}{\partial t} = \sum_{n=1}^{\infty} B_n \cdot \frac{n\pi a}{l} \sin \frac{n\pi x}{l} \cos \frac{n\pi at}{l} \quad \text{...(IV)}$$

$$\left(\frac{\partial y}{\partial t}\right)_{t=0} = \sum_{1}^{\infty} B_n \frac{n\pi a}{l} \sin \frac{n\pi x}{l} = v_0 \sin^3 \frac{\pi x}{l}$$

$$= v_0 \left[\frac{3}{4} \sin \frac{\pi x}{l} - \frac{1}{4} \sin \frac{3\pi x}{l}\right]$$

Comparing both sides

$$B_1 \cdot \frac{\pi a}{l} = \frac{3v_0}{4} \text{ and } B_3 \cdot \frac{3\pi a}{l} = -\frac{v_0}{4} \text{ and } B_n = 0 \text{ for } n \neq 1, n \neq 3$$

$$\therefore \; B_1 = \frac{3lv_0}{4\pi a}$$

$$B_3 = \frac{-v_0 l}{12\pi a}$$

$$B_n = 0, n \neq 1, n \neq 3.$$

Using in (3),

$$y(x, t) = \frac{3lv_0}{4\pi a} \sin \frac{\pi x}{l} \sin \frac{\pi at}{l} - \frac{v_0 l}{12\pi a} \sin \frac{3\pi x}{l} \sin \frac{3\pi at}{l}$$

Example 7. *A tightly stretched string with end points $x = 0$ and $x = l$ is initially in a position given by $y(x, 0) = y_0 \sin \frac{\pi x}{l}$. If it is released from rest from this position, find the displacement $y(x, t)$ at any point of the string.*

The boundary conditions are

$$y(0, t) = 0 \text{ for all } t \geq 0 \quad \text{...(i)}$$

$$y(l, t) = 0 \text{ for } t \geq 0 \quad \text{...(ii)}$$

$$\left(\frac{\partial y}{\partial t}\right)_{t=0} = \text{ for } 0 \leq x \leq l \quad \text{...(iii)}$$

$$y(x, 0) = y_0 \sin \frac{\pi x}{l} \text{ for } 0 \leq x \leq l \quad \text{...(iv)}$$

Looking into example 2, and using the first three conditions, we have

$$y(x, t) = \sum_{n=1}^{\infty} B_n \sin \frac{n\pi x}{l} \cos \frac{n\pi at}{l} \quad \text{...(IV)}$$

Using boundary conditions (iv), in IV,

$$y(x, 0) = \Sigma B_n \sin \frac{n\pi x}{l} = y_0 \sin \frac{\pi x}{l}$$

Comparing both sides, $B_1 = y_0$ and $B_n = 0 \;\; n \neq 1$

Using B_1 in IV,

$$y(x, t) = y_0 \sin \frac{\pi x}{l} \cos \frac{\pi at}{l}.$$

Example 8. *A string is stretched and fastened to two points l apart. Motion is started by displacing the string into the form $y = k(lx - x^2)$ from which it is released at time $t = 0$. Find the displacement of any point of the string at a distance x from one end at any time t.* **(M.S. 1990 Nov.)**

The boundary conditions are

$$y(0, t) = 0, \;\; t \geq 0 \quad \text{...(i)}$$

$$y(l, t) = 0, \;\; t \geq 0 \quad \text{..(ii)}$$

$$\left(\frac{\partial y}{\partial t}\right)_{t=0} = 0, \;\; 0 \leq x \leq l \quad \text{...(iii)}$$

$$y(x, 0) = k(lx - x^2), \; 0 \leq x \leq l \quad \text{...(iv)}$$

Using boundary conditions (*i*), (*ii*) and (*iii*) as in example 2, we get

$$y(x, t) = \sum_{n=1}^{\infty} B_n \sin \frac{n\pi x}{l} \cdot \cos \frac{n\pi at}{l} \quad \text{...(IV)}$$

using boundary condition (*iv*),

$$y(x, 0) = \sum_{n=1}^{\infty} B_n \sin\frac{n\pi x}{l} = k\,(lx - x^2)$$

This shows that this is the half range Fourier sine series of $k(lx - x^2)$. Using the formula for Fourier coefficients,

$$B_n = b_n = \frac{2}{l}\int_0^l k(lx - x^2)\sin\frac{n\pi x}{l}\,dx$$

$$= \frac{2k}{l}\left[(lx - x^2)\left(-\frac{\cos\frac{n\pi x}{l}}{\frac{n\pi}{l}}\right) - (l - 2x)\left(-\frac{\sin\frac{n\pi x}{l}}{\frac{n^2\pi^2}{l^2}}\right) + (-2)\left(\frac{\cos\frac{n\pi x}{l}}{\frac{n^3\pi^3}{l^3}}\right)\right]_0^l$$

$$= \frac{2k}{l}\left[\frac{-2l^3}{n^3\pi^3}\{(-1)^n - 1\}\right]$$

$$= \frac{4kl^2}{n^3\pi^3}\left[1 - (-1)^n\right]$$

$= 0$ if n is even

$= \dfrac{8kl^2}{n^3\pi^3}$ if n is odd

Substituting in IV,

$$y(x, t) = \frac{8kl^2}{\pi^3}\sum_{n=1}^{\infty}\frac{1}{(2n-1)^3}\sin\frac{(2n-1)\pi x}{l}\cos\frac{(2n-1)\pi a t}{l}$$

Example 9. *A taut string of length 2l is fastened at both ends. The mid point of the string is taken to a height b and then released from the rest in that position. Find the displacement of the string.* **(*MS. 1992 Ap.*)**

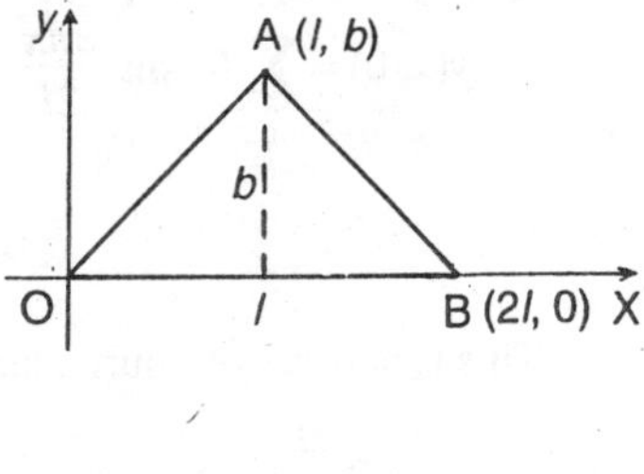

Taking an end as origin, the boundary conditions are

$$y(0, t) = 0,\ t \geq 0 \qquad \text{...(i)}$$

$$y(2l, t) = 0,\ t \geq 0 \qquad \text{...(ii)}$$

$$\left(\frac{\partial y}{\partial t}\right)_{t=0} = 0,\ 0 \le x \le 2l \qquad \text{...(iii)}$$

$$y(x, 0) = \frac{b}{l}x,\ 0 \le x \le l$$

$$= -\frac{b}{l}(x - 2l),\ l \le x \le 2l$$

$$\left[\text{since, equation of } OA \text{ is } y = \frac{b}{l}x \text{ and equation of } AB \text{ is } \frac{y-0}{x-2l} = \frac{b-0}{l-2l}\right]$$

Starting with the solution

$$y(x, t) = (A \cos \lambda x + B \cos \lambda x)(C \cos \lambda at + D \sin \lambda at) \qquad \text{...I}$$

using the first boundary condition,

$$y(o, t) = A\,(C \cos \lambda at + D \sin \lambda ab) = 0$$

$$\therefore \quad A = 0,$$

using $y(2l, t) = 0$ we get

$$B \sin 2l\lambda\,(C \cos \lambda at + D \sin \lambda at) = 0$$

$$B \ne 0;\ 2l\lambda = n\pi\ ;\ \lambda = \frac{n\pi}{2l}$$

Using $\left(\frac{\partial y}{\partial t}\right)_{t=0}$; $D = 0$.

$$\therefore \quad y(x, t) = \sum_{n=1}^{\infty} B_n \sin \frac{n\pi x}{2l} \cos \frac{n\pi at}{2l} \qquad \text{...IV}$$

Using boundary condition (*iv*) in IV,

$$y(x, 0) = \sum_{1}^{\infty} B_n \sin \frac{n\pi x}{2l} = \frac{b}{l}x,\ 0 \le x \le l$$

$$= -\frac{b}{l}(x - 2l),\ l \le x \le 2l$$

This is half-range Fourier sine series

$$\therefore \quad B_n = \frac{2}{2l}\int_0^{2l} f(x) \sin \frac{n\pi x}{2l}\,dx$$

$$= \frac{1}{l}\left[\int_0^{l} \frac{b}{l}x \sin \frac{n\pi x}{2l}\,dx - \frac{b}{l}\int_l^{2l} (x - 2l) \sin \frac{n\pi x}{2l}\,dx\right]$$

$$= \frac{b}{l^2}\left[\left\{(x)\left(-\frac{\cos\frac{n\pi x}{2l}}{\frac{n\pi}{2l}}\right)-\left(-\frac{\sin\frac{n\pi x}{2l}}{\frac{n^2\pi^2}{4l^2}}\right)\right\}_0^l\right.$$

$$\left.-\left\{(x-2l)\left(-\frac{\cos\frac{n\pi x}{2l}}{\frac{n\pi}{2l}}\right)-\left(-\frac{\sin\frac{n\pi x}{2l}}{\frac{n^2\pi^2}{4l^2}}\right)\right\}_l^{2l}\right]$$

$$= \frac{b}{l^2}\left[-\frac{2l^2}{n\pi}\cos\frac{n\pi}{2}+\frac{4l^2}{n^2\pi^2}\left(\sin\frac{n\pi}{2}\right)+\frac{2l^2}{n\pi}\cos\frac{n\pi}{2}+\frac{4l^2}{n^2\pi^2}\sin\frac{n\pi}{2}\right]$$

$$= \frac{b}{l^2}\left[\frac{8l^2}{n^2\pi^2}\sin\frac{n\pi}{2}\right]$$

$$= \frac{8b}{n^2\pi^2}\sin\frac{n\pi}{2}$$

$$= 0 \text{ for } n \text{ even}$$

$$= \frac{8b}{n^2\pi^2}\sin\frac{n\pi}{2} \text{ for odd } n.$$

Substituting in IV,

$$y(x,t) = \frac{8b}{\pi^2}\sum_{n=1}^{\infty}\frac{1}{(2n-1)^2}\sin(2n-1)\frac{\pi}{2}\cdot\sin\frac{(2n-1)\pi x}{2l}\cos\frac{(2n-1)\pi at}{2l}$$

$$= \frac{8b}{\pi^2}\sum_{n=1}^{\infty}\frac{(-1)^{n-1}}{(2n-1)^2}\sin\frac{(2n-1)\pi x}{2l}\cos\frac{(2n-1)\pi at}{2l}$$

Example 10. *Show that the solution of the equation of a vibrating string of length l, satisfying the conditions y(0, t) = y(l, t) = 0 and*

$$y = f(x),\ \frac{\partial y}{\partial t} = g(0) \text{ at } t = 0 \text{ is}$$

$$y = \sum_{1}^{\infty} a_n \sin\frac{n\pi x}{l}\cos\frac{n\pi at}{l}$$

where $$a_n = \frac{2}{l}\int_0^l f(x)\sin\frac{n\pi x}{l}\,dx$$

The four boundary and initial conditions are given in the problem itself

As in problem 2, using the first two conditions,

$$y(x, t) = B \sin \frac{n\pi x}{l}\left(C \cos \frac{n\pi at}{l} + D \sin \frac{n\pi at}{l}\right) \quad ...IV$$

Again using the boundary condition $\frac{\partial y}{\partial t}(x, 0) = 0$

We get,

$$\frac{\partial y}{\partial t}(x, 0) = B \sin \frac{n\pi x}{l} \cdot \left(D \frac{n\pi a}{l}\right) = 0$$

Since, $B \neq 0 \,; D = 0$

Therefore, the most general solution is

$$y(x, t) = \underset{n=1}{\overset{\infty}{Sum}}\, B_n \sin \frac{n\pi x}{l} \cdot \cos \frac{n\pi at}{l} \quad ...(3)$$

Using $y(x, 0) = f(x)$, we get

$$y(x, 0) = \sum_1^{\infty} B_n \sin \frac{n\pi x}{l} = f(x)$$

Hence,

$$B_n = \frac{2}{l}\int_0^l f(x) \sin \frac{n\pi x}{l}\, dx \quad ...(4)$$

Hence, the solution is

$$y(x, t) = \sum_1^{\infty} B_n \sin \frac{n\pi x}{l} \cdot \cos \frac{n\pi at}{l} \text{ where}$$

B_n is given by (4)

Example 11. *Solve the problem of the vibrating string for the following boundary conditions.*

(*i*) $y(0, t) = 0$, (*ii*) $y(l, t) = 0$

(*iii*) $\frac{\partial y}{\partial t}(x, 0) = x(x - l), 0 < x < l$

(*iv*) $y(x, 0) = x$ in $0 < x < l/2$

$= l - x$ in $\frac{l}{2} < x < l.$ **(*MI. 1978*)**

This is same as the problem (Example 1) where $f(x)$ and $g(x)$ are clearly specified.

$g(x) = x(x - l)$

$f(x) = x$ in $0 < x < l/2$

$$= l - x \text{ in } l/2 < x < l$$

Hence $C_n = \dfrac{2}{l}\displaystyle\int_0^l f(x)\sin\frac{n\pi x}{l}\,dx$

$$= \frac{2}{l}\left[\int_0^{l/2} x\sin\frac{n\pi x}{l}\,d_n + \int_{l/2}^{l}(l-x)\sin\frac{n\pi x}{l}\,dx\right]$$

$$= \frac{2}{l}\left[\left\{(x)\left(-\frac{\cos\frac{n\pi x}{l}}{\frac{n\pi}{l}}\right) - \left(-\frac{\sin\frac{n\pi x}{l}}{\frac{n^2\pi^2}{l^2}}\right)\right\}_0^{l/2}\right.$$

$$\left. + \left\{(l-x)\left(-\frac{\cos\frac{n\pi x}{l}}{\frac{n\pi}{l}}\right) - (-1)\left(-\frac{\sin\frac{n\pi x}{l}}{\frac{n^2\pi^2}{l^2}}\right)\right\}_{l/2}^{l}\right]$$

$$= \frac{2}{l}\left[-\frac{l^2}{2n\pi}\cos\frac{n\pi}{2} + \frac{l^2}{n^2\pi^2}\sin\frac{n\pi}{2} + \frac{l^2}{2n\pi}\cos\frac{n\pi}{2} + \frac{l^2}{l^2\pi^2}\sin\frac{n\pi}{2}\right]$$

$$= \frac{2}{l}\left[\frac{2l^2}{n^2\pi^2}\sin\frac{n\pi}{2}\right]$$

$$= 0 \text{ if } n \text{ is even}$$

$$= \frac{4l}{n^2\pi^2}\sin\frac{n\pi}{2} \text{ for } n \text{ odd}$$

Again, $D_n = \dfrac{2}{n\pi a}\displaystyle\int_0^l x(x-l)\sin\frac{n\pi x}{l}\,dx$

$$= \frac{2}{n\pi a}\left[(x^2 - lx)\left(-\frac{\cos\frac{n\pi x}{l}}{\frac{n\pi}{l}}\right) - (2x-l)\left(-\frac{\sin\frac{n\pi x}{l}}{\frac{n^2\pi^2}{l^2}}\right)\right.$$

$$\left. + (2)\left(\frac{\cos\frac{n\pi x}{l}}{\frac{n^3\pi^3}{l^3}}\right)\right]_0^l$$

$$= \frac{2}{n\pi a}\left[\frac{2l^3}{n^3\pi^3}\left((-1)^n - 1\right)\right]$$

$$= 0 \text{ if } n \text{ is even}$$

$$= \frac{-8l^3}{n^4\pi^4 a} \text{ if } n \text{ is odd}$$

Hence, using the result of the Example 1, Page 180,

$$y(x, t) = \sum_{n=1,3,5\ldots}^{\infty} \frac{4l}{n^2\pi^2} \sin\frac{n\pi}{2} \sin\frac{n\pi x}{l} \cdot \cos\frac{n\pi at}{l}$$

$$- \sum_{n=1,3,5\ldots}^{\infty} - \frac{8l^3}{n^4\pi^4 .a} \sin\frac{n\pi x}{l} \cdot \sin\frac{n\pi at}{l}$$

$$= \sum_{n=1}^{\infty} \sin\frac{(2n-1)\pi x}{l}\left[\frac{(-1)^{n-1}.4l}{(2n-1)^2\pi^2}\cos\frac{(2n-1)\pi at}{l} - \frac{8l^3}{(2n-1)^4\pi^4 a} \cdot \sin\frac{(2n-1)\pi at}{l}\right]$$

Exercises 3(b)

1. (*a*) A tightly stretched string with fixed end points $x = 0$ and $x = l$ is initially in a position given by $y(x, 0) = y_0 \sin\frac{2\pi x}{l}$. If it is released from rest from this position, find the displacement y at any distance x from one end at any time t.

(*b*) In the above problem, if $y(x, 0) = k\left(\text{Sin}\frac{\pi x}{l} - \text{Sin}\frac{2\pi x}{l}\right)$ find $y(x, t)$

2. If a string of length l is initially at rest in equilibrium position and each of its points is given the velocity

$\left(\frac{\partial y}{\partial t}\right)_{t=0} = v_0 \sin\frac{\pi x}{l}$, $0 < x < l$, determine the displacement $y(x, t)$.

(***Madras, 63 B.E.***)

3. A string is stretched tightly between $x = 0$ and $x = l$ and both ends are given displacement $y = a \sin pt$ perpendicular to the string. If the string satisfies the differential equation

$\frac{\partial^2 y}{\partial t^2} = c^2 \frac{\partial^2 y}{\partial x^2}$, show that the oscillations of the string are given by

$$y = a \sec\frac{pl}{2c} \cos\left(\frac{px}{c} - \frac{pl}{2c}\right) \sin pt$$

4. A string is stretched and fastened to two points l apart. Motion is started by displacing the string into the form $y = 3(lx - x^2)$ from which it is released at time $t = 0$. Find the displacement of any point on the string at a distance of x from one end at any time t. **(*Madras, 67 B.E. 75, 78*)**

5. A tightly stretched string with fixed end points $x = 0$ and $x = l$ is initially at rest in equilibrium position. If it is set vibrating giving each point a velocity $\lambda x(l - x)$, show that

$$y(x, t) = \frac{8\lambda l^3}{a\pi^4} \sum_{n=1}^{\infty} \frac{1}{(2n-1)^4} \sin\frac{(2n-1)\pi x}{l} \sin\frac{(2n-1)\pi at}{l}.$$

(*MS. 1989 A.*)

6. A uniform elastic string of length 60 cms. is subjected to a constant tension of 2 kg. If the ends are fixed and the initial displacement $y(x, 0) = 60x - x^2$ for $0 < x < 60$, while the initial velocity is zero, find the displacement function $y(x, t)$. **(*Madras, 64 ; 66 B.E.*)**

7. A taut string of length $2l$ is fastened at both ends. The mid point of the string is taken to a height b and then released from rest in that position. Show that the displacement is

$$y(x, t) = \frac{8b}{\pi^2} \sum_{n=1}^{\infty} \frac{(-1)^{n-1}}{(2n-1)^2} \sin\frac{(2n-1)\pi x}{2l} \cos\frac{(2-1)\pi at}{2l}.$$

8. If a string of length l is initially at rest in its equilibrium position and each of its points is given a velocity v such that

$$v = cx, \text{ for } 0 < x \le l/2$$

$= c(l - x)$, for $\frac{l}{2} < x \le l$, show that the displacement $y(x, t)$ at any time t is given by

$$y(x, t) = \frac{4l^2 c}{\pi^3 a}\left[\sin\frac{\pi x}{l}\sin\frac{\pi at}{l} - \frac{1}{3^3}\sin\frac{3\pi x}{l}\sin\frac{3\pi at}{l} + \ldots\right]$$ **(*S.V.U. 67 B.E.*)**

9. Find the displacement if a string of length a is vibrating between fixed end points with initial velocity zero and initial displacement given by

$$y(x, 0) = \frac{2px}{a} \text{ for } 0 < x < a/2$$

$$= 2p - \frac{2px}{a} \text{ for } a/2 < x < a.$$ **(*Madras, 64 B.E.*)**

10. An elastic string is stretched between two points at a distance l apart. One end is taken as the origin and at a distance $\frac{2l}{3}$ from this end, the string is displaced a distance d transversely and is released from rest when it is in this position. Find the equation of the subsequent motion.

11. A uniform string of line density ρ is stretched to tension ρa^2 and executes a small transverse vibration in a plane through the undisturbed line of the string. The

ends $x = 0$ and $x = l$ of the string are fixed. The string is at rest, with the point $x = b$ drawn aside through a small distance d and released at time $t = 0$. Show that at any subsequent time t the transverse displacement y is given by

$$y(x, t) = \frac{2dl^2}{\pi^2 b(l-b)} \sum_{n=1}^{\infty} \frac{1}{n^2} \sin\frac{n\pi b}{l} \sin\frac{n\pi x}{l} \cos\frac{n\pi at}{l}$$

12. A uniform string of density ρ is stretched to tension ρc^2 and executes a small transverse vibration in a plane through the undisturbed line of the string. The ends $x = 0, l$ of the string are fixed. The string is released from rest in the position $y = \frac{4\epsilon}{l^2} x(l-x)$. Show that the motion is given by

$$y = \frac{32\epsilon}{3} \sum_{n=0}^{\infty} \frac{1}{(2n+1)^3} \sin\frac{(2+1)\pi x}{l} \cos\frac{(2n+1)\pi ct}{l}$$ **(*Madras, 72 B.E.*)**

Also find the total energy of the string in terms of Fourier coefficients.

(*Madras, 65 B.E.*)

13. A taut string of length l, fastened at both ends, is disturbed from its position of equilibrium by imparting to each of its points an initial velocity of magnitude $f(x)$. Show that the solution of the problem is

$$y = \frac{2}{\pi a} \sum_{=1}^{\infty} \left[\frac{1}{n} \sin\frac{n\pi x}{l} \sin\frac{n\pi at}{l} \int_0^l f(\theta) \sin\frac{n\pi\theta}{l}\, d\theta \right]$$

14. If the string is released from rest in the position $y = f(x)$, show that the total energy of the string is

$$\frac{\pi^2 T}{4l} \sum_{n=1}^{\infty} n^2 b_n^2 \qquad \text{where}$$

$$b_n = \frac{2}{l} \int_0^l f(x) \sin\frac{n\pi x}{l}\, dx.$$

15. The differential equation of a vibrating string that is viscously damped is

$$\frac{\partial^2 y}{\partial t^2} = a^2 \frac{\partial^2 y}{\partial x^2} - 2b \frac{\partial y}{\partial t}.$$

If the string is of length l and is fastened at both ends and has initial displacement $f(x)$ when the initial velocity is zero, show that,

$$v = \sum_{n=0}^{\infty} a_n e^{-bt} \sin\frac{n\pi x}{l} \left(\cos\alpha_n t + \frac{b}{\alpha_n} \sin\alpha_n t \right)$$

where $\alpha_n{}^2 = \frac{n^2\pi^2 a^2}{l^2} - b^2$ and $\alpha_n = \frac{2}{l} \int_0^l f(x) \sin\frac{n\pi x}{l}\, dx.$

Discuss the case when $l = \pi$. **(*Madras, 63 B.E.*)**

16. Show that the total energy of a string which is fixed at the points $x = 0$, $x = l$ and is executing small transverse vibrations is

$$W = \frac{1}{2} T \int_0^l \left\{ \left(\frac{\partial y}{\partial x} \right)^2 + \frac{1}{a^2} \left(\frac{\partial y}{\partial t} \right)^2 \right\} dx.$$

Show that if $y = f(x - at)$, $0 \le x \le l$, then the energy of the wave is equally divided between potential and kinetic energy.

3·5. One dimensional heat flow. In this article, we shall consider the flow of heat and the accompanying variation of temperature with position and with time in conducting solids. The following empirical laws are taken as the basis of investigation.

1. Heat flows from a higher to lower temperature.
2. The amount of heat required to produce a given temperature change in a body is proportional to the mass of the body and to the temperature change. This constant of proportionality is known as the specific heat (c) of the conducting material.
3. The rate at which heat flows through an area is proportional to the area and to the temperature gradient normal to the area. This constant of proportionality is known as the thermal conductivity (k) of the material.

Consider a bar or rod of homogeneous material of density ρ(gr./cm^3) and having a constant cross-sectional area A (cm^2). We suppose that the sides of the bar are insulated so that the streamlines of heat flow are all parallel and perpendicular to the area A. Take an end of the bar as the origin and the direction of heat flow as the positive x-axis.

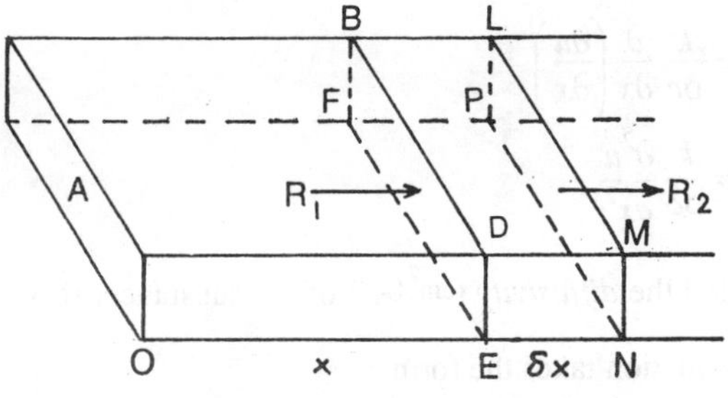

Let c be the specific heat and k the thermal conductivity of the material.

Consider an element got between two parallel sections $BDEF$ and $LMNP$ at distances x and $x + \delta x$ from the origin O, the sections being perpendicular to the x-axis. The mass of the element $= A\rho\delta x$.

Let $u(x, t)$ be the temperature at a distance x at time t. By the second law enunciated above, the rate of increase of heat in the element $= A\rho\delta x c \dfrac{\partial u}{\partial t} \cdot$ If

R_1 and R_2 are respectively the rates (cal./sec.) of inflow and outflow, for the sections $x = x$ and $x = x + \delta x$, then

$$R_1 = -kA\left(\frac{\partial u}{\partial x}\right)_x$$

and $R_2 = -kA\left(\frac{\partial u}{\partial x}\right)_{x+\delta x}$, the negative sign being due to the fact that heat flows from higher to lower temperature.

$$\left(i.e., \frac{\partial u}{\partial x} \text{ is negative}\right)$$

Equating the rates of increase of heat from the two empirical laws,

$$A\rho c\delta x \cdot \frac{\partial u}{\partial t} = R_1 - R_2$$

$$= kA\left[\left(\frac{\partial u}{\partial x}\right)_{x+\delta x} - \left(\frac{\partial u}{\partial x}\right)_x\right]$$

$$\therefore \quad \frac{\partial u}{\partial t} = \frac{k}{\rho c}\left[\frac{\left(\frac{\partial u}{\partial x}\right)_{x+\delta x} - \left(\frac{\partial u}{\partial x}\right)_x}{\delta x}\right]$$

Taking the limit as $\delta x \to 0$ *i.e.* when $x + \delta x \to x$.

$$\frac{\partial u}{\partial t} = \frac{k}{\rho c} \underset{\delta x \to 0}{\text{Lt}} \left[\frac{\left(\frac{\partial u}{\partial x}\right)_{x+\delta x} - \left(\frac{\partial u}{\partial x}\right)_x}{\delta x}\right]$$

$$= \frac{k}{\rho c}\frac{\partial}{\partial x}\left(\frac{\partial u}{\partial x}\right)$$

i.e.,
$$\frac{\partial u}{\partial t} = \frac{k}{\rho c}\frac{\partial^2 u}{\partial x^2}$$

$\frac{k}{pc}$ is called the *diffusivity* ($\text{cm}^2/\text{sec.}$) of the substance. If we denote it by α^2, the above equation takes the form

$$\frac{\partial u}{\partial t} = \alpha^2 \frac{\partial^2 u}{\partial x^2}.$$

3.6 Solution of heat equation by the method of separation of variables

We have to solve the equation

$$\frac{\partial u}{\partial t} = \alpha^2 \frac{\partial^2 u}{\partial x^2} \qquad \text{... (1)}$$

Assume a solution of the form

$$u(x, t) = X(x) \,.\, T(t),$$

where X is a function of x alone and T is a function of t along.

Then (1) becomes,

$$XT' = \alpha^2 X''T,$$

where $X'' = \dfrac{d^2X}{dx^2}$ and $T' = \dfrac{dT}{dt}$

i.e., $$\frac{X''}{X} = \frac{T'}{\alpha^2 T} \qquad ...(2)$$

The left hand side is a function of x alone and the right hand side is a function of t alone when x and t are independent variables. (2) can be true only if each expression is equal to a constant.

$\therefore$ Let $\dfrac{X''}{X} = \dfrac{T'}{\alpha^2 T} = k$ (constant)

$$\therefore\ X'' - kX = 0, \text{ and } T' - \alpha^2 kT = 0 \qquad ...(3)$$

The nature of solutions of (3) depends upon the values of k.

Case 1. Let $k = \lambda^2$, a positive number.

Then (3) becomes,

$$X'' - \lambda^2 X = 0, \text{ and } T' - \alpha^2\lambda^2 T = 0.$$

Solving, we get

$$X = A_1 e^{\lambda x} + B_1 e^{-\lambda x} \text{ and } T = C_1 e^{\alpha^2\lambda^2 t}.$$

Case 2. Let $k = -\lambda^2$, a negative number. Then (3) becomes $X'' + \lambda^2 X = 0$, and $T' + \alpha^2\lambda^2 T = 0$.

Solving, we obtain

$$X = A_2 \cos \lambda x + B_2 \sin \lambda x, \text{ and } T = C_2 e^{-\alpha^2\lambda^2 t}.$$

Case 3. Let $k = 0$.

Then $X'' = 0$ and $T' = 0$.

Solving, we arrive at ,

$$X = A_3 x + B_3 \text{ and } T = C_3.$$

Hence the possible solutions of (1) are

$$u(x, t) = (A_1 e^{\lambda x} + B_1 e^{-\lambda x})\, C_1 e^{\alpha^2\lambda^2 t} \qquad ...(I)$$

$$u(x, t) = (A_2 \cos \lambda x + B_2 \sin \lambda x)\, C_2\, e^{-\alpha^2\lambda^2 t} \qquad ...(II)$$

$$u(x, t) = (A_3x + B_3)\, C_3 \qquad ...(III)$$

Out of these three possible solutions, we have to select that solution which will suit the physical nature of the problem and the given boundary conditions. As we are concerned with heat conduction, $u(x, t)$ must decrease with increase of time. Therefore, out of the three solutions, we select the second solution to suit the physical nature of the problem. In the steady-state conditions, when the temperature no longer varies with time, the solution of the diffusion equation (1) will be the last solution (III).

Example 12. *A rod l cm. with insulated lateral surface is initially at temperature f (x) at an inner point distant x cm. from one end. If both the ends are kept at zero temperature, find the temperature at any point of the rod at any subsequent time.* **(Calicut, 72 Engg.)**

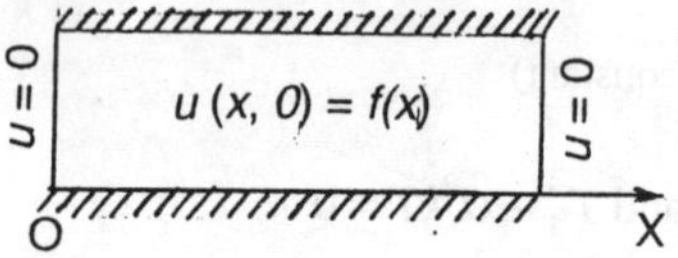

Let $u(x, t)$ be the temperature at any point distant x from one end at any time t seconds. Then u satisfies the partial differential equation

$$\frac{\partial^2 u}{\partial x^2} = \frac{1}{\alpha^2}\frac{\partial u}{\partial t} \qquad ...(1)$$

The boundary conditions, here, are

$$u(0, t) = 0 \text{ for all } t \geq 0 \qquad ...(i)$$

$$u(l, t) = 0 \text{ for all } t \geq 0 \qquad ...(ii)$$

and the initial condition is

$$u(x, 0) = f(x), \text{ for } 0 < x < l \qquad ...(iii)$$

Solving the equation (1) by the method of separation of variables and selecting the suitable solution to suit the physical nature of the problem as explained in the method § 3·6, we get

$$u(x, t) = (A \cos \lambda x + B \sin \lambda x)\, e^{-\alpha^2\lambda^2 t} \qquad ...(2)$$

Substituting the boundary condition (*i*) in (2), we get,

$$u(0, t) = Ae^{-\alpha^2\lambda^2 t} = 0, \text{ for all } t \geq 0$$

$$\therefore\ A = 0$$

Employing the boundary condition (*ii*) in (2), we obtain,

$$u(l, t) = B \sin \lambda l\, e^{-\alpha^2\lambda^2 t} = 0, \text{ for all } t \geq 0$$

i.e., $B \sin \lambda l = 0$.

If $B = 0$, (2) will be a trivial solution. Hence

$\sin \lambda l = 0$

$\therefore$ $\lambda l = n\pi$, where n is any integer.

i.e., $\lambda = \frac{n\pi}{l}$, where n is any integer.

Then (2) reduces to

$$u(x, t) = B_n \sin \frac{n\pi x}{l} e^{-\frac{\alpha^2 n^2 \pi^2}{l^2}} \quad ...(3),$$

where B_n is any constant.

Since the equation (1) is linear, its most general solution is obtained by a linear combination of solutions given by (3).

Hence the most general solution is

$$u(x, t) = \sum_{n=1}^{\infty} B_n \sin \frac{n\pi x}{l} e^{-\frac{\alpha^2 n^2 \pi^2 t}{l^2}} \quad ...(4).$$

(4) should satisfy the initial condition (*iii*).

Using (*iii*) in (4),

$$u(x, 0) = \sum_{n=1}^{\infty} B_n \sin \frac{n\pi x}{l} = f(x), \text{ for } 0 < x < l \text{ (given)} \quad ...(5).$$

If $u(x, 0)$, for $0 < x < l$, is expressed in a half-range Fourier sine series in $0 < x < l$, we know that

$$u(x, 0) = f(x) = \sum_{n=1}^{\infty} b_n \sin \frac{n\pi x}{l}, \text{ where}$$

$$b_n = \frac{2}{l} \int_0^l f(x) \sin \frac{n\pi x}{l} dx.$$

Comparing this with (5), we get

$$B_n = b_n = \frac{2}{l} \int_0^l f(x) \sin \frac{n\pi x}{l} dx.$$

Therefore the temperature function $u(x, t)$, is

$$u(x, t) = \sum_{n=1}^{\infty} \left(\frac{2}{l} \int_0^l f(x) \sin \frac{n\pi x}{l} dx \right) \sin \frac{n\pi x}{l} e^{-\frac{\alpha^2 n^2 \pi^2 t}{l^2}}.$$

Example 13. *Find the temperature u(x, t) in a silver bar (of length 10 cm, constant cross-section of 1 cm² area, density 10·6 gm/cm³, thermal conductivity 1·04 cal/cm deg. sec; specific heat 0·056 cal/gm.deg.) which is perfectly insulated laterally, if the ends are kept at 0°C. and if, initially, the temperature is 5°C. at the centre of the bar and falls uniformly to zero at its ends.*

In this problem, $\alpha^2 = \dfrac{k}{\rho c} = \dfrac{1.04}{(10.6)\,(0.056)} = 1.75.$

Initial temperature distribution is given by the graph given below.

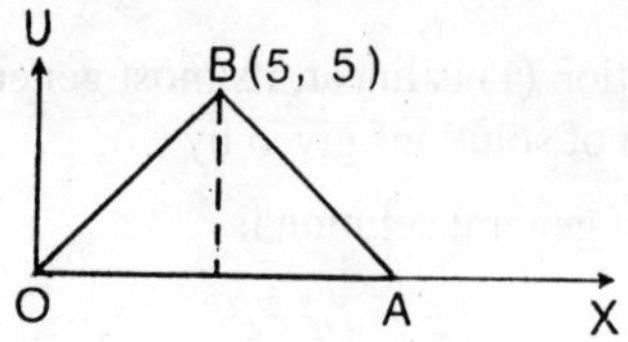

The equation to OB is $u = x$.

The equation to BA is $u = -(x-10) = 10 - x$

Therefore, the initial temperature

$$f(x) = \begin{cases} x, & \text{in } 0 < x < 5 \\ 10 - x, & \text{in } 5 < x < 10. \end{cases}$$

We can directly use the result of the previous example 12, for $l = 10$.

$$B_n = \tfrac{2}{10} \int_0^{10} f(x) \sin \frac{n\pi x}{10}\, dx$$

$$= \tfrac{1}{5}\left[\int_0^5 x \sin \frac{n\pi x}{10}\, dx + \int_5^{10} (10 - x) \sin \frac{n\pi x}{10}\, dx\right]$$

$$= \tfrac{1}{5}\left[\frac{200}{n^2\pi^2} \sin \frac{n\pi}{2}\right]$$

$$= \frac{40}{n^2\pi^2} \sin \frac{n\pi}{2}$$

$\therefore\ B_n = 0$, if n is even

$$= \frac{40}{n^2\pi^2}, \text{ for } n = 1, 5, 9, ...$$

$$= -\frac{40}{n^2\pi^2}, \text{ for } n = 3, 7, 11, \ldots$$

Hence the temperature $u(x,t)$

$$= \frac{40}{\pi^2}\left[\frac{1}{1^2}\sin\frac{\pi x}{10}\, e^{-0.0175\pi^2 t} - \frac{1}{3^2}\sin\frac{3\pi x}{10}\, e^{-0.0175(3\pi)^2 t} + \cdots\right]$$

Example 14. *A rod, 30 cm. long, has its ends A and B kept at 20°C. and 80°C., respectively, until steady state conditions prevail. The temperature at each end is then suddenly reduced to 0°C. and kept so. Find the resulting temperature function u (x, t)) taking x = 0 at A.*

The P.D.E.. of one dimensional heat flow is

$$\frac{\partial u}{\partial t} = \alpha^2 \frac{\partial^2 u}{\partial x^2} \qquad \ldots(1)$$

In steady-state conditions, the temperature at any particular point does not vary with time. That is, u depends only on x and not on time t. Hence the P.D.E. (1) in steady-state becomes

$$\frac{d^2 u}{dx^2} = 0 \qquad \ldots(2)$$

Solving (2), we get,

$$u = ax + b \qquad \ldots(3)$$

The initial conditions, in steady-state, are

$u = 20$, when $x = 0$;

and $u = 80$, when $x = 30$

Using these conditions in (3), we obtain,

$b = 20,\ a = 2.$

$$\therefore \quad u(x) = 2x + 20 \qquad \ldots(4)$$

When the temperatures at A and B are reduced to zero, the temperature distribution changes and the state is no more steady-state. For this transient state, the boundary conditions are

$$u(0, t) = 0 \ \forall \ t \geq 0 \qquad \ldots(i)$$

$$u(30, t) = 0 \ \forall \ t \geq 0 \qquad \ldots(ii).$$

The initial temperature of this state is the temperature in the previous steady-state. Hence the initial condition is

$$u(x, 0) = 2x + 20 \text{ for } 0 < x < 30 \qquad \ldots(iii)$$

Now, we have to find $u(x, t)$ satisfying the conditions (i), (ii) and (iii) and the P.D.E. (1). The suitable solution of (1) is of the form

$$u(x, t) = (A \cos \lambda x + B \sin \lambda x)e^{-\alpha^2 \lambda^2 t} \quad \text{...(5)}$$

Using (i) in (5), we get that $A = 0$.

Using (ii) in (5), $\quad B \sin 30\lambda = 0$.

Since $B \neq 0$, $\quad \sin 30\lambda = 0$.

i.e., $\quad 30\lambda = n\pi$

i.e., $\quad \lambda = \frac{n\pi}{30}$, where n is any integer.

Therefore (5) reduces to

$$u(x,t) = B_n \sin \frac{n\pi x}{30} e^{-\frac{\alpha^2 n^2 \pi^2 t}{900}} \quad \text{...(6)}$$

The most general solution of (1) is obtained by a linear combination of terms given by (6).

$$\therefore \; u(x, t) = \sum_{n=1}^{\infty} B_n \sin \frac{n\pi x}{30} e^{-\frac{\alpha^2 n^2 \pi^2 t}{900}} \quad \text{...(7)}$$

Using (iii),

$$u(x, 0) = \sum_{n=1}^{\infty} B_n \sin \frac{n\pi x}{30} = 2x + 20, \text{ for } 0 < x < 30.$$

$\sum_{n=1}^{\infty} B_n \sin \frac{n\pi x}{30}$ is the Fourier sine series for $f(x) = 2x + 20$, in $0 < x < 30$.

$$\text{Hence } B_n = \frac{2}{30} \int_0^{30} (2x + 20) \sin \frac{n\pi x}{30}\, dx$$

$$= \frac{1}{15}\left[(2x + 20)\left(-\frac{\cos \frac{n\pi x}{30}}{\frac{n\pi}{30}}\right) - (2)\left(-\frac{\sin \frac{n\pi x}{30}}{\frac{n^2\pi^2}{30^2}}\right)\right]_0^{30}$$

$$= \frac{40}{n\pi}\left[1 - 4(-1)^n\right]$$

Substituting in (7),

$$u(x,t) = \sum_{n=1}^{\infty} \frac{40}{n\pi}\left[1 - 4(-1)^n\right] \sin \frac{n\pi x}{30}\, e^{-\frac{\alpha^2 n^2 \pi^2 t}{900}} \text{ degrees.}$$

Example 15. *A bar, 10 cm. long, with insulated sides, has its ends A and B kept at 20° and 40°C., respectively, until steady-state conditions prevail, that is, until the temperature at any interior point no longer changes with time. The temperature at A is then suddenly raised to 50°C. and at the same instant that at B is lowered to 10°C. Find the subsequent temperature function u (x, t) at any time.* **(*MS 1990 Nov.*)**

The partial differential equation satisfied by $u(x, t)$ is

$$\frac{\partial u}{\partial t} = \alpha^2 \frac{\partial^2 u}{\partial x^2} \quad \text{...(1)}$$

In steady-state, this equation reduces to

$$\frac{d^2 u}{dx^2} = 0.$$

Solving this, $u = ax + b$, where a and b are arbitrary constants.

$$u = 20 \text{ at } x = 0, \text{ and } u = 40 \text{ at } x = 10.$$

$\therefore\ b = 20$ and $40 = 10a + b$.

Solving, $a = 2$.

Thus the temperature function in steady-state is

$$u(x) = 2x + 20 \quad \text{...(2)}$$

When the temperatures at A and B are changed, the state is no longer steady. Then the temperature function $u(x, t)$

satisfies (1).

The boundary conditions in the second state are

$$u(0, t) = 50 \ \forall \ t > 0 \quad \text{...}(i)$$

$$u(10, t) = 10 \ \forall \ t > 0 \quad \text{...}(ii)$$

and the initial condition is

$$u(x,0) = 2x + 20 \text{ for } 0 < x < 10 \quad \text{...}(iii),$$

as explained in the previous problem.

Now we have non-zero boundary values and the procedure adopted in the previous problem has to be modified.

So we break up the required function $u(x, t)$ into two parts and write

$$u(x, t) = u_s(x) + u_t(x, t) \qquad ...(3)$$

where $u_s(x)$ is a solution of (1) involving x only and satisfying the boundary conditions (i) and (ii) $u_t(x, t)$ is a function defined by (3) and satisfying the equation (1).

Thus $u_s(x)$ is a *steady-state* solution of (1) and $u_t(x, t)$ may then be regarded as a *transient* solution which decreases with increase of t; $u_s(x)$ satisfies (1).

$$\therefore \quad \frac{d^2u_s}{dx^2} = 0, \text{ where } u_s(0) = 50 \text{ and } u_s(10) = 10.$$

Solving, $u_s(x) = ax + b$.

$u_s(0) = b = 50$, using (*i*);

and $\quad u_s(10) = 10a + 50 = 10$, using (*ii*).

Hence $a = -4$.

Thus $u_s(x) = 50 - 4x \qquad ...(4)$

Consequently,

$$u_t(0, t) = u(0, t) - u_s(0) = 50 - 50 = 0 \qquad (iv)$$

$$u_t(10, t) = u(10, t) - u_s(10) = 10 - 10 = 0 \qquad (v)$$

and $u_t(x, 0) = u(x, 0) - u_s(x) = (2x + 20) - (50 - 4x)$

i.e., $\quad u_t(x, 0) = 6x - 30 \qquad (vi)$

Now, $u_t(x, t)$ also satisfies (1) and (iv), (v), (vi).

Solving (1) and selecting a suitable solution

$$u_t(x, t) = (A \cos \lambda x + B \sin \lambda x)e^{-\alpha^2\lambda^2 t} \qquad ...(5)$$

Using the boundary condition (iv) in (5),

$$u_t(0, t) = Ae^{-\alpha^2\lambda^2 t} = 0 \;\forall \quad t > 0$$

$\therefore \quad A = 0$

Using (v) in (5),

$$B \sin 10\lambda \,.\, e^{-\alpha^2\lambda^2 t} = 0, \text{ for } \forall \quad t > 0$$

Since $B \neq 0$, $\sin 10\lambda = 0$.

Hence $10\lambda = n\pi$

i.e., $\lambda = \frac{n\pi}{10}$, where n is any integer.

Therefore, (5) becomes,

$$u_t(x,t) = B_n \sin \frac{n\pi x}{10} e^{-\frac{\alpha^2 n^2 \pi^2 t}{100}}$$

The most general solution of (1) is,

$$u_t(x, t) = \sum_{n=1}^{\infty} B_n \sin \frac{n\pi x}{10} \cdot e^{-\frac{\alpha^2 n^2 \pi^2 t}{100}} \qquad ...(6)$$

Using the intial condition (vi) in (6),

$$u_t(x,0) = \sum_{n=1}^{\infty} B_n \sin \frac{n\pi x}{10} = 6x - 30, \text{ for } 0 < x < 10.$$

Thus is itself Fourier half-range sine series for $6x - 30$ in $0 < x < 10$ if

$$B_n = \tfrac{2}{10} \int_0^{10} (6x - 30) \sin \frac{n\pi x}{10}\, dx$$

$$= \tfrac{1}{5} \left[(6x - 30) \left(-\frac{\cos \frac{n\pi x}{10}}{\frac{n\pi}{10}} \right) - (6) \left(-\frac{\sin \frac{n\pi x}{10}}{\frac{n^2\pi^2}{10^2}} \right) \right]_0^{10}$$

$$= \tfrac{1}{5} \left[-\frac{300}{n\pi} \cos n\pi - \frac{300}{n\pi} \right]$$

$$= \frac{-60}{n\pi} [1 + (-1)^n]$$

$$= \frac{-120}{n\pi}, \quad \text{for } n \text{ even}$$

$$= 0, \quad \text{for } n \text{ odd.}$$

Substituting this value of B_n in (6),

$$u_t(x, t) = \sum_{n=2, 4, 6 \ldots}^{\infty} \left(\frac{-120}{n\pi} \right) \sin \frac{n\pi x}{10} e^{-\frac{\alpha^2 n^2 \pi^2 t}{100}}$$

$$= -\frac{60}{\pi} \sum_{n=1}^{\infty} \frac{1}{n} \sin \frac{n\pi x}{5} e^{-\frac{\alpha^2 n^2 \pi^2 t}{25}}$$

$\therefore \quad u(x, t) = u_s(x) + u_t(x, t)$

$$= 50 - 4x - \frac{60}{\pi} \sum_{n=1}^{\infty} \frac{1}{n} \sin \frac{n\pi x}{5} e^{-\frac{\alpha^2 n^2 \pi^2 t}{25}}$$

This series converges for $0 < x < 10$ and $t > 0$. We note that $u(x, t)$ tends to $u_s(x)$ as $t \to \infty$.

Example 16. *Solve* $\frac{\partial u}{\partial t} = \alpha^2 \frac{\partial^2 u}{\delta x^2}$ *subject to the conditions*

(*i*) u is not infinite as $t \to \infty$

(*ii*) $u = 0$ for $x = 0$ and $x = \pi$ for all t

(*iii*) $u = \pi x - x^2$ for $t = 0$ in $(0, \pi)$

Solving the differential equation by the method of separation of variables, we get

$$u(x, t) = (Ae^{\lambda x} + Be^{-\lambda x})\, e^{\alpha^2 \lambda^2 t} \quad \text{...(I)}$$

$$= (A \cos \lambda x + B \sin \lambda x)\, e^{-\alpha^2 \lambda^2 t} \quad \text{...(II)}$$

$$= (Ax + B) \quad \text{...(III)}$$

as three possible solutions

As $t \to \infty$, $u \to \infty$ in solution I. Hence we reject solution as u is not infinite as $t \to \infty$ as per condition (1).

Further solution III is independent of t (steady state solution). Hence we select solution II as possible one.

Let $u(x, t) = (A \cos \lambda x + B \sin \lambda x)\, e^{-\alpha^2 \lambda^2 t}$ ...(II)

$u(0, t) = 0$. using this,

$A \,.\, e^{-\alpha^2 \lambda^2 t} = 0 \quad \therefore \; A = 0$

$u(\pi, t) = 0$. using in II,

$B \sin \lambda \pi \,.\, e^{-\alpha^2 \lambda^2 t} = 0$

$B \neq 0$ (otherwise $u \equiv 0$).

Hence, $\sin \lambda \pi = 0 \quad \therefore \; \lambda = n$, any integer

$\therefore \; u(x, t) = B_n \sin nx\, e^{-\alpha^2 n^2 t}$, n any integer.

Hence, the most general solution is

$$u(x, t) = \sum_{n=1}^{\infty} B_n \sin nx\, e^{-\alpha^2 n^2 t} \quad \text{...(IV)}$$

Now, we will use the initial condition to find B_n.

$$u(x,0) = \sum_{n=1}^{\infty} B_n \sin nx = \pi x - x^2$$

This is half range Fourier sine series for $\pi x - x^2$.

$$B_n = b_n = \frac{2}{\pi}\int_0^{\pi} (\pi x - x^2) \sin nx \, dx$$

$$= \frac{2}{\pi}\left[(\pi x - x^2)\left(-\frac{\cos nx}{n}\right) - (\pi - 2x)\left(-\frac{\sin nx}{n^2}\right) + (-2)\left(\frac{\cos nx}{n^3}\right)\right]_0^{\pi}$$

$$= \frac{2}{\pi}\left[\frac{2}{n^3}\left(1 - (-1)^n\right)\right]$$

$$= \frac{4}{\pi n^3}\left[1 - (-1)^n\right]$$

$= 0$ if n is even

$= \dfrac{8}{\pi n^3}$ if n is odd

Substituting B_n value in IV,

$$\therefore u(x, t) = \frac{8}{\pi}\sum_{n=1}^{\infty} \frac{1}{(2n-1)^3} \sin (2n-1)\, x \, . \, e^{-\alpha^2 (2n-1)^2 t}$$

Example 17. *Solve* $\dfrac{\partial u}{\partial t} = \alpha^2 \dfrac{\partial^2 u}{\partial x^2}$ *subject to*

$u(o, t) = 0$ (*i*) for $t \geq 0$

$u(l, t) = 0$ (*ii*) for $t \geq 0$ **(*M.S. 1990 Ap.*)**

$u(x, 0) = x$ for $0 \leq x \leq l/2$

$= l - x$ for $\dfrac{l}{2} \leq x \leq l$ (*iii*)

Solving the given differential equation by the method of separation of variables we get three possible types of solutions, out of which we neglect the solution I and III for reasons explained in the previous question. Hence, we select.

$$u(x, t) = (A \cos \lambda x + B \sin \lambda x)\, e^{-\alpha^2 \lambda^2 t} \quad \text{...II}$$

using $u(0, t) = 0$ in II, we get $A = 0$

using $u(l, t) = 0$ in II, $B \sin \lambda l = 0$

$B \neq 0; \quad \therefore \; \lambda l = n\pi$

$$\lambda = \frac{n\pi}{l}, \quad n \text{ any integer.}$$

Hence, $u(x, t) = B_n \sin \frac{n\pi x}{l} \cdot e^{-\frac{\alpha^2 n^2 \pi^2}{l^2} t}$ n any integer.

Hence, the most general solution is

$$u(x, t) = \sum_{n=1}^{\infty} B_n \text{ Sin} \frac{n\pi x}{l} \cdot e^{-\frac{\alpha^2 n^2 \pi^2 t}{l^2}} \qquad \text{...IV}$$

Now we use the third intial condition in IV

$$u(x, 0) = \sum_{1}^{\infty} B_n \sin \frac{n \pi x}{l} = \text{given function of } x.$$

$$\therefore \; B_n = \frac{2}{l}\left[\int_{o}^{l/2} x \sin \frac{n \pi x}{l} \text{dn} + \int_{l/2}^{l} (l - x) \sin \frac{n \pi x}{l} dx\right]$$

$$= \frac{2}{l}\left[\left\{(x)\left(-\frac{\cos \frac{n\pi x}{l}}{\frac{n\pi}{l}}\right) - (1)\left(-\frac{\sin \frac{n\pi x}{l}}{\frac{n^2\pi^2}{l^2}}\right)\right\}_0^{l/2}\right.$$

$$\left. + \left\{(l - x)\left(-\frac{\cos \frac{n\pi x}{l}}{\frac{n\pi}{l}}\right) - (-1)\left(-\frac{\sin \frac{n\pi x}{l}}{\frac{n^2\pi^2}{l^2}}\right)\right\}_{l/2}^{l}\right]$$

$$= \frac{2}{l}\left[-\frac{l^2}{2n\pi} \cos \frac{n\pi}{2} + \frac{l^2}{n^2\pi^2} \sin \frac{n\pi}{2} + \frac{l^2}{2n\pi} \cos \frac{n\pi}{2} + \frac{l^2}{n^2\pi^2} \sin \frac{n\pi}{2}\right]$$

$$= \frac{4l}{n^2\pi^2} \sin \frac{n\pi}{2}$$

$\therefore$ using the value of B_n in IV,

$$u(x, t) = \frac{4l}{\pi^2} \sum_{n=1}^{\infty} \frac{1}{n^2} \sin \frac{n \pi}{2} \sin \frac{n\pi x}{l} e^{-\frac{\alpha^2 n^2 \pi^2 t}{l^2}}$$

Example 18. *Solve* $\frac{\partial u}{\partial t} = \alpha^2 \frac{\partial^2 u}{\partial x^2}$ *subject to the conditions.*

(*i*) u is not infinite as $t \to \infty$

(*ii*) $\frac{\partial u}{\partial x} = 0$ for $x = 0$ and $x = l$ (two ends thermally insulated)

(*iii*) $u = lx - x^2$ for $t = 0, 0 < x < l$ (***MS, 1987 Ap.***)

As in the previous problems, we select

$$u(x, t) = (A \cos \lambda x + B \sin \lambda x)\, e^{-\alpha^2 \lambda^2 t} \quad \text{...II}$$

$$\frac{\partial u}{\partial x} = (-A \lambda \sin \lambda x + B \lambda \cos \lambda x)\, e^{-\alpha^2 \lambda^2 t}$$

$$\frac{\partial u}{\partial x}(0, t) = 0 \;\therefore\; B \lambda e^{-\alpha^2 \lambda^2 t} = 0 \;\therefore\; B = 0$$

$$\frac{\partial u}{\partial x}(l, t) = 0 \;\therefore\; -A \lambda \sin \lambda l\, e^{-\alpha^2 \lambda^2 t} = 0$$

$$\therefore A \neq 0 \text{ Hence } \lambda l = n\pi$$

$$\lambda = \frac{n\pi}{l}, \; n \text{ any integer}$$

$$\therefore \; u(x, t) = A_n \cos \frac{n \pi x}{l}\, e^{-\frac{\alpha^2 n^2 \pi^2 t}{l^2}}$$

Hence, the most general solution is

$$u(x, t) = \sum_{n=0}^{\infty} A_n \cos \frac{n\pi x}{l}\, e^{-\frac{\alpha^2 n^2 \pi^2 t}{l^2}} \quad \text{...IV}$$

Using $u(x,0) = lx - x^2$ for $0 < x < l$ in IV

$$u(x,0) = \sum_{n=0}^{\infty} A_n \cos \frac{n\pi x}{l} = lx - x^2$$

$$= A_0 + \sum_{n=1}^{\infty} A_n \cos \frac{n \pi x}{l} = lx - x^2$$

$$\therefore \quad A_0 = \frac{a_0}{2} = \frac{1}{l} \int_0^l (lx - x^2)\, dx$$

$$= \frac{1}{l}\left(\frac{lx^2}{2} - \frac{x^3}{3}\right)_0^l$$

$$= \frac{1}{l}\left(\frac{l^3}{6}\right)$$

$$= \frac{l^3}{6}.$$

$$A_n = \frac{2}{l}\int_0^l (lx - x^2)\cos\frac{n\pi x}{l}\,dx$$

$$= \frac{2}{l}\left[(lx - x^2)\left(\frac{\sin\frac{n\pi x}{l}}{\frac{n\pi}{l}}\right) - (l - 2x)\left(-\frac{\cos\frac{n\pi x}{l}}{\frac{n^2\pi^2}{l^2}}\right) + (-2)\left(-\frac{\sin\frac{n\pi x}{l}}{\frac{n^3\pi^3}{l^3}}\right)\right]_0^l$$

$$= \frac{2}{l}\left[-\frac{l^3}{n^2\pi^2}(-1)^n - \frac{l^3}{n^2\pi^2}\right]$$

$$= -\frac{2l^2}{n^2\pi^2}[1 + (-1)^n]$$

$= 0$ if n is odd

$= \dfrac{-4\,l^2}{n^2\,\pi^2}$ if n is even

$$\therefore\; u(x, t) = \frac{l^2}{6} - \frac{4l^2}{\pi^2}\sum_{n=2,4,6}^{\infty}\frac{1}{n^2}\cos\frac{n\pi x}{l}\,.\,e^{\frac{-\alpha^2 n^2\pi^2 t}{l^2}}$$

$$u(x, t) = \frac{l^2}{6} - \frac{l^2}{\pi^2}\sum_{n=1}^{\infty}\frac{1}{n^2}\cos\frac{2n\pi x}{l}\,.\,e^{\frac{-4\alpha^2 n^2\pi^2 t}{l^2}}$$

Example 19. *Solve* $\dfrac{\partial\theta}{\partial t} = \dfrac{\partial^2\theta}{\partial x^2}$ for $t > 0$

and $0 < x < 1$, θ *being the temperature. The initial and boundary conditions are*

$\theta(0, t) = 0 \quad t > 0$

$\dfrac{\partial\theta}{\partial x}(x, t) = 0$ at $x = 1, t > 0$

$\theta(x, 0) = x$ for $0 < x < 1$

In this problem, the end $x = 1$ is insulated whereas the end $x = 0$ is not .

In the differential equation $\alpha^2 = 1$

Hence, starting with the solution (II type)

$$\theta(x, t) = (A\cos\lambda x + B\sin\lambda x)\, e^{-\lambda^2 t} \qquad ...(II)$$

Since $\theta(0, t) = 0$, $A = 0$

$$\frac{\partial\theta}{\partial x} = B\lambda\cos\lambda x\, e^{-\lambda^2 t}$$

$$\frac{\partial\theta}{\partial x}(1, t) = 0 \;\therefore\; B\lambda\cos\lambda = 0\,;\, B \neq 0\,.$$

$$\cos\lambda = 0 \;\therefore\; \lambda = (2n - 1)\frac{\pi}{2}$$

Hence $\theta(x, t) = B_n \sin\dfrac{(2n-1)\pi x}{2}\, e^{-\frac{(2n-1)^2\pi^2 t}{4}}$, n any integer

The most general solution is

$$\theta(x, t) = \sum_{n=1}^{\infty} B_n \sin\frac{(2n-1)\pi x}{2}\, e^{-\frac{(2n-1)^2\pi^2}{4}t} \qquad ...(IV)$$

At $t = 0$,

$$\theta(x, 0) = \sum_{1}^{\infty} B_n \sin\frac{(2n-1)\pi x}{2} = x$$

$$B_n = \frac{2}{1}\int_0^1 x\sin\frac{(2n-1)\pi x}{2}\,dx$$

$$= 2\left[(x)\left(-\frac{\cos\frac{(2n-1)\pi x}{2}}{\frac{(2n-1)\pi}{2}}\right) - \left(-\frac{\sin\frac{(2n-1)\pi x}{2}}{\frac{(2n-1)^2\pi^2}{4}}\right)\right]_0^1$$

$$= \frac{8(-1)^{n-1}}{(2n-1)^2\pi^2}$$

$$\theta(x, t) = \frac{8}{\pi^2}\sum_{n=1}^{\infty}\frac{(-1)^{n-1}}{(2n-1)^2\pi^2}\sin\frac{(2n-1)\pi x}{2}\, e^{-\frac{(2n-1)^2\pi^2}{4}t}$$

Example 20. *A uniform rod of length l whose surface is thermally insulated is initially at a constant temperature $\theta = \theta_0$. At time t = 0, one end is* *suddenly cooled to zero temperature while the other end remains thermally insulated. This state is maintained subsequently. Find the temperature at the end x = l.* **(BR. 1995 Ap.)**

The boundary conditions are

(*i*) $\theta(o, t) = o$ for $t \geq 0$

(*ii*) $\frac{\partial \theta}{\partial x}(l, t) = 0$ for $t \geq 0$ since insulated

(*iii*) $\theta(x, 0) = \theta_0$ for $0 < x < l$.

(*iv*) θ is finite.

Since θ is finite, we start with the solution

$$\theta(x, t) = (A \cos \lambda x + B \sin \lambda x)\, e^{-\alpha^2 \lambda^2 t} \quad \text{...II}$$

Using boundary condition (1), A = 0

$\frac{\partial \theta}{\partial x}(l, t) = 0$ implies

$B \lambda \cos \lambda l = 0$

$B \neq 0, \cos \lambda l = 0; \quad \lambda = \frac{(2n-1)\pi}{2l}$

$$\therefore \quad \theta(x, t) = B_n \sin \frac{(2n-1)\pi x}{2l}\, e^{-\alpha^2 \lambda^2 t}$$

Hence, most general solution is

$$\theta(x, t) = \sum_{n=1}^{\infty} B_n \sin \frac{(2n-1)\pi x}{2l}\, e^{-\frac{\alpha^2 (2n-1)^2 \pi^2}{4l^2} t}$$

At $t = 0, \theta = \theta_o$

$$\therefore \quad \sum_{1}^{\infty} B_n \sin \frac{(2n-1)\pi x}{2l} = \theta_0$$

$$\therefore \quad B_n = \frac{2}{l} \int_o^l \theta_o \sin \frac{(2n-1)\pi x}{2l}\, dx$$

$$= \frac{2\theta_o}{l} \left[- \frac{\cos \frac{(2n-1)\pi x}{2l}}{\frac{(2n-1)\pi}{2l}} \right]_0^l$$

$$= \frac{4\theta_o}{(2n-1)\pi}$$

$$u(x, t) = \frac{4\theta_o}{\pi} \sum_{n=1}^{\infty} \frac{1}{(2n-1)} \sin \frac{(2n-1)\pi x}{2l} \text{ Exp.} \left[\frac{-\alpha^2 (2n-1)^2 \pi^2 t}{4l^2} \right]$$

$$u(l, t) = \frac{4\theta_0}{\pi} \sum_{n=1}^{\infty} \frac{(-1)^{n-1}}{(2n-1)} \cdot \text{ Exp.} \left[\frac{-\alpha^2 (2n-1)^2 \pi^2 t}{4l^2} \right]$$

Example 21. *An insulated rod of length l has its ends A and B kept at a degree centigrade and b degree centigrade until steady state conditions prevail. The temperature at each end in suddenly reduced to zero degree centigrade and kept so. Find the resulting temperature at any point of the rod taking the end A as origin.*

Let the temperature at any point P of the rod at time t be u where AP $= x$.

Then u satisfies $\frac{\partial u}{\partial t} = \alpha^2 \frac{\partial^2 u}{\partial x^2}$. where $\alpha^2 = \frac{k}{\rho c}$.

In steady state, $\frac{\partial u}{\partial t} = 0$. Therefore $\frac{d^2 u}{dx^2} = 0$

i.e., $u = Dx + E$

At $x = 0,\ u = a$

At $x = l,\ u = b$

$\therefore\ E = a$

$b = Dl + E = Dl + a$

$\therefore\ D = \frac{b-a}{l}$

Hence, the steady state temperature is

$$u(x) = \frac{b-a}{l} x + a \quad ...(2)$$

This temperature is the initial temperature of the next unsteady state.

Let $u(x, t)$ be the temperature in unsteady state. The boundary conditions are

$$u(o, t) = 0 \text{ for } t > o \quad ...(i)$$

$$u(l, t) = 0 \text{ for } t > o \quad ...(ii)$$

$$u(x, o) = \frac{b-a}{l} x + a \text{ for } o < x < l \quad ...(iii)$$

Now, in the unsteady state,

$$u(x, t) = (A \cos\lambda x + B \sin \lambda x) e^{-\alpha^2 \lambda^2 t} \quad ...\text{II}$$

Using the first two boundaries conditions,

$$A = 0\,;\ \lambda = \frac{n\pi}{l},\ n \text{ any integer}$$

Hence, $u\,(x,t) = B_n \sin \frac{n\pi x}{l} \cdot e^{-\frac{\alpha^2 n^2 \pi^2}{l^2} t}$

The most general solution is

$$u\,(x,t) = \sum_{n=1}^{\infty} B_n \sin \frac{n\pi x}{l} \cdot e^{-\frac{\alpha^2 n^2 \pi^2}{l^2} t} \qquad \text{...IV}$$

Using boundary condition (iii),

$$u\,(x, o) = \sum_{n=1}^{\infty} B_n \sin \frac{n\pi x}{l} = \frac{b-a}{l} x + a \qquad \text{...(3)}$$

This is half-range Furrier sine series.

$$\therefore \qquad B_n = \frac{2}{l} \int_0^l \left(\frac{b-a}{l} x + a \right) \sin \frac{n\pi x}{l}\, dx$$

$$= \frac{2}{l} \left[\left(\frac{b-a}{l} x + a \right) \left(- \frac{\cos \frac{n\pi x}{l}}{\frac{n\pi}{l}} \right) - \left(\frac{b-a}{l} \right) \left(- \frac{\sin \frac{n\pi x}{l}}{\frac{n^2 \pi^2}{l^2}} \right) \right]_0^l$$

$$= \frac{2}{l} \left[- \frac{bl}{n\pi} \cos n\pi + \frac{al}{n\pi} \right]$$

$$= \frac{2}{n\pi} \left[a - b(-1)^n \right]$$

Hence

$$u\,(x, t) = \frac{2}{\pi} \sum_{n=1}^{\infty} \frac{1}{n} \left[a - b(-1)^n \right] \sin \frac{n\pi x}{l} \cdot e^{-\frac{\alpha^2 n^2 \pi^2}{l^2} t} \text{ degrees.}$$

Note: In many problems, *a*, *b* and *l* only will vary. The result can be verified by putting the values of *a*, *b*, *l* in the result of the above problem.

For example, if $a = 0°C$, $b = 100°C$ and $l = l$

$$u\,(x, t) = \frac{200}{\pi} \sum_{1}^{\infty} \frac{1}{n} (-1)^{n+1} \sin \frac{n\pi x}{l} \text{ Exp.} \left(- \frac{\alpha^2 n^2 \pi^2 t}{l^2} \right)$$

Example 22. *A rod of length l cm long, with insulated sides, has its ends A and B kept at a° centigrate and b° centigrade respectively until steady state conditions prevail. The temperature at A is then suddenly raised to c° centigrade and at the same that at B is lowered to d° centigrade. Find the temperature distribution u (x, t) subsequently.*

In the steady state, u satisfies $\frac{d^2u}{dx^2} = 0$ ...(1)

i.e., $u = Dx + E$

At $x = 0,\ u = a$

At $x = l,\ u = b$

$\therefore \quad E = a$

$Dl + a = b$

$D = \frac{b-a}{l}$

$\therefore\ u\ (x) = \frac{b-a}{l}x + a$ in steady state ...(2)

When the temperature at A and B are changed, the state is no longer steady. It becomes transient. Let $u(x, t)$ be the temperature in the next state. The boundary conditions are

$u(o, t) = c,\ t > o$ ...(*i*)

$u(l, t) = d,\ t > o$ ...(*ii*)

and $u(x, o) = \frac{b-a}{l}x + a$, for $o < x < l$ ...(*iii*)

Now, the boundary conditions are not zeros. This transient state, after a long time, will again be steady. The boundary conditions both in the transient state and subsequent steady state will be $u(o, t) = c,\ u(l, t) = d$

In other words $u\ (x, t)$ consists of two portions

$u_s\ (x)$ and $u_t\ (x, t)$ where $u_s(x)$ is the steady state function and $u_s + u_t$ the transient state function.

$\therefore \quad u\ (x, t) = u_s\ (x) + u_t\ (x, t)$ ...(3)

i.e., $u_s\ (o) = c$ and $u\ (o, t) = u_s\ (o) + u_t\ (o, t) = c$

$u_s\ (l) = d \qquad u\ (l, t) = u_s\ (l) + u_t\ (l, t) = d$

which implies, $u_t\ (o, t) = 0,\ u_t\ (l, t) = 0$

$u_s\ (x)$, the steady state function satisfies,

$\frac{d^2u_s}{dx^2} = 0$ *i.e.,* $u_s = \alpha x + \beta$ ...(4)

Now use $u_s(o) = c, u_s(l) = d$

$\therefore \ \beta = c$ and $\alpha l + \beta = d$

$$\therefore \ \alpha = \frac{d-c}{l}$$

i.e., $$u_s(x) = \frac{d-c}{l} x + c \qquad \text{...(5)}$$

Now U_t satisfies $\dfrac{\partial u}{\partial t} = \alpha^2 \dfrac{\partial^2 u}{\partial x^2}$

$$\therefore \quad u_t = (A \cos \lambda x + B \sin \lambda x)\, e^{-\alpha^2 \lambda^2 t} \qquad \text{...II}$$

subject to $u_t(o, t) = o$ ...(*iv*)

$u_t(l, t) = o$ ...(*v*)

$$u_t(x, o) = u(x,o) - u_s(x)$$

$$= \left(\frac{b-a}{l} x + a\right) - \left(\frac{d-c}{l} x + c\right)$$

$$= \frac{b+c-a-d}{l} x + (a-c) \qquad \text{...(vi)}$$

Now find A, B, in II using (*iv*), (*v*), (*vi*)

$u_t(o, t) = 0$ implies $A = 0$

$u_t(l, t) = 0$ implies $\lambda = \dfrac{n\pi}{l}$.

$$\therefore \ u_t(x, t) = B_n \sin \frac{n\pi x}{l} \cdot e^{\frac{-\alpha^2 a^2 \pi^2 t}{l^2}}.$$

The most general solution is

$$u_t(x, t) = \sum_{n=1}^{\infty} B_n \sin \frac{n\pi x}{l} \cdot e^{\frac{-\alpha^2 a^2 \pi^2 t}{l^2}} \qquad \text{...(IV)}$$

Now $$u_t(x, o) = \sum_{1}^{\infty} B_n \sin \frac{n\pi x}{l} = \frac{b+c-a-d}{l} x + (a-c).$$

$$\therefore \quad B_n = \frac{2}{l} \int_o^l \left[\frac{b+c-a-d}{l} x + a - c\right] \sin \frac{n\pi x}{l}\, dx$$

$$= \frac{2}{l}\left[\left(\frac{b+c-a-d}{l}x+a-c\right)\left(-\frac{\cos\frac{n\pi x}{l}}{\frac{n\pi}{l}}\right)-\left(\frac{b+c-a-d}{l}\right)\times\left(-\frac{\sin\frac{n\pi x}{l}}{\frac{n^2\pi^2}{l^2}}\right)\right]_0^l$$

$$= \frac{2}{l}\left[-\frac{(b-d)\,l}{n\,\pi}(-1)^n+\frac{l\,(a-c)}{n\,\pi}\right]$$

$$= \frac{2}{\pi n}\left[(a-c)-(b-d)\,(-1)^n\right]$$

using B_n in IV and then using (3), we get

$u\,(x, t) = u_s\,(x) + u_t\,(x, t)$

$$u\,(x, t) = \left(\frac{d-c}{l}x+c\right)+\frac{2}{\pi}\sum_{n=1}^{\infty}\frac{1}{n}\left[(a-c)-(b-d)(-1)^n\right]\sin\frac{n\pi x}{l}\times e^{\frac{-\alpha^2 n^2 \pi^2 t}{l^2}}$$

Note 1. Putting $a = 20, b = 40, c = 50, d = 10, l = 30$ cm, we get the worked example 15.

Note 2. Worked examples 21, and 22 give more general problem.

Exercises 3 (c)

1. Solve the one dimensional diffusion equation for the function $\theta\,(x, t)$ in the region $0 \le x \le \pi, t \ge 0$ when

   (i) θ remains finite as $t \to \infty$

   (ii) $\theta = 0$ if $x = 0$ or π for all values of $t > 0$; and

   (iii) at $t = 0, \theta = x$ for $0 \le x \le \pi/2$ and

   $\theta = \pi - x$ for $\frac{\pi}{2} \le x \le \pi$. (*Madras, '65 B.E.*)

2. Find the solution to the equation $\frac{\partial u}{\partial t} = a^2\frac{\partial^2 u}{\partial x^2}$ that satisfies the conditions

   $u\,(0, t) = 0, u(l, t) = 0$ for $t > 0$ and

   $u(x, 0) = x$ for $0 \le x \le l/2$

   $= l - x$ for $l/2 < x < l$ (*Madras, '69 B.E.; Kerala, '68 B.E.*)

3. Obtain permissible product solutions of the heat-flow equation

$$\frac{\partial^2 T}{\partial x^2} = \frac{1}{\alpha^2}\frac{\partial T}{\partial t} \text{ which satisfy the conditions}$$

$$\frac{\partial T}{\partial x}(0, t) = 0 \text{ and } \frac{\partial T}{\partial x}(l, t) = 0.$$

(*Madras, '69 B.E.*)

4. Determine the solution of one dimensional heat equation

$$\frac{\partial \theta}{\partial t} = a^2 \frac{\partial^2 \theta}{\partial x^2} \text{ under the boundary conditions}$$

$\theta(0, t) = \theta(l, t) = 0$ for $t > 0$ and the initial condition

$\theta(x, 0) = x$ for $0 < x < l$, l being the length of the bar.

5. Solve the equation $\frac{\partial u}{\partial t} = a^2 \frac{\partial^2 u}{\partial x^2}$ subject to the conditions given below:

(*i*) u is not infinite as $t \to \infty$

(*ii*) $u = 0$ for $x = 0$ or $x = \pi$ for all values of t

(*iii*) $u = \pi x - x^2$ if $t = 0$ in $(0, \pi)$.

6. Show that the solution of the equation

$$\frac{\partial \theta}{\partial t} = k \frac{\partial^2 \theta}{\partial x^2} \text{ subject to the conditions}$$

(*i*) θ is finite as $t \to \infty$

(*ii*) $\frac{\partial \theta}{\partial x} = 0$ when $x = 0$ for all $t > 0$

(*iii*) $\theta = 0$ when $x = l$ for all $t > 0$

(*iv*) $\theta = \theta_0$ when $t = 0$ for $0 < x < l$ is

$$\theta = 4\frac{\theta_0}{\pi}\sum_{n=0}^{\infty}\frac{(-1)^n}{(2n+1)} \cos\frac{(2n+1)\pi x}{2l} \cdot e^{-\frac{k(2n+1)^3 \cdot \pi^2 t}{4l^2}}$$

(*MS. 1989 Nov.*)

7. Solve $\frac{\partial u}{\partial t} = \alpha^2 \frac{\partial^2 u}{\partial x^2}$ with adiabatic boundary conditions:

(*i*) $\frac{\partial u}{\partial x}(0, t) = 0$, (*ii*) $\frac{\partial}{\partial x} u(l, t) = 0$, (*iii*) $u(x, 0) = x$.

8. Solve $\frac{\partial u}{\partial t} = \alpha^2 \frac{\partial^2 u}{\partial x^2}$ for the conduction of heat along a rod, without radiation, subject to the following conditions.

(*i*) u is not infinite as $t \to \infty$

(*ii*) $\frac{\partial u}{\partial x} = 0$ for $x = 0$ or $x = l$

(*iii*) $u = lx - x^2$ for $t = 0$ for $0 < x < l$.

(*Madras, '65 B.E.*)

9. Solve $\dfrac{\partial u}{\partial t} = \alpha^2 \dfrac{\partial^2 u}{\partial x^2} = 0$ for $0 < x < \pi$, $t > 0$

$u_x(0, t) = 0$, $u_x(\pi, t) = 0$ and $u(x, 0) = \sin x$.

10. Show that the solution of $\dfrac{\partial u}{\partial t} = \alpha^2 \dfrac{\partial^2 u}{\partial x^2}$ subject to the conditions

$$\left.\begin{aligned} &(i)\ \frac{\partial u}{\partial x} = 0 \text{ when } x = 0 \\ &(ii)\ \frac{\partial u}{\partial x} = 0 \text{ when } x = l \end{aligned}\right\} \text{ for all } t > 0$$

(*iii*) $u = f(x)$ when $t = 0$ is

$$u = \frac{a_0}{2} + \sum_{n=1}^{\infty} a_n \cos \frac{n\pi x}{l} \text{ Exp } [-\alpha^2 n^2 \pi^2 t/l^2]$$

where $a_n = \dfrac{2}{l} \displaystyle\int_0^l f(x) \cos \frac{n\pi x}{l} dx.$

11. A bar with insulated sides is initially at temperature 0°C. throughout. The end $x = 0$ is kept at 0°C. and the heat is suddenly applied at the end $x = l$ at a constant rate, so that

$\dfrac{\partial u}{\partial x} = K$ for $x = l$, K being a constant. Show that

$$u(x, t) = Kx + \frac{8Kl}{\pi^2} \sum_{n=1}^{\infty} \frac{(-1)^n}{(2n-1)^2} \sin \frac{(2n-1)\pi x}{2l} \cdot \text{ Exp } [-(2n-1)^2 \pi^2 \alpha^2 t/4l^2]$$

(*MS. 1987 Nov.*)

12. The temperatures at the ends $x = 0$ and $x = 100$ of a rod 100 cm. in length are held at 0° and 100°C respectively until steady-state conditions prevail. Then at $t = 0$, the two ends are suddenly insulated. Find the resultant temperature distribution in the rod. **(*Madras, 69 B.E.*)**

13. (*a*) A rod of length l has its ends A and B kept at 0°C. and 100°C. respectively until steady-state conditions prevail. If the temperature at B is reduced suddenly to 0°C. and kept so, while that of A is maintained, find the temperature $u(x, t)$ at a distance x from A and at time t. **(*Madras, B.E. 78.*)**

(*b*) Solve the above problem, if the change consists of raising the temperature of A to 25°C. and reducing that of B to 75°C. **(*O.U., 65 B.E.*)**

14. Solve the heat equation

$\dfrac{\partial \theta}{\partial t} = k \dfrac{\partial^2 \theta}{\partial x^2}$ in the range $0 \le x \le 2\pi$, $t > 0$ subject to the conditions

$$\theta(x, 0) = \sin^3 x, 0 \le x \le 2\pi$$

and $\theta(0, t) = \theta(2\pi, t) = 0$ for $t > 0$. **(*Calicut, 71 B.Sc. Engg.*)**

15. In a bar of 10 cm. long with its sides impervious to heat, the heat flow is one-dimensional and the two ends A and B are kept at 50°C. and 100°C. respectively, until steady-state conditions prevail. The temperature at A is then suddenly raised to 90°C. and at the same instant that at B is lowered to 60°C.; these end temperatures are maintained thereafter. Find an expression for the temperature at a distance x from A and at any time t subsequent to the changes in temperatures at the ends. **(*S.V.U., 66 B.E.*)**

16. A rod AB of length 10 cm. has the ends A and B kept at temperatures 30°C. and 100°C, respectively until the steady-state conditions prevail. At sometime later, the temperature at A is lowered to 20°C, and that at B to 40°C, and then these temperatures are maintained. Find the subsequent temperature distribution. **(*64 B.E.*)**

17. A bar, 10 cm. long, with insulated sides, has its ends A and B kept at 20°C, and 40°C, respectively, until steady-state conditions prevail. The temperature at A is then suddenly raised to 50°C, and at the same time that at B is lowered to 10°C. Find the subsequent temperature distribution $u(x, t)$ and show that the temperature at the middle point of the bar remains unaltered for all time, regardless of the material of the bar. **(*Madras, 63 B.E.*)**

18. A bar 40 cm. long has originally a temperature of 0°C, along all its length. At time $t = 0$ sec., the temperature at the end $x = 0$ is raised to 50°C, while that at the other end is raised 100°C. Determine the resulting temperature distribution. **(*Madras, 63 B.E.*)**

19. A rod of length l has its ends A and B kept at 0°C. and 100°C. respectively until steady-state conditions prevail. If the temperature of A is suddenly raised to 50°C. and that of B to 150°C., find the temperature distribution at any point of the rod and at any time. **(*Madras, 71 B.E.*)**

20. A metal bar 50 cms. long whose surface is insulated is at temperature 60°C. At $t = 0$, a temperature of 30°C. is applied at one end and a temperature of 80°C. to the other end and these temperature are maintained. Determine the temperature of the bar at any time assuming the diffusivity $K = 0{\cdot}15$ cgs unit. **(*Madras, 70 B.E.*)**

21. The ends A and B of a rod of 30 cms. long have their temperatures kept at 20°C. and the other at 80°C., until steady-state conditions prevail. The temperature of the end B is suddenly reduced to 60°C. and kept so while the end A is raised to 40°C. Find the temperature distribution in the rod after time t. **(*67 B.E.*)**

22. The temperature at one end of a bar, 50 cm. long and with insulated sides, is kept at 0°C. and the other end is kept at 100°C. until steady-state conditions prevail. The two ends are then suddenly insulated, so that the temperature gradient is zero at each end thereafter. Find the temperature distribution. **(*'63, '66 B.E.*)**

23. A uniform rod of length a whose surface is thermally insulated, is initially at temperature $\theta = \theta_0$. At time $t = 0$, one end is suddenly cooled to temperature $\theta = 0$ and subsequently maintained at this temperature. The other end remains thermally insulated; show that the temperature at this end at time t is given by

$$\theta = \frac{4\theta_0}{\pi} \sum_{n=0}^{\infty} \frac{(-1)^n}{(2n+1)} \exp\left[-\frac{(2n+1)^2 K\pi^2 t}{4a^2}\right]$$ where K is the thermometric conductivity. **(BR. 1995 April)**

24. The equation for the conduction of heat along the bar of length l is $\frac{\partial\theta}{\partial t} = k\frac{\partial^2\theta}{\partial x^2}$, neglecting radiation. Find an expression for θ, if the ends of the bar are maintained at zero temperature and if, initially, the temperature is T at the centre of the bar and falls uniformly to zero at its ends.

25. A rod of length l is at a uniform temperature of 50°C. Suddenly, the end $x = 0$ is cooled to 0°C. by an application of ice, and the end $x = l$ is heated to 100°C. by an application of steam and these two temperatures are maintained at the ends. Also the rod is insulated along its length so that no transfer of heat can occur from the sides. Find the temperature distribution in the rod at any subsequent time.

26. An insulated metal rod of length 100 cm has one end A kept at 0°C., and the other end B at 100°C. until steady-state conditions prevail. At $t = 0$, the temperature at A is then suddenly raised to 50°C. and thereafter maintained while at the same time $t = 0$ the end B is insulated. Find the temperature at any point of the rod at any subsequent time. **(S.V.U. '67 B.E.)**

3·7. Temperature in a slab with faces at zero temperature.

Consider a slab of homogeneous material bounded by two parallel planes $x = 0$ and $x = l$, having an initial temperature $u = f(x)$ which varies only with the distance from the faces, the two faces being kept at temperature zero. Since the heat-flow is one dimensional, the temperature is a function of x and t only. Hence the function $u(x, t)$ is the function representing the temperature at any interior point of the slab and satisfying the one-dimensional feat-flow equation

$$\frac{\partial^2 u}{\partial x^2} = \frac{1}{\alpha^2}\frac{\partial u}{\partial t}.$$

Example 23. *The faces of a slab of width l are kept at temperature zero. Its initial temperature is given by $u = f(x)$, where*

$$f(x) = A \text{ in } 0 < x < l/2$$
$$= 0 \text{ in } l/2 < x < l.$$

Find the temperature distribution.

The temperature function $u(x, t)$ satisfies

$$\frac{\partial u}{\partial t} = \alpha^2 \frac{\partial^2 u}{\partial x^2} \qquad \text{....(1)}$$

The boundary conditions are

$u(0, t) = 0$ for all $t > 0$ *(i)*

$u(l, t) = 0$ for all $t > 0$ *(ii)*

The initial conditions is

$$u(x, 0) = A, \text{ for } 0 < x < l/2$$

$$= 0, \text{ for } l/2 < x < l \qquad (iii).$$

Solving (1) and selecting the suitable solution as usual,

$$u(x, t) = (A \cos \lambda x + B \sin \lambda x)e^{-\alpha^2 \lambda^2 t} \qquad ....(2)$$

Using (*i*), we get $A = 0$.

Using (*ii*), $\lambda = \frac{n\pi}{l}$, where n is any integer.

Hence $u(x, t) = B_n \sin \frac{n\pi x}{l} e^{-\frac{\alpha^2 n^2 \pi^2 t}{l^2}}$

The most general solution of (1) is

$$u(x, t) = \sum_{n=1}^{\infty} B_n \sin \frac{n\pi x}{l} e^{-\frac{\alpha^3 n^2 \pi^2 t}{l^2}} \qquad ....(3)$$

Using (*iii*),

$$u(x, 0) = \sum_{n=1}^{\infty} B_n \sin \frac{n\pi x}{l} \qquad ....(4)$$

Expressing $u(x, 0)$ as a half-range Fourier sine series in $(0, l)$ and comparing it with (4), we find

$$B_n = b_n = \frac{2}{l} \int_0^l u(x, 0) \sin \frac{n\pi x}{l} dx$$

$$= \frac{2}{l} A \int_0^{l/2} \sin \frac{n\pi x}{l} dx, \text{ using } (iii)$$

$$= \frac{2A}{l} \left[-\frac{\cos \frac{n\pi x}{l}}{\frac{n\pi}{l}} \right]_0^{l/2}$$

$$= \frac{2A}{l} \cdot \frac{l}{n\pi} \left[1 - \cos \frac{n\pi}{2} \right]$$

$$= \frac{4A}{n\pi} \sin^2 n\pi/4.$$

Substituting in (3),

$$u(x, t) = \frac{4A}{\pi} \sum_{n=1}^{\infty} \frac{1}{n} \sin^2 \frac{n\pi}{4} \sin \frac{n\pi x}{l} \exp\left(- \frac{\alpha^2 n^2 \pi^2 t}{l^2} \right).$$

Exercises 3(d)

1. The faces $x = 0$ and $x = c$ of a slab which is initially at temperature $f(x)$ are kept at temperature zero. Derive the temperature formula

$$u(x, t) = \frac{2}{c} \sum_{n=1}^{\infty} \left(\int_0^c f(\theta) \sin \frac{n\pi\theta}{c} \, d\theta \right) \sin \frac{n\pi x}{c} \exp\,(-n^2\pi^2\alpha^2 t/c^2).$$

2. If $f(x) = \sin \dfrac{\pi x}{c}$ in problem 1, show that

$$u(x, t) = \sin \frac{\pi x}{c} \exp\,(-\alpha^2\pi^2 t/c^2).$$

3. Two slabs of iron, each of 20 cm. thick, one at temperature 100°C. and the other at 0°C. throughout, are placed face to face in perfect contact, and their outer faces are kept at 0°C. Given that $\alpha^2 = 0{\cdot}15$ centimetre-gram-second unit, show that the temperature at their common face 10 minutes after the contact was made is approximately 37°C.

4. If the slabs in question (3) above are made of material whose $\alpha^2 = 0{\cdot}005$ cgs unit, show that five hours are necessary for the common face to reach the temperature 36°C.

5. A homogeneous slabs of conducting material is bounded by the planes $x = 0$ and $x = \pi$ which are kept at temperature zero. If initially when $t = 0$, the slab is raised to temperature

$$u(x, 0) = \frac{\pi k x}{4} \text{ in } 0 \le x \le \pi/2$$

$$= \frac{\pi k}{4} (\pi - x) \text{ in } \pi/2 \le x \le \pi$$

where k is the constant in the heat equation

$\dfrac{\partial u}{\partial t} = k \dfrac{\partial^2 u}{\partial x^2}$, show that the temperature at any subsequent time is

$$k\left(e^{-kt} \sin x - \frac{1}{3^2} e^{-3^2 kt} \sin 3x + \frac{1}{5^2} e^{-5^2 kt} \sin 5x - \cdots \right).$$

6. Two slabs of the same material, one 2 feet thick and the other 1 foot thick, are placed side by side. The thicker slab is initially at constant temperature A, the thinner one initially at zero. The outer faces are held at temperature zero for $t > 0$. Show that the temperature at the centre of the two-feet slab is

$$\frac{4A}{\pi} \sum_{n=1}^{\infty} \frac{1}{n} \sin^2 \frac{n\pi}{3} \exp\left(-\alpha^2 n^2 \pi^2 \cdot t/9\right).$$

7. The faces $x = 0$ and $x = l$ of a slab which is initially at temperature $f(x)$ are insulated. Derive the formula

$$u(x, t) = \frac{1}{2} a_0 + \sum_{n=1}^{\infty} a_n \cos \frac{n\pi x}{l} \exp\left(-n^2 \pi^2 \alpha^2 t / l^2\right)$$

where $a_n = \frac{2}{l} \int_0^l f(x) \cos \frac{n\pi x}{l}\, dx.$

3·8. Two dimensional heat-flow

When the heat-flow is along curves instead of along straight lines, all the curves lying in parallel planes, then the flow is called two-dimensional. Let us consider now the flow of heat in a metal-plate in the *xoy* plane. Let the plate be of uniform thickness h, density ρ, thermal conductivity k and the specific heat c. Since the flow is two dimensional, the temperature at any point of the plate is independent of the z-coordinate. The heat-flow lies in the *xoy* plane and is zero along the direction normal to the *xoy* plane.

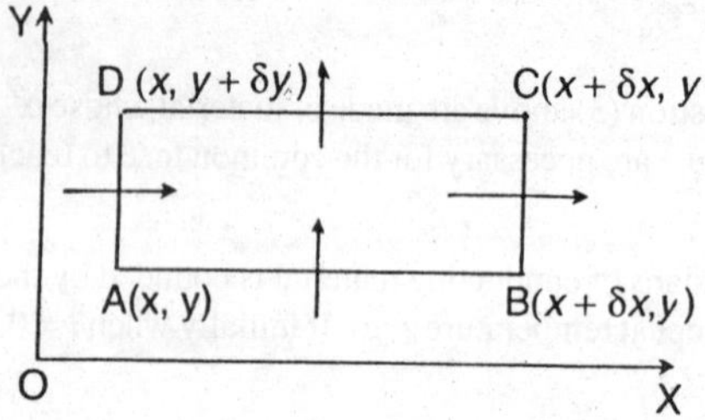

Now, consider a rectangular element *ABCD* of the plate with sides δx and δy, the edges being parallel to the coordinates axes, as shown in the figure. Then the quantity of heat *entering* the element *ABCD* per sec. through the surface *AB* is

$$= -k \left(\frac{\partial u}{\partial y}\right)_y \delta x \cdot h.$$

Similarly the quantity of heat entering the element *ABCD* per sec. through the surface *AD* is

$$= -k \left(\frac{\partial u}{\partial x}\right)_x \delta y \cdot h.$$

The amount of heat which flows *out* through the surfaces *BC* and *CD* are

$$-k\left(\frac{\partial u}{\partial x}\right)_{x+\delta x}\cdot \delta y\cdot h \text{ and } -k\left(\frac{\partial u}{\partial y}\right)_{y+\delta y}\delta x\cdot h \text{ respectively.}$$

Therefore the total gain of heat by the rectangular element *ABCD*

per sec. = inflow–outflow

$$= kh\left[\left\{\left(\frac{\partial u}{\partial x}\right)_{x+\delta x}-\left(\frac{\partial u}{\partial x}\right)_x\right\}\delta y+\left\{\left(\frac{\partial u}{\partial y}\right)_{y+\delta y}-\left(\frac{\partial u}{\partial y}\right)_y\delta x\right\}\right]$$

$$= kh\delta x\cdot\delta y\left[\frac{\left(\frac{\partial u}{\partial x}\right)_{x+\delta x}-\left(\frac{\partial u}{\partial x}\right)_x}{\delta x}+\frac{\left(\frac{\partial u}{\partial y}\right)_{y+\delta y}-\left(\frac{\partial u}{\partial y}\right)_y}{\delta y}\right] \quad \text{....(1)}$$

The rate of gain of heat by the element *ABCD* is also given by

$$\rho\delta x\cdot\delta y\cdot h\cdot c\cdot\frac{\partial u}{\partial t} \quad \text{....(2)}$$

Equating the two-expressions for gain of heat per sec. from (1) and (2), we have,

$$\rho\,\delta x\cdot\delta y\cdot h\cdot c\cdot\frac{\partial u}{\partial t}=h\,k\,\delta x\,\delta y\left[\frac{\left(\frac{\partial u}{\partial x}\right)_{x+\delta x}-\left(\frac{\partial u}{\partial x}\right)_x}{\delta x}+\frac{\left(\frac{\partial u}{\partial y}\right)_{y+\delta y}-\left(\frac{\partial u}{\partial y}\right)_y}{\delta y}\right]$$

i.e.,
$$\frac{\partial u}{\partial t}=\frac{k}{\rho c}\left[\frac{\left(\frac{\partial u}{\partial x}\right)_{x+\delta x}-\left(\frac{\partial u}{\partial x}\right)_x}{\delta x}+\frac{\left(\frac{\partial u}{\partial y}\right)_{y+\delta y}-\left(\frac{\partial u}{\partial y}\right)_y}{\delta y}\right]$$

Taking the limit as $\delta x\to 0,\ \delta y\to 0$, the above reduces to

$$\frac{\partial u}{\partial t}=\frac{k}{\rho c}\left(\frac{\partial^2 u}{\partial x^2}+\frac{\partial^2 u}{\partial y^2}\right)$$

Putting $\alpha^2=\dfrac{k}{\rho c}$ as before, the equation becomes,

$$\frac{\partial u}{\partial t}=\alpha^2\left(\frac{\partial^2 u}{\partial x^2}+\frac{\partial^2 u}{\partial y^2}\right) \quad \text{....(3)}$$

The equation (3) gives the temperature distribution of the plate in the transient state.

In the steady-state, u is independent of t, *so that* $\frac{\partial u}{\partial t} = 0$. *Hence the temperature distribution of the plate in the steady-state is* $\frac{\partial^2 u}{\partial x^2} + \frac{\partial^2 u}{\partial y^2} = 0$.

i.e., $\nabla^2 u = 0$, which is known as *Laplace's Equation* in two-dimensions.

Corollary. If the stream lines are parallel to the x-axis, then the rate of change $\frac{\partial u}{\partial y}$ of the temperature in the direction of the y-axis will be zero. Then the heat-flow equation reduces to $\frac{\partial u}{\partial t} = \alpha^2 \frac{\partial^2 u}{\partial x^2}$ which is the heat-flow equation in one-dimension.

3·9. Solution of the Equation $\frac{\partial^2 u}{\partial x^2} + \frac{\partial^2 u}{\partial y^2} = 0$.

The equation is $\frac{\partial^2 u}{\partial x^2} + \frac{\partial^2 u}{\partial y^2} = 0$. ....(1)

Assume the solution $u(x, y) = X(x) \cdot Y(y)$,
where X is a function of x alone and Y a function of y alone.

$\therefore \quad \frac{\partial^2 u}{\partial x^2} = X'' Y$, and $\frac{\partial^2 u}{\partial y^2} = XY''$.

The Laplace equation $\nabla^2 u = 0$ becomes $X''Y + Y'' X = 0$

i.e., $\frac{X''}{X} = \frac{-Y''}{Y}$ ....(2)

The left hand side of (2) is a function of x alone and the right hand side is a function of y alone. Also x and y are independent variables. Hence, this is possible only if each quantity is equal to a constant k.

$\therefore$ Let $\frac{X''}{X} = \frac{-Y''}{Y} = k$ ....(3)

i.e., $X'' - kX = 0$, and $Y'' + kY = 0$. ....(4)

Case 1. Let $k = \lambda^2$, a positive number.

Then $X'' - \lambda^2 X = 0$, and $Y'' + \lambda^2 Y = 0$.

Solving, $X = A_1 e^{\lambda x} + B_1 e^{-\lambda x}$ and $Y = C_1 \cos \lambda y + D_1 \sin \lambda y$.

Case 2. Let $k = -\lambda^2$, a negative number.

Then (4) becomes $X'' + \lambda^2 X = 0$ and $Y'' - \lambda^2 Y = 0$.

Solving these equations, we have,

$X = A_2 \cos \lambda x + B_2 \sin \lambda x$ and $Y = C_2 e^{\lambda y} + D_2 e^{-\lambda y}$.

Case 3. Let $k = 0$. Then (4) reduces to

$X'' = 0$ and $Y'' = 0$.

On solving these equations,

$X = A_3 x + B_3$ and $Y = C_3 y + D_3$.

Therefore, the possible solutions of (1) are

$$u(x, y) = (A_1 e^{\lambda x} + B_1 e^{-\lambda x})(C_1 \cos \lambda y + D_1 \sin \lambda y) \quad \text{...(I)}$$

$$u(x, y) = (A_2 \cos \lambda x + B_2 \sin \lambda x)\ (C_2 e^{\lambda y} + D_2 e^{-\lambda y}) \quad \text{...(II)}$$

$$u(x, y) = (A_3 x + B_3)(C_3 y + D_3) \quad \text{...(III)}$$

In problems where the boundary conditions are given, we have to select a suitable solution or a linear combination of solutions to satisfy (1) and the boundary conditions.

Example 24. *An infinitely long plane uniform plate is bounded by two parallel edges $x = 0$ and $x = l$, and an end at right angles to them. The breadth of this edge $y = 0$ is l and is maintained at a temperature $f(x)$. All the other three edges are at temperature zero. Find the steady-state temperature at any interior point of the plate.*

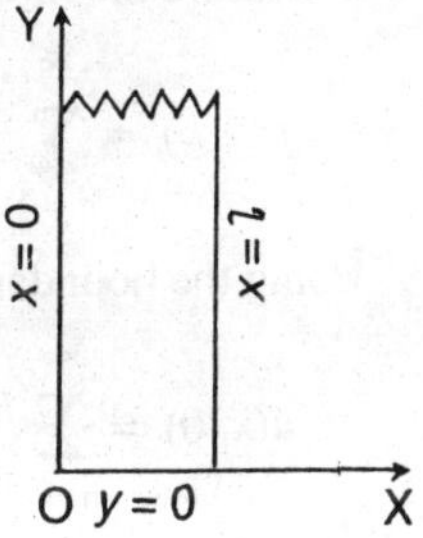

Let $u(x, y)$ be the temperature at any point (x, y) of the plate.

Then u satisfies

$$\frac{\partial^2 u}{\partial x^2} + \frac{\partial^2 u}{\partial y^2} = 0 \quad \text{....(1)}$$

The boundary conditions are

$$u(0, y) = 0, \quad \text{for } 0 \le y \le \infty \quad \text{....}(i)$$
$$u(l, y) = 0, \quad \text{for } 0 \le y \le \infty \quad \text{....}(ii)$$
$$u(x, \infty) = 0, \quad \text{for } 0 \le x \le l \quad \text{....}(iii)$$
$$u(x, 0) = f(x), \quad \text{for } 0 < x < l \quad \text{....}(iv)$$

Solving (1), we get,

$$u(x, y) = (A_1 e^{\lambda x} + B_1 e^{-\lambda x})(C_1 \cos \lambda y + D_1 \sin \lambda y) \quad \text{....(I)}$$

$$u(x, y) = (A_2 \cos \lambda x + B_2 \sin \lambda x)(C_2 e^{\lambda y} + D_2 e^{-\lambda y}) \quad \text{....(II)}$$

$$u(x, y) = (A_3 x + B_3)(C_3 y + D_3) \quad \text{....(III)}$$

Of these solutions, we have to select a solution to suit the boundary conditions.

Since $u = 0$ as $y \to \infty$, we select the solution (*II*) as a possible solution, (rejecting the other two).

$$\therefore \ u(x, y) = (A \cos \lambda x + B \sin \lambda x)\ (Ce^{\lambda y} + De^{-\lambda y}) \qquad ....(2)$$

Using the boundary condition (1),

$$u(0, y) = A\,(Ce^{\lambda y} + De^{-\lambda y}) = 0, \text{ for } 0 \le y \le \infty. \ \therefore A = 0$$

Using the boundary condition (*ii*) in (2),

$$u(l, y) = B \sin \lambda l\,(Ce^{\lambda y} + De^{-\lambda y}) = 0, \text{ for } 0 \le y \le \infty.$$

Since $B \neq 0$, $\sin \lambda l = 0$. Hence $\lambda l = n\pi$

i.e., $\lambda = \dfrac{n\pi}{l}$, where n is any integer.

As $y \to \infty$, $u \to 0$, from (*iii*). $\therefore C = 0$.

Hence $u(x, y) = B_n \sin \dfrac{n\pi x}{l}\, e^{-\frac{n\pi y}{l}}$, where $BD = B_n$.

Therefore the most general solution of (1) is

$$u(x, y) = \sum_{n=1}^{\infty} B_n \sin \frac{n\pi x}{l}\, e^{-\frac{n\pi y}{l}} \qquad ....(3)$$

Using the boundary condition (*iv*) in (3),

$$u(x, 0) = \sum_{n=1}^{\infty} B_n \sin \frac{n\pi x}{l} = f(x), \text{ in } 0 < x < l \qquad ....(4)$$

Expressing $f(x)$ as a half-range Fourier sine series in $(0, l)$, we have

$$f(x) = \sum_{1}^{\infty} b_n \sin \frac{n\pi x}{l} \qquad ....(5),$$

where $b_n = \dfrac{2}{l} \displaystyle\int_0^l f(x) \sin \frac{n\pi x}{l}\, dx.$

Comparing (4) and (5), $B_n = b_n = \dfrac{2}{l} \displaystyle\int_0^l f(x) \sin \frac{n\pi x}{l}\, dx.$

Therefore the solution is

$$u(x, y) = \sum_{n=1}^{\infty} \left(\frac{2}{l} \int_0^l f(x) \sin \frac{n\pi x}{l}\, dx \right) \sin \frac{n\pi x}{l} \cdot e^{-\frac{n\pi y}{l}}$$

Note. If $f(x)$ is given explicitly in any problem, evaluate the value of B_n from the integral and substitute.

Example 25. *The vertices of a thin square plate are (0, 0), (l, 0), (l, l) and (0, l). The upper edge of the square is maintained at an arbitary temperature given by u(x, l) = f (x). The other three edges are kept at temperature zero. Find the steady-state temperature at any point within the plate. Explain how you will find the steady-state temperature if an arbitary temperature distribution existed along each edge.*

Suppose that $u(x, y)$ is the temperature at any point (x, y) of the plate in steady-state.

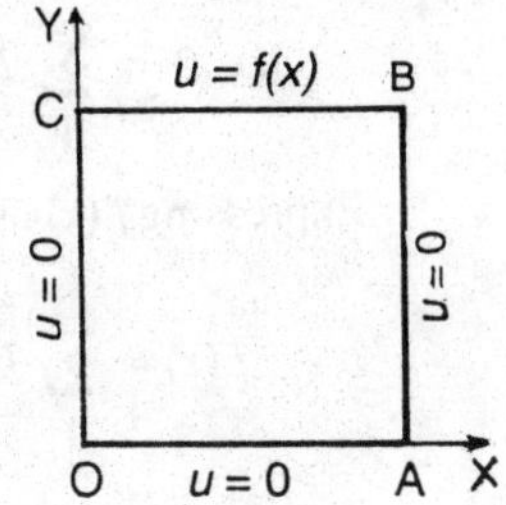

Then $\dfrac{\partial^2 u}{\partial x^2} + \dfrac{\partial^2 u}{\partial y^2} = 0$ ....(1)

The boundary conditions are

$u(0, y) = 0,\quad$ for $0 \le y < l$ (*i*)
$u(l, y) = 0,\quad$ for $0 \le y < l$ (*ii*)
$u(x, 0) = 0,\quad$ for $0 \le x \le l$ (*iii*)
$u(x, l) = f(x),\quad$ for $0 < x < l$ (*iv*)

Solving (1), we get the three possible solutions,

$$u(x, y) = (Ae^{\lambda x} + Be^{-\lambda x})(C \cos \lambda y + D \sin \lambda y) \quad \text{....(I)}$$

$$u(x, y) = (A \cos \lambda x + B \sin \lambda x)(C\, e^{\lambda y} + D\, e^{-\lambda y}) \quad \text{....(II)}$$

$$u(x, y) = (Ax + B)(Cy + D) \quad \text{....(III)}$$

where *A*, *B*, *C*, *D* are different arbitary constants in each solution.

Now we shall select the solution II.

i.e., $u(x, y) = (A \cos \lambda x + B \sin \lambda x)(Ce^{\lambda y} + De^{-\lambda y})$ ....(II)

Using the boundary condition (*i*) in (*II*),

$A\,(Ce^{\lambda y} + De^{-\lambda y}) = 0$, for $0 \le y < l$. $\therefore A = 0$

Using the condition (*ii*) in (*II*),

$u(l, y) = B \sin \lambda l\; (Ce^{\lambda y} + De^{-\lambda y}) = 0.$ But $B \ne 0; \sin \lambda l = 0$

i.e., $\lambda l = n\pi$

i.e., $\lambda = \dfrac{n\pi}{l}$ where *nn* is any integer.

Using (*iii*) in *II*,

$u(x, 0) = (C + D)(B \sin \lambda x) = 0$, for $0 \le x \le l$.

$B \ne 0$ Hence $C + D = 0$. $\therefore D = -C$.

Hence (*II*) reduces to,

$$u(x, y) = BC \sin \frac{n\pi x}{l}\left(e^{\frac{n\pi y}{l}} - e^{-\frac{n\pi y}{l}}\right)$$

$$= B_n \sin \frac{n\pi x}{l} \sinh \frac{n\pi y}{l} \text{ where } B_n = 2BC.$$

Hence the most general solution is

$$u(x, y) = \sum_{n=1}^{\infty} B_n \sin \frac{n\pi x}{l} \sinh \frac{n\pi y}{l} \qquad ....(2)$$

Using the boundary condition (*iv*) in (2),

$$u(x, l) = \sum_{1}^{\infty} B_n \sin \frac{n\pi x}{l} \sinh n\pi = f(x) \text{ for } 0 < x < l \qquad ....(3)$$

Expressing $f(x)$ as a half range Fourier sine series in $(0, l)$, we get

$$f(x) = \sum_{1}^{\infty} b_n \sin \frac{n\pi x}{l} \qquad ....(4)$$

where $b_n = \frac{2}{l} \int_0^l f(x) \sin \frac{n\pi x}{l} \, dx$

Comparing (3) and (4), $B_n \sinh n\pi = b_n$

$$\therefore \quad B_n = \frac{b_n}{\sinh n\pi} = \frac{2}{l \sinh n\pi} \int_0^l f(x) \sin \frac{n\pi x}{l} \, dx.$$

Hence putting this value in (2),

$$u(x, y) = \sum_{n=1}^{\infty} \left(\frac{2}{l \sinh n\pi} \int_0^l f(x) \sin \frac{n\pi x}{l} \, dx \right) \sin \frac{n\pi x}{l} \sinh n \frac{\pi y}{l}$$

Suppose that arbitrary temperature distribution exists along each edge, we can get four solutions each corresponding to a problem in which zero temperatures are prescribed along three of the four edges. Then by superimposing these four solutions, we get the solution of the given problem. In this case, note that you are doing four different problems of the type worked out in the above example.

Example 26. *A rectangular plate is bounded by the lines $x = 0$, $y = 0$, $x = a$ and $y = b$. Its surfaces are insulated and the temperature along two adjacent edges are kept at 100°C. while the temperature along the other two edges are at 0°C. Find the steady-state temperature at any point in the plate.*

Also find the steady-state temperature at any point of a square plate of side a if two adjacent edges are kept at 100°C. and the others at 0°C.

(***Madras B.E.***)

If $u(x, y)$ be the temperature at any interior point of the rectangular plate, then

$$\frac{\partial^2 u}{\partial x^2} + \frac{\partial^2 u}{\partial y^2} = 0 \qquad ....(1)$$

The boundary conditions are

$$u(0, y) = 0, \quad \text{for } 0 \leq y < b \qquad (i)$$
$$u(x, 0) = 0, \quad \text{for } 0 \leq x < a \qquad (ii)$$
$$u(a, y) = 100, \quad \text{for } 0 < y < b \qquad (iii)$$
$$u(x, b) = 100, \quad \text{for } 0 < x < a \qquad (iv)$$

Now we will split the solution into two solutions.

i.e., Let $u(x, y) = u_1(x, y) + u_2(x, y)$ ....(2)

where $u_1(x, y)$ and $u_2(x, y)$ are solutions of (1) and further $u_1(x, y)$ is the temperature at any point P with the edge BC maintained at 100°C. and the other three edges at 0°C. while $u_2(x, y)$ is the temperature at P with the edge AB maintained at 100°C. and the other three edges at 0°C.

(Refer to figure on page 234)

Therefore the boundary conditions for the functions $u_1(x, y)$ and $u_2(x, y)$ are as follows.

$u_1(0, y) = 0$	(*v*)	$u_2(0, y) = 0$	(*ix*)
$u_1(x, 0) = 0$	(*vi*)	$u_2(x, 0) = 0$	(*x*)
$u_1(a, y) = 0$	(*vii*)	$u_2(a, y) = 100$	(*xi*)
$u_1(x, b) = 100$	(*viii*)	$u_2(x, b) = 0$	(*xii*)

u_1, u_2 each satisfies (1).

Therefore u_1 will have three possible values when the equation (1) is solved for u.

i.e., $u_1(x, y) = (Ae^{\lambda x} + Be^{-\lambda x})(C\cos\lambda y + D\sin\lambda y)$ ...I

$u_1(x, y) = (A\cos\lambda x + B\sin\lambda x)(Ce^{\lambda y} + De^{-\lambda y})$ ...II

$u_1(x, y) = (Ax + B)(Cy + D)$ ...III

Select II solution and make use of the boundary conditions (*v*), (*vi*), (*vii*) as in example 25. We get,

$$u_1(x, y) = \sum_{n=1}^{\infty} B_n \sin\frac{n\pi x}{a} \sinh\frac{n\pi y}{a} \qquad ...(3)$$

Using boundary condition (*viii*) in (3)

$$u_1(x, b) = \sum_{n=1}^{\infty} B_n \sin\frac{n\pi x}{a} \sinh\frac{n\pi b}{a} = 100 \qquad ...(4)$$

Expressing 100 as a half-range Fourier sine series,

$$100 = \sum_{1}^{\infty} b_n \sin\frac{n\pi x}{a} \qquad ...(5)$$

where $b_n = \frac{2}{a}\int_0^a 100 \sin\frac{n\pi x}{a}\, dx = \frac{200}{n\pi}(1 - \cos n\pi)$

Comparing (5) and (4) $B_n \sinh\frac{n\pi b}{a} = b_n$.

$$\therefore \quad B_n = \frac{200}{n\pi \sinh\frac{n\pi b}{a}}(1 - \cos n\pi) = 0 \text{ if } n \text{ is even}$$

$$= \frac{400}{n\pi \sinh\frac{n\pi b}{a}} \text{ if } n \text{ is odd}$$

Substituting this value in (3),

$$u_1(x, y) = \sum_{n=1,3,5...}^{\infty} \frac{400}{n\pi} \frac{1}{\sinh\frac{n\pi b}{a}} \cdot \sin\frac{n\pi x}{a} \cdot \sinh\frac{n\pi y}{a} \qquad ...(6)$$

Similarly considering the boundary conditions of $u_2(x, y)$, we can easily obtain

$$u_2(x, y) = \frac{400}{\pi} \sum_{n=1,3,5....} \frac{1}{n} \frac{1}{\sinh\frac{n\pi a}{b}} \sinh\frac{n\pi x}{b} \sin\frac{n\pi y}{b} \qquad ...(7)$$

Since $u(x, y) = u_1(x, y) + u_2(x, y)$,

$$u(x, y) = \frac{400}{\pi} \sum_{n=1}^{\infty} \frac{1}{2n-1} \left[\frac{\sin\frac{(2n-1)\pi x}{a} \sinh\frac{(2n-1)\pi y}{a}}{\sinh\frac{(2n-1)\pi b}{a}} + \frac{\sin\frac{(2n-1)\pi y}{b} \cdot \sinh\frac{(2n-1)\pi x}{b}}{\sinh\frac{(2n-1)\pi a}{b}} \right]$$

Putting $b = a$, the rectangular plate becomes a square plate.

Hence the temperature at any point (x, y) in the square plate of side a is

$$u(x, y) = \frac{400}{\pi} \sum_{n=1}^{\infty} \frac{1}{(2n-1)\sinh(2n-1)\pi}$$

$$\times \left[\sin\frac{(2n-1)\pi x}{a} \sinh\frac{(2n-1)\pi y}{a} + \sin\frac{(2n-1)\pi y}{a} \sinh\frac{(2n-1)\pi x}{a} \right]$$

Note. While selecting the solution for $u_2(x, y)$, we choose solution *I* and not *II*.

Example 27. *An infinite long plate is bounded by two parallel edges and an end at right angles to them. The breadth is* π. *This end is maintained at a constant temperature* u_0 *at all points and the other edges are at zero temperature. Find the steady-state temperature at any point* (x, y) *of the plate.*

If $u(x, y)$ is the temperature at any point (x, y),

then $\dfrac{\partial^2 u}{\partial x^2} + \dfrac{\partial^2 u}{\partial y^2} = 0$ (steady-state) ...(1)

Solving, we get, three solutions

$$u(x, y) = (Ae^{\lambda x} + Be^{-\lambda x})(C\cos\lambda y + D\sin\lambda y) \quad \text{...I}$$

$$= (A\cos\lambda x + B\sin\lambda x)(Ce^{\lambda y} + De^{-\lambda y}) \quad \text{....II}$$

$$= (Ax + B)(Cy + D) \quad \text{...III}$$

where A, B, C, D are different arbitrary constants.

The boundary conditions are

$$\left.\begin{array}{l} u(0, y) = 0 \\ u(\pi, y) = 0 \end{array}\right\} \begin{array}{l} (i)\ 0 < y < \infty \\ (ii) \end{array}$$

$$\left.\begin{array}{l} u(x, 0) = u_0 \\ u(x, \infty) = 0 \end{array}\right\} \begin{array}{l} (iii)\ 0 < x < \pi \\ (iv) \end{array}$$

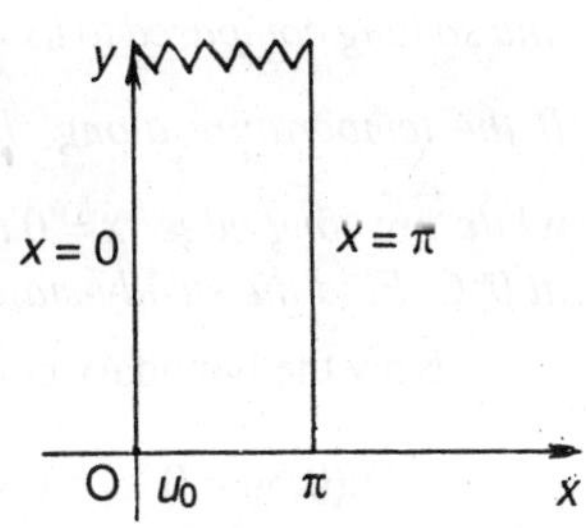

The non-zero condition occurs when x varies (*iii*) select that solution in which x occurs as Trigonometric function. That is, select solution *II*.

Let

$$u(x, y) = (A\cos\lambda x + B\sin\lambda x)(Ce^{\lambda y} + De^{-\lambda y}) \quad \text{...II}$$

use boundary condition (*i*) in *II*, we get $A=0$

use boundary condition (*iv*) in *II*, $C = 0$ (otherwise u becomes infinite)

using boundary condition (*ii*) in *II*,

$$B\sin\lambda\pi\,(De^{-\lambda y}) = 0$$

$\therefore \quad B \neq 0, \sin \lambda \pi = 0 \therefore \lambda = n$, any integer.

Hence, $u(x, y) = B_n \sin nx\, e^{-ny}$, n any integer.

The most general solution is

$$u(x, y) = \sum_{n=1}^{\infty} B_n \sin nx\; e^{-ny} \qquad \text{...(IV)}$$

using boundary condition (*iii*) in *IV*,

$$u(x, 0) = \sum_{1}^{\infty} B_n \sin nx = u_0. \qquad 0 < x < \pi$$

$$\therefore \qquad B_n = \frac{2}{\pi} \int_0^{\pi} u_0 \sin nx\, dx$$

$$= \frac{2u_0}{\pi} \left[-\frac{\cos nx}{n} \right]_0^{\pi}$$

$$= \frac{2u_0}{\pi} [1 - (-1)^n]$$

$$= 0 \text{ for } n \text{ even and } = \frac{4u_0}{n\pi} \text{ for } n \text{ odd.}$$

Hence, *IV* becomes,

$$u(x, y) = \frac{4u_0}{\pi} \sum_{n=1}^{\infty} \frac{1}{(2n-1)} \sin (2n-1)x\, e^{-(2n-1)y}$$

Example 28. *A rectangular plate with insulated surfaces is 8 cm. wide and so long compared to its width that it may be considered as an infinite plate. If the temperature along short edge y = 0 is u(x, 0) = 100 sin $\frac{\pi x}{8}$, $0 < x < 8$ while two long edges x = 0 and x = 8 as well as the other short edge are kept at 0°C. Find the steady-state temperature at any point of the plate.*

Now the boundary conditions are

$$u(0, y) = 0 \qquad 0 < y < \infty \qquad (i)$$

$$u(8, y) = 0 \qquad 0 < y < \infty \qquad (ii)$$

$$u(x, \infty) = 0 \qquad 0 < x < 8 \qquad (iii)$$

$$u(x, 0) = 100 \quad \sin\frac{\pi x}{8}\,;\ 0 < x < 8 \qquad (iv)$$

Non zero condition comes with x variable.

Select that solution in which x occurs in trigonometric function.

Let $u(x, y) = (A \cos \lambda x + B \sin \lambda x)(Ce^{\lambda y} + De^{-\lambda y})$ II

using (i), $A = 0$ using (iii), $C = 0$

using (ii), $B \sin 8\lambda = 0$; $8\lambda = n\pi$

$$\lambda = \frac{n\pi}{8}, \; n \text{ any integer}$$

$$\therefore \quad u(x, y) = B_n \sin \frac{n\pi x}{8} e^{-\frac{n\pi y}{8}}, \; n \text{ any integer}$$

Hence, most general solution is,

$$u(x, y) = \sum_{n=1}^{\infty} B_n \sin \frac{n\pi x}{8} e^{-\frac{n\pi y}{8}} \qquad \text{....(IV)}$$

using boundary condition, (iv) in IV,

$$u(x, 0) = \sum_{1}^{\infty} B_n \sin \frac{n\pi x}{8} = 100 \sin \frac{\pi x}{8}$$

$$\therefore \qquad B_1 = 100 \quad B_n = 0, \quad n \neq 1$$

Put in IV,

$$\therefore \quad u(x, y) = 100 \sin \frac{\pi x}{8} e^{\frac{-\pi y}{8}} \text{ is temperature function}$$

at any point of the plate.

Example 29. *A rectangular plate with insulated surface is 10 cm. wide and so long compared to its width that it may be considered infinite length without introducing appreciable error. If the temperature at short edge y = 0 is given by* **(*Anna Ap. 2005*)**

$$u = 20\,x \text{ for } 0 \le x \le 5$$
$$= 20\,(10 - x) \text{ for } 5 \le x \le 10$$

and all the other three edges are kept at 0°C. Find the steady-state temperature at any point of the plate.

Select

$$u(x, y) = (A \cos \lambda x + B \sin \lambda x)(Ce^{\lambda y} + De^{-\lambda y}) \qquad \text{....II}$$

The boundary conditions are

$$u(0, y) = 0, \quad 0 < y < \infty \quad (i)$$
$$u(10, y) = 0, \quad 0 < y < \infty \quad (ii)$$
$$u(x, \infty) = 0, \quad 0 < x < 10 \quad (iii)$$
$$u(x, 0) = 20x, \quad \text{for } 0 \le x \le 5$$
$$= 20(10 - x) \text{ for } 5 \le x \le 10 \quad (iv)$$

using (i) and (iii), in II, we get $A = 0, C = 0$

using (ii), $\lambda = \frac{n\pi}{10}$, n any integer.

Hence, $u(x, y) = B_n \sin \frac{n\pi x}{10} e^{\frac{-n\pi y}{10}}$, n any integer.

Therefore, the most general solution is

$$u(x, y) = \sum_{n=1}^{\infty} B_n \sin \frac{n\pi x}{10} e^{\frac{-n\pi y}{10}} \quad \text{....(IV)}$$

using $u(x, 0)$ condition in IV,

$$u(x, 0) = \sum_{1}^{\infty} B_n \sin \frac{n\pi x}{10} = \text{given function of } x.$$

$$\therefore \quad B_n = \frac{2}{10} \int_0^{10} f(x)\, dx$$

$$= \frac{1}{5} \left[\int_0^5 20\, x \sin \frac{n\pi x}{10} dx + \int_5^{10} 20(10 - x) \sin \frac{n\pi x}{10} dx \right]$$

$$= 4 \left[\left\{ (x) \left(-\frac{\cos \frac{n\pi x}{10}}{\frac{n\pi}{10}} \right) - \left(-\frac{\sin \frac{n\pi x}{10}}{\frac{n^2\pi^2}{100}} \right) \right\}_0^5 \right.$$

$$\left. + \left\{ (10 - x) \left(-\frac{\cos \frac{n\pi x}{10}}{\frac{n\pi}{10}} \right) - (-1) \left(-\frac{\sin \frac{n\pi x}{10}}{\frac{n^2\pi^2}{100}} \right) \right\}_5^{10} \right]$$

$$= 4 \left[\frac{-50}{n\pi} \cos \frac{n\pi}{2} + \frac{100}{n^2\pi^2} \sin \frac{n\pi}{2} + \frac{50}{n\pi} \cos \frac{n\pi}{2} + \frac{100}{n^2\pi^2} \sin \frac{n\pi}{2} \right]$$

$$= \frac{800}{n^2\pi^2} \sin \frac{n\pi}{2}$$

$= 0$ for n even

$= \dfrac{800}{n^2\pi^2} \sin \dfrac{n\pi}{2}$ for n odd. Substituting in *IV*,

$$\therefore\ u(x, y) = \frac{800}{\pi^2} \sum_{n=1,3,5\ldots}^{\infty} \frac{1}{n^2} \sin n\pi/2 \sin \frac{n\pi x}{10}\, e^{-\frac{n\pi y}{10}}$$

$$= \frac{800}{\pi^2} \sum_{n=1}^{\infty} \frac{(-1)^{n-1}}{(2n-1)^2} \sin \frac{(2n-1)\pi x}{10}\, e^{\frac{-(2n-1)\pi y}{10}}$$

Example 30. *A rectangular plate is bounded by lines $x = 0$, $x = a$, $y = 0$ and $y = b$ and the edge temperatures are $u(0, y) = 0$, $u(x, b) = 0$ $u(a, y) = 0$, and $u(x, 0) = 5 \sin \dfrac{5\pi x}{a} + 3 \sin \dfrac{3\pi x}{a}$. Find the steady-state temperature distribution at any point of the plate.*

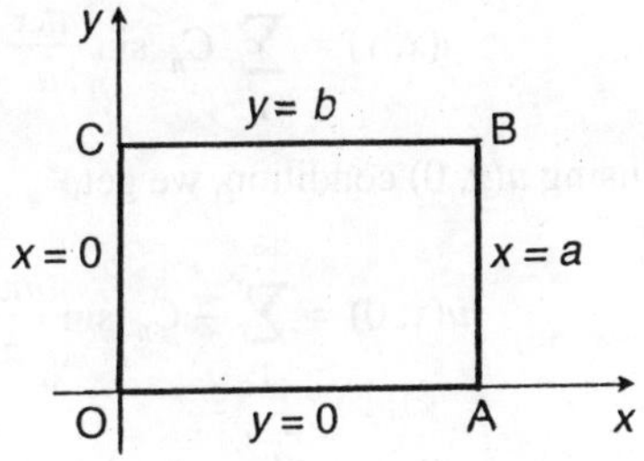

The non-zero boundary condition is in the variable x.

Hence, select that solution in which x occurs in trigonometric function.

Let $u(x, y) = (A \cos \lambda x + B \sin \lambda x)(Ce^{\lambda y} + De^{-\lambda y})$ ...II

Boundary conditions are

$u(0, y) = 0$ if $0 < y < b$ (*i*)

$u(a, y) = 0$ if $0 < y < b$ (*ii*)

$u(x, b) = 0$ if $0 < x < a$ (*iii*)

$u(x, 0) = 5 \sin \dfrac{5\pi x}{a} + 3 \sin \dfrac{3\pi x}{a}$, $0 < x < a$ (*iv*)

using (*i*) in *II*, $A = 0$

using (*ii*) in *II*, $B \sin \lambda a = 0$ $\therefore$ $B \neq 0$;

$$\lambda a = n\pi \therefore \lambda = \frac{n\pi}{a},\ n \text{ any integer.}$$

use (*iii*);

$$u(x, b) = B \sin \lambda x\ (Ce^{\lambda b} + De^{-\lambda b}) = 0$$

$\therefore \quad B \neq 0; \quad Ce^{\lambda b} = -De^{-\lambda b} \quad \therefore \; D = -Ce^{2\lambda b}$

Hence II reduces to

$$u(x, y) = B_n \sin \frac{n\pi x}{a} \left(e^{\frac{n\pi y}{a}} - e^{2\lambda b} \; e^{\frac{-n\pi y}{b}} \right)$$

$$= B'_n \sin \frac{n\pi x}{a} \left[e^{\frac{n\pi}{a}(y-b)} - e^{\frac{-n\pi}{a}(y-b)} \right]$$

$$= C_n \sin \frac{n\pi x}{a} \sin h \frac{n\pi}{a}(y-b)$$

Most general solution is

$$u(x, y) = \sum_{1}^{\infty} C_n \sin \frac{n\pi x}{a} \sinh \frac{n\pi}{a}(y-b) \qquad \text{....IV}$$

using $u(x, 0)$ condition, we get,

$$u(x, 0) = \sum_{1}^{\infty} -C_n \sin \frac{n\pi x}{a} \cdot \sinh \frac{n\pi b}{a} = 5 \sin \frac{5\pi x}{a} + 3 \sin \frac{3\pi x}{a}$$

$$\therefore \; -C_5 \sinh \frac{5\pi b}{a} = 5$$

$$-C_3 \sinh \frac{3\pi b}{a} = 3 \; ; \; C_n = 0; \; n \neq 3, \; n \neq 5$$

use in IV

$$u(x, y) = \frac{3}{\sinh \frac{3\pi b}{a}} \sin \frac{3\pi x}{a} \cdot \sinh \frac{3\pi}{a}(b-y)$$

$$+ \frac{5}{\sinh \frac{5\pi b}{a}} \cdot \sin \frac{5\pi x}{a} \cdot \sinh \frac{5\pi}{a}(b-y)$$

[**Hint.** $\sinh(-\theta) = -\sinh\theta$]

Example 31. *Solve for the steady-state temperature at any point of a rectangular plate of sides a and b insulated on the lateral surface and satisfying $u(0, y) = 0$, $u(a, y) = 0$, $u(x, b) = 0$ and $u(x, 0) = x(a - x)$.*

Select

$$u(x, y) = (A \cos \lambda x + B \sin \lambda x)(Ce^{\lambda y} + De^{-\lambda y}) \qquad \text{....II}$$

using $u(0, y) = 0, \; A = 0$

using $u(a, y) = 0$ we get $B \sin \lambda a = 0 \quad \therefore \; \lambda a = n\pi$

$$\lambda = \frac{n\pi}{a}, \; n \text{ any integer.}$$

use $\quad u(x, b) = 0$

$$\therefore B \sin \lambda x \; (Ce^{\lambda b} + De^{-\lambda b}) = 0$$

$$\therefore Ce^{\lambda b} = -De^{-\lambda b}$$

$$D = -Ce^{2\lambda b}$$

$$\therefore \quad u(x, y) = B' \sin \lambda x \cdot (e^{\lambda y} - e^{2\lambda b} e^{-\lambda y})$$

$$= C \sin \lambda x \; (e^{\lambda(y-b)} - e^{-\lambda(y-b)})$$

$$= K \sin \lambda x \; \sinh \lambda (b - y)$$

Most general solution

$$u(x, y) = \sum_{n=1}^{\infty} K_n \sin \frac{n\pi x}{a} \sinh \frac{n\pi}{a} (b - y) \qquad \text{....IV}$$

Now use the last boundary condition in *IV*

$$u(x, 0) = \sum_{1}^{\infty} K_n \sin \frac{n\pi x}{a} \cdot \sinh \frac{n\pi b}{a} = x(a - x).$$

This is half-range Fourier sine series of $f(x)$.

$$\therefore \; K_n \sinh \frac{n\pi b}{a} = \frac{2}{a} \int_0^a (ax - x^2) \sin \frac{n\pi x}{a} \, dx$$

$$= \frac{2}{a} \left[(ax - x^2) \left(- \frac{\cos \frac{n\pi x}{a}}{\frac{n\pi}{a}} \right) - (a - 2x) \left(- \frac{\sin \frac{n\pi x}{a}}{\frac{n^2\pi^2}{a^2}} \right) + (-2) \left(+ \frac{\cos \frac{n\pi x}{a}}{\frac{n^3\pi^3}{a^3}} \right) \right]_0^a$$

$$= \frac{2}{a} \left[\frac{-2a^3}{n^3\pi^3} \left((-1)^n - 1 \right) \right]$$

$$= 0 \text{ if } n \text{ is even and } = \frac{8a^2}{n^3\pi^3} \text{ if } n \text{ is odd}$$

Putting in *IV*,

$$u(x, y) = \frac{8a^2}{\pi^3} \sum_{n=1}^{\infty} \frac{1}{(2n-1)^3} \frac{\sin \frac{(2n-1)\pi x}{a} \sinh \frac{(2n-1)\pi}{a}(b-y)}{\sinh \frac{(2n-1)\pi b}{a}}$$

Example 32. *A square plate has its faces and the edge $y = 0$ insulated. Its edges $x = 0$ and $x = \pi$ are kept at zero temperature and its fourth edge $y = \pi$ is kept as temperature $f(x)$. Find the steady-state temperature at any point of the plate.*

The boundary conditions are

$$u(0, y) = 0, \quad 0 < y < \pi, \quad (i)$$

$$u(\pi, y) = 0, \quad 0 < y < \pi \quad (ii)$$

$$\left(\frac{\partial u}{\partial y}\right)_{y=0} = 0, \quad 0 < x < \pi \quad (iii), \text{ since } y = 0 \text{ is insulated.}$$

$$u(x, \pi) = f(x), \quad 0 < x < \pi \quad (iv)$$

Let $u(x, y) = (A \cos \lambda x + B \sin \lambda x)(A' e^{\lambda y} + B' e^{-\lambda y})$

This can also be rewritten as

$$u(x, y) = (A \cos \lambda x + B \sin \lambda x)(C \cosh \lambda y + D \sinh \lambda y) \quad \text{....II}$$

using $u(0, y) = 0$; we get $A = 0$

using $u(\pi, y) = 0$, $B \sin \lambda\pi = 0 \quad \therefore \lambda = n$, any integer

since $B \neq 0$.

$$u(x, y) = B \sin n x \,(C \cosh xy + D \sinh xy)$$

$$\frac{\partial u}{\partial y} = B \sin nx \,[Cn \sinh ny + nD \cosh ny]$$

$$\frac{\partial u}{\partial y}(x, 0) = 0 \text{ implies, } D = 0 \quad (\because B \neq 0)$$

$\therefore \quad u(x, y) = B_n \sin n x \cosh ny$, n any integer

The most general solution is

$$u(x, y) = \sum_{n=1}^{\infty} B_n \sin nx \cosh ny \quad \text{....IV}$$

using $u(x, \pi) = f(x)$,

$$u(x, \pi) = \sum_{1}^{\infty} B_n \cosh n\pi \sin nx = f(x)$$

$$B_n \cosh n\pi = \frac{2}{\pi}\int_0^{\pi} f(x)\sin nx\,dx$$

$$\therefore \quad B_n = \frac{2}{\pi\cosh\pi}\int_0^{\pi} f(x)\sin nx\,dx \qquad \text{....V}$$

$$u(x,y) = \frac{2}{\pi}\sum_{n=1}^{\infty}\frac{\sin nx\cosh ny}{\cosh n\pi}\cdot\left(\int_0^{\pi} f(x)\sin nx\,dx\right)\cdot$$

Exercises 3(e)

1. An infinitely long plane uniform plate is bounded by two parallel edges and an end at right angles to them. The breadth is π; this end is maintained at a temperature 100°C at all points and other edges are at zero temperature. Show that, in the steady-state, the temperature is given by

$$u(x,y) = \frac{400}{\pi}\left[e^{-y}\sin x + \frac{1}{3}e^{-3y}\sin 3x + \frac{1}{5}e^{-5y}\sin 5x + ...\right]$$

2. A rectangular plate with insulated surfaces is l cm. wide and so long compared to its width that it may be considered infinite in the length without introducing an appreciable error. If the temperature along one short edge $y = 0$ is given by $u(x, 0) = 100\sin\frac{\pi x}{l}$, $0 < x < l$, while the two long edges $x = 0$ and $x = l$ as well as the other short edge, are kept at , how that the steady-state temperature

$$u(x,y) = 100\sin\frac{\pi x}{l}\,e^{-\frac{\pi y}{l}}$$

3. An infinitely long metal plate in the form of an area is enclosed between the lines $y = 0$ and $y = l$ for $x > 0$. The temperature is zero along the edges $y = 0$ and $y = l$ and at infinity. If the edge $x = 0$ is kept at a constant temperature u_0, show that the steady-state temperature distribution is given by

$$u(x,y) = \frac{4u_0}{\pi}\left[e^{-\lambda x}\sin\lambda y + \frac{1}{3}e^{-3\lambda x}\sin 3\lambda y + \frac{1}{5}e^{-5\lambda x}\sin 5\lambda y + \cdots\right]$$

where $\lambda l = \pi$. ***(Madras, 70 B.E.)***

4. A thin iron plate bounded by $x = 0, y = 0, x = 5$ and $y = \infty$ has the temperature at the end $y = 0$ and is given by $\phi(x)$, and has the temperature at other edges always zero. Assuming that the heat cannot escape from either surface of the plate, the temperature $u(x, y)$ at any point in the steady-state conditions is the given by

$$u(x,y) = \sum_{1}^{\infty} B_n \sin\frac{n\pi x}{5}\,e^{-\frac{n\pi y}{5}}$$

where $B_n = \frac{2}{5}\int_0^5 \phi(x) \sin \frac{n\pi x}{5}\, dx.$

5. The temperature T is maintained at 0°C. along three edges of a square plate of length 100 cm. and the fourth edge is maintained at 100°C. until the steady-state conditions prevail. Find the expression for the temperature at any point on the Plate. Also calculate the temperature at the centre of the plate.

(Madras, '69 B.E.)

6. A rectangular plate with insulated surfaces is 10 cm. wide and so long compared to its width that it may be considered infinite in length without introducing an appreciable error. If the temperature along one short edge ($y = 0$) is $T(x, 0) = 4\,(10x - x^2)$ degree, $0 < x < 10$ while the two long edges $x = 0$? and $x = 10$ as well as the other short edge are kept at 0°C. find the steady-state temperature function $T(x, y)$. ***(Madras, 64 B.E.)***

7. A long rectangular plate has its surface insulated and the two long sides, as well as one of the short sides are maintained at 0°C. Find an expression for the steady-state temperature $T(x, y)$ if the short side $y = 0$ is 30 cm long and kept at 40°C. ***(Madras, 64 B.E.)***

8. A rectangular plate with insulated surface is 10 cm. wide and so long compared to its width that it may be considered infinite in length without introducing an appreciable error. If the temperature of the short edge $y = 0$ is given by $u = x$ for $0 \le x \le 5$ and $(10 - x)$ for $5 \le x \le 10$ and the two long edges $x = 0, x = 10$ as well as the other short edge are kept at 0°C., prove that the temperature $u(x,y)$ at any point (x, y) of the plate is given by

$$u(x, y) = \frac{40}{\pi^2} \sum_{n=1}^{\infty} \frac{(-1)^{n+1}}{(2n-1)^2} \sin \frac{(2n-1)\pi x}{10}\, e^{-\frac{(2n-1)n\pi}{10}}$$

9. The temperature u is maintained at 0°C. along three edges of a square plate of length 100° cm. and the fourth edge is maintained at a constant temperature u_0 until steady-state conditions prevail. Find an expression for the temperature u at any point (x, y) of the plate. Calculate the temperature at the centre of the plate.

(BR. 1995 Ap.)

10. A square plate is bounded by the lines $x = 0, y = 0\; x = 20$ and $y = 20$ and its faces are insulated. The temperature along the upper horizontal edge is given by $u(x, 20) = x(20 - x)$ for $0 < x < 20$, while the other three edges are maintained at 0°C. Find the steady-state temperature distribution in the plate.

(S.V.U., 67 B.E.)

11. Obtain the steady-state temperature in a rectangular plate of length a and width b, the sides of which are kept at temperature zero, the lower horizontal length is kept at temperature $f(x)$ while the upper edge is kept insulated.

12. Obtain the steady-state temperature distribution in a rectangular metal plate of length a and width b, the sides of which are kept at 0°C., the lower edge is kept at 0°C. while the upper edge is kept insulated.

13. If a rectangular plate is bounded by the lines $x = 0, x = a, \ y = 0, \ y = b$ and the edge temperature are $u(0, y) = 0, \ u(a, y) = 0, \ u(x, b) = 0, \ u(x, 0) = 5 \sin \frac{4\pi x}{a} + 3 \sin \frac{3\pi x}{a}$, find the temperature distribution. *(Madras, 64 B.E.)*

14. A square metal plate of side a has edges represented by the lines $x = a$ and $y = a$ insulated. The edge $x = 0$ is kept at 0°C. and the edge $y = 0$ at T°C.. where T is a constant. Show that the temperature distribution in the steady-state is

$$u(x, y) = \frac{4T}{\pi} \sum_{1}^{\infty} \frac{\operatorname{sech} (2n-1) \frac{\pi}{2}}{(2n-1)} \times \cosh \frac{(2n-1)\pi(a-y)}{2a} \cdot \sin \frac{(2n-1)\pi x}{2a}.$$

15. A square iron plate of side l is bounded by $x = 0, y = 0, x = l, y = l$. The edges $x = 0, y = l$ are kept at zero temperature while the edge $y = 0$ is insulated and the edge $x = l$ is kept at T°C. Show that, in the steady-state, the temperature at any point of the plate is

$$u(x, y) = \frac{4T}{\pi} \sum_{n=1}^{\infty} \frac{(-1)^{n-1} \sinh \frac{(2n-1)\pi x}{2a} \cdot \cos \frac{(2n-1)\pi y}{2a}}{(2n-1) \sinh (2n-1) \frac{\pi}{2}}.$$

16. Find the steady-state temperature distribution in a rectangular plate of side a and b insulated on the lateral surfaces and satisfying the following boundary conditions.

$u(0, y) = u(a, y) = u(x, b) = 0$ and $u(x, 0) = x(a - x)$.

17. A rectangular metal plate has sides of lengths a and b. Taking the side of length a as ox and one of the length b as oy and the sides to be $x = a$ and $y = b$, the sides $x = 0, x = a, y = b$ are insulated and the edge $y = 0$ is kept at temperature $u_0 \cos \frac{\pi x}{a}$. Show that the temperature in the steady-state is

$$u(x, y) = u_0 \operatorname{sech} \frac{\pi b}{a} \cosh \frac{\pi(b-y)}{a} \cdot \frac{\cos \pi x}{a}.$$

18. A long rectangular plate has its surface insulated and the two long sides, as well as one of the short sides are kept at 0°C. while the other short side $u(x, 0) = 3x$ and the length being 5 cm. Find an expression for the steady-state temperature.

(Os. U., 64 B.E.)

19. A long rectangular plate of width l cm. with insulated surface has its temperature u equal to zero on both the long sides and one of the shorter sides so that $u(0, y) = u(l, y) = u(x, \infty) = 0$ and $u(x, 0) = kx$. Show that the steady-state temperature is

$$u(x, y) = \frac{2lk}{\pi} \sum_{n=1}^{\infty} \frac{(-1)^{n+1}}{n} \sin \frac{n\pi x}{l} e^{-\frac{n\pi y}{l}}$$

20. A square plate has its faces and its edges $x = 0$ and $x = \pi \ (0 < y < \pi)$ insulated. Its edge $y = 0$ and $y = \pi$ are kept temperatures zero and $f(x)$ respectively. Derive this formula for steady temperatures

$$u(x, y) = \frac{1}{2\pi} a_0 y + \sum_{n=1}^{\infty} a_n \frac{\sinh ny}{\sinh n\pi} \cos nx$$

where $a_n = \frac{2}{\pi} \int_0^{\pi} f(x) \cos nx \, dx$ for $n = 0, 1, 2, \ldots$

21. Solve $\frac{\partial^2 u}{\partial x^2} + \frac{\partial^2 u}{\partial y^2} = 0$ for $0 < x < \pi$; $0 < y < \pi$

$$u(x, 0) = x^2, \; u(x, \pi) = 0$$

$$\frac{\partial u}{\partial x}(0, y) = \frac{\partial u}{\partial x}(\pi, y) = 0.$$

3·10. Equation of heat-flow in two dimension in polar co-ordinates.

Consider a sheet of conducting material of uniform density ρ, uniform thickness h, thermal conductivity k and specific heat c. Let O, the pole and OX, the initial line, be taken on the plane of sheet. As we are dealing with two dimensional heat-flow, the temperature function $u(r, \theta, t)$ at point (r, θ) of the plate is a function of r, θ and time t.

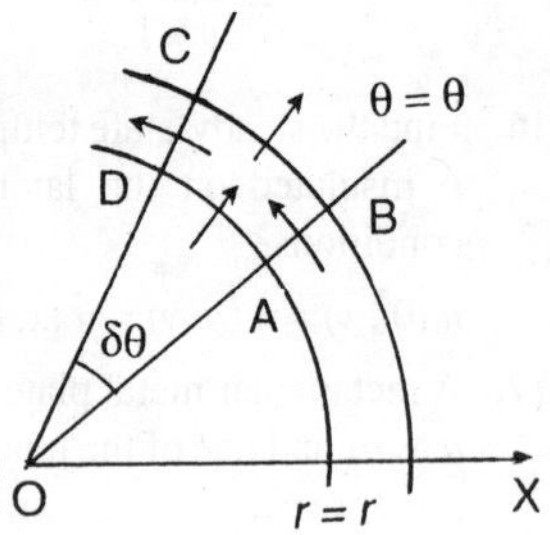

Consider an element of the sheet included between the circles $r = r$, $r = r + \delta r$ and the straight lines $\theta = \theta$ and $\theta = \theta + \delta\theta$ through the pole. Heat-flow directions are assumed to be positive in the positive directions associated with r and θ. The mass of the element $ABCD = \rho h (r\delta r\delta\theta)$

Let δu denote the temperature rise in the element during a short interval of time δt succeeding the time t.

$\therefore$ Rate of increase of heat content in the element

$$= \lim_{\delta t \to 0} \rho h (r\delta r \delta\theta) c \frac{\delta u}{\delta t}$$

$$= \rho h c r \delta r \delta\theta \frac{\partial u}{\partial t} \qquad \ldots(1)$$

If R_1, R_2, R_3, R_4 are the rates of flow of heat across the sides of the element through the edges AB, CD, AD and BC respectively, then

$$R_1 = -k \left(\frac{1}{r} \frac{\partial u}{\partial \theta} \right)_{\theta} h\delta r$$

$$R_2 = -k\left(\frac{1}{r}\frac{\partial u}{\partial \theta}\right)_{\theta+\delta\theta} h\delta r$$

$$R_3 = -k\left(\frac{\partial u}{\partial r}\right)_{r} hr\delta\theta$$

$$R_4 = -k\left(\frac{\partial u}{\partial r}\right)_{r+\delta r} h\,(r+\delta r)\,\delta\theta$$

The rate of increase of heat in the element $= R_1 - R_2 + R_3 - R_4$ which is equal to the expression in (1)

$$\therefore\ \rho hcr\delta r\delta\theta\,\frac{\partial u}{\partial t} = kh\left[\left\{\left(\frac{1}{r}\frac{\partial u}{\partial \theta}\right)_{\theta+\delta\theta} - \left(\frac{1}{r}\frac{\partial u}{\partial \theta}\right)_{\theta}\right\}\delta r + \left\{\left(\frac{\partial u}{\partial r}\right)_{r+\delta r} - \left(\frac{\partial u}{\partial r}\right)_{r}\right\} r\delta\theta + \delta r\delta\theta\left(\frac{\partial u}{\partial r}\right)_{r+\delta r}\right]$$

Dividing by $\delta r \cdot \delta\theta \cdot hr\rho c$,

$$\frac{\partial u}{\partial t} = \frac{k}{\rho c}\left[\frac{1}{r^2}\,\frac{\left\{\left(\frac{\partial u}{\partial \theta}\right)_{\theta+\delta\theta} - \left(\frac{\partial u}{\partial \theta}\right)_{\theta}\right\}}{\delta\theta} + \frac{\left(\frac{\partial u}{\partial r}\right)_{r+\delta r} - \left(\frac{\partial u}{\partial r}\right)_{r}}{\delta r} + \frac{1}{r}\left(\frac{\partial u}{\partial r}\right)_{r+\delta r}\right]$$

Taking the limit as $\delta\theta, \delta r \to 0$,

$$\frac{\partial u}{\partial t} = \frac{k}{pc}\left[\frac{1}{r^2}\frac{\partial^2 u}{\partial \theta^2} + \frac{\partial^2 u}{\partial r^2} + \frac{1}{r}\frac{\partial u}{\partial r}\right]$$

Therefore the equation of heat-flow in polar coordinates is

$$\frac{\partial^2 u}{\partial r^2} + \frac{1}{r}\frac{\partial u}{\partial r} + \frac{1}{r^2}\frac{\partial^2 u}{\delta \theta^2} = \frac{1}{\alpha^2}\frac{\partial u}{\partial t} \text{ where } \alpha^2 = \frac{k}{\rho c}$$

In steady-state, $\frac{\partial u}{\partial t} = 0$. Hence the equation of heat-flow in steady-state is

$$\nabla^2 u \equiv \frac{\partial^2 u}{\partial r^2} + \frac{1}{r}\frac{\partial u}{\partial r} + \frac{1}{r^2}\frac{\partial^2 u}{\partial \theta^2} = 0$$

$$i.e.,\ r^2\frac{\partial^2 u}{\partial r^2} + r\frac{\partial u}{\partial r} + \frac{\partial^2 u}{\partial \theta^2} = 0.$$

3·11. Solution of Laplace's equation in polar coordinates

We have to solve $r^2 \frac{\partial^2 u}{\partial r^2} + r\frac{\partial u}{\partial r} + \frac{\partial^2 u}{\partial \theta^2} = 0$ ...(1)

Assume a solution $u(r, \theta) = R(r)\,\phi(\theta)$ ...(2)

where R is a function of r alone and ϕ is a function of θ alone.

Therefore (1) becomes,

$$r^2 R''\phi + rR'\phi + R\phi'' = 0.$$

i.e., $$\frac{r^2R'' + rR'}{R} = \frac{-\phi''}{\phi} \quad ...(3)$$

The left hand side is a function of r only while the right hand side is a function of θ only and r, θ are independent variables. This is possible only if each ratio is a constant, k say.

$$\therefore \quad \frac{r^2R'' + rR'}{R} = -\frac{\phi''}{\phi} = k, \ k \text{ being a constant.}$$

$$\therefore \quad r^2R'' + rR' - kR = 0 \quad ...(4)$$

$$\phi'' + k\phi = 0 \quad ...(5)$$

Now, equation (4) is a differential equation with variable coefficients.

Put $r = e^z$ *i.e.,* $z = \log r$.

$$R' = \frac{dR}{dr} = \frac{dR}{dz}\cdot\frac{dz}{dr} = \frac{dR}{dz}\cdot\frac{1}{r}$$

$$\therefore \quad rR' = \frac{dR}{dz}$$

Once again differentiating with respect to r,

$$R' + rR'' = \frac{d^2R}{dz^2}\cdot\frac{dz}{dr}$$

$$= \frac{d^2R}{dz^2}\cdot\frac{1}{r}$$

i.e., $$r^2R'' + rR' = \frac{d^2R}{dz^2}$$

Hence the equation (4) reduces to

$$\frac{d^2R}{dz^2} - kR = 0 \quad ...(6)$$

Also $$\frac{d^2\phi}{d\theta^2} + k\phi = 0 \quad ...(5)$$

Case 1. Let $k = \lambda^2 = a$ positive quantity.

$\therefore$ The two equations become, $\dfrac{d^2R}{dz^2} - \lambda^2 R = 0$ and $\phi'' + \lambda^2 \phi = 0.$

Solving, we have, $R = A_1 e^{\lambda z} + B_1 e^{-\lambda z} = A_1 (e^z)^\lambda + B_1 (e^z)^{-\lambda}$

i.e., $\quad R = A_1 r^\lambda + B_1 r^{-\lambda}$

and $\quad \phi = C_1 \cos \lambda\theta + D_1 \sin \lambda\theta.$

Case 2. Let $k = -\lambda^2$, a negative number.

Equations (5) and (6) become

$$\phi'' - \lambda^2 \varphi = 0 \text{ and } \frac{d^2R}{dz^2} + \lambda^2 R = 0.$$

Hence, $\quad \phi = A_2 e^{\lambda\theta} + B_2 e^{-\lambda\theta}$

and $\quad R = C_2 \cos(\lambda z) + D_2 \sin(\lambda z)$

i.e., $\quad R = C_2 \cos(\lambda \log r) + D_2 \sin(\lambda \log r)$

Case 3. Let $k = 0$. Then (5) and (6) become,

$$\frac{d^2R}{dz^2} = 0 \text{ and } \phi'' = 0.$$

Solving, we get,

$$R = A_3 z + B_3 = A_3 \log r + B_3$$

and $\quad \phi = C_3\theta + D_3.$

Therefore the various possible solutions of the equation (1) are

$$u(r, \theta) = (A_1 r^\lambda + B_1 r^{-\lambda})(C_1 \cos \lambda\theta + D_1 \sin \lambda\theta) \qquad \text{...I}$$

$$u(r, \theta) = (A_2 e^{\lambda\theta} + B_2 e^{-\lambda\theta})\,[C_2 \cos(\lambda \log r) + D_2 \sin(\lambda \log r)] \qquad \text{...II}$$

$$u(r, \theta) = (A_3 \log r + B_3)(C_3\theta + D_3) \qquad \text{...III}$$

Out of these solutions, we have to select that solution which suits the physical nature of the problem and the given boundary conditions in the problems.

Example 33. *A thin semicircular plate of radius a has its bounding diameter kept at temperature zero and its circumference at $f(\theta)$, $0 < \theta < \pi$. Find the steady-state temperature distribution at any point of the plate.*

If $f(\theta) = k$, a constant, find the temperature. **(Calicut, 72 Engg.)**

Take the centre of the semicircle as pole and the bounding diameter AB as initial line.

Let $u(r, \theta)$ be the temperature at any point $P(r, \theta)$ of the plate.

Then u satisfies $\nabla^2 u = 0$

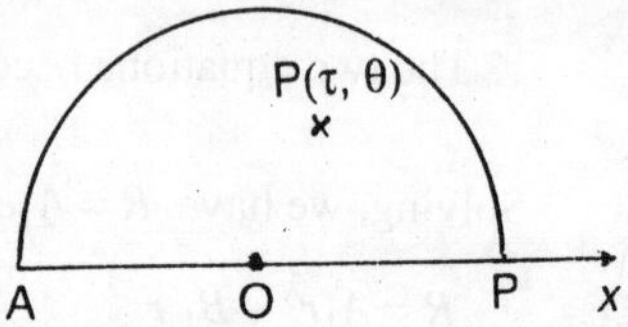

i.e., $$\frac{\partial^2 u}{\partial r^2} + \frac{1}{r}\frac{\partial u}{\partial r} + \frac{1}{r^2}\frac{\partial^2 u}{\partial \theta^2} = 0$$

i.e., $$r^2 \frac{\partial^2 u}{\partial r^2} + r\frac{\partial u}{\partial r} + \frac{\partial^2 u}{\partial \theta^2} = 0 \qquad \text{...(1)}$$

The boundary conditions are

$u(r, 0) = 0$, for $0 \le r \le a$ (*i*)

$u(r, \pi) = 0$, for $0 \le r \le a$ (*ii*)

$u(a, \theta) = f(\theta)$, $0 < \theta < \pi$ (*iii*)

Solving (1), we get the possible solutions

$$u(r, \theta) = (A_1 r^{\lambda} + B_1 r^{-\lambda})(C_1 \cos \lambda\theta + D_1 \sin \lambda\theta) \qquad \text{...(I)}$$

$$u(r, \theta) = [A_2 \cos(\lambda \log r) + B_2 \sin(\lambda \log r)] \times [C_2 e^{\lambda\theta} + D_2 e^{-\lambda\theta}] \qquad \text{...(II)}$$

$$u(r, \theta) = (A_3 \log r + B_3)(C_3 \theta + D_3) \qquad \text{...(III)}$$

As $r \to 0$, u must tend to zero, a finite value. Hence solutions (II) and (III) are rejected.

Selecting (I), we have

$$u(r, \theta) = (Ar^{\lambda} + Br^{-\lambda})(C \cos \lambda\theta + D \sin \lambda\theta) \qquad \text{...(2)}$$

Using the boundary condition (*i*),

$u(r, 0) = C(Ar^{\lambda} + Br^{-\lambda}) = 0$ for $0 < r < a$.

$\therefore \quad C = 0.$

Applying the boundary condition (*ii*) in (2),

$u(r, \pi) = D \sin \lambda\pi \,(Ar^{\lambda} + Br^{-\lambda}) = 0$, for $0 < r < a$.

Since $D \ne 0$; $\sin \lambda\pi = 0$.

$\therefore \quad \lambda = n$, an integer.

$\therefore$ (2) becomes,

$$u(r, \theta) = (A_n r^n + B_n r^{-n}) \sin n\theta \qquad \text{...(3)}$$

where $A_n = AD,$

$B_n = BD$

As $r \to 0$, u is finite. $\therefore B_n = 0$.

Hence $u(r, \theta) = A_n r^n \sin n\theta$.

The most general solution of (1) is

$$u(r, \theta) = \sum_{n=1}^{\infty} A_n r^n \sin n\theta \qquad ...(4)$$

Using the boundary condition (*iii*) in (4),

$$u(a, \theta) = \sum_{n=1}^{\infty} A_n a^n \sin n\theta = f(\theta),\ 0 < \theta. < \pi$$

$$\therefore \quad A_n a^n = \frac{2}{\pi}\int_0^{\pi} f(\theta) \sin n\theta.$$

Substituting the value of A_n in (4), we have,

$$u(r, \theta) = \sum_{n=1}^{\infty} b_n \left(\frac{r}{a}\right)^n \sin n\theta, \text{ where}$$

$$b_n = \frac{2}{\pi}\int_0^{\pi} f(\theta) \sin n\theta \, d\theta \qquad ...(5)$$

(*ii*) If $f(\theta) = k$, then

$$b_n = \frac{2k}{\pi}\int_0^{\pi} \sin n\theta \, d\theta$$

$$= \frac{2k}{\pi n}(-\cos n\theta)_0^{\pi}$$

$$= \frac{2k}{\pi n}(1 - \cos n\pi)$$

$$\therefore \quad b_n = 0 \text{ if } n \text{ is even}$$

$$= \frac{4k}{\pi n}. \text{ if } n \text{ is odd}$$

$$\text{Hence, } u(r, \theta) = \frac{4k}{\pi} \sum_{n=1,3,5\cdots}^{\infty} \frac{1}{n}\left(\frac{r}{a}\right)^n \sin n\theta$$

i.e., $$u(r,\theta)=\frac{4k}{\pi}\sum_{n=1}^{\infty}\frac{1}{(2n-1)}\left(\frac{r}{a}\right)^{2n-1}\sin(2n-1)\theta.$$

Example 34. *Semicircular plate of radius 10 cm. has insulated faces and heat flows in plane curves. The bounding diameter is maintained at 0°C. and on the circumference the temperature distribution is*

$$u(10,\theta)=\frac{400}{\pi}(\pi\theta-\theta^2),\ 0\le\theta\le\pi.$$

Determine the temperature distribution u(r, θ) in the steady-state.

(Madras, 63, B.E.)

In the above example, put $f(\theta)=\frac{400}{\pi}(\pi\theta-\theta^2)$ and $a=10$.

Hence $$b_n=\frac{2}{\pi}\int_0^{\pi}\frac{400}{\pi}(\pi\theta-\theta^2)\sin n\theta\,d\theta$$

$$=\frac{800}{\pi^2}\left[(\pi\theta-\theta^2)\left(-\frac{\cos n\theta}{n}\right)-(\pi-2\theta)\left(-\frac{\sin n\theta}{n^2}\right)+(-2)\left(\frac{\cos n\theta}{n^2}\right)\right]_0^{\pi}$$

$$=\frac{800}{\pi^2}\left[\frac{2}{n^3}(1-\cos n\pi)\right]$$

$$=\frac{1600}{\pi^2 n^3}[1-(-1)^n]$$

$\therefore$ $b_n=0$, if n is even

$=\frac{3200}{\pi^2 n^3}$ if n is odd.

Hence $$u(r,\theta)=\frac{3200}{\pi^2}\sum_{n=1}^{\infty}\frac{1}{(2n-1)^3}\left(\frac{r}{10}\right)^{2n-1}\sin(2n-1)\theta.$$

Example 35. *Determine the steady-state temperature at points of the sector $0\le\theta\le\alpha$, $0\le r\le a$ of a circular plate if the temperature is kept at zero along the straight edges and at a prescribed distribution $u(a,\theta)=f(\theta)$ when $0<\theta<\alpha$, along the curved edge.* **(BR. 1995 Ap.)**

$u(r,\theta)$ satisfies $$r^2\frac{\partial^2 u}{\partial r^2}+r\frac{\partial u}{\partial r}+\frac{\partial^2 u}{\partial\theta^2}=0 \quad ...(1)$$

The boundary, conditions are

$$u(r, 0) = 0, \qquad \text{for } 0 < r < a \qquad (i)$$

$$u(r, \alpha) = 0, \qquad \text{for } 0 < r < a \qquad (ii)$$

and $$u(a, \theta) = f(\theta), 0 < \theta < \alpha \qquad (iii)$$

Solving (1) and selecting the suitable solution as in example 33, we have,

$$u(r, \theta) = (Ar^{\lambda} + Br^{-\lambda})(C \cos \lambda\theta + D \sin \lambda\theta) \qquad ...(2)$$

Using the boundary condition (*i*), $C = 0$.

Applying the condition (*ii*) in (2), $D \sin \lambda\alpha = 0$.

Since $D \neq 0$; $\lambda\alpha = n\pi$.

i.e., $\lambda = \frac{n\pi}{\alpha}$ *where* n is any integer.

As $r \to 0$, u is finite.

Hence $B = 0$.

$$\therefore \quad u(r, \theta) = A_n r^{\frac{n\pi}{a}} \sin \frac{n\pi\theta}{\alpha} \text{ where } n \text{ is any integer.}$$

The most general solution is

$$u(r, \theta) = \sum_{n=1}^{\infty} A_n r^{\frac{n\pi}{\alpha}} \sin \frac{n\pi\theta}{\alpha} \qquad ...(3)$$

Using the boundary condition (*iii*),

$$u(a, \theta) = \sum_{n=1}^{\infty} A_n a^{\frac{n\pi}{\alpha}} \sin \frac{n\pi\theta}{\alpha} = f(\theta) \qquad ...(4)$$

for $0 < \theta < \alpha$.

Hence, $$A_n a^{\frac{n\pi}{a}} = \frac{2}{\alpha} \int_0^{\alpha} f(\theta) \sin \frac{n\pi\theta}{\alpha} d\theta.$$

Substituting A_n in (3), we have,

$$u(r, \theta) = \sum_{n=1}^{\infty} b_n \left(\frac{r}{a}\right)^{\frac{n\pi}{\alpha}} \sin \frac{n\pi\theta}{\alpha}$$

where $$b_n = \frac{2}{\alpha} \int_0^{\alpha} f(\theta) \sin \frac{n\pi\theta}{\alpha} d\theta.$$

Example 36. *A semicircular plate of radius a is kept at temperature* u_0 *along the bounding diameter and at* u_1 *along the circumference. Find the steady-state temperature at any point of the plate.*

If $u(r, \theta)$ is the temperature at (r, θ), then *u satisfies*

$$r^2 \frac{\partial^2 u}{\partial r^2} + r\frac{\partial u}{\partial r} + \frac{\partial^2 u}{\partial \theta^2} = 0 \qquad ...(1).$$

The boundary conditions are

$$\left.\begin{aligned} u(r, 0) &= u_0 \text{ for } 0 < r < a \\ u(r, \pi) &= u_0 \text{ for } 0 < r < a \\ u(a, \theta) &= u_1 \text{ for } 0 < \theta < \pi \end{aligned}\right\} \qquad ...(2)$$

Assume $u(r, \theta) = v(r, \theta) + u_0$ ...(3)

Now $v(r, \theta)$ also satisfies (1) and the boundary conditions on v are

$$\left.\begin{aligned} v(r, 0) &= 0, \text{ for } 0 < r < a \\ v(r, \pi) &= 0, \text{ for } 0 < r < a \\ v(a, \theta) &= u_1 - u_0 \text{ for } 0 < \theta < \pi \end{aligned}\right\} \qquad ...(4)$$

making use of (2) and (3),

Repeating the proof of example 13 (*ii*),

$$u(r, \theta) = \frac{4(u_1 - u_0)}{\pi} \sum_{n=1}^{\infty} \frac{1}{(2n-1)} \left(\frac{r}{a}\right)^{2n-1} \sin(2n-1)\theta.$$

$$\text{Hence } u(r, \theta) = u_0 + \frac{4(u_1 - u_0)}{\pi} \sum_{n=1}^{\infty} \frac{1}{(2n-1)} \left(\frac{r}{a}\right)^{2n-1} \sin(2n-1)\theta.$$

Example 37. *In a semicircular plate of radius 'a' with bounding diameter at 0°C and the circumference at 100°C find the steady-state temperature.* ***(MS. 1989 Ap.)***

Let $u(r, \theta)$ be the temperature at any point (r, θ) of the plate

$$u(r, \theta) \text{ satisfies } r^2 \frac{\partial^2 u}{\partial r^2} + r\frac{\partial u}{\partial r} + \frac{\partial^2 u}{\partial \theta^2} = 0$$

The boundary conditions are

$u(r, 0) = 0, 0 \le r \le a$ (*i*)

$u(r, \pi) = 0, 0 \le r \le a$ (*ii*)

$u(a, \theta) = 100, 0 < \theta < \pi$ (*iii*)

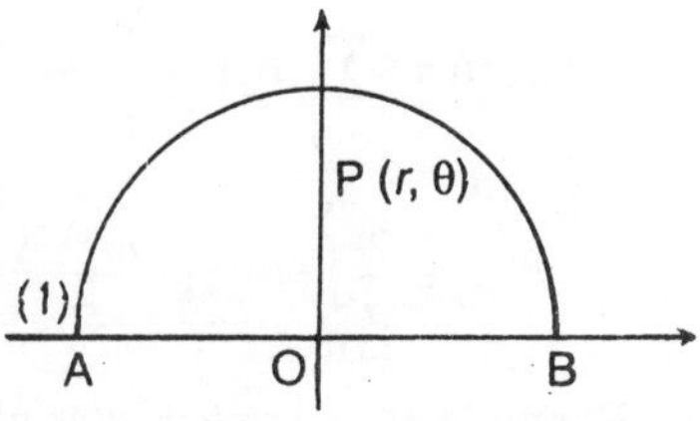

Solving equation (1), we get three possible solutions, out of which we select that solution which is periodic in θ

$$\therefore \quad u(r, \theta) = (Ar^{\lambda} + Br^{-\lambda})(C \cos \lambda\theta + D \sin \lambda\theta) \qquad ...II$$

using boundary condition (*i*),

$$u(r, 0) = (Ar^{\lambda} + Br^{-\lambda})C = 0 \text{ for all } r$$

$$\therefore \quad C = 0$$

$$u(r, \pi) = (Ar^{\lambda} + Br^{-\lambda})\, D \sin \lambda\pi = 0$$

$$D \neq 0; \therefore\ \lambda\pi = n\pi \ \therefore\ \lambda = n \text{ any integer}$$

Since $r = 0$ is a point of the plate, u cannot become infinite at $r = 0$

That is $u \to$ finite as $r \to 0$

$\therefore\ B = 0$

$\therefore\ u(r, \theta) = B_n r^n \sin n\theta$, n any integer.

Hence, the most general solution is

$$u(r, \theta) = \sum_{n=1}^{\infty} B_n r^n \sin n\theta \qquad \text{...IV}$$

Using bounding condition (*iii*),

$$u(a, \theta) = \sum_{1}^{\infty} B_n a^n \sin n\,\theta = 100, 0 < \theta < \pi$$

$$\therefore \quad B_n a^n = \frac{2}{\pi}\int_0^{\pi} 100 \sin n\theta\, d\theta$$

$$= \frac{200}{\pi}\left[-\frac{\cos n\theta}{n}\right]_0^{\pi}$$

$$= -\frac{200}{\pi n}\left[(-1)^n - 1\right]$$

$$= 0 \text{ if } n \text{ is even}$$

$$= \frac{400}{n\pi} \text{ for } n \text{ odd.}$$

Hence, substituting in IV,

$$u(r, \theta) = \frac{400}{\pi}\sum_{n=1}^{\infty}\frac{1}{(2n-1)}\left(\frac{r}{a}\right)^{2n-1}\sin(2n-1)\,\theta.$$

Example 38. *A semicircular plate of radius a cm has insulated faces and heat flows in plane curves. The bounding diameter is kept at 0°C. and the semi-circumference is maintained at temperature given by*

$$u(a, \theta) = \frac{k\theta}{\pi}, \quad 0 \le \theta \le \pi/2$$

$$= \frac{k}{\pi}(\pi - \theta), \quad \pi/2 \le \theta < \pi$$

Find the steady-state temperature distribution. ***(MS. 1989 Ap.)***

All conditions except the non-zero boundary conditions are same as in the previous problem.

i.e.

$$u(r, 0) = 0, \qquad 0 \le r \le a$$

$$u(r, \pi) = 0, \qquad 0 \le r \le a$$

$$u(a, \theta) = \frac{k\theta}{\pi}, \qquad 0 \le \theta \le \frac{\pi}{2}$$

$$= \frac{k}{\pi}(\pi - \theta), \quad \frac{\pi}{2} \le \theta \le \pi$$

Hence, from the previous example,

$$\therefore \; u(r, \theta) = \sum_{n=1}^{\infty} B_n r^n \sin n\theta \qquad \text{...IV}$$

Now use the non-zero boundary condition.

$$\therefore \; u(a, \theta) = \sum_{n=1}^{\infty} B_n a^n \sin n\theta = f(\theta), \text{ given}$$

$$\therefore \qquad B_n a^n = \frac{2}{\pi}\int_0^{\pi} f(\theta) \sin n\theta \, d\theta$$

$$= \frac{2}{\pi}\left[\int_0^{\pi/2} \frac{k\theta}{\pi} \sin n\theta \, d\theta + \int_{\pi/2}^{\pi} \frac{k}{\pi}(\pi - \theta) \sin n\theta \, d\theta\right]$$

$$= \frac{2k}{\pi^2}\left[\left\{\theta\left(-\frac{\cos n\theta}{n}\right) - \left(-\frac{\sin n\theta}{n^2}\right)\right\}_0^{\pi/2}\right.$$

$$\left. + \left\{(\pi - \theta)\left(-\frac{\cos n\theta}{n}\right) - (-1)\left(-\frac{\sin n\theta}{n^2}\right)\right\}_{\pi/2}^{\pi}\right]$$

$$= \frac{2k}{\pi^2}\left[-\frac{\pi}{2n}\cos\frac{n\pi}{2} + \frac{1}{n^2}\sin\frac{n\pi}{2} + \frac{\pi}{2n}\cos\frac{n\pi}{2} + \frac{1}{n^2}\sin\frac{n\pi}{2}\right]$$

$$= \frac{4k}{n^2\pi^2} \sin n\frac{\pi}{2}$$

$$\therefore \quad B_n = \frac{4k}{n^2\pi^2 a^n} \sin n\frac{\pi}{2}$$

$$\therefore \ u(r, \theta) = \frac{4k}{\pi^2} \sum_{n=1}^{\infty} \frac{1}{n^2} \sin\frac{n\pi}{2}\left(\frac{r}{a}\right)^n \sin n\,\theta$$

$$= \frac{4k}{\pi^2} \sum_{n=1}^{\infty} \frac{(-1)^{n-1}}{(2n-1)^2}\left(\frac{r}{a}\right)^{2n-1} \sin(2n-1)\,\theta$$

Example 39. *Find the steady-state temperature in a circular plate of radius 'a' which has one half of its circumference at 0°C. and the other half at K°C. ($k \neq 0$).* **(*MS. 1989 Ap.*)**

The bounding diameter which bisects the plate into two equal parts may be taken as initial line with pole at the centre so that 0°C. may be on one side and *K*°C. may be on the other side as detailed below.

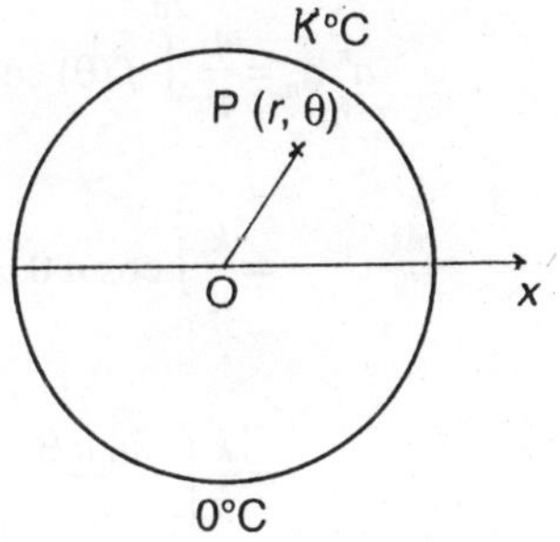

Boundary conditions are

$u(a, \theta) = k, 0 < \theta < \pi$

$= 0 \quad \pi < \theta < 2\pi$ ...(*i*)

Also as $r \to 0$, u is finite ...(*ii*)

The solution must be periodic of period 2π in θ (*iii*)

Let $u(r, \theta) = (Ar^{\lambda} + Br^{-\lambda})(C\cos\lambda\,\theta + D\sin\lambda\,\theta)$ ...II

Since u is finite as $r \to 0$, $B = 0$.

Since the solution is to be periodic of period 2π in θ, $\lambda = n$, an integer.

$\therefore \ u(r, \theta) = r^n (a_n \cos n\,\theta + b_n \sin n\,\theta)$, n any integer

The most general solution is

$$u(r, \theta) = \sum_{n=0}^{\infty} r^n (a_n \cos n\,\theta + b_n \sin n\,\theta)$$

$$= a_0 + \sum_{n=1}^{\infty} r^n (a_n \cos n\,\theta + b_n \sin n\,\theta) \qquad \text{...IV}$$

Using $u(a, \theta)$ = given value

$$u(a, \theta) = a_0 + \sum_{1}^{\infty} a^n (a_n \cos n\theta + b_n \sin n\theta) \qquad \text{...V}$$

$$= \begin{cases} k & 0 < \theta < \pi \\ 0 & \pi < \theta < 2\pi \end{cases}$$

This is a Fourier of periodicity 2π in $(0, 2\pi)$

$$\therefore \quad a_0 = \frac{1}{2\pi} \int_0^{2\pi} f(\theta)\, d\theta$$

$$= \frac{1}{2\pi} \left[\int_0^{\pi} k\, d\theta + \int_{\pi}^{2\pi} 0\, d\theta \right]$$

$$= \frac{k}{2\pi} [\theta]_0^{\pi} = \frac{k}{2}$$

$$a^n a_n = \frac{1}{\pi} \int_0^{2\pi} f(\theta) \cos n\theta\, d\theta$$

$$= \frac{k}{\pi} \int_0^{\pi} \cos n\theta\, d\theta + \frac{1}{\pi} \int_{\pi}^{2\pi} 0\, d\theta$$

$$= \frac{k}{\pi} \left(\frac{\sin n\theta}{n} \right)_0^{\pi} = 0 \qquad \therefore \boxed{a_n = 0}$$

$$a^n \cdot b_n = \frac{1}{\pi} \int_0^{2\pi} f(\theta) \sin n\theta d\theta$$

$$= \frac{1}{\pi} \left[\int_0^{\pi} k \sin n\theta\, d\theta + \int_{\pi}^{2\pi} 0\, d\theta \right]$$

$$= \frac{k}{\pi} \left[-\frac{\cos n\theta}{n} \right]_0^{\pi}$$

$$= -\frac{k}{\pi n} [(-1)^n - 1]$$

$= 0$ if n is even

$$= \frac{2k}{\pi n} \text{ if } n \text{ is odd} \qquad \boxed{\therefore \quad b_n = \frac{2k}{\pi n a^n}.}$$

Using IV,

$$\therefore \; u(r, \theta) = \frac{k}{2} + \frac{2k}{\pi} \sum_{n=1,3,5\cdots}^{\infty} \frac{1}{n\,a^n} \cdot r^n \cdot \sin n\theta$$

$$u(r, \theta) = \frac{k}{2} + \frac{2k}{\pi} \sum_{n=1}^{\infty} \frac{1}{(2n-1)} \left(\frac{r}{a}\right)^{2n-1} \sin(2n-1)\,\theta \quad \text{degrees.}$$

Example 40. *The bounding radii $\theta = 0$ and $\theta = \pi/2$ of a circular quadrant are kept at 0°C. and the temperature along the curve is kept at $50(\pi\theta - 2\theta^2)$ °C*

Find the temperature distribution at any point of the plate and also at $\left(\frac{a}{2}, \frac{\pi}{4}\right)$ if the radius is 'a'.

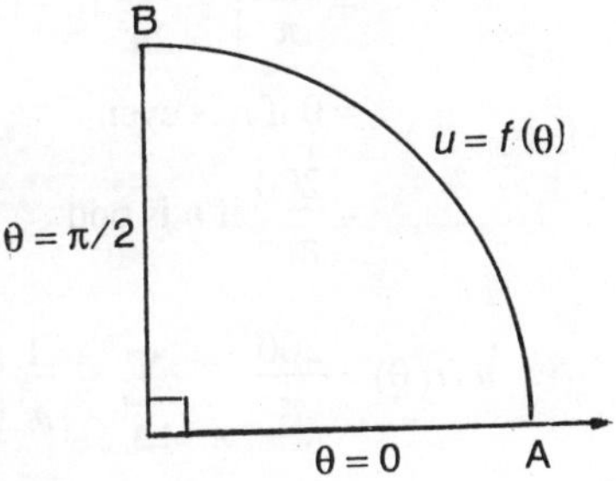

The boundary condition are

$u(r, 0) = 0;\ 0 < r < a$ (*i*)

$u\left(r, \frac{\pi}{2}\right) = 0,\ 0 < r < a$ (*ii*)

u is finite as $r \to 0$ (*iii*)

$u(a, \theta) = 50(\pi\theta - 2\theta^2),\ 0 < \theta < \pi/2$ (*iv*)

Select $u(r, \theta) = (Ar^{\lambda} + Br^{-\lambda})(C\cos\lambda\theta + D\sin\lambda\theta)$ ...II

Using (*i*), $C = 0$

using (*ii*) $D \sin\frac{\pi}{2}\lambda = 0$

$$\therefore \; \frac{\pi\lambda}{2} = n\pi$$

$\lambda = 2n$, n any integer.

Also $B = 0$ by condition (*iii*)

$\therefore \; u(r, \theta) = B_n r^{2n} \sin 2n\,\theta$, n any integer.

Hence most general solution is,

$$u(r, \theta) = \sum_{n=1}^{\infty} B_n r^{2n} \sin 2n\,\theta \qquad \text{...IV}$$

$$u(a, \theta) = \sum_{1}^{\infty} B_n a^{2n} \cdot \sin 2n\theta = 50(\pi\theta - 2\theta^2)\ 0 < \theta < \frac{\pi}{2}$$

$$\therefore\ B_n a^{2n} = \frac{2}{\pi/2} \int_0^{\pi/2} 50(\pi\theta - 2\theta^2) \sin 2n\,\theta\, d\theta$$

$$= \frac{200}{\pi}\left[(\pi\theta - 2\theta^2)\left(-\frac{\cos 2n\,\theta}{2n}\right) - (\pi - 4\theta)\left(-\frac{\sin 2n\,\theta}{4n^2}\right) + (-4)\left(\frac{\cos 2n\,\theta}{8n^3}\right)\right]_0^{\pi/2}$$

$$= \frac{200}{\pi}\left[-\frac{1}{2n^3}((-1)^n - 1)\right]$$

$$= 0 \text{ if } n \text{ is even}$$

$$= \frac{200}{\pi n^3} \text{ if } n \text{ is odd}$$

$$\therefore\ u(r, \theta) = \frac{200}{\pi} \sum_{n=1,3,5\cdots}^{\infty} \frac{1}{n^3}\left(\frac{r}{a}\right)^{2n} \sin 2n\theta$$

$$u\left(\frac{a}{2}, \frac{\pi}{4}\right) = \frac{200}{\pi} \sum_{n=1,3,5,\cdots}^{\infty} \frac{1}{n^3}\left(\frac{1}{2}\right)^{2n} \sin\frac{n\pi}{2}.$$

$$= \frac{200}{\pi}\left[\frac{1}{2^2} - \frac{1}{2^6 \cdot 3^3} + \frac{1}{2^{10} \cdot 5^3} \cdots\right]$$

The result may be calculated depending upon the accuracy, leaving the terms which do not influence the result.

Exercises 3(f)

1. In a semicircular plate of radius a with bounding diameter at 0°C and the circumference at T°C, show that the steady-state temperature distribution is given by

$$u(r, \theta) = \frac{4T}{\pi} \sum_{n=1}^{\infty} \frac{1}{(2n-1)}\left(\frac{r}{a}\right)^{2n-1} \sin(2n-1)\,\theta.$$

2. A semicircular plate of radius 10 cm has insulated faces and heat flows in plane curves. The bounding diameter is kept at 0°C.., so that we have, measuring θ from one of the bounding radii,

$$u(r, 0) = 0 \text{ for } 0 \le r < 10$$

$u(r, \pi) = 0$ for $0 < r < 10$.

On the semi-circumference, the maintained temperature distribution is given by

$$u(10, \theta) = \frac{200}{\pi}\theta \text{ for } 0 \leq \theta \leq \pi/2$$

$$= \frac{200}{\pi}(\pi - \theta) \text{ for } \pi/2 < \theta \leq \pi.$$

After steady-state conditions prevail, what is the temperature at any point (r, θ) of the plate.

3. The bounding diameter of a semicircular plate of radius 10 cm is kept at 0°C. and the temperature along the semicircular boundary is given by

$$u(10, \theta) = 50\theta \text{ for } 0 < \theta < \pi/2$$

$$= 50(\pi - \theta) \text{ for } \pi/2 \leq \theta < \pi.$$

Find the steady-state temperature function $u(r, \theta)$ at any point of the plate.

4. A semi-circular plate of radius a has its circumference maintained at $u(a, \theta) = k\theta(\pi - \theta)$ for $0 < \theta < \pi$ while the bounding diameter is maintained at 0°C. Assuming the lateral surface of the plate to be insulated, find the temperature distribution $u(r, \theta)$ in the steady-state.

5. A semi-circular plate of radius a cm has insulated faces and heat flows in plane curves. The bounding diameter is kept at 0°C and the semi-circumference is maintained at temperature given by

$$u(a, 0) - \frac{100\theta}{\pi} \text{ in } 0 \leq \theta \leq \frac{\pi}{2}$$

$$= \frac{100}{\pi}(\pi - \theta) \text{ in } \pi/2 \leq \theta \leq \pi.$$

Find the steady-state temperature distribution.

6. Find the steady-state temperature in a circular plate of radius a which has one half of its circumference at 0°C and the other half at 100°C.

(Madras, 63 B.E.; Calicut, 72 B.Sc. Engg.)

7. A plate in the shape of a truncated quadrant of a circle is bounded by $r = a$, $r = b$ and $\theta = 0$, $\theta = \pi/2$. It has its faces insulated and heat flows in plane curves. It is kept at temperature 0°C. along three of the edges while along the edge $r = a$, it is kept at the temperature $\theta(\pi/2 - \theta)$. Determine the steady-state temperature distribution. ***(Madras, 71 B.E.)***

8. A plate in the form of a circular sector is bounded by the lines $\theta = 0$, $\theta = \alpha$, $r = a$. Its surface are insulated and the temperatures along the boundary are $u(r, 0) = u(r, \alpha) = 0$ and $u(a, \theta) = f(\theta)$, $0 < \theta < \alpha$. Find the steady-state temperature distribution if $\alpha = \frac{\pi}{3}$ and $f(\theta) = 100\theta\left(\frac{\pi}{3} - \theta\right)$ ***(Madras, 64 B.E.)***

9. Show that Laplace's equation in plane polar coordinates (r, θ) has solutions of the form $(Ar^n + Br^{-n})e^{\pm in\theta}$ where A, B, n are constants. Determine the potential function V in the region $0 \leq r \leq 9$, $0 \leq \theta \leq 2\pi$ if V satisfies the conditions (*i*) V is finite as $r \to 0$ and (*ii*) $V = \sum_n C_n \cos n\theta$ on $r = a$. ***(Madras, 72 B.E.)***

10. A plate with insulated surfaces has the shape of a quadrant of a circle of radius 10 cm. The bounding radii $\theta = 0$ and $\theta = \pi/2$ are kept at 0°C. and the temperature along the circular quadrant is kept at $100\,(\pi\theta - 2\theta^2)$°C. for $0 < \theta < \frac{\pi}{2}$ until steady-state conditions prevail. Find the temperature at the point $\left(\frac{40\sqrt{2}}{3\pi}, \frac{\pi}{4}\right)$.

11. The bounding radii $\theta = 0$ and $\theta = \pi/2$ are kept at 0°C. and the temperature along the circular quadrant is kept at $50\,(\pi\theta - 2\theta^2)$°C.

Find the steady-state temperature at the point $\left(\frac{a}{2}, \frac{\pi}{4}\right)$.

3·12. Steady-state temperature distribution in a circular annulus.

If the temperature distributions along the inner radius $r = r_1$ and the outer radius $r = r_2$ of a circular annulus are maintained as $f_1(\theta)$ and $f_2(\theta)$ respectively until steady-state conditions prevail, find the steady-state temperature at an internal point (r, θ) of the annulus.

Laplace's equation in polar coordinates is

$$\nabla^2 u = \frac{\partial^2 u}{\partial r^2} + \frac{1}{r}\frac{\partial u}{\partial r} + \frac{1}{r^2}\frac{\partial^2 u}{\partial \theta^2} = 0 \qquad ...(1)$$

Assuming a product solution and solving we get the various possible solutions.

$$u(r, \theta) = (A_1 r^{\lambda} + B_1 r^{-\lambda})(C_1 \cos \lambda\theta + D_1 \sin \lambda\theta) \qquad ...I$$

$$u(r, \theta) = [A_2 \cos(\lambda \log r) + B_2 \sin(\lambda \log r)]\,[C_2 e^{\lambda\theta} + D_2 e^{-\lambda\theta}] \qquad ...II$$

$$u(r, \theta) = (A_3 \log r + B_3)(C_3 \theta + D_3) \qquad ...III$$

Out of these solutions, we have to select the solutions which suit the physical nature of the problem and the boundary conditions. Since we are dealing with an annulus, the temperature function has to be periodic in θ of period 2π. Hence neglecting the solution II, we can take a linear combination

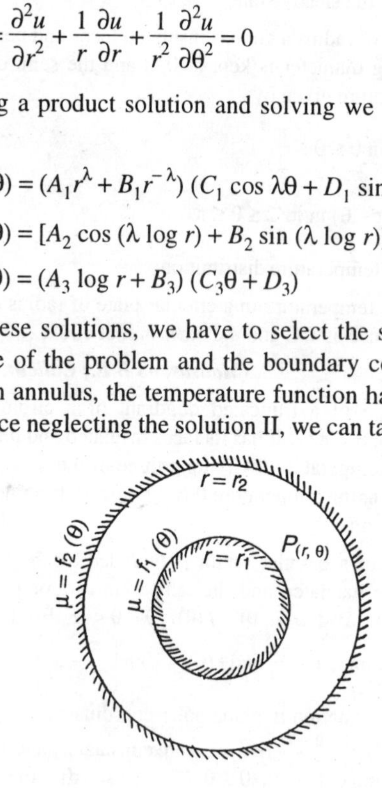

of I and III. Even in III, the coefficient of θ, namely C_3, must be zero because of the periodic nature of the solution in θ. Hence

$$u(r, \theta) = \sum_{\lambda} (A_1 r^{\lambda} + B_1 r^{-\lambda})(C_1 \cos \lambda\theta + D_1 \sin \lambda\theta) + (A_3 \log r + B_3) D_3 \quad ...(2)$$

is a general solution of (1), where λ assumes non-zero arbitrary values. Since this solution is periodic in θ of periodicity 2π,

$$\lambda = n, \text{ any integer } (n = 1, 2, 3 \cdots)$$

Again rewriting the equation (2), we have,

$$u(r, \theta) = (a_0 \log r + b_0) + \sum_{n=1}^{\infty} [(a_n r^n + b_n r^{-n}) \cos n\theta + (c_n r^n + d_n r^{-n}) \sin n\theta] \quad ...(3)$$

for $r_1 \le r \le r_2$.

The boundary conditions are

$u(r_1, \theta) = f_1(\theta)$ and $u(r_2, \theta) = f_2(\theta)$, $\quad 0 \le \theta \le 2\pi$.

Making use of the boundary conditions in (3),

$$f_1(\theta) = (a_a \log r_1 + b_0) + \sum_{n=1}^{\infty} [(a_n r_1^{\ n} + b_n r_1^{\ -n}) \cos n\theta + (c_n r_1^{\ n} + d_n r_1^{\ -n}) \sin n\theta] \quad ...(4)$$

and

$$f_2(\theta) = (a_0 \log r_2 + b_0) + \sum_{n=1}^{\infty} [(a_n r_2^{\ n} + b_n r_2^{\ -n}) \cos n\theta + (c_n r_2^{\ n} + d_n r_2^{\ -n}) \sin n\theta] \quad ...(5)$$

Expressing $f_1(\theta)$ and $f_2(\theta)$ as Fourier series in $0 \le \theta \le 2\pi$ and comparing them with (4) and (5), we have,

$$a_0 \log r_1 + b_0 = \frac{1}{2\pi} \int_0^{2\pi} f_1(\theta)\, d\theta \quad ...(6)$$

$$a_0 \log r_2 + b_0 = \frac{1}{2\pi} \int_0^{2\pi} f_2(\theta)\, d\theta \quad ...(7)$$

$$a_n r_1^{\ n} + b_1 r_1^{\ -n} = \frac{1}{\pi} \int_0^{2\pi} f_1(\theta) \cos n\theta d\theta \quad ...(8)$$

$$a_n r_2^{\ n} + b_n r_2^{\ -n} = \frac{1}{\pi} \int_0^{2\pi} f_2(\theta) \cos n\theta d\theta \quad ...(9)$$

$$c_n r_1^{\,n} + d_n r_n^{\,-n} = \frac{1}{\pi} \int_0^{2\pi} f_1(\theta) \sin n\theta d\theta \qquad ...(10)$$

$$c_n r_2^{\,n} + d_n r_2^{\,-n} = \frac{1}{\pi} \int_0^{2\pi} f_2(\theta) \sin n\theta d\theta \qquad ...(11)$$

(6) and (7) determine a_0, b_0

(8) and (9) determine a_n, b_n

(10) and (11) determine c_n, d_n

Substituting these values in (3), we get the solution given by (3).

Example 41. *A plate in the form of a ring is bounded by the lines $r = 2$ and $r = 4$. Its surfaces are insulated and* the temperature $u(r, \theta)$ *along the boundaries are $10 \sin\theta + 6 \cos\theta$ when $r = 2$ and $17 \sin\theta + 15 \cos\theta$ when $r = 4$. Find the steady-state temperature $u(r, \theta)$, in the ring.*

(*Madras, 65 B.E.*)

In the above article, put $r_1 = 2$, $r_2 = 4$

$f_1(\theta) = 10 \sin\theta + 6 \cos\theta$ and $f_2(\theta) = 17 \sin\theta + 15 \cos\theta$.

Equation (3) becomes (when $r = 2$ and $r = 4$)

$$u(2, \theta) = a_0 \log 2 + b_0 + \sum_{n=1}^{\infty} [(a_n 2^n + b_n 2^{-n}) \cos n\theta$$
$$+ (c_n 2^n + d_n 2^{-n}) \sin n\theta] = 10 \sin\theta + 6 \cos\theta \qquad ...(12)$$

$$u(4, \theta) = a_0 \log 4 + b_0 + \sum_{n=1}^{\infty} [(a_n 4^n + b_n 4^{-n}) \cos n\theta$$
$$+ (c_n 4^n + d_n 4^{-n}) \sin n\theta] = 17 \sin\theta + 15 \cos\theta \qquad ...(13)$$

From (12) and (13),

$a_0 \log 2 + b_0 = 0$,

$a_0 \log 4 + b_0 = 0$, giving $a_0 = b_0 = 0$.

$$\left.\begin{aligned} a_n 2^n + b_n 2^{-n} &= 0 \\ a_n 4^n + b_n 4^{-n} &= 0 \\ c_n 2^n + d_n 2^{-n} &= 0 \\ c_n 4^n + d_n 4^{-n} &= 0 \end{aligned}\right\} \text{ for } n \neq 1.$$

$\therefore \quad a_n = b_n = c_n = d_n = 0$, for $n \neq 1$.

$a_1 2^1 + b_1 2^{-1} = 6$

$c_1 4^1 + b_1 4^{-1} = 15$

$$c_1 2^1 + d_1 2^{-1} = 10$$
$$c_1 4^1 + d_1 4^{-1} = 17$$

Solving, we get $a_1 = 4$, $b_1 = -4$, $c_1 = 4$, $d_1 = 4$. Hence the solution, putting in (3), is

$$u(r, \theta) = 4\left(r - \frac{1}{r}\right)\cos\theta + 4\left(r + \frac{1}{r}\right)\sin\theta.$$

Example 42. *A plate in the form of a ring is bounded by the circles of radii r = 5 and r = 10. Its surfaces are insulated and the temperature along the boundaries are*

$$u(5, \theta) = 10\cos\theta + 6\sin\theta$$
$$u(10, \theta) = 17\cos\theta + 15\sin\theta. \qquad \textbf{(MS. 1991 Nov.)}$$

Find the steady state temperature in the plate

In the article 3·12, put $r_1 = 5$, $r_2 = 10$

$$f_1(\theta) = 10\cos\theta + 6\sin\theta$$
$$f_2(\theta) = 17\cos\theta + 15\sin\theta$$

Equation (3) becomes,

$$u(5, \theta) = a_0 \log 5 + b_0 + \sum_{n=1}^{\infty} [(a_n 5^n + b_n 5^{-n})\cos n\theta + (c_n 5^n + d_n 5^{-n})\sin n\theta] = 10\cos\theta + 6\sin\theta$$

$$u(10, \theta) = a_0 \log 10 + b_0 + \sum_{n=1}^{\infty} [(a_n 10^n + b_n 10^{-n})\cos n\theta \qquad ...(12)$$

$$+ (c_n 10^n + d_n 10^{-n})\sin n\theta] = 17\cos\theta + 15\sin\theta \qquad ...(13)$$

From (12) and (3)

$$\left.\begin{aligned} a_0 \log 5 + b_0 &= 0 \\ a_0 \log 10 + b_0 &= 0 \end{aligned}\right\} \text{giving } a_0 = 0,\ b_0 = 0$$

$$\left.\begin{aligned} a_n 5^n + b_n 5^{-n} &= 0 \\ c_n 5^n + d_n 5^{-n} &= 0 \\ a_n 10^n + b_n 10^{-n} &= 0 \\ c_n 10^n + d_n 10^{-n} &= 0 \end{aligned}\right\} \text{for } n \neq 1$$

$\therefore a_n = b_n = c_n = d_n = 0$ if $n \neq 1$

For $n = 1$,

$$a_1 5 + b_1 5^{-1} = 10 \qquad\qquad 5a_1 + \frac{1}{5} b_1 = 10 \quad a_1 = 8/5$$

$$c_1 5 + d_1 5^{-1} = 6 \qquad\qquad 10a_1 + \frac{1}{10} b_1 = 17 \quad b_1 = 10$$

$$a_1 10 + b_1 10^{-1} = 17 \qquad\qquad 5c_1 + \frac{1}{5} d_1 = 6 \quad c_1 = 8/5$$

$$c_1 10 + d_1 10^{-1} = 15 \qquad\qquad 10c_1 + \frac{1}{10} d_1 = 15 \quad d_1 = -10$$

$$\therefore \quad u(r, \theta) = \left(\frac{8}{5} r + \frac{10}{r}\right) \cos\theta + \left(\frac{8}{5} r - \frac{10}{r}\right) \sin\theta$$

Exercises 3 (g)

1. Along the inner boundary of a circular annulus of radii 10 cm and 20 cm. the temperature is maintained as $T(10,\theta) = 15 \cos\theta$, and along the outer boundary the distribution $T(20,\theta) = 30 \sin\theta$ is maintained. Find the steady-state temperature at an arbitrary point (r, θ) in the annulus. **(*Madras, 70 B.E.*)**
2. A plate in the form of a ring is bounded by the lines $r = 2$ and $r = 4$. Its surfaces are insulated and the temperature $u(r, \theta)$ along the boundaries are $u(2, \theta) = 10 \cos\theta + 6 \sin\theta$ and $u(4,\theta) = 17 \cos\theta + 15 \sin\theta$. Find the steady-state temperature in the ring. **(*Madras, 66 B.E.*)**

CHAPTER IV

FOURIER TRANSFORMS

Definition. If $f(x)$ is defined in (a, b), the integral transform of $f(x)$ with the Kernel $K(s, x)$ is defined by

$$F(s) = \bar{f}(s) = \int_a^b f(x)\, K(s, x)\, dx, \text{ if the integral exists.}$$

Here, $K(s, x)$ is called the Kernel of the transform while a, b, are fixed limits. If a, b are finite, the transform is finite and if a, b are infinite, it is an infinite transform.

$$\text{If } \quad K(s, x) = e^{-sx} \quad \text{for } x \geq 0$$
$$= 0 \quad \text{for } x < 0,$$

the above infinite transform becomes the well known Laplace transform.

$$\text{If } \quad F(s) = \int_a^b f(x)\, K(s, x)\, dx \qquad \text{...(1)}$$

it may be possible to get $f(x)$ as

$$f(x) = \int_c^d F(s)\, H(s, x)\, ds \qquad \text{...(2)}$$

In such a case, (2) is called the inversion formula for (1). In (1), if $F(s)$ is known while $f(x)$ is unknown, it may be regarded as an integral equation.

From the general integral transform definition, we can get various integral transforms by properly defining the kernel.

Some of the well known integral transforms are listed below together with the corresponding inversion formula.

1. Infinite Fourier Transform (complex form)

$$F(s) = \frac{1}{\sqrt{2\pi}} \int_{-\infty}^{\infty} f(x)\, e^{isx}\, dx$$

$$f(x) = \frac{1}{\sqrt{2\pi}} \int_{-\infty}^{\infty} F(s)\, e^{-isx}\, ds \quad \text{(inversion formula)}$$

2. Infinite Fourier Cosine Transform

$$F_c(s) = \bar{f}_c(x) = \sqrt{\frac{2}{\pi}} \int_0^\infty f(x) \cos sx \, dx$$

$$f(x) = \sqrt{\frac{2}{\pi}} \int_0^\infty F_c(s) \cos sx \, ds$$

3. Infinite Fourier Sine Transform

$$F_s(s) = \sqrt{\frac{2}{\pi}} \int_0^\infty f(x) \sin sx \, dx$$

$$f(x) = \sqrt{\frac{2}{\pi}} \int_0^\infty F_s(s) \sin sx \, ds$$

4. Laplace Transform

$$F(s) = \int_0^\infty e^{-sx} f(x) \, dx$$

$$f(x) = \frac{1}{2\pi i} \int_{c-i\infty}^{c+i\infty} F(s) e^{sx} \, ds$$

5. Hankel Transform

$$F_n(s) = \int_0^\infty x f(x) J_n(sx) \, dx$$

$$f(x) = \int_0^\infty s F_n(s) J_n(sx) \, ds$$

where $J_n(x)$ is the Bessel function of the first kind of order n.

6. Mellin Transform

$$F(s) = \int_0^\infty f(x) x^{s-1} \, dx$$

$$f(x) = \frac{1}{2\pi i} \int_{c-i\infty}^{c+i\infty} F(s) x^{-s} \, ds$$

Theorem 1. *If $f(x)$ satisfies Dirichlet's conditions in $(-l, l)$, then the complex Fourier series of $f(x)$ is*

$$f(x) = \sum_{n=-\infty}^{\infty} C_n e^{\frac{in\pi x}{l}} \text{ where} \qquad ...(1)$$

$$C_n = \frac{1}{2l} \int_{-l}^{l} f(x)\, e^{\frac{-in\pi x}{l}} dx \qquad ...(2)$$

(For proof, refer to chapter 'Fourier series', Page 92, Note 3)

Theorem 2. (FOURIER INTEGRAL THEOREM)

If f(x) is piece-wise continuously differentiable and absolutely integrable in $(-\infty, \infty)$*, then*

$$f(x) = \frac{1}{2\pi} \int_{-\infty}^{\infty} \int_{-\infty}^{\infty} f(t)\, e^{i(x-t)s}\, dt\, ds.$$

Proof. In theorem 1, substituting the value of C_n from (2) in (1), and changing the variable of integration in (2) as t, we get,

$$f(x) = \sum_{n=-\infty}^{\infty} \left(\frac{1}{2l} \int_{-l}^{l} f(t)\, e^{\frac{in\pi(x-t)}{l}} dt \right)$$

Let $\frac{\pi}{l} = \delta s.$ $\qquad (-l \le x \le l)$

$$\therefore \qquad f(x) = \sum_{n=-\infty}^{\infty} \left(\frac{1}{2l} \int_{-l}^{l} f(t)\, e^{i(x-t)n\delta s} dt \right)$$

$$= \frac{1}{2\pi} \sum_{n=-\infty}^{\infty} \delta s \left(\int_{-l}^{l} f(t)\, e^{i(x-t)n\delta s} dt \right) \qquad \text{since } l = \frac{\pi}{\delta s}$$

Let $l \to \infty$, so that the range $(-l, l)$ becomes $(-\infty, \infty)$. As $l \to \infty$, $\delta s \to 0$. Changing the summation to definite integral, we get,

$$f(x) = \frac{1}{2\pi} \int_{-\infty}^{\infty} \int_{-\infty}^{\infty} f(t)\, e^{i(x-t)s}\, dt\, ds.$$

Note 1. Using Euler's theorem, $e^{i\theta} = \cos\theta + i \sin\theta$, we get

$$f(x) = \frac{1}{2\pi} \int_{-\infty}^{\infty} \int_{-\infty}^{\infty} f(t)\, [\cos(x-t)s + i \sin(x-t)s]\, dt\, ds.$$

Equating real and imaginary parts on both sides ($f(x)$ real) we get,

$$f(x) = \frac{1}{2\pi} \int_{-\infty}^{\infty} \int_{-\infty}^{\infty} f(t) \cos (x - t)\, s \, dt \, ds$$

and
$$\frac{1}{2\pi} \int_{-\infty}^{\infty} \int_{-\infty}^{\infty} f(t) \sin (x - t) s \, dt \, ds = 0.$$

Complex Fourier Transform (Infinite)

Let $f(x)$ be a function defined in $(-\infty, \infty)$ and be piece-wise continuous in each finite partial interval and absolutely integrable in $(-\infty, \infty)$. Then the complex Fourier Transform of $f(x)$ is defined by

$$\mathcal{F}(s) = F(s) = \frac{1}{\sqrt{2\pi}} \int_{-\infty}^{\infty} f(x)\, e^{isx} \, dx.$$

Theorem 3. Inversion Theorem for Complex Fourier Transform

If $f(x)$ satisfies the Dirichlet's conditions in every finite interval $(-l, l)$ and if it is absolutely integrable in the range, and if

$F(s)$ denotes the complex Fourier transform of $f(x)$, then at every point of continuity of $f(x)$, we have

$$f(x) = \frac{1}{\sqrt{2\pi}} \int_{-\infty}^{\infty} F(s)\, e^{-isx} \, ds$$

Proof. From Fourier integral theorem

$$f(x) = \frac{1}{2\pi} \int_{-\infty}^{\infty} \int_{-\infty}^{\infty} f(t)\, e^{i(x-t)s} \, dt \, ds$$

$$= \frac{1}{2\pi} \int_{-\infty}^{\infty} e^{ixs} \left(\int_{-\infty}^{\infty} f(t)\, e^{-its} \, dt \right) ds$$

$$= \frac{1}{2\pi} \int_{-\infty}^{\infty} e^{-ixv} \left(\int_{-\infty}^{\infty} f(t)\, e^{itv} \, dt \right) dv \qquad \text{putting } s = -v$$

$$= \frac{1}{\sqrt{2\pi}} \int_{-\infty}^{\infty} e^{-ixv} \left(\frac{1}{\sqrt{2\pi}} \int_{-\infty}^{\infty} f(t)\, e^{itv} \, dt \right) dv$$

$$= \frac{1}{\sqrt{2\pi}} \int_{-\infty}^{\infty} e^{-ixv} \left(F(v)\right) dv$$

$$= \frac{1}{\sqrt{2\pi}} \int_{-\infty}^{\infty} F(s) e^{-isx} \, ds \quad \text{changing } v \text{ as } s.$$

Note 1. The parameter s is taken as p by some authors.

Note 2. Some authors define the Complex Fourier transforms in different forms, changing the Kernels of the transform. Various forms are listed below.

1. $$\bar{f}(s) = \int_{-\infty}^{\infty} e^{-isx} f(x) \, dx$$

$$f(x) = \frac{1}{2\pi} \int_{-\infty}^{\infty} e^{isx} \bar{f}(s) \, ds$$

2. $$F(s) = \int_{-\infty}^{\infty} e^{isx} f(x) \, dx$$

$$f(x) = \frac{1}{2\pi} \int_{-\infty}^{\infty} e^{-isx} F(s) \, ds$$

3. $$F(s) = \frac{1}{\sqrt{2\pi}} \int_{-\infty}^{\infty} e^{-isx} f(x) \, dx$$

$$f(x) = \frac{1}{\sqrt{2\pi}} \int_{-\infty}^{\infty} e^{isx} F(s) \, ds$$

Properties of Fourier Transforms

Theorem 1. *Fourier transform is linear.*

i.e. $F[af(x) + bg(x)] = a F[f(x)] + bF[g(x)]$

where F stands for Fourier transform.

$$F[af(x) + bg(x)] = \frac{1}{\sqrt{2\pi}} \int_{-\infty}^{\infty} (af(x) + bg(x)) e^{isx} \, dx$$

$$= a \cdot \frac{1}{\sqrt{2\pi}} \int_{-\infty}^{\infty} f(x) e^{isx} \, dx + b \cdot \frac{1}{\sqrt{2\pi}} \int_{-\infty}^{\infty} g(x) e^{isx} \, dx$$

$$= a F[f(x)] + b F[g(x)].$$

Theorem 2. Shifting theorem

If $F\{f(x)\} = F(s)$, *then* $F\{f(x-a)\} = e^{isa} F(s)$

$$F\{f(x-a)\} = \frac{1}{\sqrt{2\pi}} \int_{-\infty}^{\infty} f(x-a)\, e^{isx}\, dx$$

$$= \frac{1}{\sqrt{2\pi}} \int_{-\infty}^{\infty} f(t)\, e^{i(a+t)s}\, dt \quad \text{putting } x-a=t$$

$$= e^{ias} \cdot \frac{1}{\sqrt{2\pi}} \int_{-\infty}^{\infty} f(t)\, e^{its}\, dt$$

$$= e^{ias}\, F(s).$$

Theorem 3. Change of scale property

If $F\{f(x)\} = F(s)$, *then* $F\{f(ax)\} = \frac{1}{|a|} F\left(\frac{s}{a}\right)$ *where* $a \neq 0$

$$F\{f(ax)\} = \frac{1}{\sqrt{2\pi}} \int_{-\infty}^{\infty} e^{isx} f(ax)\, dx$$

$$= \frac{1}{\sqrt{2\pi}} \int_{-\infty}^{\infty} e^{i\left(\frac{s}{a}\right)t} f(t)\, \frac{dt}{a} \qquad \text{putting } ax = t \text{ and } a > 0.$$

$$= \frac{1}{a} F\left(\frac{s}{a}\right) \quad \text{if } a > 0.$$

$$F\{f(ax) = \frac{1}{\sqrt{2\pi}} \int_{+\infty}^{-\infty} e^{i\left(\frac{s}{a}\right)t} f(t)\, \frac{dt}{a} \text{ if } a < 0$$

$$= -\frac{1}{a} F\left(\frac{s}{a}\right) \text{if } a < 0$$

Hence $F\{f(ax)\} = \frac{1}{|a|} F\left(\frac{s}{a}\right)$

Theorem 4. $F\{e^{iax} f(x)\} = F(s+a)$ ***(BR 1995 Ap.)***

$$F\{e^{iax} f(x)\} = \frac{1}{\sqrt{2\pi}} \int_{-\infty}^{\infty} e^{iax} f(x)\, e^{isx}\, dx$$

$$= \frac{1}{\sqrt{2\pi}} \int_{-\infty}^{\infty} e^{i(s+a)x} f(x)\, dx$$

$$= F(s+a).$$

Theorem 5. Modulation theorem

If $F\{f(x)\} = F(s)$, then

$$F\{f(x)\cos ax\} = \tfrac{1}{2}[F(s-a) + F(s+a)]$$

Proof :

$$F\{f(x)\cos ax\} = \frac{1}{\sqrt{2\pi}} \int_{-\infty}^{\infty} e^{isx} f(x) \cos ax \, dx$$

$$= \frac{1}{\sqrt{2\pi}} \int_{-\infty}^{\infty} e^{isx} f(x) \, . \, \frac{e^{iax} + e^{-iax}}{2} \, dx$$

$$= \frac{1}{2}\left[\frac{1}{\sqrt{2\pi}} \int_{-\infty}^{\infty} e^{i(s+a)x} f(x) \, dx + \frac{1}{\sqrt{2\pi}} \int_{-\infty}^{\infty} e^{i(s-a)x} f(x) \, dx \right]$$

$$= \tfrac{1}{2}[F(s+a) + F(s-a)]$$

Theorem 6. *If* $F\{f(x)\} = F(s)$, *then* $F\{x^n f(x)\} = (-i)^n \dfrac{d^n}{ds^n} F(s)$

$$F(s) = \frac{1}{\sqrt{2\pi}} \int_{-\infty}^{\infty} e^{isx} f(x) \, dx$$

Differentiating w.r.t. s both sides, n times,

$$\frac{d^n F(s)}{ds^n} = \frac{1}{\sqrt{2\pi}} \int_{-\infty}^{\infty} (ix)^n e^{isx} f(x) \, dx$$

$$= (i)^n \frac{1}{\sqrt{2\pi}} \int_{-\infty}^{\infty} x^n e^{isx} f(x) \, dx$$

$$= (i)^n F\{x^n f(x)\}$$

$$F\{x^n f(x)\} = (-i)^n \frac{d^n}{ds^n} \{F(s)\}$$

Theorem 7. $F\{f'(x)\} = -is\,F(s)$ if $f(x) \to 0$ as $x \to \pm\infty$

$$F\{f'(x)\} = \frac{1}{\sqrt{2\pi}} \int_{-\infty}^{\infty} e^{isx} f'(x) \, dx$$

$$= \frac{1}{\sqrt{2\pi}} \int_{-\infty}^{\infty} e^{isx} \, d\{f(x)\}$$

$$= \frac{1}{\sqrt{2\pi}}\left[\left\{e^{isx} f(x)\right\}_{-\infty}^{\infty} - is \int_{-\infty}^{\infty} f(x)\, e^{isx}\, dx\right]$$

$$= - is\, F(s) \text{ if } f(x) \to 0 \text{ as } x \to \pm\infty$$

Cor.: $F\{f^n(x)\} = (-is)^n F(s)$ if $f, f', f'', \ldots f^{(n-1)} \to 0$ as $x \to \pm\infty$

Theorem 8. $F\left\{\int_a^x f(x)\, dx\right\} = \frac{F(s)}{(-is)}$

Let $$\phi(x) = \int_a^x f(x)\, dx$$

Then, $$\phi'(x) = f(x)$$

$$F\{\phi'(x)\} = (-is)\, \bar{\phi}(s)$$

$$= (-is)\, F(\phi(x))$$

$$= (-is)\, F \int_a^x f(x)\, dx$$

$$F\left(\int_a^x f(x)\, dx\right) = \frac{1}{(-is)}\, F\{\phi'(x)\}$$

$$= \frac{1}{(-is)}\, F(f(x)) = \frac{F(s)}{(-is)}$$

Example 1. *Find the Complex Fourier transform of*

$$f(x) = x \text{ for } |x| \le a$$

$$= 0 \text{ for } |x| > a$$

$$F\{f(x)\} = \frac{1}{\sqrt{2\pi}} \int_{-\infty}^{\infty} e^{isx} f(x)\, dx$$

$$= \frac{1}{\sqrt{2\pi}} \int_{-a}^{a} x\, e^{isx}\, dx, \quad \text{the other integrals vanish.}$$

$$= \frac{1}{\sqrt{2\pi}}\left[\frac{x\, e^{isx}}{is} - \frac{e^{isx}}{(is)^2}\right]_{-a}^{a}$$

$$= \frac{1}{\sqrt{2\pi}} \left[\frac{1}{is} \left\{ a\, e^{isa} + a\, e^{-isa} \right\} + \frac{1}{s^2} \left\{ e^{isa} - e^{-isa} \right\} \right]$$

$$= \frac{1}{\sqrt{2\pi}} \left[\frac{-2ai}{s} \cos sa + \frac{2i}{s^2} (\sin sa) \right]$$

$$= \frac{2i}{s^2} \cdot \frac{1}{\sqrt{2\pi}} [\sin sa - as \cos sa]$$

Example 2. *Find the Fourier transform (complex) of*

$$f(x) = e^{ikx}, a < x < b$$
$$= 0, x < a \text{ and } x > b$$

$$F\{f(x)\} = \frac{1}{\sqrt{2\pi}} \int_a^b e^{ikx}\, e^{isx}\, dx$$

$$= \frac{1}{\sqrt{2\pi}} \left[\frac{e^{i(k+s)x}}{i(k+s)} \right]_a^b$$

$$= \frac{i}{\sqrt{2\pi}\,(k+s)} \left[e^{i(k+s)a} - e^{i(k+s)b} \right]$$

Example 3. *Find the Fourier transform of*

$$f(x) = \begin{cases} 1 - x^2 & \text{if } |x| < 1 \\ 0 & \text{if } |x| > 1 \end{cases}$$ ***(Anna Ap 2005)***

Hence evaluate $\int_0^\infty \left(\frac{x \cos x - \sin x}{x^3} \right) \cos \frac{x}{2}\, dx$

$$F\{f(x)\} = \frac{1}{\sqrt{2\pi}} \int_{-\infty}^{\infty} e^{isx} f(x)\, dx$$

$$= \frac{1}{\sqrt{2\pi}} \int_{-1}^{1} e^{isx} (1 - x^2)\, dx$$

$$= \frac{1}{\sqrt{2\pi}} \left[(1 - x^2) \left(\frac{e^{isx}}{is} \right) - (-2x) \left(\frac{e^{isx}}{i^2 s^2} \right) + (-2) \left(\frac{e^{isx}}{i^3 s^3} \right) \right]_{-1}^{1}$$

$$= \frac{1}{\sqrt{2\pi}} \left[\frac{-2}{s^2} \{ e^{is} + e^{-is} \} - \frac{2i}{s^3} \{ e^{is} - e^{-is} \} \right]$$

$$= -\frac{2}{s^3}\frac{1}{\sqrt{2\pi}}[2\,s\cos s - 2\sin s]$$

$$= \frac{-4}{\sqrt{2\pi}}\left[\frac{s\cos s - \sin s}{s^3}\right]$$

Using inversion formula

$$f(x) = \frac{1}{\sqrt{2\pi}}\int_{-\infty}^{\infty}\left[\frac{-4}{\sqrt{2\pi}}\left(\frac{s\cos s - \sin s}{s^3}\right)e^{-isx}\,ds\right.$$

$$= -\frac{2}{\pi}\int_{-\infty}^{\infty}\left(\frac{s\cos s - \sin s}{s^3}\right)e^{-isx}ds$$

$$\int_{-\infty}^{\infty}\left(\frac{s\cos s - \sin s}{s^3}\right)(\cos sx - i\sin sx)\,ds$$

$$= -\frac{\pi}{2}(1-x^2) \quad \text{if } |x| < 1$$

$$= 0 \quad \text{if } |x| > 1$$

Equating real parts,

$$\int_{-\infty}^{\infty}\left(\frac{s\cos s - \sin s}{s^3}\right)\cos sx\,ds = -\frac{\pi}{2}(1-x^2) \text{ if } |x| < 1$$

$$= 0 \text{ if } |x| > 1$$

Set $x = \frac{1}{2}$

$$\therefore \quad \int_{-\infty}^{\infty}\left(\frac{s\cos s - \sin s}{s^3}\right)\cos\frac{s}{2}\,ds = -\frac{\pi}{2}\left(1-\frac{1}{4}\right) = -\frac{3\pi}{8}$$

$$2\int_{0}^{\infty}\left(\frac{s\cos s - \sin s}{s^3}\right)\cos\frac{s}{2}\,ds = -\frac{3\pi}{8}$$

$$\therefore \quad \int_{0}^{\infty}\left(\frac{s\cos s - \sin s}{s^3}\right)\cos\frac{s}{2}\,ds = -\frac{3\pi}{16}$$

Example 4. *Show that the transform of $e^{-\frac{x^2}{2}}$ is $e^{-\frac{s^2}{2}}$ by finding the Fourier transform of $e^{-a^2x^2}$, $a > 0$*

$$F\{e^{-a^2x^2}\} = \frac{1}{\sqrt{2\pi}}\int_{-\infty}^{\infty} e^{-a^2x^2}e^{isx}\,dx$$

$$= \frac{1}{\sqrt{2\pi}} \int_{-\infty}^{\infty} e^{-\left(ax - \frac{is}{2a}\right)^2} e^{-\frac{s^2}{4a^2}} dx$$

$$= e^{-\frac{s^2}{4a^2}} \cdot \frac{1}{a\sqrt{2\pi}} \int_{-\infty}^{\infty} e^{-t^2} dt, \text{ putting } ax - \frac{is}{2a} = t \text{ and } a > 0$$

$$= e^{-\frac{s^2}{4a^2}} \cdot \frac{1}{a\sqrt{2\pi}} \times \sqrt{\pi} \text{ since } \int_{-\infty}^{\infty} e^{-t^2} dt = \sqrt{\pi}$$

$$= \frac{1}{\sqrt{2}\,a} e^{-\frac{s^2}{4a^2}}$$

Setting $a = \frac{1}{\sqrt{2}}$,

$$F\left\{e^{-\frac{x^2}{2}}\right\} = e^{-\frac{s^2}{2}}$$

That is, $e^{-\frac{x^2}{2}}$ is self-reciprocal.

Example 5. *Find the Fourier transform of* $f(x)$ given by

$$f(x) = \begin{cases} 1 \text{ for } |x| < a \\ 0 \text{ for } |x| > a > 0 \end{cases}$$ **(*Anna. Univ. 2002*)**

and hence evaluate $\int_0^{\infty} \frac{\sin x}{x} dx$

and $\int_{-\infty}^{\infty} \frac{\sin as \cos sx}{s} ds$

$$F\{f(x)\} = \frac{1}{\sqrt{2\pi}} \int_{-\infty}^{\infty} f(x)\, e^{ixs} dx$$

$$= \frac{1}{\sqrt{2\pi}} \int_{-a}^{a} e^{ixs} dx$$

$$= \frac{1}{\sqrt{2\pi}} \left(\frac{e^{isx}}{is}\right)_{-a}^{a}$$

$$= \frac{1}{\sqrt{2\pi}} \cdot \frac{1}{is}\left[e^{ias} - e^{-ias}\right]$$

$$= \sqrt{\frac{2}{\pi}} \frac{\sin as}{s}$$

Using inversion formula, we get,

$$\frac{1}{\sqrt{2\pi}} \int_{-\infty}^{\infty} \left(\sqrt{\frac{2}{\pi}} \frac{\sin as}{s} \right) e^{-ixs}\, ds = f(x) = \begin{cases} 1 \text{ for } |x| < a \\ 0 \text{ for } |x| > a \end{cases}$$

$$\frac{1}{\sqrt{2\pi}} \int_{-\infty}^{\infty} \left(\sqrt{\frac{2}{\pi}} \frac{\sin as}{s} \right) (\cos xs - i \sin xs)\, ds = f(x)$$

Equating real parts,

$$\frac{1}{\pi} \int_{-\infty}^{\infty} \frac{\sin as}{s} \cos xs\, ds = f(x) = \begin{cases} 1 \text{ for } |x| < a \\ 0 \text{ for } |x| > a \end{cases}$$

$$\frac{2}{\pi} \int_{0}^{\infty} \frac{\sin as}{s} \cos xs\, ds = f(x)$$

$$\int_{0}^{\infty} \frac{\sin as}{s} \cos xs\, ds = \frac{\pi}{2} \text{ for } |x| < a$$

$$= 0 \text{ for } |x| > a$$

$$= \frac{1}{2}\left(\frac{\pi}{2} + 0 \right) = \frac{\pi}{4} \text{ for } |x| = a$$

Setting $x = 0$,

$$\int_{0}^{\infty} \frac{\sin as}{s}\, ds = \frac{\pi}{2}$$

$$\therefore \quad \int_{0}^{\infty} \frac{\sin ax}{x}\, dx = \frac{\pi}{2}$$

Setting $ax = \theta$, we get,

$$\int_{0}^{\infty} \frac{\sin \theta}{\theta}\, d\theta = \frac{\pi}{2}$$

Convolution Theorem or Faltung Theorem

Def. The convolution of two functions $f(x)$ and $g(x)$ is defined as

$$f * g = \frac{1}{\sqrt{2\pi}} \int_{-\infty}^{\infty} f(t)\, g(x-t)\, dt$$

Theorem. *The Fourier transform of the convolution of f (x) and g(x) is the product of their Fourier transforms.*

That is, $F\{f(x) * g(x)\} = F(s) \,.\, G(s) = F\{f(x)\} \,.\, F\{g(x)\}$

$$F\{f * g\} = \frac{1}{\sqrt{2\pi}} \int_{-\infty}^{\infty} (f * g)\, e^{ixs}\, dx$$

$$= \frac{1}{\sqrt{2\pi}} \int_{-\infty}^{\infty} \left(\frac{1}{\sqrt{2\pi}} \int_{-\infty}^{\infty} f(t)\, g(x-t)\, dt \right) e^{ixs}\, dx$$

$$= \frac{1}{2\pi} \int_{-\infty}^{\infty} f(t) \left(\int_{-\infty}^{\infty} g(x-t)\, e^{ixs}\, dx \right) dt$$

by changing the order of integration

$$= \frac{1}{\sqrt{2\pi}} \int_{-\infty}^{\infty} f(t)\, F\{g(x-t)\}\, dt$$

$$= \frac{1}{\sqrt{2\pi}} \int_{-\infty}^{\infty} f(t)\, e^{its}\, G(s)\, dt, \quad \text{using shifting theorem}$$

$$= G(s) \,.\, \frac{1}{\sqrt{2\pi}} \int_{-\infty}^{\infty} f(t)\, e^{its}\, dt$$

$$= G(s) \,.\, F(s)$$

$$= F(s) \,.\, G(s)$$

By inversion,

$$F^{-1}\{F(s)\, G(s)\} = f * g = F^{-1}\{F(s)\} * F^{-1}\{G(s)\}$$

Parseval's identity. If $F(s)$ is the Fourier transform of $f(x)$.

then, $$\int_{-\infty}^{\infty} |f(x)|^2\, dx = \int_{-\infty}^{\infty} |F(s)|^2\, ds$$

Proof. Firstly let us prove

$\overline{F\{f(-x)\}} = \overline{F(s)}$ where $\overline{F(s)}$ indicates the complex conjugate of $F(s)$

$$\overline{F(s)} = \frac{1}{\sqrt{2\pi}} \int_{-\infty}^{\infty} \overline{f(x)}\, e^{-isx}\, dx$$

$$= \frac{1}{\sqrt{2\pi}} \int_{-\infty}^{\infty} \overline{f(-v)}\, e^{isv}\, dv \quad \text{putting } x = -v$$

$$= F\,\overline{\{f(-v)\}}$$

$$= F\,\overline{\{f(-x)\}}, \text{ changing the dummy variable} \qquad ...(1)$$

By convolution Theorem,

$$F\{f(x) * g(x)\} = F(s)\, G(s)$$

$$f * g = F^{-1}\{F(s)\, G(s)\}$$

$$\frac{1}{\sqrt{2\pi}} \int_{-\infty}^{\infty} f(t)\, g(x-t)\, dt = \frac{1}{\sqrt{2\pi}} \int_{-\infty}^{\infty} F(s)\, G(s)\, e^{-isx}\, ds$$

Putting $x = 0$, we get

$$\int_{-\infty}^{\infty} f(t)\, g(-t)\, dt = \int_{-\infty}^{\infty} F(s)\, G(s)\, ds \qquad ...(2)$$

Since it is true for all $g(t)$, take $g(t) = \overline{f(-t)}$

$\therefore \quad g(-t) = \overline{f(t)}$

$G(s) = F\{g(t)\} = F\{\overline{f(-t)}\} = \overline{F(s)}$ using (1). Use this in (2).

$$\int_{-\infty}^{\infty} f(t)\, \overline{f(t)}\, dt = \int_{-\infty}^{\infty} F(s)\, \overline{F(s)}\, ds$$

$$\therefore \int_{-\infty}^{\infty} |f(t)|^2\, dt = \int_{-\infty}^{\infty} |F(s)|^2\, ds.$$

Example 6. *Using Parseval's identity, prove* $\int_0^{\infty} \left(\frac{\sin t}{t}\right)^2 dt = \frac{\pi}{2}.$

In example 5, if $f(x) = \begin{cases} 1 \text{ for } |x| < a \\ 0 \text{ for } |x| > a > 0 \end{cases}$

We have proved $F(s) = \sqrt{\frac{2}{\pi}}\, \frac{\sin as}{s}$.

Using Parseval's identity.

$$\int_{-\infty}^{\infty} |f(t)|^2\, dt = \int_{-\infty}^{\infty} |F(s)|^2\, ds$$

$$\therefore \quad \int_{-a}^{a} 1\, dt = \int_{-\infty}^{\infty} \frac{2}{\pi}\left(\frac{\sin as}{s}\right)^2 ds$$

$$2a = \frac{2}{\pi}\int_{-\infty}^{\infty}\left(\frac{\sin as}{s}\right)^2 ds.$$

Setting $as = t$, we get,

$$\int_{-\infty}^{\infty}\left(\frac{\sin t}{t}\right)^2 dt = \pi$$

$$\therefore \quad 2\int_{0}^{\infty}\left(\frac{\sin t}{t}\right)^2 dt = \pi$$

$$\therefore \quad \int_{0}^{\infty}\left(\frac{\sin t}{t}\right)^2 dt = \pi/2$$

Example 7. *Find the Fourier transform of* $f(x) = 1 - |x|$ *if* $|x| < 1$

and hence find the value $\int_0^\infty \frac{\sin^4 t}{t^4}\, dt.$ $= 0$ *for* $|x| > 1$

(*Anna Ap. 2005*)

$$F\{f(x)\} = \frac{1}{\sqrt{2\pi}}\int_{-\infty}^{\infty}(1 - |x|)\, e^{isx}\, dx$$

$$= \frac{1}{\sqrt{2\pi}}\int_{-1}^{1}(1 - |x|)(\cos sx + i \sin sx)\, dx$$

$$= \frac{1}{\sqrt{2\pi}} \cdot 2\int_{0}^{1}(1 - x)\cos sx\, dx \quad \text{since } (1 - |x|) \text{ is even}$$

$$= \sqrt{\frac{2}{\pi}}\left(\frac{1 - \cos s}{s^2}\right)$$

Using Parseval's identity,

$$\frac{2}{\pi}\int_{-\infty}^{\infty}\frac{(1 - \cos s)^2}{s^4}\, ds = \int_{-1}^{1}(1 - |x|)^2\, dx$$

$$\frac{4}{\pi}\int_0^\infty \frac{(1-\cos s)^2}{s^4}\,ds = 2\int_0^1 (1-x)^2\,dx = 2/3$$

$$\frac{16}{\pi}\int_0^\infty \frac{\sin^4 s/2}{s^4}\,ds = 2/3$$

Setting $\frac{s}{2} = x$, we get

$$\int_0^\infty \frac{\sin^4 x}{x^4}\,dx = \frac{\pi}{3}$$

Infinite Fourier cosine transform and sine transform:

Infinite Fourier cosine transform.

Let $f(x)$ be defined for all $x \geq 0$. Now we shall extend the function $f(x)$ to the negative side of x axis. That is, let the extended function $F(x)$ be defined as

$$F(x) = f(x) \text{ for } x \geq 0$$
$$= f(-x) \text{ for } x < 0$$ so that $F(x)$ is even in $(-\infty, \infty)$. Now,

$$F\{F(x)\} = \frac{1}{\sqrt{2\pi}}\int_{-\infty}^{\infty} F(x)\,e^{isx}\,dx$$

$$= \frac{1}{\sqrt{2\pi}}\int_{-\infty}^{\infty} F(x)\,[\cos sx + i\sin sx]\,dx$$

$$= \frac{1}{\sqrt{2\pi}} \times 2\int_0^\infty F(x)\cos sx\,dx \qquad \text{since } F(x) \text{ is even}$$

$$= \sqrt{\frac{2}{\pi}}\int_0^\infty f(x)\cos sx\,dx \text{ since } F(x) = f(x) \text{ for } x > 0$$

The R.H.S. is defined as the infinite Fourier cosine transform of $f(x)$ denote by $F_c(s)$

$\therefore$ Infinite Fourier cosine transform of $f(x)$ is defined to be

$$F_c(s) = \sqrt{\frac{2}{\pi}}\int_0^\infty f(x)\cos sx\,dx \qquad ...(1)$$

which is an even function of s

Then by inversion theorem,

$$F(x) = \frac{1}{\sqrt{2\pi}} \int_{-\infty}^{\infty} F_c(s)\, e^{-isx}\, ds$$

$$= \frac{1}{\sqrt{2\pi}} \int_{-\infty}^{\infty} F_c(s)\{\cos sx - i \sin sx\} ds$$

$$= \frac{1}{\sqrt{2\pi}} \times 2 \int_{0}^{\infty} F_c(s) \cos sx\, dx \text{ since } F_c(s) \text{ is even}$$

Since $F(x) = f(x)$ *in* $(0, \infty)$,

$$f(x) = \sqrt{\frac{2}{\pi}} \int_{0}^{\infty} Fc(s) \cos sx \, ds \quad ...(2)$$

Equation (1) defines the Fourier cosine transform of $f(x)$ and (2) gives the Inversion Theorem for Fourier cosine transform.

Infinite Fourier Sine Transform

Let $f(x)$ be defined for $x \geq 0$. We extend $f(x)$ to the negative side of x axis. Thus, let,

$$F(x) = f(x) \text{ for } x \geq 0$$
$$= -f(-x) \text{ for } x < 0$$

so that $F(x)$ is odd in $(-\infty, \infty)$

$$F\{F(x)\} = \frac{1}{\sqrt{2\pi}} \int_{-\infty}^{\infty} F(x)\, e^{isx}\, dx$$

$$= \frac{1}{\sqrt{2\pi}} \int_{-\infty}^{\infty} F(x)\, [\cos sx + i \sin sx]\, dx$$

$$= \frac{i}{\sqrt{2\pi}} \int_{-\infty}^{\infty} F(x) \sin sx\, dx \text{ since } F(x) \text{ is odd}$$

$$= i\sqrt{\frac{2}{\pi}} \int_{0}^{\infty} f(x) \sin sx\, dx$$

Now, we define Fourier sine transform of $f(x)$ as

$$F_s(s) = \text{ imaginary point of } F\{F(x)\}$$

$$F_s\{f(x)\} = F_s(s) = \sqrt{\frac{2}{\pi}} \int_0^{\infty} f(x) \sin sx \, dx \qquad ...(3)$$

By inversion theorem,

$$F(x) = F^{-1}\{i F_s(s)\}$$

$$= i \frac{1}{\sqrt{2\pi}} \int_{-\infty}^{\infty} F_s(s)\, e^{-isx}\, ds$$

$$f(x) = \sqrt{\frac{2}{\pi}} \int_0^{\infty} F_s(s) \sin sx \, ds \qquad ...(4)$$

since $F_s(s)$ is odd function of s and $F(x) = f(x)$ in $(0, \infty)$

Equation (3) defines infinite Fourier sine transform of $f(x)$ and (4) gives the inversion theorem for Fourier sine transform.

Hence the transform pairs are

$$F_c\{f(x)\} = \sqrt{\frac{2}{\pi}} \int_0^{\infty} f(x) \cos sx \, dx \,;\, f(x) = \sqrt{\frac{2}{\pi}} \int_0^{\infty} F_c(s) \cos sx \, ds$$

$$F_s\{f(x)\} = \sqrt{\frac{2}{\pi}} \int_0^{\infty} f(x) \sin sx \, dx \,;\, f(x) = \sqrt{\frac{2}{\pi}} \int_0^{\infty} F_s(s) \sin sx \, ds$$

Properties regarding cosine and sine transforms

1. Cosine and sine transforms are linear.

   That is,

   $F_c\{af(x) + bg(x)\} = a\, F_c\{f(x)\} + b\, F_c[g(x)]$

   The proof is obvious.

2. $F_s[af(x) + bg(x)] = a\, F_s\{f(x)\} + b\, F_s\{g(x)\}$

3. $F_s[f(x) \sin ax] = \frac{1}{2}[F_c(s-a) - F_c(s+a)]$

4. $F_s[f(x) \cos ax] = \frac{1}{2}[F_s(s+a) + F_s(s-a)]$

5. $F_c[f(x) \sin ax] = \frac{1}{2}[F_s(a+s) + F_s(a-s)]$

6. $F_c[f(x) \cos ax] = \frac{1}{2}[F_c(s+a) + F_c(s-a)]$

Proof. $F_s\,[f(x)\sin ax] = \sqrt{\dfrac{2}{\pi}}\displaystyle\int_0^\infty f(x)\sin ax \sin sx\, dx$

$$= \frac{1}{2}\sqrt{\frac{2}{\pi}}\int_0^\infty f(x)\,.\,[\cos(s-a)\,x - \cos(a+s)\,x\,]\,dx$$

$$= \tfrac{1}{2}[F_c(s-a) - F_c(s+a)\,]$$

$$F_s\,[f(x)\cos ax\,] = \sqrt{\frac{2}{\pi}}\int_0^\infty f(x)\cos ax \sin sx\, dx$$

$$= \frac{1}{2}\sqrt{\frac{2}{\pi}}\int_0^\infty f(x)\,.\,[\sin(s+a)\,x + \sin(s-a)\,x\,]\,dx$$

$$= \tfrac{1}{2}\,[F_s(s+a) + F_s(s-a)\,]$$

Similarly, we can prove the results (5) and (6).

7. $F_c\,\{f(ax)\,\} = \dfrac{1}{a}F_c\left(\dfrac{s}{a}\right)$

8. $F_s\,\{f(ax)\,\} = \dfrac{1}{a}F_s\left(\dfrac{s}{a}\right)$

Hint. put $ax = u$ and proceed.

Identities

If $F_c(s)$, $G_c(s)$ are the Fourier cosine transforms and $F_s(s)$, $G_s(s)$ are the Fourier sine transforms of $f(x)$ and $g(x)$ respectively, then

1. $\displaystyle\int_0^\infty f(x)\,g(x)\,dx = \int_0^\infty F_c(s)\,G_c(s)\,ds$

2. $\displaystyle\int_0^\infty f(x)\,g(x)\,dx = \int_0^\infty F_s(s)\,G_s(s)\,ds$

3. $\displaystyle\int_0^\infty |\,f(x)\,|^2\,dx = \int_0^\infty |\,F_c(s)\,|^2\,ds = \int_0^\infty |\,F_s(s)\,|^2\,ds.$

Proof. $\displaystyle\int_0^\infty F_s(s)\,G_s(s)\,ds = \int_0^\infty F_s(s)\left\{\sqrt{\frac{2}{\pi}}\int_0^\infty g(x)\sin sx\,dx\right\}ds$

$$= \int_0^\infty g(x) \left\{ \sqrt{\frac{2}{\pi}} \int_0^\infty F_s(s) \sin sx \, ds \right\} dx$$

$$= \int_0^\infty g(x) f(x) \, dx$$

Similarly, we can prove the first identity and the third follows by setting $g(x) = f(x)$.

Example 8. *Find Fourier cosine and sine transforms of e^{-ax}, $a > 0$ and hence deduce the inversion formula.*

$$F_s(e^{-ax}) = \sqrt{\frac{2}{\pi}} \int_0^\infty e^{-ax} \sin sx \, dx$$

$$= \sqrt{\frac{2}{\pi}} \left\{ \frac{e^{-ax}}{a^2 + s^2} (-a \sin sx - s \cos sx) \right\}_0^\infty$$

$$= \sqrt{\frac{2}{\pi}} \frac{s}{a^2 + s^2}, \text{ if } a > 0 \quad ...(1)$$

$$F_c(e^{-ax}) = \sqrt{\frac{2}{\pi}} \int_0^\infty e^{-ax} \cos sx \, dx$$

$$= \sqrt{\frac{2}{\pi}} \left[\frac{e^{-ax}}{a^2 + s^2} (-a \cos sx + s \sin sx) \right]_0^\infty$$

$$= \sqrt{\frac{2}{\pi}} \cdot \frac{a}{a^2 + s^2} \text{ if } a > 0 \quad ...(2)$$

By inversion formula of (1),

$$e^{-ax} = \sqrt{\frac{2}{\pi}} \int_0^\infty \left(\sqrt{\frac{2}{\pi}} \frac{s}{a^2 + s^2} \cdot \sin sx \right) ds$$

$$\therefore \int_0^\infty \frac{s}{a^2 + s^2} \sin sx \, ds = \frac{\pi}{2} e^{-ax}, a > 0$$

Changing the variables,

$$\int_0^\infty \frac{x \sin \alpha x}{a^2 + x^2}\, dx = \frac{\pi}{2} e^{-\alpha a}, a > 0 \qquad ...(3)$$

Again, by inversion formula of (2)

$$e^{-ax} = \sqrt{\frac{2}{\pi}} \int_0^\infty \left(\sqrt{\frac{2}{\pi}} \frac{a}{a^2 + s^2} \right) \cos sx\, ds$$

$$\therefore \quad \int_0^\infty \frac{\cos sx}{a^2 + s^2}\, ds = \frac{\pi}{2a} e^{-ax}$$

Changing the variables,

$$\int_0^\infty \frac{\cos \alpha x}{a^2 + x^2}\, dx = \frac{\pi}{2a} e^{-\alpha a}, a > 0 \qquad ...(4)$$

Example 9. *Find Fourier sine transform of* $\frac{x}{a^2 + x^2}$ *and Fourier cosine transform of* $\frac{1}{a^2 + x^2}$.

$$F_s\left(\frac{x}{a^2 + x^2}\right) = \sqrt{\frac{2}{\pi}} \int_0^\infty \frac{x}{a^2 + x^2} \sin sx\, dx$$

$$= \sqrt{\frac{2}{\pi}} \left[\frac{\pi}{2} e^{-as} \right] \text{ using (3) of Example 8.}$$

$$= \sqrt{\frac{\pi}{2}}\, e^{-as}$$

$$F_c\left(\frac{1}{a^2 + x^2}\right) = \sqrt{\frac{2}{\pi}} \int_0^\infty \frac{1}{a^2 + x^2} \cos sx\, dx$$

$$= \sqrt{\frac{2}{\pi}} \left[\frac{\pi}{2a} e^{-as} \right] \text{ using (4) of Example 8.}$$

$$= \sqrt{\frac{\pi}{2}} \cdot \frac{1}{a} e^{-as}.$$

Example 10. *Using Parseval's identity evaluate*

$$\int_0^\infty \frac{dx}{(a^2 + x^2)^2} \text{ and } \int_0^\infty \frac{x^2}{(a^2 + x^2)^2}\, dx \textit{ if } a > 0.$$

By example 8, If $f(x) = e^{-ax}$, $F_s(s) = \sqrt{\frac{2}{\pi}} \frac{s}{a^2 + s^2}$

and $F_c(s) = \sqrt{\frac{2}{\pi}} \frac{a}{a^2 + s^2}$

Using Parseval's identity,

$$\int_0^\infty \frac{2}{\pi} \frac{s^2}{(a^2 + s^2)^2} ds = \int_0^\infty e^{-2ax} dx$$

$$= \left(\frac{e^{-2ax}}{-2a}\right)_0^\infty$$

$$= \frac{1}{2a}$$

$$\therefore \quad \int_0^\infty \frac{x^2}{(a^2 + x^2)^2} dx = \frac{\pi}{4a}. \text{ if } a > 0$$

Also $\displaystyle\int_0^\infty \frac{2}{\pi} \frac{a^2}{(a^2 + s^2)^2} ds = \int_0^\infty e^{-2ax} dx = \frac{1}{2a}$

$$\int_0^\infty \frac{dx}{(a^2 + x^2)^2} = \frac{\pi}{4a^3} \text{ if } a > 0$$

Example 11. *Evaluate* $\displaystyle\int_0^\infty \frac{dx}{(a^2 + x^2)(b^2 + x^2)}$ *using transform methods.*

Let $f(x) = e^{-ax}$, $g(x) = e^{-bx}$

$$F_c(s) = \sqrt{\frac{2}{\pi}} \int_0^\infty e^{-ax} \cos sx dx$$

$$= \sqrt{\frac{2}{\pi}} \frac{a}{a^2 + s^2} \text{ by example 8}$$

Similarly,

$$G_c(s) = \sqrt{\frac{2}{\pi}} \frac{b}{b^2 + s^2}$$

$\therefore$ Using, $\displaystyle\int_0^\infty F_c(s)\, G_c(s)\, ds = \int_0^\infty f(x)\, g(x)\, dx$

$$\frac{2}{\pi}\int_0^\infty \frac{ab}{(a^2+s^2)(b^2+s^2)}\,ds = \int_0^\infty e^{-(a+b)x}\,dx$$

$$= \frac{1}{a+b}$$

$$\int_0^\infty \frac{dx}{(a^2+x^2)(b^2+x^2)} = \frac{\pi}{2ab(a+b)},\ \text{if } a, b > 0.$$

Example 12. *Find Fourier cosine transform of*

$$f(x) = \begin{cases} \cos x \text{ in } 0 < x < a \\ 0 \qquad\quad x \geq a \end{cases}$$

$$F_c(s) = \sqrt{\frac{2}{\pi}}\int_0^\infty f(x)\cos sx\,dx$$

$$= \sqrt{\frac{2}{\pi}}\int_0^a \cos x\cos sx\,dx$$

$$= \sqrt{\frac{2}{\pi}}\,\frac{1}{2}\int_0^a [\cos(s+1)x + \cos(s-1)x]\,dx$$

$$= \frac{1}{\sqrt{2\pi}}\left[\frac{\sin(s+1)a}{s+1} + \frac{\sin(s-1)a}{s-1}\right] \text{ if } s \neq 1, -1$$

Example 13. *Find Fourier sine transform of* $\frac{1}{x}$ **(*Anna Ap. 2005*)**

$$F_s\left(\frac{1}{x}\right) = \sqrt{\frac{2}{\pi}}\int_0^\infty \frac{\sin sx}{x}\,dx$$

$$= \sqrt{\frac{2}{\pi}}\int_0^\infty \frac{\sin\theta}{\theta}\,d\theta,\ \text{putting } sx = \theta$$

$$= \sqrt{\frac{2}{\pi}} \times \frac{\pi}{2}$$

$$= \sqrt{\frac{\pi}{2}}$$

Example 14. *Find Fourier sine and cosine transform of* x^{n-1}.

We know, $\Gamma(n) = \int_0^\infty e^{-x}x^{n-1}\,dx,\ n > 0$

$$= \int_0^\infty e^{-ax} (ax)^{n-1} \, adx \text{ setting } x \text{ as } ax, a > 0$$

$$\therefore \quad \int_0^\infty e^{-ax} x^{n-1} \, dx = \frac{\Gamma(n)}{a^n}, n > 0, a > 0$$

We can prove the above result even if a is complex.

Setting $a = is$,

$$\int_0^\infty e^{-isx} x^{n-1} \, dx = \frac{\Gamma(n)}{(is)^n}$$

$$= \frac{(-i)^a \, \Gamma(n)}{s^n}$$

$$= \frac{e^{-\frac{\pi}{2} ni}}{s^n} \Gamma(n) \text{ since } -i = e^{-\frac{\pi}{2} i}$$

Equating real and imaginary parts on both sides, we get

$$\int_0^\infty x^{n-1} \cos sx \, dx = \frac{\Gamma(n)}{s^n} \cos \frac{n\pi}{2}$$

$$\int_0^\infty x^{n-1} \sin sx \, dx = \frac{\Gamma(n)}{s^n} \sin \frac{n\pi}{2}$$

$$\therefore \quad F_c(x^{n-1}) = \sqrt{\frac{2}{\pi}} \frac{\Gamma(n)}{s^n} \cos \frac{n\pi}{2}$$

$$F_s(x^{n-1}) = \sqrt{\frac{2}{\pi}} \frac{\Gamma(n)}{s^n} \sin \frac{n\pi}{2}$$

Taking $n = 1/2$,

$$F_c\left(\frac{1}{\sqrt{x}}\right) = \frac{1}{\sqrt{s}} \text{ since } \Gamma\left(\frac{1}{2}\right) = \sqrt{\pi}$$

$$F_s\left(\frac{1}{\sqrt{x}}\right) = \frac{1}{\sqrt{s}}$$

Note. : $\frac{1}{\sqrt{x}}$ is self reciprocal under Fourier sine and cosine transforms.

Example 15. *Show that*

(*i*) $F_s[xf(x)] = -\frac{d}{ds} F_c(s)$ (*ii*) $F_c[xf(x)] = \frac{d}{ds} F_s(s)$

and hence find Fourier cosine and sine transform of xe^{-ax}

$$F_c(s) = \sqrt{\frac{2}{\pi}} \int_0^\infty f(x) \cos sx \, dx$$

$$\frac{d}{ds} F_c(s) = -\sqrt{\frac{2}{\pi}} \int_0^\infty xf(x) \sin sx \, dx$$

$$= -F_s\{xf(x)\}$$

Similarly,

$$F_s(s) = \sqrt{\frac{2}{\pi}} \int_0^\infty f(x) \sin sx \, dx$$

$$\frac{d}{ds} F_s(s) = \sqrt{\frac{2}{\pi}} \int_0^\infty xf(x) \cos sx \, dx$$

$$= F_c\{xf(x)\}$$

$$F_c(xe^{-ax}) = \frac{d}{ds} F_s(e^{-ax})$$

$$= \frac{d}{ds}\left(\sqrt{\frac{2}{\pi}} \frac{s}{a^2+s^2}\right) \text{ using example 8}$$

$$= \sqrt{\frac{2}{\pi}} \frac{a^2-s^2}{(a^2+s^2)^2}$$

$$F_s(xe^{-ax}) = -\frac{d}{ds} F_c(e^{-ax})$$

$$= -\frac{d}{ds}\left(\sqrt{\frac{2}{\pi}} \frac{a}{a^2+s^2}\right) \text{ (Refer example 8)}$$

$$= \sqrt{\frac{2}{\pi}} \frac{2as}{(a^2+s^2)^2}$$

Example 16. *Find Fourier cosine transform of* $e^{-a^2x^2}$ *and hence evaluate Fourier sine transform of* $xe^{-a^2x^2}$

$$F_c(e^{-a^2x^2}) = \sqrt{\frac{2}{\pi}} \int_0^\infty e^{-a^2x^2} \cdot \cos sx \, dx$$

$$= \text{Real part of } \sqrt{\frac{2}{\pi}} \int_0^{\infty} e^{-a^2x^2+isx}\, dx$$

$$= \frac{1}{a\sqrt{2}}\, e^{-\frac{s^2}{4a^2}} \quad \text{(Refer example 4)}$$

$$F_s\,(xe^{-a^2x^2}) = -\frac{d}{ds} F_c\,(e^{-a^2x^2})$$

$$= \frac{1}{a\sqrt{2}}\, e^{-\frac{s^2}{4a^2}}\, \frac{s}{2a^2}$$

$$= \frac{s}{2a^3\sqrt{2}}\, e^{-\frac{s^2}{4a^2}}$$

Example 17. *Solve for f (x) from the integral equation*

$$\int_0^{\infty} f(x) \cos \alpha x\, dx = e^{-\alpha}$$

Proof. Multiplying by $\sqrt{\frac{2}{\pi}}$, we get

$$\sqrt{\frac{2}{\pi}} \int_0^{\infty} f(x) \cos \alpha x\, dx = \sqrt{\frac{2}{\pi}}\, e^{-\alpha}$$

$$F_c\,\{f(x)\} = \sqrt{\frac{2}{\pi}}\, e^{-\alpha}, \ \alpha \text{ parameter}$$

$$\therefore \qquad f(x) = F_c^{-1}\left(\sqrt{\frac{2}{\pi}}\, e^{-\alpha}\right)$$

$$= \frac{2}{\pi} \int_0^{\infty} e^{-\alpha} \cos \alpha x\, d\alpha$$

$$= \frac{2}{\pi} \cdot \frac{1}{1+x^2} \text{ on integration}$$

Example 18. *Solve for f(x) from the integral equation*

$$\int_0^{\infty} f(x) \sin sx\, dx = \begin{cases} 1 \text{ for } 0 \le s < 1 \\ 2 \text{ for } 1 \le s < 2 \\ 0 \text{ for } s \ge 2 \end{cases}$$

Proof. Multiplying by $\sqrt{\frac{2}{\pi}}$, both sides,

$$F_s(f(x)) = \begin{cases} \sqrt{\dfrac{2}{\pi}} & \text{for } 0 \le s < 1 \\ 2\sqrt{\dfrac{2}{\pi}} & \text{for } 1 \le s < 2 \\ 0 & \text{for } s \ge 2 \end{cases}$$

$$\therefore \quad f(x) = F_s^{-1} \text{ (R.H.S.)}$$

$$= \frac{2}{\pi}\int_0^1 \sin sx\, ds + \frac{4}{\pi}\int_1^2 \sin sx\, ds$$

$$= \frac{2}{\pi}\left(\frac{1-\cos x}{x}\right) + \frac{4}{\pi}\left(\frac{\cos x - \cos 2x}{x}\right)$$

$$f(x) = \frac{2}{\pi x}(1 + \cos x - 2\cos 2x)$$

Example 19. *Find the Fourier cosine transform of* e^{-x^2} *(second method).*

$$I = F_c(e^{-x^2}) = \sqrt{\frac{2}{\pi}}\int_0^\infty e^{-x^2}\cos sx\, dx$$

$$\frac{dI}{ds} = -\sqrt{\frac{2}{\pi}}\int_0^\infty xe^{-x^2}\sin sx\, dx$$

$$= +\sqrt{\frac{2}{\pi}}\int_0^\infty \sin sx\, d\left(\frac{e^{-x^2}}{2}\right)$$

$$= \sqrt{\frac{2}{\pi}}\left[\left(\frac{e^{-x^2}}{2}\sin sx\right)_0^\infty - \frac{1}{2}\int_0^\infty se^{-x^2}\cos sx\, dx\right]$$

$$= -\frac{s}{2}I$$

$$\therefore \quad \frac{dI}{I} = -\frac{s}{2}ds$$

$$\therefore \quad \log I = -\frac{s^2}{4} + \log c$$

$$I = ce^{-\frac{s^2}{4}} \qquad \text{...(1)}$$

when $s = 0$, $\quad I = \sqrt{\dfrac{2}{\pi}}\displaystyle\int_0^\infty e^{-x^2}\, dx = \dfrac{1}{\sqrt{2}}$

Using in (1),

$$\frac{1}{\sqrt{2}} = c$$

$$\therefore \qquad I = F_c\,(e^{-x^2}) = \frac{1}{\sqrt{2}}\, e^{-\frac{s^2}{4}}$$

Example 20. *Find the complex Fourier transform of dirac delta function* $\delta\,(t-a)$.

$$F\{\delta\,(t-a)\} = \frac{1}{\sqrt{2\pi}} \int_{-\infty}^{\infty} e^{ist}\,\delta\,(t-a)\,dt$$

$$= \frac{1}{\sqrt{2\pi}} \operatorname*{Lt}_{h\to 0} \int_a^{a+h} \frac{1}{h}\, e^{ist}\,dt$$

$$= \frac{1}{\sqrt{2\pi}} \operatorname*{Lt}_{h\to 0} \frac{1}{h} \left(\frac{e^{ist}}{is}\right)_a^{a+h}$$

$$= \frac{1}{\sqrt{2\pi}} \operatorname*{Lt}_{h\to 0} e^{isa} \left(\frac{e^{ish}-1}{ish}\right)$$

$$= \frac{e^{isa}}{\sqrt{2\pi}} \text{ since } \operatorname*{Lt}_{\theta\to 0} \frac{e^{\theta}-1}{\theta} = 1$$

Note. Dirac delta function $\delta\,(t-a)$ is defined as

$$\delta\,(t-a) = \operatorname*{Lt}_{h\to 0} I(h, t-a) \text{ where}$$

$$I\,(h, t-a) = \frac{1}{h} \text{ for } a < t < a+h$$

$$= 0 \text{ for } t < a \text{ and } t > a+h$$

Example 21. *Find the function if its sine transform is* $\frac{e^{-as}}{s}$.

Let $F_s\,(f\,(x)) = \dfrac{e^{-as}}{s}$

Then, $f\,(x) = \sqrt{\dfrac{2}{\pi}} \displaystyle\int_0^{\infty} \frac{e^{-as}}{s} \sin sx\,ds$ ...(1)

$$\therefore \qquad \frac{df}{dx} = \sqrt{\frac{2}{\pi}} \int_0^{\infty} e^{-as} \cos sx\,ds$$

$$= \sqrt{\frac{2}{\pi}} \cdot \frac{a}{a^2 + x^2}$$

$$\therefore \qquad f(x) = \sqrt{\frac{2}{\pi}} \cdot a \int \frac{dx}{a^2 + x^2}$$

$$= \sqrt{\frac{2}{\pi}} \tan^{-1} \frac{x}{a} + c \qquad \text{...(2)}$$

At $x = 0, f(0) = 0$ using (1)

Using this in (2), $c = 0$

Hence, $\quad f(x) = \sqrt{\frac{2}{\pi}} \tan^{-1} \frac{x}{a}$.

Setting $a = 0$,

$$F_s^{-1}\left(\frac{1}{s}\right) = \sqrt{\frac{2}{\pi}} \cdot \frac{\pi}{2} = \sqrt{\frac{\pi}{2}}$$

Example 22. *Prove* (*i*) $F\{x^n f(x)\} = (-i)^n \dfrac{d^n F(s)}{ds^n}$

and (*ii*) $F\{f^n(x)\} = (-is)^n F(s)$ (*iii*) *Hence solve for* $f(x)$ *if*

$$\int_{-\infty}^{\infty} f(t)\, e^{-|x-t|}\, dt = \phi(x) \text{ where } \phi(x) \text{ is known.}$$

Proof. (*i*) $F(s) = \dfrac{1}{\sqrt{2\pi}} \displaystyle\int_{-\infty}^{\infty} e^{isx} f(x)\, dx$

$$\frac{d^n}{ds^n} F(s) = \frac{1}{\sqrt{2\pi}} \int_{-\infty}^{\infty} (ix)^n e^{isx} f(x)\, dx$$

$$(-i)^n \frac{d^n}{ds^n} F(s) = \frac{1}{\sqrt{2\pi}} \int_{-\infty}^{\infty} x^n e^{isx} f(x)\, dx$$

$$= F\{x^n f(x)\}$$

(*ii*) Similarly,

$$F\{f^n(x)\} = \frac{1}{\sqrt{2\pi}} \int_{-\infty}^{\infty} e^{isx} \frac{d^n}{dx^n} f(x)\,.\, dx$$

$$= (is)^n \mathrm{F}(s),$$

Using integration by parts successively and making assumptions that $f, f', ..., f^{(n-1)} \to 0$ as $x \to \pm\infty$.

(*iii*) $\frac{1}{\sqrt{2\pi}} \phi(x) = \frac{1}{\sqrt{2\pi}} \int_{-\infty}^{\infty} f(t)\, e^{-|x-t|}\, dt$, from the given equation

$$= f(x) * e^{-|x|}$$

By convolution theorem,

$$\frac{1}{\sqrt{2\pi}} \bar{\phi}(s) = F(s) \, . \sqrt{\frac{2}{\pi}} \frac{1}{1+s^2}$$

$$F(s) = \tfrac{1}{2} (1+s^2)\, \bar{\phi}(s)$$

$$= \tfrac{1}{2} [\bar{\phi}(s) - (-is)^2\, \bar{\phi}(s)]$$

$\therefore \; f(x) = \frac{1}{2} \phi(x) - \frac{1}{2} \phi''(x)$ using the result derived in (*ii*)

Exercises 4 (a)

1. Show that the Fourier transform of

$$f(x) = a - |x| \text{ for } |x| < a$$
$$= 0 \quad \text{for } |x| > a > 0$$

is $\sqrt{\frac{2}{\pi}} \frac{1 - \cos as}{s^2}$. Hence show that $\int_0^{\infty} \left(\frac{\sin t}{t}\right)^2 dt = \pi/2$

2. Show that the Fourier transform of

$$f(x) = 0 \text{ for } x < \alpha$$
$$= 1 \text{ for } \alpha < x < \beta$$
$$= 0 \text{ for } x > \beta$$

is $\frac{1}{\sqrt{2\pi}} \left(\frac{e^{i\beta s} - e^{i\alpha s}}{is}\right)$

3. Show that the Fourier transform of

$$f(x) = \begin{cases} \frac{\sqrt{2\pi}}{2a}, & \text{for } |x| \le a \\ 0 & \text{for } |x| > a \end{cases}$$

is $\frac{\sin sa}{sa}$

4. Find the Fourier sine and cosine transforms of e^{-x} and hence using the inversion formulae, show that

$$\int_0^{\infty} \frac{x \sin \alpha x}{1+x^2} dx = \frac{\pi}{2} e^{-\alpha} = \int_0^{\infty} \frac{\cos \alpha x}{1+x^2} dx$$

5. Show that the Fourier sine transform of

$$f(x) = \begin{cases} x & \text{for } 0 < x < 1 \\ 2-x & \text{for } 1 < x < 2 \\ 0 & \text{for } x > 2 \end{cases}$$

is $\dfrac{2\sqrt{2}}{\sqrt{\pi}} \sin s\,(1-\cos s)/s^2$. ***(1987 Ap. Barathadasan)***

6. Find the Fourier sine and cosine transform of $\cosh x - \sinh x$ (same as question 4).

7. Find the Fourier sine and cosine transform of $ae^{-\alpha x} + be^{-\beta x}$, $\alpha, \beta > 0$.

8. Find the Fourier transform of $f(x) = \begin{cases} 1 \text{ for } |x| \le a \\ 0 \text{ for } |x| > a \end{cases}$

9. Show that the Fourier cosine transform of $\dfrac{1}{1+x^2}$ is $\sqrt{\dfrac{\pi}{2}}\, e^{-s}$

10. Show that the Fourier sine transform of $\dfrac{x}{1+x^2}$ is $\sqrt{\dfrac{\pi}{2}}\, e^{-s}$

11. Show that the Fourier transform of $e^{\frac{-x^2}{2}}$ is self-reciprocal.

12. Find Fourier transform of $e^{-a|x|}$ if $a > 0$.

13. Find Fourier transform of $\dfrac{1}{\sqrt{|x|}}$.

Fourier Transform of Derivatives

We have already seen (refer Example 22) that,

$$F\{f^n(x)\} = (-is)^n F(s)$$

(*i*) $\therefore F\left(\dfrac{\partial^2 u}{\partial x^2}\right) = (-is)^2 F\{u(x)\} = -s^2 \bar{u}$ where $\bar{u}$ is Fourier transform of u w.r.t. x

(*ii*) $F_c\{f'(x)\} = -\sqrt{\dfrac{2}{\pi}}\, f(o) + s\, F_s(s)$

$$\text{L.H.S.} = \sqrt{\frac{2}{\pi}} \int_0^\infty f'(x) \,.\, \cos sx\, dx$$

$$= \sqrt{\frac{2}{\pi}} \int_0^\infty \cos sx\, d\,\{f(x)\}$$

$$= \sqrt{\frac{2}{\pi}} \left[\{f(x) \cos sx\}_0^\infty + s \int_0^\infty f(x) \sin sx\, dx \right]$$

$$= s\, F_s(s) - \sqrt{\frac{2}{\pi}}\, f(0) \text{ assuming } f(x) \to 0 \text{ as } x \to \infty$$

$$(iii)\ F_s\{f'(x)\} = \sqrt{\frac{2}{\pi}} \int_0^{\infty} \sin sx\, d\,[f(x)]$$

$$= \sqrt{\frac{2}{\pi}} \left[(f(x) \sin sx)_0^{\infty} - s \int_0^{\infty} f(x) \cos sx\, dx \right]$$

$$= -s\, F_c(s)$$

$$(iv)\ F_c\{f''(x)\} = \sqrt{\frac{2}{\pi}} \int_0^{\infty} \cos sx\, d\,[f'(x)]$$

$$= \sqrt{\frac{2}{\pi}} \left[\{f'(x) \cos sx\}_0^{\infty} + s \int_0^{\infty} f'(x) \sin sx\, dx \right]$$

$$= -\sqrt{\frac{2}{\pi}}\, f'(0) + s\, F_s\{f'(x)\}$$

$$= -s^2 F_c(s) - \sqrt{\frac{2}{\pi}}\, f'(0)$$

assuming $f(x), f'(x) \to 0$ as $x \to \infty$

$$= \sqrt{\frac{2}{\pi}} \left[\int_0^{\infty} \sin sx\, d\,[f'(x)] \right]$$

$$= \sqrt{\frac{2}{\pi}} \left[\left(f'(x) \sin sx \right)_0^{\infty} - s \int_0^{\infty} f'(x) \cos sx\, dx \right]$$

$$= -s\, F_c\{f'(x)\}$$

$$= -s \left[s\, F_s(s) - \sqrt{\frac{2}{\pi}}\, f(0) \right]$$

$$= -s^2 F_s(s) + \sqrt{\frac{2}{\pi}}\, sf(0)$$

assuming $f(x), f'(x) \to 0$ as $x \to \infty$.

Relationship between Fourier and Laplace Transforms

Consider

$$f(t) = \begin{cases} e^{-xt}\, g(t) & \text{for } t > 0 \\ 0 & \text{for } t < 0 \end{cases} \qquad \ldots(1)$$

Then the Fourier transform of $f(t)$ is given by

$$F\{f(t)\} = \frac{1}{\sqrt{2\pi}} \int_{-\infty}^{\infty} e^{ist} f(t)\, dt$$

$$= \frac{1}{\sqrt{2\pi}} \int_{0}^{\infty} e^{ist} f(t)\, dt$$

$$= \frac{1}{\sqrt{2\pi}} \int_{0}^{\infty} e^{(is-x)\,t}\, g(t)\, dt$$

$$= \frac{1}{\sqrt{2\pi}} \int_{0}^{\infty} e^{-pt}\, g(t)\, dt \text{ where } p = x - is$$

$$= \frac{1}{\sqrt{2\pi}} L\{g(t)\}$$

$\therefore$ Fourier transform of $f(t) = \frac{1}{\sqrt{2\pi}} \times$ Laplace transform of $g(t)$ defined by (1).

APPLICATIONS TO BOUNDARY VALUE PROBLEMS

Example 1. *Solve the diffusion equation*

$$\frac{\partial u}{\partial t} = K \frac{\partial^2 u}{\partial x^2},\ -\infty < x < \infty,\ t > 0 \text{ with the conditions,}$$

$u(x, 0) = f(x)$, *and* $\frac{\partial u}{\partial x}$, *u tend to zero as x tend to* $\pm\infty$.

Proof. Here, $u = u(x, t),\ -\infty < x < \infty,\ t > 0$

Defining Fourier transform of $u(x, t)$ as

$\bar{u}(s, t) = \frac{1}{\sqrt{2\pi}} \int_{-\infty}^{\infty} u(x, t)\, e^{isx}\, dx$, take Fourier transform of the given differential equation.

$$KF\left\{\frac{\partial^2 u}{\partial x^2}\right\} = F\left\{\frac{\partial u}{\partial t}\right\}$$

i.e., $$K(-s^2 \bar{u}(s, t)) = \frac{1}{\sqrt{2\pi}} \int_{0}^{\infty} \frac{\partial u}{\partial t} e^{isx}\, dx$$

$$= \frac{d}{dt}\{F(u)\}$$

$$= \frac{d\bar{u}}{dt}$$

$\therefore \quad \frac{d\bar{u}}{dt} + s^2 . k\bar{u} = 0$ where $\bar{u}$ = Fourier transform of $\bar{u}$ ...(1)

Solving (1), $\bar{u}(s, t) = ce^{-s^2 kt}$ ...(2)

Since $u(x, 0) = f(x)$, taking transform w.r.t. x,

$$\bar{u}(s, o) = F(s)$$

Using (3) in (2), we get, $c = F(s)$

$$\bar{u}(s, t) = F(s)\, e^{-s^2 kt}$$

Taking inverse transform,

$$u(x, t) = F^{-1}\{F(s)\, e^{-s^2 kt}\}$$

$$= f(x) * F^{-1}(e^{-s^2 kt})$$

$$= f(x) * \frac{e^{-\frac{x^2}{4kt}}}{\sqrt{2kt}}$$

$$\left(\text{since } F^{-1}\left(e^{-\frac{s^2}{4a^2}}\right) = \sqrt{2}a\, e^{-a^2x^2} \text{ by example 4}\right)$$

$$= \frac{1}{\sqrt{4\pi kt}} \int_{-\infty}^{\infty} f(\theta)\, e^{-\frac{(x-\theta)^2}{4kt}}\, d\theta$$

Putting, $\frac{x-\theta}{2\sqrt{kt}} = \phi$, we get $\theta = x - 2\sqrt{kt}\,\phi$

$$\therefore \quad u(x, t) = \frac{1}{\sqrt{\pi}} \int_{-\infty}^{\infty} f(x - 2\sqrt{kt}\,\phi)\, e^{-\phi^2}\, d\phi$$

Example 2. *Solve* $\frac{\partial^2 u}{\partial t^2} = \alpha^2 \frac{\partial^2 u}{\partial x^2}, -\infty < x < \infty, t \geq 0$ *with conditions* $u(x, 0) = f(x), \frac{\partial u}{\partial t}(x, 0) = g(x)$ *and assuming* $u, \frac{\partial u}{\partial x} \to 0$ *as* $x \to \pm\infty$.

Taking Fourier transform on both sides of the differential equation,

$\frac{d^2\bar{u}}{dt^2} = \alpha^2(-s^2\bar{u})$ where $\bar{u}$ is Fourier transform of u with respect to x.

$$\frac{d^2\bar{u}}{dt^2} + \alpha^2 s^2 \bar{u} = 0$$

Auxiliary equation is $m^2 + \alpha^2 s^2 = 0$

$$m = \pm i\,\alpha\, s$$

$$\therefore \bar{u}(s,t) = Ae^{i\alpha st} + Be - i\alpha st \quad \text{...(1)}$$

Since $u(x, 0) = f(x)$ and $\frac{\partial u}{\partial t}(x, 0) = g(x)$,

$\bar{u}(s, 0) = F(s)$ and $\frac{d\bar{u}}{dt}(s, o) = G(s)$ on taking transforms.

Using these conditions in (1),

$$\bar{u}(s, 0) = A + B = F(s) \quad \text{...(2)}$$

$$\frac{d\bar{u}}{dt}(s, 0) = i\,\alpha s(A - B) = G(s) \quad \text{...(3)}$$

$$\text{Solving } A = \tfrac{1}{2}\left[F(s) + \frac{G(s)}{i\alpha s}\right]$$

$$B = \tfrac{1}{2}\left[F(s) - \frac{G(s)}{i\alpha s}\right]$$

Using these values in (1),

$$\bar{u}(s, t) = \tfrac{1}{2}\left[F(s) + \frac{G(s)}{i\alpha s}\right]e^{i\alpha st} + \tfrac{1}{2}\left[F(s) - \frac{G(s)}{i\alpha s}\right]e^{-i\alpha st} \quad \text{...(4)}$$

By inversion theorem, (4) reduces to,

$$u(x, t) = \tfrac{1}{2}\left[f(x - \alpha t) - \frac{1}{\alpha}\int_{\alpha}^{x-\alpha t} g(\theta)\, d\theta\right]$$

$$+ \tfrac{1}{2}\left[f(x + \alpha t) + \frac{1}{\alpha}\int_{\alpha}^{x+\alpha t} g(\theta)\, d\theta\right]$$

Using the result

$$F\left(\int_{\alpha}^{x} f(t)\, dt\right) = \frac{F(s)}{(-is)}$$

Boundary Value Problems using sine and cosine Transforms

Example 3. *Solve* $\frac{\partial u}{\partial t} = \frac{\partial^2 u}{\partial x^2}$ *for* $x > 0$, $t > 0$ *given that*

(*i*) $u(0, t) = 0$ *for* $t > 0$

(*ii*) $u(x, 0) = \begin{cases} 1 \text{ for } 0 < x < 1 \\ 0 \text{ for } x \geq 1 \end{cases}$

and (*iii*) $u(x, t)$ *is bounded.*

(**Note.** If u at $x = 0$ is given, take Fourier sine transform and if $\frac{\partial u}{\partial x}$ at $x = 0$ is given, take Fourier cosine transform,)

In this problem, u at $x = 0$ is given. Therefore, take fourier (infinite) sine transform.

$$F_s\left(\frac{\partial u}{\partial t}\right) = F_s\left(\frac{\partial^2 u}{\partial x^2}\right)$$

$$\sqrt{\frac{2}{\pi}}\int_0^\infty \frac{\partial u}{\partial t}\sin sx\,dx = \sqrt{\frac{2}{\pi}}\int_0^\infty \frac{\partial^2 u}{\partial x^2}\sin sx\,dx$$

i.e., $\dfrac{d\bar{u}}{dt} = -s^2\bar{u} + \sqrt{\dfrac{2}{\pi}}\, s\, u\,(0, t)$ where $\bar{u}$ stands for Fourier sine transform of u.

Using boundary condition (*i*), we get

$$\frac{d\bar{u}}{dt} + s^2\bar{u} = 0$$

Solving

$$\bar{u}\,(s, t) = ce^{-s^2 t} \qquad \ldots(iv)$$

Now, $u\,(x, 0) = \begin{cases} 1 \text{ for } 0 < x < 1 \\ 0 \text{ for } x \geq 1 \end{cases}$

Taking Fourier sine transform of this initial condition,

$$\bar{u}\,(s, 0) = \sqrt{\frac{2}{\pi}}\int_0^1 \sin sx\,dx$$

$$= \sqrt{\frac{2}{\pi}}\left(\frac{1-\cos s}{s}\right) \qquad \ldots(v)$$

Using (*v*) in (*iv*),

$$c = \sqrt{\frac{2}{\pi}}\left(\frac{1-\cos s}{s}\right)$$

Hence (*iv*) reduces to,

$$\bar{u}\,(s, t) = \sqrt{\frac{2}{\pi}}\left(\frac{1-\cos s}{s}\right)\cdot e^{-s^2 t}$$

By inversion theorem, this becomes,

$$u\,(x, t) = \frac{2}{\pi}\int_0^\infty \left(\frac{1-\cos s}{s}\right) e^{-s^2 t}.\ \sin sx\,ds.$$

Example 4. *Solve* $\dfrac{\partial u}{\partial t} = K\dfrac{\partial^2 u}{\partial x^2}$ for $0 \leq x < \infty$, $t > 0$ *given the conditions*

(*i*) $u\,(x, 0) = 0$ for $x \geq 0$

(ii) $\frac{\partial u}{\partial x}(0, t) = -a$ (constant)

(iii) $u(x, t)$ is bounded.

In this problem, $\frac{\partial u}{\partial x}$ at $x = 0$ is given. Hence, take Fourier cosine transform on both sides of the given equation.

$$F_c\left(\frac{\partial u}{\partial t}\right) = F_c\left(K\frac{\partial^2 u}{\partial x^2}\right)$$

$$\frac{d\bar{u}}{dt} = K\left(-s^2\bar{u} - \sqrt{\frac{2}{\pi}} \cdot \frac{\partial u}{\partial x}(0, t)\right)$$

$$= -ks^2\bar{u} + \sqrt{\frac{2}{\pi}}\, ka \quad \text{using condition } (ii)$$

$$\frac{d\bar{u}}{dt} + ks^2\bar{u} = \sqrt{\frac{2}{\pi}}\, ka$$

This is linear in $\bar{u}$. Therefore, solving

$$\bar{u}e^{ks^2 t} = \int \sqrt{\frac{2}{\pi}}\, ka\, e^{ks^2 t}\, dt$$

$$= \sqrt{\frac{2}{\pi}}\, ka \frac{e^{ks^2 t}}{ks^2} + c$$

$$\bar{u}(s, t) = \sqrt{\frac{2}{\pi}} \frac{a}{s^2} + c\, e^{-ks^2 t} \qquad \ldots(iv)$$

Since $u(x, 0) = 0$ for $x \geq 0$,

$\bar{u}(s, 0) = 0$.

Using this in (iv), we get

$$\bar{u}(s, 0) = c + \sqrt{\frac{2}{\pi}} \frac{a}{s^2} = 0$$

$$\therefore \qquad c = -\sqrt{\frac{2}{\pi}} \frac{a}{s^2}$$

Substituting this in (iv)

$$\bar{u}(s, t) = \sqrt{\frac{2}{\pi}} \frac{a}{s^2}\left(1 - e^{-ks^2 t}\right)$$

By inversion theorem,

$$u(x, t) = \frac{2}{\pi} \cdot a \int_0^\infty \frac{1 - e^{-ks^2 t}}{s^2} \cos sx \, ds.$$

Example 5. *Solve* $\frac{\partial u}{dt} = k \frac{\partial^2 u}{\partial x^2}$ *for* $x \geq 0, t \geq 0$ *under the given conditions* $u = u_0$ *at* $x = 0, t > 0$ *with initial condition* $u(x, 0) = 0, x \geq 0$

Taking Fourier sine transforms

$$F_s\left(\frac{\partial u}{\partial t}\right) = F_s\left(k \frac{\partial^2 u}{\partial x^2}\right)$$

$$\frac{d}{dt} \bar{u} = k\left[-s^2 \bar{u} + \sqrt{\frac{2}{\pi}} \, su(0, t)\right]$$

$$= -ks^2 \bar{u} + \sqrt{\frac{2}{\pi}} \, ksu_0 \quad \text{where } \bar{u} \text{ is the Fourier sine transform of } u.$$

$$\frac{d\bar{u}}{dt} + ks^2 \bar{u} = \sqrt{\frac{2}{\pi}} \, ksu_0$$

This is linear in $\bar{u}$.

$$\therefore \quad \bar{u} \, e^{ks^2 t} = \sqrt{\frac{2}{\pi}} \, ku_0 \int s \, e^{ks^2 t} \, dt$$

$$= \sqrt{\frac{2}{\pi}} \frac{u_0}{s} e^{ks^2 t} + c \qquad (1)$$

Since, $u(x, 0) = 0$, $\bar{u}(s, 0) = 0$. Using this in (1)

$$0 = \sqrt{\frac{2}{\pi}} \frac{u_0}{s} + c$$

$$\therefore \quad c = -\sqrt{\frac{2}{\pi}} \frac{u_0}{s}$$

$$e^{ks^2 t} \bar{u} = \sqrt{\frac{2}{\pi}} \frac{u_0}{s} \left(e^{ks^2 t} - 1\right)$$

$$\therefore \quad \bar{u} = \sqrt{\frac{2}{\pi}} \frac{u_0}{s} \left(1 - e^{ks^2 t}\right)$$

By inversion theorem,

$$u(x, t) = \frac{2 u_0}{\pi} \int_0^\infty \left(\frac{1 - e^{-ks^2 t}}{s}\right) \sin sx \, ds.$$

Exercises 4 (b)

1. Show that the solution of $\frac{\partial u}{\partial t} = 2\frac{\partial^2 u}{\partial x^2}$,

subject to $u(0, t) = 0$, for $t > 0$ and $u(x, 0) = e^{-x}$ for $x > 0$ and $u(x, t)$ is bounded,

is $\frac{2}{\pi}\int\limits^{\infty} \frac{s\, e^{-2s^2 t}}{1+s^2} \sin sx\, ds$

2. Solve $\frac{\partial u}{\partial t} = \frac{\partial^2 u}{\partial x^2}$ if

(*i*) $\frac{\partial u}{\partial x}(0, t) = 0$ for $t > 0$.

(*ii*) $u(x, 0) = \begin{cases} x \text{ for } 0 \le x \le 1 \\ 0 \text{ for } x > 1 \end{cases}$

(*iii*) and $u(x, t)$ is bounded for $x > 0, t > 0$.

$$\left[\textbf{Ans. } u(x, t) = \frac{2}{\pi}\int_0^{\infty}\left(\frac{\sin s}{s} + \frac{\cos s - 1}{s^2}\right) e^{-s^2 t} \cos sx ds\right]$$

FINITE FOURIER TRANSFORMS

Let $f(x)$ denote a function which is sectionally continuous over the range $(0, l)$. Then the **finite Fourier sine transform** of $f(x)$ on this interval is defined as

$$F_s(p) = \bar{f}_s(p) = \int_0^l f(x) \sin \frac{p\pi x}{l}\, dx$$

where p is an integer (Instead of s, we take p as a parameter).

Inversion formula for sine transform

If $\bar{f}_s(p) = F_s(p)$ is the finite Fourier sine transform of $f(x)$ in $(0, l)$ then the inversion formula for sine transform is

$$f(x) = \frac{2}{l}\sum_{p=1}^{\infty} \bar{f}_s(p) \sin \frac{p\pi x}{l}$$

Proof. For the given function $f(x)$ in $(0, l)$, if we find the half range Fourier sine series, we get,

$$f(x) = \sum_{n=1}^{\infty} b_n \sin \frac{n\pi x}{l} \qquad ...(1)$$

where $$b_n = \frac{2}{l}\int_0^l f(x)\sin\frac{n\pi x}{l}\,dx$$

$$\therefore \quad b_p = \frac{2}{l}\int_0^l f(x)\sin\frac{p\pi x}{l}\,dx$$

$$= \frac{2}{l}\bar{f}_s(p) \text{ by definition}$$

Substituting in (1),

$$\therefore \quad f(x) = \frac{2}{l}\sum_{p=1}^{\infty}\bar{f}_s(p)\sin\frac{p\pi x}{l}$$

Finite Fourier Cosine Transform

Let $f(x)$ denote a sectionally continuous function in $(0, l)$.

Then the Finite Fourier cosine transform of $f(x)$ over $(0, l)$ is defined as

$$f_c(p) = \bar{f}_c(p) = \int_0^l f(x)\cos\frac{p\pi x}{l}\,dx \text{ where } p \text{ is an integer.}$$

Inversion formula for cosine transform

If $\bar{f}_c(p)$ is the finite Fourier cosine transform of $f(x)$ in $(0, l)$, then the inversion formula for cosine transform is

$$f(x) = \frac{1}{l}\bar{f}_c(0) + \frac{2}{l}\sum_{p=l}^{\infty}\bar{f}_c(p)\cos\frac{p\pi x}{l}$$

where $$\bar{f}_c(0) = \int_0^l f(x)\,dx.$$

Proof. If we find half range Fourier cosine series for $f(x)$ in $(0, l)$, we obtain,

$$f(x) = \frac{a_0}{2} + \sum_{n=1}^{\infty} a_n\cos\frac{n\pi x}{l} \quad ...(2) \text{ where}$$

$$a_n = \frac{2}{l}\int_0^l f(x)\cos\frac{n\pi x}{l}\,dx$$

$$\therefore \quad a_p = \frac{2}{l}\bar{f}_c(p)$$

$$a_0 = \frac{2}{l} \int_0^l f(x)\, dx$$

$$= \frac{2}{l} \mathcal{F}_c(0).$$

Substituting in (2), we get,

$$f(x) = \frac{1}{l} \mathcal{F}_c(0) + \frac{2}{l} \sum_{p=1}^{\infty} \mathcal{F}_c(p) \cos \frac{p\pi x}{l}$$

Example 1. *Find the finite Fourier sine and cosine transforms of*

(*i*) $f(x) = 1$ in $(0, \pi)$

(*ii*) $f(x) = x$ in $(0, l)$

(*iii*) $f(x) = x^2$ in $(0, l)$

(*iv*) $f(x) = 1$ in $0, < x < \pi/2$

$= -1$ in $\pi/2 < x < \pi$

(*v*) $f(x) = x^3$ in $(0, l)$

(*vi*) $f(x) = e^{ax}$ in $(0, l)$

(*i*) $\mathcal{F}_s(p) = F_s(1) = \int_0^{\pi} 1 \cdot \sin \frac{p\pi x}{\pi}\, dx$

$$= \left(-\frac{\cos px}{p} \right)_0^{\pi}$$

$$= \frac{1 - \cos p\pi}{p} \qquad \text{if } p \neq 0$$

$$\mathcal{F}_c(p) = \int_0^{\pi} 1 \cdot \cos px\, dx$$

$$= \left(\frac{\sin px}{p} \right)_0^{\pi} = \frac{1}{p}(0 - 0) = 0$$

(*ii*) $\mathcal{F}_s(p) = F_s(p) = \int_0^{l} x \sin \frac{p\pi x}{l}\, dx$

$$= \left[(x)\left(\frac{-\cos\frac{p\pi x}{l}}{\frac{p\pi}{l}}\right) - (1)\left(-\frac{\sin\frac{p\pi x}{l}}{\frac{p^2\pi^2}{l^2}}\right)\right]_0^l$$

$$= \frac{-l}{p\pi}(l\cos p\pi)$$

$$= \frac{-l^2}{p\pi}(-1)^p \qquad \text{if } p \neq 0$$

$$\bar{f}_c(p) = F_c(x) = \int_0^l x\cos\frac{p\pi x}{l}\,dx$$

$$= \left[(x)\left(\frac{\sin\frac{p\pi x}{l}}{\frac{p\pi}{l}}\right) - (1)\left(-\frac{\cos\frac{p\pi x}{l}}{\frac{p^2\pi^2}{l^2}}\right)\right]_0^l$$

$$= \frac{l^2}{p^2\pi^2}[(-1)^p - 1] \qquad \text{if } p \neq 0$$

(iii) $$\bar{f}_s(p) = F_s(x^2) = \int_0^l x^2\sin\frac{p\pi x}{l}\,dx$$

$$= \left[(x^2)\left(-\frac{\cos\frac{p\pi x}{l}}{\frac{p\pi}{l}}\right) - (2x)\left(-\frac{\sin\frac{p\pi x}{l}}{\frac{p^2\pi^2}{l^2}}\right) + (2)\left(\frac{\cos\frac{p\pi x}{l}}{\frac{p^3\pi^3}{l^3}}\right)\right]_0^l$$

$$= \frac{-l^3}{p\pi}[(-1)^p] + \frac{2l^3}{p^3\pi^3}[(-1)^p - 1] \qquad \text{if } p \neq 0$$

$$\bar{f}_c(p) = \int_0^l (x^2)\cos\frac{p\pi x}{l}\,dx =$$

$$= \left[(x^2)\left(\frac{\sin\frac{p\pi x}{l}}{\frac{p\pi}{l}}\right) - (2x)\left(-\frac{\cos\frac{p\pi x}{l}}{\frac{p^2\pi^2}{l^2}}\right) + (2)\left(\frac{\sin\frac{p\pi x}{l}}{\frac{p^3\pi^3}{l^3}}\right)\right]_0^l$$

$$= \frac{2l^3}{p^2\pi^2}[(-1)^p] \qquad \text{if } p \neq 0$$

$$(iv)\ \ F_s\{f(x)\} = \int_0^{\pi/2} \sin px\, dx + \int_{\pi/2}^{\pi} (-1)\sin px\, dx$$

$$= \left(-\frac{\cos px}{p}\right)_0^{\pi/2} + \left(\frac{\cos px}{p}\right)_{\pi/2}^{\pi}$$

$$= -\frac{1}{p}\left(\cos\frac{p\pi}{2} - 1\right) + \frac{1}{p}(\cos p\pi - \cos p\pi/2)$$

$$= \frac{1}{p}\left(\cos p\pi - -2\cos\frac{p\pi}{2} + 1\right) \qquad \text{if } p \neq 0$$

$$F_c(f(x)) = \int_0^{\pi/2} \cos px\, dx - \int_{\pi/2}^{\pi} \cos px\, dx$$

$$= \left(\frac{\sin px}{p}\right)_0^{\pi/2} - \left(\frac{\sin px}{p}\right)_{\pi/2}^{\pi} = \frac{2}{p}\sin\frac{p\pi}{2} \qquad \text{If } p \neq 0$$

$$(v)\ \ F_s(x^3) = \int_0^l x^3 \sin\frac{p\pi x}{l}\, dx$$

$$= \left[(x^3)\left(-\frac{\cos\frac{p\pi x}{l}}{\frac{p\pi}{l}}\right) - (3x^2)\left(-\frac{\sin\frac{p\pi x}{l}}{\frac{p^2\pi^2}{l^2}}\right) + (6x)\left(\frac{\cos\frac{p\pi x}{l}}{\frac{p^3\pi^3}{l^3}}\right) - (6)\left(-\frac{\sin\frac{p\pi x}{l}}{\frac{p^4\pi^4}{l^4}}\right)\right]_0^l$$

$$= -\frac{l^4}{p\pi}\left[(-1)^p\right] + \frac{6l^4}{p^3\pi^3}(-1)^p \qquad \text{if } p \neq 0$$

$$F_c(x^3) = \int_0^l x^3 \cos\frac{p\pi x}{l}\,dx$$

$$= \left[(x^3)\left(\frac{\sin\frac{p\pi x}{l}}{\frac{p\pi}{l}}\right) - (3x^2)\left(-\frac{\cos\frac{p\pi x}{l}}{\frac{p^2\pi^2}{l^2}}\right) + (6x)\left(\frac{\sin\frac{p\pi x}{l}}{\frac{p^3\pi^3}{l^3}}\right) - (6)\left(\frac{\cos\frac{p\pi x}{l}}{\frac{p^4\pi^4}{l^4}}\right)\right]_0^l$$

$$= \frac{3l^4}{\pi^2 p^2}(-1)^p - \frac{6l^4}{p^4\pi^4}\left[(-1)^p - 1\right] \qquad \text{if } p \neq 0$$

(vi) $$F_s(e^{ax}) = \int_0^l e^{ax}\sin\frac{p\pi x}{l}\,dx$$

$$= \left\{\frac{e^{ax}}{a^2 + \frac{p^2\pi^2}{l^2}}\left[a\sin\frac{p\pi x}{l} - \frac{p\pi}{l}\cos\frac{p\pi x}{l}\right]\right\}_0^l$$

$$= \frac{e^{ax}}{a^2 + \frac{p^2 a^2}{l^2}}\left(-\frac{p\pi}{l}(-1)^p\right) + \frac{1}{a^2 + \frac{p^2\pi^2}{l^2}}\left(\frac{p\pi}{l}\right)$$

$$F_c(e^{ax}) = \left\{\frac{e^{ax}}{a^2 + \frac{p^2\pi^2}{l^2}}\left[a\cos\frac{p\pi x}{l} + \frac{p\pi}{l}\sin\frac{p\pi x}{l}\right]\right\}_0^l$$

$$= \frac{e^{al}}{a^2 + \frac{p^2\pi^2}{l^2}}a(-1)^p - \frac{1}{a^2 + \frac{p^2\pi^2}{l^2}}\left(a\right)$$

Note. In all the above problems, if $p = 0$, do the integration separately.

Example 2. *Find $f(x)$ if its finite Fourier sine transform is $\frac{2\pi}{p^3}(-1)^{p-1}$ for $p = 1, 2, \ldots\ldots, 0 < x < \pi$.*

By inversion Theorem,

$$f(x) = \frac{2}{\pi} \sum_{p=1}^{\infty} \frac{2\pi}{p^3} (-1)^{p-1} \sin px$$

$$= 4 \sum_{p=1}^{\infty} \frac{(-1)^{p-1}}{p^3} \sin px$$

Example 3. *Find f(x) if its finite Fourier sine transform is given by*

(*i*) $F_s(p) = \dfrac{1 - \cos p\pi}{p^2 \pi^2}$ for $p = 1, 2, 3, \ldots\ldots$ and $0 < x < \pi$

(*ii*) $F_s(p) = \dfrac{16(-1)^{p-1}}{p^3}$ for $p = 1, 2, 3, \ldots.$ where $0 < x < 8$

(*iii*) $F_s(p) = \dfrac{\cos \dfrac{2\pi p}{3}}{(2p+1)^2}$ for $p = 1, 2, 3, \ldots.$ and $0 < x < 1$.

Solution :

By inversion theorem

(*i*) $$f(x) = \frac{2}{\pi} \sum_{p=1}^{\infty} \left(\frac{1 - \cos p\pi}{p^2 \pi^2} \right) . \sin px$$

$$= \frac{2}{\pi^3} \sum_{p=1}^{\infty} \left(\frac{1 - \cos p\pi}{p^2} \right) . \sin px$$

(*ii*) $$f(x) = \frac{2}{l} \sum_{p=1}^{\infty} F_s(p) \sin \left(\frac{p\pi x}{l} \right)$$

$$= \frac{2}{8} \sum_{p=1}^{\infty} \frac{16(-1)^{p-1}}{p^3} \sin \left(\frac{p\pi x}{8} \right) \quad \text{since } l = 8$$

$$= 4 \sum_{p=1}^{\infty} \frac{(-1)^{p-1}}{p^3} \sin \left(\frac{p\pi x}{8} \right)$$

(*iii*) $$f(x) = \frac{2}{l} \sum_{p=1}^{\infty} F_s(p) \sin \left(\frac{p\pi x}{l} \right)$$

$$= 2 \sum_{p=1}^{\infty} \frac{\cos\left(\dfrac{2\pi p}{3} \right)}{(2p+1)^2} \sin(p\pi x) \quad \text{since } l = 1$$

Example 4. *Find f(x) if its finite Fourier cosine transform is*

(i) $F_c(p) = \frac{1}{2p} \sin\left(\frac{p\pi}{2}\right)$ for $p = 1, 2, 3, ...$

$= \frac{\pi}{4}$ for $p = 0$ given $0 < x < 2\pi$

(ii) $F_c(p) = \frac{6 \sin \frac{p\pi}{2} - \cos p\pi}{(2p+1)\pi}$ for $p = 1, 2, 3, ...$

$= \frac{2}{\pi}$ for $p = 0$ given $0 < x < 4$

(iii) $F_c(p) = \frac{\cos\left(\frac{2p\pi}{3}\right)}{(2p+1)^2}$ for $p = 1, 2, 3, ...$

$= 1$ for $p = 0$ given $0 < x < 1$

Solution :

By inversion theorem,

$$f(x) = \frac{1}{l} F_c(0) + \frac{2}{l} \sum_{p=1}^{\infty} F_c(p) \cdot \cos \frac{p\pi x}{l}.$$

(i) Here $F_c(0) = \pi/4$ and $l = 2\pi$

$$\therefore f(x) = \frac{1}{2\pi}\left(\frac{\pi}{4}\right) + \frac{2}{2\pi} \sum_{p=1}^{\infty} \frac{1}{2p} \sin\left(\frac{p\pi}{2}\right) \cos\left(\frac{p\pi x}{2\pi}\right)$$

$$= \frac{1}{8} + \frac{1}{2\pi} \sum_{p=1}^{\infty} \frac{1}{p} \sin\left(\frac{p\pi}{2}\right) \cos\left(\frac{px}{2}\right)$$

(ii) Here, $F_c(0) = \frac{2}{\pi}$ and $l = 4$

$$f(x) = \frac{1}{4}\left(\frac{2}{\pi}\right) + \frac{2}{4} \sum_{p=1}^{\infty} \left(\frac{6 \sin \frac{p\pi}{2} - \cos p\pi}{(2p+1)\pi}\right) \cos\left(\frac{p\pi x}{4}\right)$$

$$= \frac{1}{2\pi} + \frac{1}{2\pi} \sum_{p=1}^{\infty} \left(\frac{6 \sin \frac{p\pi}{2} - \cos p\pi}{(2p+1)}\right) \cdot \cos\left(\frac{p\pi x}{4}\right)$$

(iii) Here $F_c(0) = 1$, $l = 1$

$$\therefore f(x) = \frac{1}{1} + \frac{2}{1} \sum_{p=1}^{\infty} \frac{1}{(2p+1)^2} \cos\left(\frac{2p\pi}{3}\right) \cdot \cos(p\pi x)$$

$$= 1 + 2 \sum_{p=1}^{\infty} \frac{\cos\left(\frac{2p\pi}{3}\right)}{(2p+1)^2} \cos(p\pi x).$$

Example 5. *Find the finite Fourier sine transform of in f(x) = 1 (0 , π). Use the inversion theorem and find Fourier sine series for f(x) = 1 in (0 , π) Hence prove*

(i) $1 - \frac{1}{3} + \frac{1}{5} - \frac{1}{7} + ... = \pi/4$

(ii) $\frac{1}{1^2} + \frac{1}{3^2} + \frac{1}{5^2} + ... = \pi^2/8$

Solution.

$$F_s(1) = \int_0^{\pi} 1.\ \sin\left(\frac{p\pi x}{\pi}\right) dx$$

$$\bar{f}_s(p) = \frac{1 - \cos p\pi}{p} \text{ if } p \neq 0$$

By inversion theorem,

$$f(x) = \frac{2}{l} \sum_{p=1}^{\infty} F_s(p) \sin \frac{p\pi x}{l}$$

$$1 = \frac{2}{\pi} \sum_{p=1}^{\infty} \frac{1-(-1)^p}{p} .\ \sin px \text{ since } l = \pi$$

$$1 = \frac{4}{\pi} \left[\sin x + \frac{1}{3} \sin 3x + \frac{1}{5} \sin 5x + ... \right]$$

This is the half range Fourier sine series for $f(x) = 1$ in $(0, \pi)$ getting $x = \pi/2$,

$$\frac{4}{\pi} \left[1 - \frac{1}{3} + \frac{1}{5} - \frac{1}{7} + ... \right] = 1$$

$$\therefore \quad 1 - \frac{1}{3} + \frac{1}{5} - \frac{1}{7} + ... = \pi/4$$

In the half Fourier sine series $l_n = \frac{4}{\pi} \cdot \frac{1}{n}$ for n odd
By using Parseval's Theorem.

(range) $$\left[\frac{1}{2} \sum b_n^2\right] = \int_0^{\pi} (1)^2 dx$$

$$\pi \left[\frac{1}{2} \cdot \frac{16}{\pi^2} \sum_{n=1,3,5....}^{\infty} \frac{1}{n^2} \right] = \pi$$

i.e., $$\sum_{n=1,3,5....}^{\infty} \frac{1}{n^2} = \frac{\pi^2}{8}$$

Example 6. *Find f (x) if its finite Fourier cosine transform is* $\frac{2l^3}{p^2\pi^2}(-1)^p$ *for p = 1, 2, 3, ... and is* $\frac{l^3}{3}$ *for p = 0; 0 < x < l.*

Proof: By inversion formula,

$$f(x) = \frac{1}{l}\bar{f}_c(0) + \frac{2}{l}\sum_{p=1}^{\infty}\bar{f}_c(p)\cos\left(\frac{p\pi x}{l}\right)$$

$$= \frac{1}{l}\cdot\frac{l^3}{3} + \frac{2}{l}\sum_{p=1}^{\infty}\frac{2l^3(-1)^p}{p^2\pi^2}\cos\frac{p\pi x}{l}$$

$$= \frac{l^2}{3} + \frac{4l^2}{\pi^2}\sum_{p=1}^{\infty}\frac{(-1)^p}{p^2}\cos\left(\frac{p\pi x}{l}\right)$$

Exercises 4 (c)

Find the finite Fourier sine and cosine transforms of

1. $f(x) = 2x$ in $(0, 4)$
2. $f(x) = x$ in $(0, \pi)$ ***(BR 1995 April.)***
3. $f(x) = \cos ax$ in $(0, \pi)$
4. $f(x) = 1 - \frac{x}{\pi}$ in $(0, \pi)$
5. $f(x) = \begin{cases} x \text{ in } (0, \pi/2) \\ \pi - x \text{ in } (\pi/2, \pi) \end{cases}$
6. $f(x) = e^{-ax}$ in $(0, l)$
7. Find finite Fourier cosine transform of $\left(1 - \frac{x}{\pi}\right)^2$, $0 < x < \pi$
8. Find $f(x)$ if $\bar{f}_c(p) = \frac{\sin\left(\frac{p\pi}{2}\right)}{2p}$, $p = 1, 2, 3....$

   and $= \frac{\pi}{4}$ if $p = 0$ given $0 < x < 2\pi$.
9. Find finite Fourier cosine and sine transform of

   $f(x) = \frac{\pi}{3} - x + \frac{x^2}{2\pi}$. ***(BR 1995 April)***

Finite Fourier sine and cosine transform of derivatives

Using the definition and the integration by parts, we can easily prove the following results. For $0 \le x \le l$,

$$F_s\ \{f''(x)\} = -\frac{p^2\pi^2}{l^2}\bar{f}_s\ (p) + \frac{p\pi}{l}\left[f(0) - (-1)^p f(l)\right]$$

$$F_c\ \{f''(x)\} = -\frac{p^2\pi^2}{l^2}\bar{f}_c\ (p) + f'(l)\ (-1)^p - f'(0)$$

$$F_s\ \{f'(x)\} = -\frac{p\pi}{l}\bar{f}_c(p)$$

$$F_c\ \{f'(x)\} = f(l)\ (-1)^p - f(0) + \frac{p\pi}{l}\bar{f}_s(p)$$

Proof: (*i*) $F_s\ \{f'(x)\} = \int_0^l f'(x)\sin\frac{p\pi x}{l}\,dx$

$$= \int_0^l \sin\frac{p\pi x}{l}\cdot d[f(x)]$$

$$= \left(f(x)\sin\frac{p\pi x}{l}\right)_0^l - \int_0^l f(x)\cdot\cos\frac{p\pi x}{l}\cdot\frac{p\pi}{l}\,dx$$

$$= -\frac{p\pi}{l}\bar{f}_c(p)$$

(*ii*) $F_c\ \{f'(x)\} = \int_0^l f'(x)\cos\frac{p\pi x}{l}\,dx$

$$= \left(f(x)\cos\frac{p\pi x}{l}\right)_0^l - \int_0^l f(x)\cdot\frac{p\pi}{l}\sin\frac{p\pi x}{l}\,dx$$

$$= (-1)^p f(l) - f(0) + \frac{p\pi}{l}\bar{f}_s\ (p)$$

(*iii*) $F_s\ \{f''(x)\} = \int_0^l \sin\frac{p\pi x}{l}\,d\,[f'(x)]$

$$= \left(f'(x)\sin\frac{p\pi x}{l}\right)_0^l - \frac{p\pi}{l}\int_0^l f'(x)\cos\frac{p\pi x}{l}\,dx$$

$$= -\frac{p\pi}{l}\left[(-1)^p f(l) - f(0) + \frac{p\pi}{l}\bar{f}_s(p)\right]$$

$$= -\frac{p^2\pi^2}{l^2}\bar{f}_s(p) + \frac{p\pi}{l}\left[f(0) - (-1)^p f(l)\right]$$

(iv) $F_c\{f''(x)\} = \int_0^l \cos\frac{p\pi x}{l}\, d[f'(x)]$

$$= \left[f'(x)\cos\frac{p\pi x}{l}\right]_0^l + \frac{p\pi}{l}\int_0^l f'(x)\sin\frac{p\pi x}{l}\,dx$$

$$= (-1)^p f'(l) - f'(0) + \frac{p\pi}{l}\left[-\frac{p\pi}{l}\bar{f}_c(p)\right]$$

$$= -\frac{p^2\pi^2}{l^2}\bar{f}_c(p) + f'(l)(-1)^p - f'(0)$$

Note. If $u = u(x, t)$, then

$$F_s\left[\frac{\partial u}{\partial x}\right] = -\frac{p\pi}{l}F_c(u)$$

$$F_c\left[\frac{\partial u}{\partial x}\right] = \frac{p\pi}{l}F_s(u) - u(0, t) + (-1)^p u(l, t)$$

$$F_s\left[\frac{\partial^2 u}{\partial x^2}\right] = -\frac{p^2\pi^2}{l^2}F_s(u) + \frac{p\pi}{l}[u(0, t) - (-1)^p u(l, t)]$$

$$F_c\left[\frac{\partial^2 u}{\partial x^2}\right] = -\frac{p^2\pi^2}{l^2}F_c(u) + \frac{\partial u}{\partial x}(l, t)\cos p\pi - \frac{\partial u}{\partial x}(0, t)$$

Example 1. *Using finite Fourier transform, solve*

$$\frac{\partial u}{\partial t} = \frac{\partial^2 u}{\partial x^2} \text{ given } u(0, t) = 0 \text{ and } u(4, t) = 0$$

and $u(x, 0) = 2x$ *where* $0 < x < 4,\ t > 0$

Proof. Since $u(0, t)$ is given, take finite Fourier sine transform.

$$\int_0^4 \frac{\partial u}{\partial t}\sin\frac{p\pi x}{4}\,dx = \int_0^4 \frac{\partial^2 u}{\partial x^2}\sin\frac{p\pi x}{4}\,dx$$

$$\frac{d}{dt}\bar{u}_s = F_s\left(\frac{\partial^2 u}{\partial x^2}\right)$$

$$= -\frac{p^2\pi^2}{16}\bar{u}_s + \frac{p\pi}{4}[u(0, t) - (-1)^p u(4, t)]$$

$$= -\frac{p^2\pi^2}{16}\bar{u}_s \quad \text{using } u(0, t) = 0,\ u(4, t) = 0$$

$$\frac{d\bar{u}_s}{\bar{u}_s} = -\frac{p^2\pi^2}{16}\,dt$$

Integrating

$$\log \bar{u}_s = -\frac{p^2\pi^2}{16}\,t + c$$

$$\bar{u}_s = A\,e^{-\frac{\pi^2 p^2}{16}t} \qquad ...(1)$$

Since $\quad u(x, 0) = 2x$

$$\bar{u}_s(p, 0) = \int_0^4 (2x)\sin\left(\frac{p\pi x}{4}\right) dx$$

$$= -\frac{32}{p\pi}\cos p\pi \qquad ...(2)$$

Using (2) in (1),

$$\bar{u}_s(p, 0) = A = -\frac{32}{p\pi}\cos p\pi.$$

Substituting in (1),

$$\therefore \qquad \bar{u}_s = -\frac{32}{p\pi}(-1)^p\,e^{-\frac{p^2\pi^2}{16}t}$$

By inversion theorem,

$$u(x, t) = \frac{2}{4}\sum_{p=1}^{\infty}\frac{32}{p\pi}(-1)^{p+1}\,e^{-\frac{p^2\pi^2}{16}.t}\sin\left(\frac{p\pi x}{4}\right)$$

Example 2. *Solve* $\frac{\partial u}{\partial t} = \frac{\partial^2 u}{\partial x^2}$, *$0 < x < 6$, $t > 0$,*

given $\frac{\partial u}{\partial x}(0, t) = 0$, $\frac{\partial u}{\partial x}(6, t) = 0$ and $u(x, 0) = 2x$

Proof. Since $\frac{\partial u}{\partial x}(0, t)$ is given, use finite Fourier cosine transform.

$$\int_0^6 \frac{\partial u}{\partial t}\cos\frac{p\pi x}{6}\,dx = \int_0^6 \frac{\partial^2 u}{\partial x^2}\cos\frac{p\pi x}{6}\,dx$$

$$\frac{d}{dt}\bar{u}_c = -\frac{p^2\pi^2}{36}\bar{u}_c + \frac{\partial u}{\partial x}(6, t)\cos p\pi - \frac{\partial u}{\partial x}(0, t)$$

$$= -\frac{p^2\pi^2}{36}\bar{u}_c$$

$$\frac{d\bar{u}_c}{\bar{u}_c} = -\frac{p^2\pi^2}{36}\,dt$$

$$\log \bar{u}_c = -\frac{p^2\pi^2}{36}\,t + c$$

$$\bar{u}_c = A\,e^{-\frac{p^2\pi^2}{36}t} \qquad \text{...(1)}$$

$$u(x, 0) = 2x.$$

$\therefore$ At $t = 0$

$$\bar{u}_c(p, 0) = \int_0^6 (2x)\cos\frac{p\pi x}{6}\,dx$$

$$= \frac{72}{p^2\pi^2}(\cos p\pi - 1) \qquad \text{...(2)}$$

Using this in (1),

$$\bar{u}_c(p, 0) = A = \frac{72}{p^2\pi^2}(\cos p\pi - 1)$$

Substituting in (1),

$$\bar{u}_c(p, t) = \frac{72}{p^2\pi^2}(\cos p\pi - 1)\,e^{-\frac{p^2\pi^2}{36}t}$$

By inversion theorem,

$$u(x, t) = \frac{1}{l}\bar{f}_c(0) + \frac{2}{l}\sum_{p=1}^{\infty}\bar{f}_c(p)\cos\left(\frac{p\pi x}{l}\right)$$

$$= \frac{1}{6}\int_0^6 (2x)\,dx + \frac{2}{6}\sum_{p=1}^{\infty}\frac{72}{p^2\pi^2}(\cos p\pi - 1)\,e^{-\frac{p^2\pi^2}{36}t}\,.\cos\left(\frac{p\pi x}{6}\right)$$

$$= 6 + \frac{24}{\pi^2}\sum_{p=1}^{\infty}\frac{(\cos p\pi - 1)}{p^2}\,e^{-\frac{p^2\pi^2}{36}t}\cos\left(\frac{p\pi x}{6}\right).$$

Example 3. *Solve* $\frac{\partial u}{\partial t} = 2\frac{\partial^2 u}{\partial x^2}$, $0 < x < 4, t > 0$

given $u(0, t) = 0$, $u(4, t) = 0$; $u(x, 0) = 3\sin \pi x - 2\sin 5\pi x$.

Proof. Since $u(0, t)$ is given, take finite Fourier sine transform. The equation becomes (as in example 1)

$$\frac{d}{dt}\bar{u}_c = 2\left[-\frac{p^2\pi^2}{16}\bar{u}_s + \frac{p\pi}{4}\{u(0, t) - (-1)^p\,u(4, t)\}\right]$$

$$= -\frac{p^2\pi^2}{8}\bar{u}_s$$

Solving we get,

$$\bar{u}_s = A\, e^{-\frac{p^2\pi^2}{8}t} \qquad \text{...(1)}$$

$$u(x, 0) = 3 \sin \pi x - 2 \sin 5\pi x$$

Taking sine transform,

$$\bar{u}_s(p, 0) = \int_0^4 (3 \sin \pi x - 2 \sin 5\pi x) \sin \frac{p\pi x}{4}\, dx$$

$$= 0 \text{ if } p \neq 4, p \neq 20.$$

If $p = 4$, $\bar{u}_s(4, 0) = 6$

If $p = 20$, $\bar{u}_s(20, 0) = -4$

$$u(x, t) = \frac{2}{4}\sum_{p=1}^{\infty} \bar{u}_s(p, t) \sin\left(\frac{p\pi x}{4}\right)$$

$$= \frac{1}{2}\left[6\, e^{-\frac{p^2\pi^2}{8}t} \sin \pi x - 4\, e^{-\frac{p^2\pi^2}{8}t} \sin 5\pi x\right]$$

where p in the first term is 4 and p in the second term is 20

$$= 3\, e^{-2\pi^2 t} \sin \pi x - 2\, e^{-50\pi^2 t} \sin 5\pi x.$$

Example 4. *Solve* $\frac{\partial u}{\partial t} = \frac{\partial^2 u}{\partial x^2}$, $0 < x < \pi$, $t > 0$ *given* $u(0, t) = u(\pi, t) = 0$ *for* $t > 0$ *and* $u(x, 0) = \sin^3 x$.

Solution. Since $u(0, t)$ and $u(\pi, t)$ are given, take finite Fourier sine transforms on both sides of the given equation with respect to x in $(0, \pi)$.

$$\int_0^\pi \frac{\partial u}{\partial t} \sin \frac{p\pi x}{\pi}\, dx = \int_0^\pi \frac{\partial^2 u}{\partial x^2} \sin \frac{p\pi x}{\pi}\, dx$$

$$\frac{d}{dt}\bar{u}_s = F_s\left(\frac{\partial^2 u}{\partial x^2}\right)$$

$$= -p^2\bar{u}_s + p\,[u(0, t) - (-1)^p u(\pi, t)]$$

$$= -p^2\bar{u}_s \text{ using } u(0, t) = u(\pi, t) = 0.$$

$$\frac{d\,\bar{u}_s}{\bar{u}_s} = -p^2 dt$$

Integrating

$$\log(\bar{u}_s) = -p^2 t + k$$

$$\therefore \qquad \bar{u}_s(p, t) = c\,e^{-p^2 t} \qquad \text{... (1)}$$

Since, $\quad u(x, 0) = \sin^3 x$

$$\bar{u}_s(p, 0) = \int_0^{\pi} \sin^3 x \sin px\, dx$$

$$= \int_0^{\pi} \left(\frac{3}{4}\sin x - \frac{1}{4}\sin 3x\right) \sin px\, dx$$

$$= 0 \text{ for } p \neq 1,\ p \neq 3$$

when $\quad p = 1,\ \bar{u}_s(1,0) = \int_0^{\pi} \sin^4 x dx$

$$= 2\int_0^{\pi/2} \sin^4 x\, dx$$

$$= 2\left(\frac{3}{4} \cdot \frac{1}{2} \cdot \pi/2\right)$$

$$= \frac{3}{8}\pi \qquad \text{.... (2)}$$

when $\quad p = 3,\ \bar{u}_s(3, 0) = \int_0^{\pi} \left(-\frac{1}{4}\sin^2 3x\right) dx$

$$= -\frac{1}{4}\int_0^{\pi} \frac{1 - \cos 6x}{2}\, dx$$

$$= -\frac{1}{8}\left[x - \frac{\sin 6x}{6}\right]_0^{\pi}$$

$$= -\pi/8 \qquad \text{... (3)}$$

Using (2) and (3) in (1),

when $\quad p = 1,\ \bar{u}_s(1, 0) = C = \dfrac{3\pi}{8}$

when $\quad p = 3,\ \bar{u}_s(3, 0) = C = -\pi/8$

For all other values of p, $c = 0$

By inversion formula, $u(x, t) = \dfrac{2}{\pi} \sum_{p=1}^{\infty} \bar{u}_s(p, t) \sin nx$

ie. $$u(x, t) = \frac{2}{\pi}\left[\frac{3\pi}{8} e^{-t} \sin x - \frac{\pi}{8} e^{-9t} \sin 3x\right]$$

$$= \frac{3}{4} e^{-t} \sin x - \frac{1}{4} e^{-9t} \sin 3x$$

Example 5. *Solve* $\dfrac{\partial u}{\partial t} = \dfrac{\partial^2 u}{\partial x^2}$, $0 < x < 10$ *given* $u(0, t) = u(10, t) = 0$ *for* $t > 0$ *and* $u(x, 0) = 10x - x^2$ *for* $0 < x < 10$.

Solution. As the previous problem, take finite Fourier sine transform on both sides of the heat equation with respect to x in (0, 10).

$$\therefore \quad \frac{d}{dt}\bar{u}_s = -\frac{p^2\pi^2}{100}\bar{u}_s + \frac{p\pi}{10}[u(0,t) - (-1)^p u(10, t)]$$

$$= -\frac{p^2\pi^2}{100}\bar{u}_s \text{ using } u(0, t) = u(10, t) = 0$$

Solving for $\bar{u}_s$, we get '

$$\bar{u}_s = C e^{\frac{-p^2\pi^2}{100}t} \qquad \ldots (1)$$

Since $u(x, 0) = 10x - x^2$,

$$\bar{u}_s(p, 0) = \int_0^{10} (10x - x^2) \sin\frac{p\pi x}{10}\, dx$$

$$= \left\{(10x - x^2)\left[\frac{-\cos\frac{p\pi x}{10}}{\frac{p\pi}{10}}\right] - (10 - 2x)\left[-\frac{\sin\frac{p\pi x}{10}}{\frac{p^2\pi^2}{100}}\right] + (-2)\left[\frac{\cos\frac{p\pi x}{10}}{\frac{p^3\pi^3}{1000}}\right]\right\}_0^{10}$$

$$= -\frac{2000}{p^3\pi^3}\left[(-1)^p - 1\right]$$

$$= \begin{cases} \dfrac{4000}{p^3\pi^3} & \text{if p is odd} \\ 0 & \text{if p is even} \end{cases}$$

Using (1), $\bar{u}_s(p, 0) = C$.. using this again in (1), C is eliminated. Hence, taking inversion formula, (use (1))

$$u(x, t) = \frac{2}{10}\sum \bar{u}_s \cdot \sin\frac{p\pi x}{10}$$

$$= \frac{800}{\pi^3} \sum_{p = 1, 3, 5 \ldots}^{\infty} \frac{1}{p^3} \sin \frac{p\pi x}{10} e^{\frac{-p^2\pi^2}{100}t}$$

Exercises 4 (d)

1. Solve $\frac{\partial u}{\partial t} = \frac{\partial^2 u}{\partial x^2}$, $0 < x < 6$, $t > 0$ given that

$u(0, t) = 0 = u(6, t)$ and $u(x, 0) = \begin{cases} 1 \text{ for } 0 < x < 3 \\ 0 \text{ for } 3 < x < 6 \end{cases}$

$$\left[\textbf{Ans. } u(x, t) = \frac{2}{\pi} \sum_{p=1}^{\pi} \left(\frac{1 - \cos \frac{p\pi}{2}}{p} \right) e^{-\frac{p^2\pi^2 t}{36}} \sin\left(\frac{p\pi x}{6}\right)\right]$$

2. Solve $\frac{\partial v}{\partial t} = \frac{\partial^2 v}{\partial x^2}$ subject to conditions $v(0, t) = 1$, $v(\pi, t) = 3$

$v(x, 0) = 1$ for $0 < x < \pi$, $t > 0$.

$$\left[\textbf{Ans. } v(x, t) = \frac{4}{\pi} \sum_{p=1}^{\infty} \frac{\cos p\pi}{p} e^{-p^2 t} \sin px + 1 + \frac{2x}{\pi}.\right]$$

3. Solve $\frac{\partial \theta}{\partial t} = \frac{\partial^2 \theta}{\partial x^2}$ given

$\theta(0, t) = 0$, $\theta(\pi, t) = 0$, $\theta(x, 0) = 2x$ for $0 < x < \pi$, $t > 0$

[Refer example 3]

4. Solve $\frac{\partial u}{\partial t} = \frac{\partial^2 u}{\partial x^2}$ for $0 < x < l$, $t > 0$ given $\frac{\partial u}{\partial x}(0, t) = 0$, $\frac{\partial u}{\partial x}(l, t) = 0$ for $t > 0$

and $u(x, 0) = lx - x^2$ for $0 < x < l$

$$\left[\textbf{Ans. } u(x, t) = \frac{l^2}{6} - \frac{l^2}{\pi^2} \sum \frac{1}{p^2} \cos \frac{2p\pi x}{l} e^{\frac{-4p^2\pi^2 t}{l^2}}\right.$$

5. Solve $\frac{\partial u}{\partial t} = 2 \frac{\partial^2 u}{\partial x^2}$ for $0 < x < 1$, $t > 0$ given $u(0, t) = u(1, t) = 0$ for $t > 0$ and

$u(x, 0) = \sin 3\pi x + \sin \pi x$

$$\left[\textbf{Ans: } u(x, t) = e^{-2\pi^2 t} \sin \pi x + e^{-18\pi^2 t} \sin 3\pi x\right]$$

6. Solve $\frac{\partial u}{\partial t} = \alpha^2 \frac{\partial^2 u}{\partial x^2}$ subject to

$u(0, t) = u(l, t) = 0$ for $t > 0$ and $u(x, 0) = x$ for $0 \le x \le l/2$

$= l - x$ for $l/2 \le x =< l$

$$\left[\textbf{Ans: } u(x, t) = \frac{4l}{\pi^2} \sum \frac{1}{p^2} \sin \frac{p\pi}{2} \sin \frac{p\pi x}{l}\, e^{\frac{-\alpha^2 p^2 \pi^2 t}{l^2}}\right]$$

SHORT ANSWER QUESTIONS

1. Define Periodic function.
2. If $f(x)$ is expressed in Fourier series of period 2π in $(c, c + 2\pi)$ write down the Fourier series and formulae for Euler's coefficients.
3. If $f(x)$ is expressed in Fourier series of period $2l$ in $(c, c + 2l)$ write down the Fourier series and formulae for Euler's coefficients.
4. If $f(x)$ is expressed in half range Fourier cosine series or sine series in $(0, l)$ of periodicity $2l$, write down two series and the corresponding integrals for coefficients.
5. State Dirichlet's conditions
6. Draw the graph of $f(x) = |x|$
7. Define $|x|$ in $(-\infty, \infty)$
8. Find half range Fourier Cosine series or sine series in $(0, \pi)$ of periodicity 2π if $f(x) = k$.
9. Define odd and even functions.
10. State any property you know regarding $\int_{-l}^{l} f(x)\, dx$ if $f(x)$ is odd or even.
11. Evaluate $\int_{-1}^{1} (1 - |x|) \cos n\pi x\, dx$, n any integer
12. Find the half range sine series in $0 < x < \pi$ if $f(x) = x^2$.
13. Find the half range cosine series in $0 < x < \pi$ if $f(x) = x + 1$
14. Write down the analytic expression for $f(x)$ if $y = f(x)$ has the graph

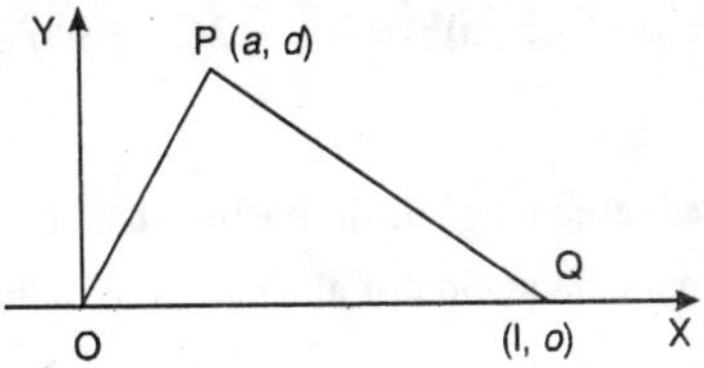

15. Define root mean square value of $f(x)$ over the range (a, b)
16. Using the R.M.S. value fill up the blanks in terms of Fourier coefficients

$$\int_{c}^{c+2\pi} [f(x)]^2\, dx = \text{.......}$$

17. Write down the complex Fourier series, stating the formula for coefficient C_n.

Write true or false, with reason, if possible, (18 to 19)

18. In $(-l, l), f(x) = (1 - |x|) \sin nx$ is odd

19. $\int_c^{c+2\pi} \sin mx \cos nx \, dx = 0$, for all integral values of m, n.

20. Find the Fourier series of periodicity 2π for $f(x) = x$ in $-\pi < x < \pi$

21. $f(x) = 1 + \frac{2x}{\pi}$ in $-\pi < x < 0$

$= 1 - \frac{2x}{\pi}$ in $0 < x < \pi$ is even. State true of false

22. If Fourier series is found out for $f(x) = x^2$ in $-1 < x < 1$, the series will contain only cosine terms. State true or false.

23. If $f(x) = x^3$ in $-\pi < x < \pi$, find the constant term of its Fourier series.

24. If $f(x) = \sin \alpha x$, *in* $-\pi < x < \pi$, α not an integer, find the constant term of its Fourier expansion.

25. If $f(x) = x^2$ in $(-l, l)$ is expressed as a Fourier series of periodicity $2l$, find the constant term of the series.

By eliminating the arbitrary constants, form the partial differential equations (26 to 30).

26. $z = (x^2 + a)(y^2 + b)$

27. $\frac{x}{a} + \frac{y}{a} + \frac{z}{c} = 1$

28. $z = ax + by + ab$

29. $(x - a)^2 + (y - b)^2 + z^2 = 1$

30. $\frac{x^2}{a^2} + \frac{y^2}{b^2} + \frac{z^2}{c^2} = 1$

By eliminating the arbitrary functions form the partial differential equation (31 to 37)

31. $z = F(x^2 + y^2)$

32. $z = f(x^2 + y^2 + z^2)$

33. $xyz = \phi(x + y + z)$

34. $z = f\left(\frac{xy}{z}\right)$

35. $f(x^2 + y^2 + z^2, xyz) = 0$

36. $z = e^y f(x + y)$

37. $z = f(2x + 3y) + F(y + 2x)$

38. Find the partial differential equation of all spheres whose centres lie on the z-axis.

39. Find the partial differential equation of all spheres of radius K with centres on the xy plane.

40. Define complete solution of a p.d.e.

41. If $\frac{\partial^2 z}{\partial x \partial y} = \sin x$ find z

42. Solve $\frac{\partial^2 z}{\partial x^2} = xy$

43. Solve for $z \cdot \frac{\partial z}{\partial x} = 3x - y$; $\frac{\partial z}{\partial y} = \cos y - x$.

Solve for z. (44 to 54)

44. $\sqrt{p} + \sqrt{q} = 1$ **45.** $z = px + qy + \sqrt{1 + p^2 + q^2}$ **46.** $p = 2qx$

47. $p + q = x + y$ **48.** $p + q = pq$ **49.** $q + \sin p = 0$

50. $(z - px - qy)(p + q) = 1$ **51.** $(y - z)p + (z - x)q = x - y$

52. $px^2 + qy^2 = (x + y)z$ **53.** $x(y - z)p + y(z - x)q = z(x - y)$

54. $px + qy = nz$

Find the complimentary function of the following Partial differential equation.

55. $(D^2 - 6DD' + 5D'^2)z = 0$ **56.** $(D^3 - 7DD'^2 - 6D'^3)z = x^2y$

57. $(D^2 - DD')z = \cos x \cos 2y$ **58.** $(D^2 - D'^2)z = e^{x + 2y}$

Find the Particular integrals of the following

59. $(D^2 - D'^2)z = e^{x + 2y}$ **60.** $(D^2 - D'^2)z = e^{x + y}$

61. Evaluate $\dfrac{e^{x + 2y}}{D^3 - 3D^2D' + 4D'^3}$ **62.** Solve: $(2D^2 + 5DD' + 2D'^2)z = 0$

Write true or false

63. If $(D - D')(D + D')z = 0$, the general solution is $z = f_1(y + x) + f_2(y - x)$

64. If $(D^2 - 2DD' + D'^2)z = 0$, its general solution is $z = f_1(y + x) + xf_2(y + x)$.

65. Solve $2xp - 3yq = 0$ by the method of separation of variables.

66. Write down one dimensional wave equation.

67. What are the various solutions of $\dfrac{\partial^2 y}{\partial t^2} = a^2 \dfrac{\partial^2 y}{\partial x^2}$.

68. A tightly stretched string with fixed end points $x = 0$ and $x = l$ is initially at rest in equilibrium position. If it is set vibrating giving each point a velocity $3x(l - x)$, write down the boundary conditions of the problem.

69. Write the analytic expression for the function given by graph

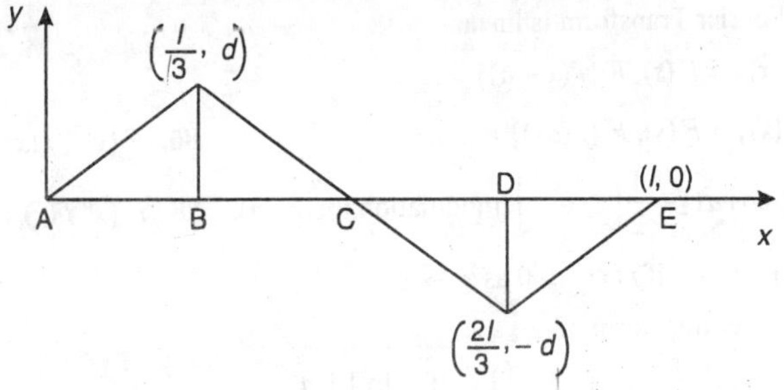

70. Write down the partial differential equation that governs the temperature flow in two dimension (cartesian).

71. Write down the steady-state differential equation that governs the temperature function. Write also the various types of solutions you get in solving it.

72. A rectangular plate is bounded by $x = 0, y = 0, x = a$ and $y = b$. Its surfaces are insulated and temperature along two edges $x = a$ and $y = b$ are 100° C while the temperature along the other edges are 0° C. Write down the Boundary conditions in mathematical form.

73. Express $\frac{\partial^2 u}{\partial x^2} + \frac{\partial^2 u}{\partial y^2}$ in its equivalent in polar form.

74. What are the solutions of the equation $\nabla^2 u = 0$ in polar form?

75. Evaluate $\int_0^{\pi} \frac{400}{\pi} (\pi\,\theta - \theta^2) \sin n\,\theta\, d\,\theta$; n positive integer.

76. A semi-circular plate of radius a is kept at temperature u_0 along the bounding diameter and at u_1 along the circumference. In solving the above problem for $u(r, \theta)$ in steady-state; write down the boundary conditions in u. Further, making a transformation $u(r, \theta) = v(r, \theta) + u_0$ what are the boundary conditions for $v(r, \theta)$

77. Define steady-state solution.

78. An infinitely long plane uniform plate is bounded by edges $x = 0, x = l$ and an end at right angles to them. The breadth of this edge $y = 0$ is l and is maintained at $f(x)$. All the other three edges, are kept at 0° C. Write down the boundary conditions in mathematical form.

79. In the definition of Fourier transform, $F(s) = \frac{1}{\sqrt{2\pi}} \int_{-\infty}^{\infty} f(x)\, e^{isx}\, dx$ what is the Kernel?

80. Define infinite complex Fourier transform of $f(x)$ and write its inversion formula also.

81. State Fourier integral theorem.

82. State Inversion theorem for complex Fourier Transform.

83. Prove Fourier Transform is linear.

84. If $F\{f(x)\} = F(s)$, $F\{f(x - a)\} = \ldots..$

85. If $F\{f(x)\} = F(s)$, $F\{f(ax)\} = \ldots\ldots$

86. $F\{e^{iax} f(x)\} = \ldots\ldots.$

87. $F\{f(x) \cos ax\} = \frac{1}{2}\left[\quad + \quad\right]$ (modulation theorem).

88. $F\{x^n f(x)\} = \ldots\ldots..$

89. $F\{f'(x)\} = \ldots\ldots.$ if $f(x) \to 0$ as $x \to \pm\infty$

90. Find Fourier transform of

$$f(x) = \begin{cases} 1 & \text{if} \quad |x| < a \\ 0 & \text{if} \quad |x| > a > 0. \end{cases}$$

91. Define convolution of two functions of $f(x)$ and $g(x)$.

92. Prove convolution of two functions is commutative.

93. $F^{-1}\{F(s)\, G(s)\} = \ldots\ldots$

94. State convolution Theorem.

95. State parseval's identity.

96. Define infinite Fourier sine and cosine Transform.

97. Write the corresponding inversion forms.

98. Find Fourier sine and cosine transform of $e^{-x}, x \geq 0$,

99. Find Fourier sine transform of $\frac{1}{x}$.

100. Solve $f(x)$ if $\int_0^\infty f(x) \cos \alpha x = e^{-\alpha}$.

101. The Fourier transform of $e^{-\frac{1}{2}x^2}$ is ....

102. Define Finite Fourier sine and cosine transform of $f(x)$.

103. Write down the corresponding inversion results.

104. Find Fourier sine and cosine transforms of $f'(x)$.

105. Find Fourier sine and cosine transforms of $f''(x)$.

106. Find finite Fourier sine and cosine transform of

(*i*) $f(x) = x$ in $(0, \pi)$

(*ii*) $f(x) = x^2$ in $(0, l)$

(*iii*) $f(x) = 1 - \frac{x}{\pi}$ in $(0, \pi)$

(*iv*) $f(x) = 1$ in $(0, \pi)$

107. If $F_s(p) = \frac{2\pi}{p^3}(-1)^{p-1}$ for $p = 1, 2, 3....$ $0 \leq x \leq \pi$ find $f(x)$

108. Find $f(x)$, if $F_c(p) = \frac{\sin \frac{p\pi}{2}}{2p}$ if $p = 1, 2, 3....$

$= \frac{\pi}{4}$ if $p = 0$ in $0 < x < 2\pi$

109. Prove $F_s\{f'(x)\} = -\frac{p\pi}{l}\bar{f}_c(p)$ (finite)

110. Prove $F_c(f''(x)) = -\frac{p^2\pi^2}{l^2}\bar{f}_c(p) + f'(l)(-1)^p - f'(0)$

111. Prove $F_c\{f'(x)\} = f(l)(-1)^p - f(0) + \frac{p\pi}{l}\bar{f}_s(p)$ (finite)

112. Prove $F_s(f''(x)) = \frac{-p^2\pi^2}{l^2}\bar{f}_s(p) + \frac{p\pi}{l}\left[f(0) - (-1)^p f(l)\right]$

113. Find Fourier cosine and sine transform $e^{-x}$of

114. $F_c\{f(ax)\} = ...$ **115.** $F_s\{f(ax)\} = ...$

116. If $F\{f(x)\} = F(s)$ than $F[f(x-a)] = ...$

117. $F\left[e^{-|x|}\right]$ (*i*) does not exist **(BR 1995 Ap.)**

(*ii*) is $\dfrac{2}{1+s^2}$

(*iii*) is $\dfrac{1}{1+s^2}$

(*iv*) is $\dfrac{s}{1+s^2}$

(*v*) is $-\dfrac{2}{1+s^2}$

State true or false:

118. Dirichlet's conditions are necessary for the uniform convergence of Fourier series.

119. When the number of independent variables is equal to the number of arbitrary constants, elimination of the latter leads to first order partial differential equation?

120. One dimensional heat conduction equation is

$$\frac{\partial u}{\partial t} + \frac{\partial^2 u}{\partial x^2} = 0.$$

121. Two dimensional Laplace's equation in Cartesian coordinates is

$$\frac{\partial^2 u}{\partial x^2} + \frac{\partial^2 u}{\partial y^2} = 1.$$

122. $F\{f * g\} = \int_{-\infty}^{\infty} f(u)\, g(u-t)\, du.$

Fill in the blanks:

123. Euler's formula for the Fourier coefficients in the half-range sine series of $f(x)$ in $(0, 2l)$ are............

124. The general solution in terms of arbitrary functions for the partial differential equation $2p + 3q = 1$ is............

125. The steady state temperature of a rod of length l whose ends are kept at 30° and 40° is.........

126. A separable solution of Laplace equation in polar coordinates suitable for a circular disc, which are periodic in θ is..........

127. If $F[f(t)] = \bar{f}(s)$, then $F[f(at)] =$ ....

Pick the correct answer:

128. If the complex form of the Fourier series of $f(x)$ is

$\sum_{-\infty}^{\infty} C_n e^{inx}$ in $(0, 2\pi)$, then C_n is (*a*) $\dfrac{1}{\pi}\int_0^{2\pi} f(x)\, e^{-inx}\, dx.$

(b) $\frac{2}{\pi}\int_0^{2\pi} f(x)\,e^{-inx}\,dx.$ (c) $\frac{1}{2\pi}\int_0^{2\pi} f(x)\,e^{-inx}\,dx.$

(d) $\frac{2}{\pi}\int_0^{\pi} f(x)\,e^{-inx}\,dx.$ (e) $\frac{1}{2\pi}\int_0^{2\pi} f(x)\,e^{inx}\,dx.$

129. The general solution of $\frac{\partial^2 z}{\partial x^2} = 0$ is

(a) $z = f(x) + g(y).$ (b) $z = xf(y) + g(y).$ (c) $z = yf(y) + g(y).$

(d) $z = f(x) + g(x).$ (e) $z = yf(x) + g(x).$

130. The general solution $y(x, t)$ of vibratory motion of a string of length l with fixed end points and zero initial velocity is

(a) $\Sigma\lambda_n \sin\frac{n\pi x}{l}\sin\frac{n\pi at}{l}.$ (b) $\Sigma\lambda_n \cos\frac{n\pi x}{l}\cos\frac{n\pi at}{l}.$

(c) $\Sigma\lambda_n \sin\frac{n\pi x}{l}\cos\frac{n\pi at}{l}.$ (d) $\Sigma\lambda_n \cos\frac{n\pi x}{l}\sin\frac{n\pi at}{l}.$

(e) $\Sigma\lambda_n \sin\frac{n\pi x}{l}.$

131. The equation of steady-state heat conduction in a plate is

(a) $r^2 u_{rr} + ru_r + u_{\theta\theta} = 0.$ (b) $r^2 u_{rr} + ru_r + u_{\theta\theta} = 1.$

(c) $ru_{rr} + r^2 u_r + u_{\theta\theta} = 0.$ (d) $r^2 u_{rr} + ru_r + \theta u_{\theta\theta} = 0.$

(e) $r^2 u_{rr} + 2ru_r + u_{\theta\theta} = 0.$

132. If the Fourier series of $f(x)$ and $g(x)$ are

$\frac{a_0}{2} + \Sigma\,[a_n \cos nx + b_n \sin nx]$ and $A_0 + \Sigma\,[2A_n \cos nx + 2B_n \sin nx]$ over $[-\pi, \pi]$, write down the Fourier series of $2f(x) + g(x)$ over $[-\pi, \pi]$.

133. Find the complete integral of the partial differential equation $z = px + qy + \sqrt{pq}$.

134. Express the boundary conditions in respect of insulated ends of a bar of length a and also the initial temperature distribution.

135. Write down a suitable separable solution for the partial differential equation governing steady state heat conduction in a rectangular plate when the edges $y = 0, y = b$ of a rectangular plate bounded by $x = 0, x = a, y = 0, y = b$ are insulated.

136. Find the finite cosine transform of x when $0 < x < \pi$.

Question from University Papers

State *true* or *false*.

137. Fourier series of period 2 for $|x|$ in $(0, 2)$ contains only cosine terms.

138. A solution of a p.d.e. involving arbitrary functions, is called complete integral.

139. The p.d.e. governing the transverse vibrations of an elastic string is $\frac{\partial u}{\partial t} = C^2 \frac{\partial^2 u}{\partial x^2}$.

140. A separable solution of Laplace equation in polar coordinates, suitable for a circular plate, is $u\,(r\,\theta) = r^{-p}\,(A \cos p\theta + B \sin p\theta)$.

141. The inverse finite sine transform of $F(s)$ is $\int_0^\infty F(s) \sin sx ds$.

Fill in blanks :

142. The cosine series for $f(x) = \sin^2 x$ in $(0, \pi)$ is ____________

143. The p.d.e. whose solution is $z = f(x^2 + y^2)$ (f, using an arbitrary function) is ________________

144. A uniform rod of length a is initially at temperature $\theta = \theta_0$. At time $t = 0$, one end is suddenly cooled to temperature $\theta = 0$ and subsequently maintained at this temperature. The other end remains thermally insulated. The mathematical formulation of this problem is ____________________.

145. The two-dimensional steady state heat conduction equation is ____________.

146. The Fourier transform of $f(x) = 1, |x| < a$
$= 0, |x| > a$ is ______________

Choose the correct answer :

147. Sum of the Fourier series for

$$\left\{\begin{matrix} f(x) = x\,,\ 0 < x < 1 \\ \quad = 2\ 1 < x < 2 \end{matrix}\right\} \text{ at } x = 1 \text{ is}$$

(*a*) 0 (*b*) 1 (*c*) 2 (d) $\frac{1}{2}$ (*e*) $\frac{3}{2}$.

148. The degree of the p.d.e. $\frac{\partial^2 u}{\partial x\,\partial y} = \left(\frac{\partial u}{\partial z}\right)^3$ is

(*a*) 3 (*b*) 1 (*c*) 2 (*d*) $\frac{1}{2}$ (*e*) not defined.

49. A separable solution of $\frac{\partial^2 y}{\partial t^2} = C^2 \frac{\partial^2 y}{\partial x^2}$ satisfying the end conditions $y(0, t) = y(1, t) = 0$ is

(*a*) $\sin\left(\frac{n\,\pi\,Ct}{l}\right) \sin\left(\frac{n\pi x}{l}\right)$ (*b*) $e^{\frac{-C^2 n^2 \pi^2 t}{t}} \sin\left(\frac{n\pi x}{l}\right)$

(*c*) $x(x-l)\sin\left(\frac{n\pi Ct}{l}\right)$ (*d*) $\sin\left(\frac{n\pi Ct}{l}\right)\cos\frac{n\pi x}{l}$

150. A suitable expression for the steady state temperature in a rectangular plate when its edges $x = 0$ and $x = a$ are kept at zero temperature is

(*a*) $(C_1 \cos px + C_2 \sin px)\, e^{py}$

(*b*) $(C_1 e^{px} + C_2 e^{-px}) \sin py$

(*c*) $x(x-a)\sin py$

(*d*) $\sin\left(\frac{n\pi x}{a}\right)\sin\left(\frac{n\pi y}{b}\right)$

(*e*) $e^{\frac{px}{a}}\, e^{\frac{-py}{a}}$

151. If $\bar{f}(s)$ is Fourier transform of $f(t)$, then the Fourier transform of $f(t-a)$ is :

(*a*) $\bar{f}(s-a)$ (*b*) $\bar{f}(s+a)$ (*c*) $e^{ias}\bar{f}(s)$

(*d*) $e^{-ias}\bar{f}(s)$ (*e*) $\bar{f}(s)$.

152. Write down the complex form of the Fourier series for $f(x)$ in a given interval.

153. solve : $\frac{\partial^2 z}{\partial y^2} = \sin xy$.

154. Write down one-dimensional heat e^4quation and a separable solution for the same.

155. In the case of flow of heat in a metal plate in the $x-y$ plane, how do you calculate the rate of gain of heaf by the plate.

156. State inversion theorem for a complex Fourier transform.

State *True* or *False:*

157. The application of Fourier series is not restricted to the expansion of periodic functions of period 2π.

158. The Fourier series for $f(x)$ in the interval $c \le x \le c + 2\pi$ is given by

$$f(x) = \frac{a_0}{2} + \sum_{n=1}^{\infty} a_n \cos nx$$

$$\text{where } a_0 = \frac{1}{\pi}\int_0^{2\pi} f(x)\, dx$$

$$a_n = \frac{1}{\pi}\int_c^{c+2\pi} f(x) \cos nx\, dx.$$

159. $z = px - qy$, where $p = \frac{\partial z}{\partial x}$, $q = \frac{\partial z}{\partial y}$ is the partial differential equation, obtained by eliminating arbitary constants a and b from the equation $z = ax + by$..

160. $\frac{\partial^2 u}{\partial x^2} + \frac{\partial^2 u}{\partial y^2}$ is the Laplace's equation, which arises in the conduction of heat in plate in steady state.

161. $$f(x) = \frac{1}{\sqrt{2\pi}} \int_{-\infty}^{\infty} e^{iwt}\, g(w)\, dw \text{ where}$$

$$g(w) = \frac{1}{\sqrt{2\pi}} \int_{-\infty}^{\infty} e^{-iws} f(s)\, ds$$

does not constitute the symmetrical fourier transform.

Fill in the blanks:

162. At a point $x = x_0$, where $f(x)$ has a finite discontinuity, the sum of the fourier series is . . . .

163. For half-range cosine series in $0 < x < l$ the co-efficient a_n is given by a_n . . . .

164. $Pp + Qq = R$ is known as . . . . where P, Q, R are functions of x, y, z.

165. The quantity of heat required to produce a given temperature change in a body is proportional to the mass of the body and to the

. . . .

166. $f(t) = \frac{2}{\pi} \int_0^{\infty} \int_0^{\infty} f(s) \cos ws \, . \cos wt \, ds\, dw$ is called the fourier cosine integral and it is similar to . . . .

Pick the correct answer:

167. If $f(x) = \sum_{n=1}^{\infty} a_n \cos \frac{n\pi x}{l}$ in $(0, l)$, then a_n is

(*a*) $\frac{1}{l} \int_{-l}^{l} f(x) \cos \frac{n\pi x}{l}\, dx$ (*b*) $\frac{2}{l} \int_{-l}^{l} f(x) \cos \frac{n\pi x}{l}\, dx$

(*c*) $\int_{-l}^{l} f(x) \cos \frac{n\pi x}{l}\, dx$ (*d*) Zero (*e*) $\frac{1}{l}\int_0^l f(x)\cos\frac{n\pi x}{l}\,dx.$

168. The complete integral of the P.D.E. of the form $f(z, p, q) = 0$ is

(*a*) $g(z, a) = x + ay + \varphi(a)$

(*b*) $g(z, a) = x + ay + b$

(*c*) $g'(z, a) = y + \varphi'(a)$

(*d*) $g(z, a) = ax + by + \varphi(a)$

(*e*) $g(z, a) = ax + y + \varphi'(a)$.

169. The complementary function of P.D.E. $(D_x + D_y)^2 z = e^{x-y}$ is

(*a*) $z = \varphi_1(y + x) + x\,\varphi_2(y - x)$

(*b*) $z = \varphi_1(y - x) + x\,\varphi_2(y - x)$

(*c*) $z = \varphi_1(y + x) + \varphi_2(y - x)$

(*d*) $z = \varphi_1(y + x) + \varphi_2(y - 2x)$

(*e*) $z = \varphi_1(y - 2x) + \varphi_2(y + x)$.

170. The equation of the two dimensional heat flow is

(*a*) $\frac{\partial^2 u}{\partial t^2 u} = a^2 \frac{\partial^2}{\partial x^2}$

(*b*) $\frac{\partial^2 u}{\partial x^2} + \frac{\partial^2 u}{\partial y^2} = 0$

(*c*) $\frac{\partial^2 u}{\partial x^2} + \frac{\partial^2 u}{\partial y^2} + \frac{\partial^2 u}{\partial x \partial y} = 0$

(*d*) $\frac{\partial u}{\partial t} = \alpha^2 \frac{\partial^2 u}{\partial x^2}$

171. The Fourier integral in its exponential or complex form is

(*a*) $f(t) = \frac{1}{2\pi}\int_{-\infty}^{\infty}\int_{0}^{\infty} f(s)\, e^{-iw(s-t)}\, ds\, dw$

(*b*) $f(t) = \frac{1}{2\pi}\int_{-\infty}^{\infty}\int_{-\infty}^{\infty} f(s)\, e^{-iw(s-t)}\, ds\, dw$

(*c*) $f(t) = \frac{1}{2\pi}\int_{0}^{\infty}\int_{-\infty}^{\infty} f(s)\, e^{-iw(s-t)}\, ds\, dw$

(*d*) $f(t) = \frac{1}{2\pi}\int_{-\infty}^{-\infty}\int_{-\infty}^{\infty} f(s)\, e^{iw(s+t)}\, ds\, dw$

(e) None.

172. Examine whether the function $f(x) = \sin(1/x)$ in the interval $[-\pi, \pi]$ can be expanded in a Fourier series.

173. Find the complete integral of the P.D.E. $pq = k$, where $p = \frac{\partial z}{\partial x}, q = \frac{\partial z}{\partial y}$.

174. Obtain the general solution of the equation

$$\tan x \frac{\partial z}{\partial x} + \tan y \frac{\partial z}{\partial y} = \tan z.$$

175. Solve $(D^2 + 2DD' + D'^2) z = 0$.

176. Write down the various possible solutions of the wave equation $\frac{\partial^2 y}{\partial t^2} = a^2 \frac{\partial^2 y}{\partial x^2}$.

State *True* or *False*:

177. One of the Dirichlets conditions for convergence is that a function may have finite number of maxima or minima in any one period.

178. The number of arbitary constants will be the order of the partial differential equation on their elimination to form the partial differential equation.

179. In one-dimensional heat flow equation, if the temperature function u is independent of time, then the solution is $u = ax + b$.

180. One of the possible solutions to two-dimensional heat flow equation in cartesian system of co-ordinates is $u = (c_1 x + c_2)(c_3 y + c_4)$.

181. Is $F[f(x)] = \frac{1}{2\pi} \int_{-\infty}^{\infty} e^{-sx} f(x)\, dx$. ?

Fill in the blanks:

182. If a periodic function $f(x)$ is even in $(-l, l)$ then a_n is _________.

183. The Lagranges auxilary equation to $x \frac{\partial z}{\partial x} + y \frac{\partial z}{\partial y} + t \frac{\partial z}{\partial t} = xy\,t$ are _________.

184. If $y = X(x) . T(t)$ is the solution of $\frac{\partial^2 y}{\partial t^2} = c^2 \frac{\partial^2 y}{\partial x^2}$, then $\frac{X''}{X}$ = _________.

185. In two dimensional heat flow, the temperature at any point (of the xy plane) is independent of _________ cordinate.

186. The Fourier cosine transform of $f(x)$ can be defined as _________.

Choose the correct answer:

187. In harmonic analysis upto the fundamental harmonic, the given function

(*a*) $f(x) = a_0 + a_1 \cos x$

(*b*) $f(x) = \frac{a_0}{2} + a_1 \sin x + b_1 \cos x$

(*c*) $f(x) = a_0 + a_1 \cos x + b_1 \sin x$

(*d*) $a_1 \cos x + b_1 \sin x + \frac{a_0}{2} = f(x)$.

188. The solution to $\frac{dx}{\sqrt{x^2 + a}} = \frac{dy}{y}$ is

(*a*) $\sin^{-1}\left(\frac{x}{\sqrt{a}}\right) = \log_e y + b$ (*b*) $\cos^{-1}\left(\frac{x}{\sqrt{a}}\right) = \log_e y + b$

(*c*) $\sinh^{-1}\left(\frac{x}{\sqrt{a}}\right) = \log_e y + b$ (*d*) $\cosh^{-1}\left(\frac{x}{\sqrt{a}}\right) = \log_e y + b$.

189. When the ends of a rod is non-zero for one-dimensional heat flow equation, the temperature function $u(x, t)$ is modified as the sum of steady state and transient state temperatures. The transient part of the solution which

(*a*) increases with increase of time

(*b*) decreases with increase of time

(*c*) increases with decrease of time

(*d*) decreases with decrease of time.

190. In two dimensional heat flow, the rate of heat flow across an area is proportional to the

(*a*) area and to the temperature gradient parallel to the area

(*b*) area and to the temperature normal to the area

(*c*) area and to the temperature gradient normal to the area

(*d*) area and to the temperature parallel to the area.

191. If $F[f(x)] = \phi(\lambda)$, then $F[f(x + a)]$ is

(*a*) $e^{ia\lambda} \phi(\lambda)$ (*b*) $e^{-ia\lambda} \phi(\lambda)$

(*c*) $e^{ia\lambda} \phi(i\lambda)$ (*d*) $e^{-ia\lambda} \phi(i\lambda)$.

192. What is the equivalent Fourier constant to the expression $2X$ mean value of $f(x)$ in $(l, l + 2\pi)$.

193. Form the partial differential equation by eliminating f from $u = f(x^3 - y^3)$.

194. The ends of a rod of length 5 units are maintained at 15°C and 5°C respectively until steady state prevails. Determine the temperature at that state.

195. If $u(r, \theta) = \sum_{1}^{\infty} b_n \left(\frac{r}{a}\right)^n \sin n\theta$ is the solution of two dimensional steady state heat flow equation of a thin semicircular plate of radius a units, then determine $u(r, \theta)$ when $r = a$ and $u(a, \theta) = K$.

196. Find the Fourier transform of second derivative of a function $f(t)$.

State *True* or *False*:

197. Is the function

$$f(x) = \begin{cases} 1 + \dfrac{2x}{\pi}, & -\pi < x < 0 \\ 1 - \dfrac{2x}{\pi}, & 0 < x < \pi \end{cases}$$

odd?

198. If the number of constants to be eliminated is equal to the number of independent variables, then the P.D.E' is of first order.

199. One dimensional heat flow equation is $\frac{\partial u}{\partial t} - \frac{\partial^2 u}{\partial x^2} = 0.$

200. Two dimensional Laplace's equation in polar co-ordinates is $r^2 \frac{\partial^2 u}{\partial r^2} + r \frac{\partial u}{\partial r} + \frac{\partial^2 u}{\partial \theta^2} = 0.$

201. The Fourier transform of the convolution of two functions $f(x)$ and $g(x)$ is $\frac{1}{\sqrt{2}\pi} \int_{-\infty}^{\infty} f(t)\, g(x - t)\, dt.$

Fill in the blanks :

202. Euler's formulae for the Fourier co-efficients in the half-range cosine series of $f(x)$ in (0, l) are __________.

203. The complete integral of the equation $z = px + qy + f(p, q)$ is __________.

204. D'Alembert's solution of the one dimensional wave equation is __________.

205. In steady-state, two dimensional heat-flow equation in Cartesian co-ordinates is __________

206. The Fourier transform of $\int_{a}^{x} f(x)\, dx$ is __________.

Pick the correct answer :

207. The exponential form of the Fourier series of $f(x)$ in (–1, 1) is given by $f(x) = \sum_{n=-\infty}^{\infty} c_n e^{\frac{in\pi x}{l}}$ where c_n is

(a) $\frac{2}{l}\int_{-l}^{l} f(x)\, e^{-\frac{in\pi x}{l}}\, dx$ (b) $\frac{2}{2l}\int_{-l}^{l} f(x)\, e^{-\frac{in\pi x}{l}}\, dx$

(c) $\frac{2}{l}\int_{0}^{l} f(x)\, e^{-\frac{in\pi x}{l}}\, dx$ (d) $\frac{1}{2l}\int_{0}^{l} f(x)\, e^{-\frac{in\pi x}{l}}\, dx$

(a) $\frac{1}{2l}\int_{0}^{l} f(x)\, e^{\frac{in\pi x}{l}}\, dx$.

208. The general integral of $\frac{y^2 z}{x}\frac{\partial z}{\partial x} + xz\frac{\partial z}{\partial y} = y^2$ is

(a) $F(x^3 - y^3, x^2 - z^2) = 0$ (b) $F(x^3 - y^2, x - z) = 0$
(c) $F(x^2 - z^3, x^2 - y^2) = 0$ (d) $F(Y^3 - z^3, x^2 - y^2) = 0$
(e) $F(x - y, x^2 - z^2) = 0$.

209. The P.D.E. of one dimensional heat flow in steady state is given by

(a) $\frac{\partial u}{\partial t} = 0$ (b) $\frac{\partial^2 u}{\partial t^2} = 0$

(c) $\frac{\partial^2 u}{\partial x^2} - \frac{\partial u}{\partial t} = 0$ (d) $\frac{\partial^2 u}{\partial x^2} = 0$

(e) $\frac{du}{dt} = 0$.

210. A semicircular plate of radius 'a' is kept at temperature zero along the bounding diameter and at $f(\theta)$, $0 < \theta < \pi$ along the circumference. The steady-state temperature $u(r, \theta)$ at any point of the plate is given by $\sum_{n=1}^{\infty} b_n \left(\frac{r}{a}\right)^n \sin n\theta$ where b_n is

(a) $\frac{1}{\pi}\int_0^{\pi} f(\theta) \sin n\theta\, d\theta$ (b) $\frac{2}{\pi}\int_0^{\pi} f(\theta) \sin n\theta\, d\theta$

(c) $\frac{1}{\pi}\int_0^{\pi} f(\theta) \sin \theta\, d\theta$ (d) $\frac{1}{\pi}\int_0^{\pi} f(\theta) \cos \theta\, d\theta$

(e) $\frac{2}{\pi}\int_0^{\pi} f(\theta) \cos n\theta\, d\theta$.

211. The Fourier sine transform of $xf(x)$ is

(a) $F_c'(s)$ (b) $F_s'(s)$ (c) $-F_c'(s)$
(d) $-F_s'(s)$ (e) $F_s''(s)$.

212. Find the Fourier series for

$$f(x) = \begin{cases} x - 1, & -\pi < x < 0 \\ x + 1, & 0 < x < \pi \end{cases}$$

213. Find the P.D.E. by eliminating the arbitrary function in $z = f\left(\frac{xy}{z}\right)$.

214. A tightly stretched string with fixed end points $x = 0$ and $x = 1$ is initially at rest in equilibrium position. If it is set vibrating giving each point a velocity $\lambda x(1 - x)$, write down the boundary conditions of the problem.

215. Write down a suitable separable solution for the P.D.E. governing steady state heat conduction in a square plate when the edges $Y = 0$, $Y = 10$ of a square plate bounded by $X = 0$, $X = 10$, $Y = 0$, $Y = 10$ are insulated.

216. Find the Fourier Transform of $e^{-a|x|}$, $a > 0$.

State *True* **or** *False*:

217. If a function $f(x)$ has a finite discontinuity at $x = a$, then the sum of the Fourier series is the arithmetic mean of the left and right-hand limits of $f(x)$ at $x = a$.

218. The singular integral is not contained in the complete integral whereas the particular integral is obtained from the complete integral.

219. One dimensional wave equation is $\frac{\partial^2 u}{\partial t^2} + \frac{\partial^2 u}{\partial x^2} = 0$

220. The temperature distribution of the plate in the steady-state is $\nabla^2 u = 0$

221. The Fourier transform of $e^{-ax} f(x)$ in terms of Fourier transform $F(s)$ of $f(x)$ is given by $F(s - a)$.

222. If a periodic function $f(x)$ is even, its Fourier expansion contains only ___________.

223. The particular integral of $(D^2 - 2DD' + D'^2)\, z = e^{x+2y}$ is _____.

224. In the steady-state condition, when the temperature no longer varies with time, the solution of the diffusion equation is __________.

225. Two dimensional heat flow equation in polar co-ordinates is_________.

226. The Fourier sine transform of $f(x) \sin ax$ in terms of Fourier transform $F(s)$ of $f(x)$ is_____________.

Pick the correct answer :

227. If $\frac{a_0}{2} + \sum_{n=1}^{\infty} a_n \cos \frac{n\pi x}{l}$ is the half-range consine series of $f(x)$ of period $2l$ in $(0, l)$, then $\int_0^l [f(x)]^2\, dx$ is equal to

(a) $2l\left[\frac{a_0^2}{2}+\sum_{n=1}^{\infty} a_n^2\right]$ (b) $\frac{l}{2}\left[\frac{a_0^2}{2}+\sum_{n=1}^{\infty} a_n^2\right]$

(c) $\frac{l}{2}\left[\frac{a_0^2}{2}+\sum_{n=1}^{\infty} a_n^2\right]$ (d) $l\left[\frac{a_0^2}{2}+\sum_{n=1}^{\infty} a_n^2\right]$

(e) $2l\left[a_0^2+\sum_{n=1}^{\infty} a_n^2\right]$.

228. **The complete integral of the P.D.E. of the form $F(p, q) = 0$ is given by**

(a) $Z = ax + \phi'(a)y + b$ (b) $Z = ax + \phi(a)y + b$

(c) $Z = a + \phi(a)(a)x$ (d) $Z = ax + \phi(a)$

(e) $Z = ax + \phi'(a)y$.

229. **The general solution $u(x, t)$ of vibratory motion of a string of length $2l$ with fixed end points and zero initial velocity is**

(a) $\sum_{n=1}^{\infty} B_n \sin\frac{n\pi x}{l}\sin\frac{n\pi at}{l}$

(b) $\sum_{n=1}^{\infty} B_n \sin\frac{n\pi x}{l}\cos\frac{n\pi at}{l}$

(c) $\sum_{n=1}^{\infty} B_n \sin\frac{n\pi x}{2l}\cos\frac{n\pi al}{2l}$

(d) $\sum_{n=1}^{\infty} B_n \sin\frac{n\pi x}{2l}\sin\frac{n\pi at}{2l}$

(e) $\sum_{n=1}^{\infty} B_n \sin\frac{n\pi x}{2l}\cos\frac{n\pi at}{2l}$

230. $\sum_{n=1}^{\infty} i \sin x$ **is of period**

(a) 0 (b) $\pi/2$ (c) π (d) $\frac{3\pi}{2}$ (e) 2π.

231. **If $f(x) \to 0$ as $x \to \infty$, then Fourier cosine transform of first derivative of $f(x)$ is given by**

(a) $s F_s(s)$ (b) $-s F_c(s)$

(c) $s F_s(s) - \frac{\sqrt{2}}{\pi} f(0)$ (d) $s F_c(s) - \frac{\sqrt{2}}{\pi} f(0)$

(e) $-\left[s F_s(s) + \frac{\sqrt{2}}{\pi} f(0)\right]$

232. **Find the Fourier series for $f(x) = |x|$ in $(-\pi, \pi)$.**

State *true* or *false*

233. At a point $x = x_0$, where $f(x)$ has a finite discontinuity, the sum of the Fourier series is $\frac{1}{2}\lim_{h\to 0}[hf(x_0 - h) + hf(x_0 + h)]$.

234. If the number of constants to be eliminated is just equal to the number of independent variables, the process of elimination will give rise to partial differential equation of the second order.

235. In one dimensional heat flow equation, the temperature function u is to decrease with increase of time.

236. One of the possible solutions to two dimensional heat flow equation in polar co-ordinate is
$u(r, \theta) = (C_1 \log r + C_2)(C_3 \log \theta + c)$

237. Fourier sine transform of $f(x)$ is equal to $\frac{1}{\sqrt{2\pi}}\int_0^\infty \sin sx\, f(x)\, dx$.

Fill in the blanks:

238. If a periodic function $f(x)$ is an odd function in $(-l, l)$, then b_n in the Fourier series is ________.

239. A solution which contains the maximum possible number of arbitrary function is called ________.

240. The one dimensional wave equation is ________.

241. Two dimensional Laplace equation in polar co-ordinates is ________.

242. The Fourier inverse cosine transform is defined as ________.

243. The root-mean square value of a function $y = f(x)$ over an interval from $x = a$ to $x = b$ is defined as

(*a*) $\bar{y} = \frac{1}{(b-a)^{\frac{1}{2}}}\int_a^b y^2\, dx$ (*b*) $\bar{y} = \frac{1}{(b-a)^{\frac{1}{2}}}\int_a^b f(x)\, dx$

(*c*) $\bar{y} = \frac{1}{(b-a)}\left(\int_a^b y^2\, dx\right)^{\frac{1}{2}}$ (*d*) $\bar{y} = \left[\frac{1}{(b-a)}\int_a^b y^2\, dx\right]^{\frac{1}{2}}$.

244. A rod of length l has its ends A and B kept at 0°C and 100°C respectively, until steady-state conditions prevail. Then the initial condition is given by

(*a*) $u(x, 0) = ax + b + 100l$ (*b*) $u(x,0) = \frac{100x}{l}$

(*c*) $u(x, 0) = 100xl$ (*d*) $u(x, 0) = (x + l)\, 100$.

245. The solution of the Lagrange's linear partial differential equation $Pp + Qq = R$ is of the form

(a) $\phi(u+v)=0$ where ϕ is an arbitrary function
(b) $u=\phi(v)$ where ϕ is an arbitrary function
(c) $uv=\phi(v)$ where ϕ is an arbitrary function
(d) $v=\phi(u+v)$ where ϕ is an arbitrary function

246. The solution of one dimensional wave equation is
(a) $y(x,t)=(C_1\cosh px+C_2\sinh px)(C_3\cos pat+C_4\sin pat)$
(b) $y(x,t)=(C_1\cosh px+C_2\sinh px)(C_3e^{pat}+C_4e^{-pat})$
(c) $y(x,t)=(C_1\cos px+C_2\sin px)(C_3\cos pat+C_4\sin pat)$
(d) $y(x,t)=(C_1e^{px}+C_2e^{-px})(C_3e^{py}+C_4e^{-py})$

247. $F\left(\int_a^x f(x)\,dx\right)$ is equal to

(a) $\dfrac{F(s)}{is}$ (b) $\dfrac{F(s)}{s}$ (c) $\dfrac{F(s)}{-is}$ (d) $e^{ia}\dfrac{F(s)}{-is}$

248. If $f(x)=|x|$ when $-\pi<x<\pi$, then obtain Fourier series expansion for $f(x)$.

249. Form the partial differential equation by eliminating f and ϕ and $Z=f(x+ct)+\phi(x-ct)$.

250. A rod 30 cm long, has its end A and B kept at 20°C and 80°C respectively, until steady state conditions prevail. Determine the temperature at that state.

251. For the steady state heat flow in a circular plate, is it possible to have the general solution $u(r,\theta)$ as $(A\log r+B)=(C\theta+D)$? If not, why?

252. If $f(x)=e^{-ax}$, $a>0$, find Fourier sine transform of $f(x)$.

253. Write the Fourier sine series of K in the interval $(0,\pi)$.

254. Fill in the blanks
The fourier series of $f(x)$ in $(0,l)$ is

$$f(x)=\frac{1}{2}a_0+\sum_{n=1}^{\infty}a_n\cos\frac{n\pi x}{l},$$

then $a_0=$ ______________ and $a_n=$ ______________.

255. What is the sine transform of $f(ax)$ if $f_s(s)$ is the Fourier sine transform of $f(x)$?

256. State Parseval's identity on Fourier transform.

257. Write true or false.
The non-linear equation $f(p,q)=0$ has singular integral.

258. Fill in the blanks
The general solution in terms of arbitrary functions of the PDE $2p+3q=1$ is ______________.

259. Pick the correct answer

The general solution of $\frac{\partial^2 z}{\partial x^2} = 0$ is

(*a*) $z = f(x) + g(y)$ (*b*) $z = xf(y) + g(y)$
(*c*) $z = yf(x) + g(x)$ (*d*) $z = f(x) + g(x)$

260. Solve: $(D^2 - 4DD' + 5D'^2)z = 0$.

261. Write the three possible solutions of the equation $\frac{\partial^2 u}{\partial t^2} = C^2 \frac{\partial^2 u}{\partial x^2}$.

262. Choose the correct answer

The one dimensional heat equation in steady-state is

(*a*) $\frac{\partial^2 u}{\partial x^2} + \frac{\partial^2 u}{\partial y^2} = 0$ (*b*) $\frac{\partial u}{\partial t} = c^2 \frac{\partial^2 u}{\partial x^2}$

(*c*) $\frac{d^2 u}{dx^2} = 0$ (*d*) $\frac{\partial^2 u}{\partial t^2} = c^2 \frac{\partial^2 u}{\partial x^2}$

263. Analyse the possible solution of steady state heat flow in two dimension for the following boundary conditions

(*a*) $u(0, y) = 0, 0 \le y \le \infty$
(*b*) $u(l, y) = 0, 0 \le y \le \infty$
(*c*) $u(x, \infty) = 0, 0 \le x \le l$
(*d*) $u(x, 0) = f(x), 0 < x < l$

264. Express $\nabla^2 u$ in polar form.

265. Write down the boundary conditions for the following boundary value problem.

If a string of length l is initially at rest in its equilibrium position and each of its points is given the velocity $\left(\frac{\partial y}{\partial t}\right)_{t=0} = v_0 \sin^3 \frac{\pi x}{l}$, $0 < x < l$, determine the displacement $y(x, t)$.

State *true* or *false*

266. Fourier series of period 2 for $x \sin x$ in the internal $(-1, 1)$ contains only sine terms.

267. The difference between the partial differential equation and the ordinary differential equation is that order of the differential equation equals the number of arbitrary constants eliminated.

268. In one dimensional heat flow equation, if the temperature function u is independent of time, then solution is $u = ax + b$.

269. The temperature distribution of the plate in the transient state is given by $\frac{\partial u}{\partial t} = \frac{\partial^2 u}{\partial x^2} + \frac{\partial^2 u}{\partial y^2}$.

270. Is $F\{f(x)\} = \frac{1}{2\pi}\int_{-\infty}^{\infty} e^{isx} f(x)\,dx$?

Fill in the blanks:

271. The half-range sine series for $f(x) = x$ in (0, 1) is ________.

272. The PDE whose solution on is $z = ax + by$, where a and b are arbitrary constants, is given by ________.

273. The steady state temperature of a rod of length $2l$ whose ends are kept at 20°C and 70°C is ________.

274. The two-dimensional steady state heat conduction equation is ________.

275. If $F\{f(x)\} = F(s)$, then $F\left\{\int_a^x f(x)\,dx\right\} =$ ________.

Choose the correct Answer

276. The Fourier series of $f(x) = C$, where C is a constant, the periodicity being 2π is

(*a*) 0 (*b*) $2C$ (*c*) 2π (*d*) C

277. The complete integral of the equation $p = 2qx$ is

(*a*) $z = ax^2 + ay + c$ (*b*) $z = ax + c$

(*c*) $z = ay^2 + ax + c$ (*d*) $z = ay + c$.

278. The general solution $y(x, t)$ of vibratory motion of a string of length l with fixed end points and zero initial velocity is

(*a*) $\sum \lambda_n \sin\frac{n\pi x}{l}\sin\frac{n\pi at}{l}$ (*b*) $\sum \lambda_n \sin\frac{n\pi x}{l}\cos\frac{n\pi at}{l}$

(c) $\sum \lambda_n \cos\frac{n\pi x}{l}\cos\frac{n\pi at}{l}$ (*d*) $\sum \lambda_n \cos\frac{n\pi x}{l}\sin\frac{n\pi at}{l}$.

279. In two dimensional heat flow, the rate of heat flow across an area is proportional to the

(*a*) area and to the temperature gradient parallel to the area

(*b*) area and to the temperature normal to the area

(*c*) area and to the temperature gradient normal to the area

(*d*) area and to the temperature parallel to the area.

280. Fourier sine transform of $\frac{1}{x}$ is

(*a*) π (*b*) $\frac{\pi}{2}$ (*c*) $\sqrt{\pi/2}$ (*d*) 2π

281. Find the r.m.s. value of $f(x) = x - x^2$ in the internal.

282. Obtain partial differential equation by eliminating the arbitrary functions *f* and *g* from
$z = f(2x+y) + g(3x-y)$.

283. The ends of a rod of length 20 cm are maintained at 10°C and 30°C respectively until steady-state prevails. Find the temperature at that state.

284. Write down the various possible separable solutions of Laplace's equation in two dimensions.

285. Find the Fourier transform of

$$f(x) = \begin{cases} 1-|x| & \text{if } |x| < 1 \\ 0 & \text{if } |x| > 1. \end{cases}$$

286. Find the Fourier constant a_n for $x \cos x$ in $(-\pi, \pi)$, $n = 2, 3, \ldots$ to ∞.

287. Find the sine series of $f(x) = K$ in $(0, l)$.

288. What is known as harmonic analysis?

289. Form the p.d.c. by eliminating '*f*' from $Z = f(y/x)$.

290. Find the solution of $px^2 + qy^2 = z^2$.

291. Find the complete integral of $p + q = x + y$.

292. Find the particular integral of
$(D^2 - 4DD' + 3D'^2)\, Z = e^{x+y}$

293. Write down the one dimensional wave equation.

294. Write different possible solutions of one dimensional wave equation.

295. What is meant by steady state condition in one-dimensional heat flow?

296. If the ends $x = 0$ and $x = l$ are insulated in one-dimensional problems, write the boundary conditions.

297. Write down the unsteady two-dimensional heat flow equation in Cartesian co-ordinates.

298. Write down the different solutions of Laplace equation in Cartesian co-ordinates.

299. Write the most general solution to find the steady-state temperature over annulus.

300. A quadrant plate of radius '*a*' cms has its temperature at 0°C on the bounding radii and 60°C on its circumference. Write the corresponding boundary conditions.

301. If the Fourier transform of $f(x)$ is $\bar{f}(s)$, what is the Fourier transform of $f(x-a)$?

302. Find the Fourier sine transform of e^{-2x}.

303. State Parseval's identity on Fourier transform.

304. Find the finite Fourier Cosine transform of $f(x) = x$ in $(0, \pi)$.

Say True or False (305 to 309):

305. $$f(x) = \begin{cases} 1-x & \text{in } 0 < x < \pi \\ 1+x & \text{in } \pi < x < 2\pi \end{cases}$$

is an even function and hence the Fourier series of $f(x)$ contains only cosine terms.

306. When the number of independent variables is equal to the number of arbitrary constants, the partial differential equation obtained by eliminating the arbitrary constants is of first order.

307. The one dimensional heat conduction equation is $\dfrac{\partial^2 u}{\partial t^2} = \alpha^2 \dfrac{\partial^2 u}{\partial x^2}$.

308. $e^{-x^2/2}$ is self reciprocal under Fourier transform.

309. $F_c[xf(x)] = F_s'[f(x)]$.

310. The Fourier sine series of $f(x)$ in $0 < x < \pi$ is $f(x) = \sum_1^{\infty} b_n \sin nx$, where $b_n =$ __________.

311. The complete integral of the partial differential equation $z = px + qy + pq$ is __________.

312. The general solution of $\dfrac{\partial^2 z}{\partial x^2} - \dfrac{\partial^2 z}{\partial y^2} = 0$ is __________.

313. The steady state temperature distribution in a rod of length 10 cm whose ends $x = 0$ and $x = 10$ are kept at 20°C and 50°C respectively, is __________.

314. If $F[f(x)] = F(s)$, then $F[e^{iax} f(x)] =$ __________.

Choose the correct answer (315 to 319):

315. The sum of the Fourier series of $f(x) = x + x^2$ in $-\pi < x < \pi$ at $x = \pi$ is

(*a*) π (*b*) π^2 (*c*) $\dfrac{\pi}{2}$ (*d*) $\dfrac{\pi^2}{2}$.

316. The general solution of the partial differential equation $(4D^2 - 4DD' + D'^2)z = 0$ is

316. The general solution of the partial differential equation $(4D^2 - 4DD' + D'^2)\, z = 0$ is

(a) $z = \varphi_1(y+2x) + x\varphi_2(y+2x)$

(b) $z = \varphi_1\left(y+\frac{1}{2}x\right) + \varphi_2\left(y+\frac{1}{2}x\right)$

(c) $z = \varphi_1(x+2y) + \varphi_2(x+2y)$

(d) $z = \varphi_1\left(y+\frac{1}{2}x\right) + x\varphi_2\left(y+\frac{1}{2}x\right)$.

317. The general solution $y\,(x, t)$ of the transverse vibration of a stretched string with fixed end points and zero initial velocity is

(a) $\sum B_n \sin\left(\frac{n\pi x}{l}\right)\sin\left(\frac{n\pi at}{l}\right)$

(b) $\sum B_n \cos\left(\frac{n\pi x}{l}\right)\sin\left(\frac{n\pi at}{l}\right)$

(c) $\sum B_n \sin\left(\frac{n\pi x}{l}\right)\cos\left(\frac{n\pi at}{l}\right)$

(d) $\sum B_n \cos\left(\frac{n\pi x}{l}\right)\cos\left(\frac{n\pi at}{l}\right)$

318. The two dimensional heat equation in steady state is

(a) $\frac{\partial^2 u}{\partial t^2} = \alpha^2 \frac{\partial^2 u}{\partial x^2}$ (b) $\frac{\partial u}{\partial t} = \alpha^2 \frac{\partial^2 u}{\partial x^2}$

(c) $\frac{\partial u}{\partial t} = \alpha^2\left(\frac{\partial^2 u}{\partial x^2} + \frac{\partial^2 u}{\partial y^2}\right)$ (d) $\frac{\partial^2 u}{\partial x^2} + \frac{\partial^2 u}{\partial y^2} = 0$

319. If $F[f(x)] = F(s)$, then $F[f(x+a)] =$

(a) $F(s-a)$ (b) $F(s+a)$ (c) $e^{ias} F(s)$ (d) $e^{-ias} F(s)$.

320. Write down the Parsevals formula corresponding to the Fourier cosine series of $f(x)$ in $0 < x < \pi$.

321. Form the partial differential equation by eliminating the arbitrary constants from $z = ax^2 + by^2$.

322. List all the possible solutions of the one dimensional wave equation (obtained by the method of separation of variables) and state the proper solution.

325. If $f(x) = x^2$ in the interval $-2 \le x \le 2$, what is the value of b_n ?
326. Find the value of a_n in the cosine series expansion of $f(x) = k$ in (0, 10).
327. What is known as harmonic analysis ?
328. Form the partial differential equation by eliminating the arbitrary function f from $z = f\left(\frac{y}{x}\right)$.
329. Write the complete integral of $p + q = x + y$.
330. Find the solution of $p\sqrt{x} + q\sqrt{y} = \sqrt{z}$.
331. Write the particular integral of $(D^2 + 3DD' + 4D'^2)z = e^{x-y}$.
332. Write any two solution of the transverse vibration of a string.
333. If the ends of a string of length l are fixed and the mid point of the string is drawn aside through a height h and the string is released from rest, write the intial conditions.
334. Write down steady state one dimensional heat flow equation.
335. The upper end of a rod of length l is always kept insulated. Write this condition mathematically.
336. Write the Laplace's equation in Cartesian co-ordinates.
337. A rectangular plate is bounded by $x = 0$, $y = 0$, $x = a$ and $y = b$. Its surfaces are insulated and temperature along two edges $x = a$ and $y = b$ are 100°C while the temperature along the other edges are 0°C. Write down the boundary conditions for the above problem.
338. Write down any two solution of Laplace's equation in polar coordinates.
339. A circular plate of radius 10 cm has the upper half of its circumference at 0°C and the lower half of 100°C. Write the corresponding boundary conditions.
340. Define Fourier transform of $f(x)$.
341. Define self-reciprocal function w.r.t. Fourier Cosine transform.
342. Find the Fourier sine transform of $\frac{1}{x}$.
343. Find the finite Fourier sine transform of $f(x) = x$ in $(0, l)$.

Say True or False (344-348)

344. The Fourier series of $f(x) = x\cos x$ in $-\pi < x < \pi$ contains only cosine terms.
345. The general solution of $\frac{\partial^2 z}{\partial x \partial y} = 0$ is $z = f(x) + g(y)$.

346. The two dimensional steady-state heat equation in polar coordinates is $r^2\dfrac{\partial^2 u}{\partial r^2}+\dfrac{\partial u}{\partial r}+r\dfrac{\partial^2 u}{\partial \theta^2}=0.$

347. If $f(x)$ is an even functions of x, then its Fourier transform $F(s)$ is also an even function of s.

348. $e^{-x^2/2}$ is self-reciprocal under Fourier sine transform.

Fill in the blanks (349 to 353)

349. The value of a_0 in the Fourier series expansion of $f(x) = \pi - x$ in $0 < x < 2\pi$ is ________

350. The complete solution of the partial differential equation $pq = 1$ is ________.

351. The general solution of $\dfrac{\partial^2 z}{\partial x^2}-2\dfrac{\partial^2 z}{\partial x\partial y}+\dfrac{\partial^2 z}{\partial y^2}=0$ is ________.

352. The steady-state temperature distribution in a rod of length 10 cm whose ends are kept at 10°C and 30°C is ________.

353. If $F[f(x)] = F(s)$, then $F[f(2x)] =$ ________.

Choose the correct answer (354-358)

354. The general solution of $\dfrac{\partial^2 z}{\partial y^2}=0$ is

(*a*) $z = f(x) + g(y)$ (*b*) $z = f(x) + y\,g(x)$

(*c*) $z = f(y) + xg(y)$ (*c*) $z = f(y) + xg(y)$

355. The one dimensional heat equation in steady state is

(*a*) $\dfrac{\partial^2 u}{\partial x^2}+\dfrac{\partial^2 u}{\partial y^2}=0$ (*b*) $\dfrac{\partial u}{\partial t}=0$

(*c*) $\dfrac{\partial^2 u}{\partial x^2}-\dfrac{\partial u}{\partial t}=0$ (*d*) $\dfrac{d^2 u}{dx^2}=0$

356. The general solution for the displacement $y(x, t)$ of string of length l vibrating between fixed end points with initial valocity zero and initial displacement $f(x)$ is

(*a*) $\Sigma B_n \sin\left(\dfrac{n\pi x}{l}\right)\cos\left(\dfrac{n\pi at}{l}\right)$ (*b*) $\Sigma B_n \sin\left(\dfrac{n\pi x}{l}\right)\sin\left(\dfrac{n\pi at}{l}\right)$

(*c*) $\Sigma B_n \cos\left(\dfrac{n\pi x}{l}\right)\sin\left(\dfrac{n\pi at}{l}\right)$ (*d*) $\Sigma B_n \sin\left(\dfrac{n\pi x}{l}\right)$

357. If $F(s)$ is the fourier transform of $f(x)$, then the Fourier transform of $f(x-a)$ is

(*a*) $F(s+a)$ (*b*) $F(s-a)$ (*c*) $e^{ias} F(s)$ (*d*) $e^{-ias} F(s)$

Answer the following (358 to 358)

358. Write down the complex form of Fourier series of $f(x)$ defined in $-\pi < x < \pi$.

359. Find the particular integral of $\dfrac{\partial^2 z}{\partial x^2} + \dfrac{\partial^2 z}{\partial y^2} = \cos(x-y)$

360. Write down the three possible solutions of the one dimensional heat equation and state the suitable solution.

361. If $F(s)$ is the Fourier transform of $f(x)$, what is the Fourier transform of $f''(x)$?

362. Using Parseval's theorem prove

(*i*) $\int_{-\pi}^{\pi} \cos^4 x dx = \dfrac{3\pi}{4}$ (*ii*) $\int_{-\pi}^{\pi} \cos^6 x\, dx = \dfrac{5\pi}{8}$

(*iii*) $\int_{-\pi}^{\pi} \cos^8 x\, dx = \dfrac{35\pi}{64}$.

Hint : (*i*) $f(x) = \cos^2 x = \dfrac{1}{2} + \dfrac{1}{2}\cos 2x$

$$\frac{a_0}{2} = \frac{1}{2},\ a_2 = \frac{1}{2}$$

$$\int_{-\pi}^{\pi} [f(x)]^2 dx = 2\pi \left[\frac{a_0^2}{4} + \frac{1}{2}\sum (a_n^2 + b_n^2)\right]$$

$$\int_{-\pi}^{\pi} \cos^4 x\, dx = 2\pi \left[\frac{1}{4} + \frac{1}{2}\left(\frac{1}{4}\right)\right]$$

$$= \frac{3\pi}{4}$$

(*ii*) Take $f(x) = \cos^3 x = \dfrac{3}{4}\cos x + \dfrac{1}{4}\cos 3x$.

$$a_1 = 3/4;\ a_3 = 1/4.$$

$$\int_{-\pi}^{\pi} \cos^6 x\, dx = 2\pi \left[\frac{1}{2}\left(\frac{9}{16} + \frac{1}{16}\right)\right]$$

$$= \frac{5\pi}{8}$$

(*iii*) Take $f(x) = \cos^4 x = \left(\dfrac{1+\cos 2x}{2}\right)^2$

$$= \frac{3}{8} + \frac{1}{2}\cos 2x + \frac{1}{8}\cos 4x$$

$$a_0 = \frac{3}{4},\ a_2 = \frac{1}{2},\ a_4 = \frac{1}{8}$$

ANSWERS

(FOURIER SERIES)

Exercises 1(a) Page 1

1. (*i*) π, (*ii*) $\frac{2}{3}\pi$, (*iii*) $2\pi/n$, (*iv*) 1, (*v*) k.

Exercises 1(b) Page 3

1. 9; 9; 9. **2.** $\frac{1}{2}; \frac{1}{2}; \frac{1}{2}$. **3.** $+\infty; -\infty$. **4.** 1, 0, does not exist.

5. 1, 0; $f(x)$ is not continuous at $x = 0$; jump at 0 is 1; $f(0) = 1$.

6. 1, 2; 1.

Exercises 1(c) Page 37

1. $f(x) = \dfrac{e^{2\pi} - 1}{\pi e^{\pi}} \left\{ \dfrac{1}{2} + \sum_{n=1}^{\infty} \dfrac{(-1)^n}{n^2 + 1} (\cos nx - n \sin nx) \right\}$.

2. $f_1(x) = 2\left(\sin x - \frac{1}{2}\sin 2x + \frac{1}{3}\sin 3x - \cdots\right) \quad \dfrac{\pi}{4} = 1 - \frac{1}{3} + \dfrac{1}{5} - \dfrac{1}{7} + \ldots$

3. $F(x) = f(x) + f_1(x)$, where $f(x)$ and $f_1(x)$ are given above.

4. $f(x) = \dfrac{\pi^2}{3} + 4 \sum_{n=1}^{\infty} \dfrac{(-1)^n}{n^2} \cos nx$.

5. $f(x) = 2\left\{\left(\dfrac{\pi^2}{1} - \dfrac{6}{1^3}\right)\sin x - \left(\dfrac{\pi^2}{2} - \dfrac{6}{2^3}\right)\sin 2x + \left(\dfrac{\pi^2}{3} - \dfrac{6}{3^3}\right)\sin 3x - \cdots\right\}$.

6. $f(x) = \sin x + \frac{1}{3}\sin 3x + \frac{1}{5}\sin 5x + \cdots$ to ∞.

7. $f(x) = \frac{1}{2}\pi + 2\left(\sin x + \frac{1}{3}\sin 3x + \frac{1}{5}\sin 5x + \cdots \text{ to } \infty\right)$.

8. $f(x) = \frac{1}{2}\pi + 2\left(\frac{1}{2}\sin 2x + \frac{1}{4}\sin 4x + \frac{1}{6}\sin 6x + \cdots \text{ to } \infty\right)$.

9. $f(x) = \dfrac{3\pi}{4} + \sum_{n=1}^{\infty} \dfrac{(-1)^{n-1}\cos(2n-1)x - \sin(2n-1)x + \sin 2(2n-1)x}{2n-1}$

10. $f(x) = \dfrac{4}{\pi}\left(\cos x - \dfrac{\cos 3x}{3} + \dfrac{\cos 5x}{5} - \cdots \text{ to } \infty\right)$.

11. $a_0 = \frac{1}{2}$, $a_n = \begin{cases} 0, n = 2, 4, 6, \cdots \\ 1/n\pi, n = 1, 5, 9, \cdots \\ -1/n\pi, n = 3, 7, 11, \cdots \end{cases}$ $\quad b_n = \begin{cases} 1/n\pi, n = 1, 3, 5, \cdots \\ 2/n\pi, n = 2, 6, 10, \cdots \\ 0, n = 4, 8, 12 \cdots \end{cases}$

12. $a_0 = 2$; $a_n = 0$, when $n \neq 0$; $b_n = \begin{cases} 3/n\pi, n = 1, 2, 4, 5, 7, 8, \cdots \\ 0, n = 3, 6, 9, \cdots \end{cases}$

13. $a_0 = \dfrac{\pi}{2}$; $a_n = \dfrac{-2}{n^2\pi}$ for odd n, and $= 0$ for even n; $b_n = \pm\dfrac{1}{n}$, as n is odd or even.

14. $f(x) = \dfrac{2}{\pi}\sin x + \dfrac{1}{2}\sin 2x - \dfrac{2}{9\pi}\sin 3x - \frac{1}{4}\sin 4x + \dfrac{2}{25\pi}\sin 5x + \cdots$ to ∞.

15. $a_0 = \dfrac{\pi^2}{3}$; $a_n = \dfrac{(-1)^n 2}{n^2}$, for $n \neq 0$; $b_n = \dfrac{\pi}{n} - \dfrac{4}{\pi n^3}$ for odd n, and $= -\dfrac{\pi}{n}$ for even n.

16. $a_n = 0$; $b_n = \dfrac{(-1)^{n+1}\, 8n}{\pi\,(4n^2 - 1)}$.

17. $\dfrac{\cos x}{2} + \dfrac{4}{\pi}\sum_{n=1}^{\infty}\dfrac{n \sin 2nx}{4n^2 - 1}$

18. $i = \dfrac{2I}{\pi}\left\{1 - \dfrac{\cos 2\theta}{1\cdot 3} - \dfrac{\cos 4\theta}{3\cdot 5} - \dfrac{\cos 6\theta}{5\cdot 7} - \cdots\right\}$

19. (*i*) $f(x) = \dfrac{4}{\pi}\sum \dfrac{1}{n}\sin nx$ for n odd.

(*ii*) $f(x) = \dfrac{3}{2} - \dfrac{2}{\pi}\left[\sin x + \dfrac{1}{3}\sin 3x + \dfrac{1}{5}\sin 5x + \cdots\right]$

(*iii*) $f(x) = \dfrac{2}{\pi}\left[\sin x + \dfrac{1}{3}\sin 3x + \dfrac{1}{5}\sin 5x + \cdots\right]$

(*v*) $f(x) = \dfrac{1}{2} + \dfrac{2}{\pi}\left(\sin x + \dfrac{1}{3}\sin 3x + \dfrac{1}{5}\sin 3x + \cdots\right)$

(*vi*) $f(x) = \dfrac{\pi^2}{3} + 4\left(\dfrac{\cos x}{1^2} + \dfrac{\cos 2x}{2^2} + \dfrac{\cos 3x}{3^2} + \cdots\right)$

Exercises 1(d) Page 42

1. $\dfrac{2}{\pi}\left[\left(\dfrac{\pi^2}{1} - \dfrac{4}{1^2}\right)\sin x - \dfrac{\pi^2}{2}\sin 2x + \left(\dfrac{\pi^2}{3} - \dfrac{4}{3^2}\right)\sin 3x - \dfrac{\pi^2}{4}\sin 4x + \cdots\right]$

2. $2\left[\left(\dfrac{\pi^2}{1} - \dfrac{6}{1^3}\right)\sin x - \left(\dfrac{\pi^2}{3} - \dfrac{6}{2^3}\right)\sin 2x + \left(\dfrac{\pi^2}{3} - \dfrac{6}{3^3}\right)\sin 3x - \cdots\right]$

3. $-\dfrac{1}{\pi}\sin x - \dfrac{8}{\pi}\sum_{r=1}^{\infty}\dfrac{r}{4r^2 - 1}\sin 2r\pi.$

4. $\dfrac{\sin \pi\alpha}{\pi\alpha} + \sum_{r=1}^{\infty}(-1)^n\dfrac{2\alpha \sin \pi\alpha}{\pi\,(\alpha^2 - n^2)}\cos nx.$

5. $\dfrac{2\sin \pi\alpha}{\pi}\left\{\dfrac{\sin x}{1^2 - \alpha^2} - \dfrac{2\sin 2x}{2^2 - \alpha^2} + \dfrac{3\sin 3x}{3^2 - \alpha^2}\cdots\right\}$

Exercises 1(e) Page 51

1. (*i*) $\dfrac{2}{\pi}\left\{(\pi + 2)\sin x - \dfrac{\pi \sin 2x}{2} + \dfrac{(\pi + 2)\sin 3x}{3} - \dfrac{\pi \sin 4x}{4} + \cdots\right\}$;

$$\frac{\pi}{2}+1-\frac{4}{\pi}\left\{\cos x+\frac{\cos 3x}{3^2}+\frac{\cos 5x}{5^2}+\cdots\right\}$$

(ii) $$\frac{2}{\pi}\sum_{n=1}^{\infty}\frac{n\{1-(-1)^n e^{\pi}\}}{n^2+1}\sin nx;$$

$$\frac{e^{\pi}-1}{\pi}-\frac{2}{\pi}\sum_{n=1}^{\infty}\frac{1-(-1)^n e^{\pi}}{n^2+1}\cos nx$$

(iii) The same as exercises 1 (d), ex. 1;

$$\frac{\pi^2}{3}-4\left\{\cos x-\frac{\cos 2x}{2^2}+\frac{\cos 3x}{3^2}-\frac{\cos 4x}{4^2}+\cdots\right\}.$$

2. $$\frac{\pi}{4}-\frac{2}{\pi}\left\{\cos 2x+\frac{\cos 6x}{3^2}+\frac{\cos 10x}{5^2}+\cdots+\frac{\cos 2(2r+1)x}{(2r+1)^2}+\cdots\right\}.$$

3. The sine series for sin x is sin x. This answer is consistent with the theorem. If two Fourier series in sines alone (or cosines alone) converge to the same sum for all values of x, then the corresponding coefficients are equal.

Exercises 1(f) Page 60

1. (a) The same as in 1 (b), with $k = 1$.

(b) $$\frac{k}{2}+\frac{2k}{\pi}\left\{\sin\frac{\pi x}{2}+\frac{1}{3}\sin\frac{3\pi x}{2}+\frac{1}{5}\sin\frac{5\pi x}{2}+\cdots\right\}.$$

2. $$\frac{4}{\pi}\sum_{0}^{\infty}\frac{1}{2n+1}\sin\frac{(2n+1)\pi x}{2}.$$ **3.** $$\frac{1}{2}+\frac{2}{\pi}\left\{\cos\frac{\pi x}{2}-\frac{1}{3}\cos\frac{3\pi x}{2}+\frac{1}{5}\cos\frac{5\pi x}{2}-\cdots\right\}.$$

4. $$\frac{2}{\pi}\left\{\sin\pi x-\frac{\sin 2\pi x}{2}+\frac{\sin 3\pi x}{3}-\cdots\right\}.$$ **5.** $$\frac{1}{2}-\sum_{n=1}^{\infty}\frac{4\cos(2n-1)\pi x}{\pi^2(2n-1)^2}.$$

6. $$4\sum_{n=1}^{\infty}\frac{\cos 2n\pi x}{4n^2\pi^2}.$$ **7.** $$b_n=0;\ a_0=\frac{1}{2};\ a_n=\begin{cases}\dfrac{4}{n^2\pi^2}, & n=1,3,5,\cdots\\ 0, & n=4,8,12,\cdots\\ \dfrac{8}{n^2\pi^2}, & n=2,6,10,\cdots\end{cases}$$

8. $$\frac{1}{3}-\frac{4}{\pi^2}\left\{\cos\pi x-\frac{\cos 2\pi x}{4}+\frac{\cos 3\pi x}{6}-\frac{\cos 4\pi x}{8}+\cdots\right\}.$$

9. $$\frac{2}{\pi}-\frac{4}{\pi}\left\{\frac{\cos 2x}{1\cdot 3}+\frac{\cos 4x}{3\cdot 5}+\frac{\cos 6x}{5\cdot 6}-\frac{\cos 4\pi x}{7\cdot 9}+\cdots\right\}.$$

10. $1 + (2\pi \sin 1) \sum_{n=1}^{\infty} \frac{(-1)^{n+1}\, n \sin n\pi x}{n^2\pi^2 - 1}$

11. $a_0 = \frac{2}{3}$ and $a_3 = \frac{1}{3}\cdot$ For $n = 1, 2, 4, 5, 6, 7, \cdots$

$$a_n = \frac{18}{(n^2 - 9)n\pi} \sin \frac{n\pi}{3}\text{; } b_n = 0, \text{ for all } n.$$

12. $\frac{E}{\pi} + \frac{E}{2} \sin \omega t + \frac{2E}{\pi} \left\{ \frac{\cos 2\omega t}{1\cdot 3} + \frac{\cos 4\omega t}{3\cdot 5} + \frac{\cos 6\omega t}{5\cdot 7} - \cdots \right\}$

13. $a_0 = \frac{1}{2}(3 - e^{-2})$; $a_n = \frac{2k_n}{l_n}$ where $k_n = 1 - (-1)^n e^{-2}$, $l_n = 4 + n^2\pi^2$.

For even n, $b_n = \frac{n\pi k_n}{l_n}$ and for odd n, $b_n = \frac{n\pi k_n}{l_n} - \frac{2}{n\pi}\cdot$

15. $\frac{1}{2} - \sum_{n=1}^{\infty} \frac{(-1)^n}{2n-1} \cos \frac{(2n-1)\,x}{2}\cdot$

16. $\frac{1}{6} - \frac{4}{\pi^2} \left\{ \frac{\cos 2\pi t}{4} + \frac{\cos 4\pi t}{16} + \frac{\cos 6\pi t}{9} + \frac{\cos 8\pi t}{64} + \cdots \right\}$;

$\frac{8}{\pi^2} \left\{ \frac{\sin \pi t}{1} + \frac{\sin 3\pi t}{27} + \frac{\sin 5\pi t}{125} + \cdots \right\}$

17. $\frac{k}{2} - \frac{16k}{\pi^2} \left\{ \frac{1}{2^2} \cos \frac{2\pi t}{l} + \frac{1}{6^2} \cos \frac{6\pi t}{l} + \cdots \right\}$;

$\frac{8k}{\pi^2} \left\{ \sin \frac{\pi t}{l} - \frac{1}{3^2} \sin \frac{3\pi t}{l} + \frac{1}{5^2} \sin \frac{5\pi t}{l} - \cdots \right\}$

18. $a_0 = \frac{5}{6}$; $a_n = \begin{cases} \frac{4}{n^2\pi^2} - \frac{16}{n^3\pi^3}, & n = 1, 5, 9, \cdots \\ -\frac{16}{n^2\pi^2}, & n = 2, 6, 10, \cdots \\ \frac{4}{n^2\pi^2} + \frac{16}{n^3\pi^3}, & n = 3, 7, 11, \cdots \\ \frac{8}{n^2\pi^2}, & n = 4, 8, 12, \cdots \end{cases}$

19. $-\frac{32}{\pi^3} \left\{ \sin \frac{\pi x}{2} + \frac{\pi^2}{8} \sin \frac{2\pi x}{2} + \frac{1}{3^3} \sin \frac{3\pi x}{2} + \frac{\pi^2}{16} \sin \frac{4\pi x}{2} + \cdots \right\}$

20. $\frac{2}{\pi^3} \left\{ \left(\frac{\pi^2}{1} - \frac{4}{1^3} \right) \sin \pi x + \frac{\pi^2}{2} \sin 2\pi x + \left(\frac{\pi^2}{3} - \frac{4}{3^3} \right) \sin 3\pi x + \frac{\pi^2}{4} \sin 4\pi x + \cdots \right\}$

21. For even n, $b_n = 0$ and for odd n, $b_n = \frac{4n}{(n^2 - 4)\,\pi}\cdot$

22. $\frac{8}{\pi}\left\{\frac{\sin x}{1\cdot3}+\frac{2\sin 2x}{3\cdot5}+\frac{3\sin 3x}{5\cdot7}+\frac{4\sin 4x}{7\cdot9}+\cdots\right\}$.

23. $\frac{2}{\pi}-\frac{4}{\pi}\left\{\frac{1}{1\cdot3}\cos\frac{2\pi x}{l}+\frac{1}{3\cdot5}\cos\frac{4\pi x}{l}+\frac{1}{5\cdot7}\cos\frac{6\pi x}{l}\cdots\right\}$.

24. $\sum_{n=1}^{\infty}\frac{n\pi}{1+n^2\pi^2}(1-e\cos n\pi)\sin n\pi x$.

Exercises 1(h) Page 90

1. $y=0\cdot373+0\cdot359\cos x-0\cdot085\cos 2x-0\cdot117\cos 3x+\cdots$
$+0\cdot466\sin x-0\cdot289\sin 2x-0\cdot090\sin 3x+\cdots$

2. $1\cdot45-0\cdot367\cos x-0\cdot1\cos 2x+0\cdot033\cos 3x$
$+0\cdot173\sin x-0\cdot067\sin 2x+0(\sin 3x)+\cdots$

3. $5\cdot73\cos\theta+30\cdot26\sin\theta-5\cdot10\cos 3\theta+3\cdot17\sin 3\theta$

4. $1\cdot26+0\cdot04\cos x+0\cdot53\cos 2x-0\cdot1\cos 3x$
$-0\cdot63\sin x-0\cdot23\sin 2x+0\cdot085\sin 3x$.

5. $7850\sin\theta+1500\sin 2\theta;\ T=8332$ **6.** $1\cdot071$.

8. $y=2\cdot10-0\cdot283\cos x-0\cdot18\cos 2x+1\cdot6\sin x-0\cdot49\sin 2x$

10. $y=6\cdot44-1\cdot65\cos x-0\cdot34\cos 2x+0\cdot71\sin x-0\cdot69\sin 2x$

Exercises 1(i) Page 97

1. $e^{ax}=\frac{\sinh a\pi}{a\pi}+\frac{\sinh a\pi}{\pi^2}\sum_{1}^{\infty}\frac{(-)^n 2a\cos nx}{a^2+n^2}-\frac{2\sinh a\pi}{\pi^2}\sum_{1}^{\infty}\frac{(-1)^n n\sin nx}{a^2+n^2}$.

2. (*i*) $\frac{\sin\alpha\pi}{\pi}\left\{\frac{1}{\alpha}+\sum_{1}^{\infty}\frac{(-1)^n 2\alpha\cos nx}{\alpha^2-n^2}\right\}$

(*ii*) $\frac{2\sin\alpha\pi}{\pi}\sum_{1}^{\infty}\frac{(-1)^n n\sin nx}{\alpha^2-n^2}$.

3. $e^{-x}=\sum_{n=-\infty}^{\infty}(-1)^n\frac{(1-in\pi)\sinh 1}{1+n^2\pi^2}e^{in\pi x}$.

$$=\sinh 1-2\sinh 1\left\{\frac{\cos\pi x}{1+\pi^2}-\frac{\cos 2\pi x}{1+4\pi^2}+\frac{\cos 3\pi x}{1+9\pi^2}-\cdots\right\}$$
$$-2\pi\sinh 1\left\{\frac{\sin\pi x}{1+\pi^2}-\frac{2\sin 2\pi x}{1+4\pi^2}+\frac{3\sin 3\pi x}{1+9\pi^2}-\cdots\right\}$$

4. $C_0=\frac{1}{6};\ C_n=\frac{3}{4n^2\pi^2}\left\{\left(1+\frac{2}{3}in\pi\right)e^{-\frac{2}{3}in\pi}-1\right\}$.

8. (*i*) $\sin ax = \frac{\sin a\pi}{\pi} \sum_{-\infty}^{\infty} (-1)^n \frac{n}{a^2 - n^2} e^{inx}$

(*ii*) $e^x = \frac{\sinh \pi}{\pi} \sum_{-\infty}^{\infty} (-1)^n \frac{1 + in}{1 + n^2} e^{inx}$

(Partial Differential Equations)

Exercises 2(a) Page 107

1. (*i*) $px + qy = 3z$, (*ii*) $z = px + qy + pq$, (*iii*) $p = q$,

(*iv*) $z^2 (p^2 + q^2 + 1) = 1$, (*v*) $z (px + qy) = z^2 - 1$,

(*iv*) $px + qy = q^2$, (*vii*) $z = pq$.

2. (*i*) $px + qy = nz$ (*ii*) $z^2 = x^2 + y^2 + 2z (px + qy)$

(*iii*) $p + x = qy$ (*iv*) $p + q = px + qy$

(*v*) $p^2 + q^2 = 16 p^2 q^2 (x + y)$ (*vi*) $p^2 + q^2 = \tan^2 \alpha$

(*vii*) $z = px + qy + \frac{b}{q} - q$

3. (*i*) $py = qx$, (*ii*) $py = qx$, (*iii*) $px (y - z) + qy (z - x) = z (x - y)$,

(*iv*) $xy \frac{\partial^2 z}{\partial x \partial y} = px + qy - z$, (*v*) $\frac{\partial^2 z}{\partial x^2} + \frac{\partial^2 z}{\partial y \partial x} - 6 \frac{\partial^2 z}{\partial y^2} = 0$,

(*vi*) $py - qx = y^2 - x^2$, (*vii*) $px + qy = 0$, (*viii*) $px = qy$,

(*ix*) $px - qy = x - y$, (*x*) $a^2 \frac{\partial^2 z}{\partial y^2} - 2ab \frac{\partial^2 z}{\partial x \partial y} + b^2 \frac{\partial^2 z}{\partial x^2} = 0$,

(*xi*) $(x + iy) \left(\frac{\partial^2 z}{\partial x^2} + \frac{\partial^2 z}{\partial y^2} \right) = 2 (p + iq)$,

(*xii*) $p (x + y) (x + 2z) - q (x + y) (y + 2z) = z (x - y)$,

(*xiii*) $q - p = z$, (*xiv*) $qz - pz = x - y$,

(*xv*) $(z - y) p + (x - z) q = y - x$, (*xvi*) $z = px + qy - xys - y^2 t$,

(*xvii*) $\frac{\partial^4 z}{\partial x^4} - 2 \frac{\partial^4 z}{\partial x^2 \partial y^2} + \frac{\partial^4 z}{\partial y^4} = 0$. (*xviii*) $px^2 + qy = 2y^2$

(*xix*) $(x - 2z) p + (2z - y) q = y - x$ (*xx*) $3r - 8s + 4t = 0$

4. $py - qx = 0$ **5.** $z^2 (p^2 + q^2 + 1) = k^2$

9. $z = px + qy + k (1 + p^2 + q^2)$

Exercises 2(b) Page 114

1. $z = f(y)$. **2.** $z = f(x) + y\phi (x)$. **3.** $z = -\cos x + x\phi (y) + f(y)$

4. $z = 6xy - \frac{1}{2}x^2 \log y + F(y) + \varphi(x)$.

5. $z = -\frac{1}{a^2 b}\sin(ax + by) + xF(y) + f(y) + \varphi(x)$.

6. $z = \frac{1}{4}\cos(2x - y) - x^3y^3 + xF(y) + f(x) + \varphi(y)$.

7. $z = -x - \frac{y}{3} + x^3 f(y)$.

8. $z = (1 + \cos x)\cos y$.

9. $z = e^{2y}\varphi(x) + e^{3y} f(x) + 2y + \frac{5}{3}$.

10. $z = \frac{x^3 y}{3} + \frac{y^3 x}{3} + F(y) + f(x)$.

11. $z = e^y \cosh x + e^{-y}\sinh x$.

12. $z = \frac{3x^2}{2} - xy + \sin y + k$.

13. $u = \varphi(x + iy) + f(x - iy)$.

14. $z = f(2x + y) + \phi(3x + y)$.

15. $z = axy + bx + cy + d$.

Exercises 2(c) Page 132

1. $z = ax + \frac{a}{a-1}y + b$.

2. $z = ax + \frac{1}{a}y + b$.

3. $z = ax + y\sqrt{(m^2 - a^2)} + b$.

4. $z = b^2x + by + c$.

5. $z = ax + \frac{1-2a}{3}y + c$,

6. $z = ax + \frac{ay}{2}[n \pm \sqrt{(n^2 - 4)}] + b$.

7. $z = (2 + 3b - b^2)x + by + c$.

8. $z = 3(x \tan\alpha + y \sec\alpha) + b$.

9. $z = \alpha x - y \sin\alpha + \beta$.

10. $z = ax + \frac{5 - a^3}{3 - 2a}y + b$.

11. $z = ax + by + a^2 - b^2$.

12. $z = ax + by + 3ab$.

13. $z = ax + by - 4a^2b^2$.

14. $(a + b)(z - ax - by) = 1$.

15. $z = ax + by + \sqrt{(a^2 + b^2)}$.

16. $a(\log z)^2 = (x + ay + b)^2$.

17. $4(1 + a^2)z = (x + ay + b)^2$.

18. $az\sqrt{(1 + a^2z)} - \log az + \sqrt{(1 + a^2z^2)} = 2a(ax + y + b)$.

19. $4(az - a^2 - 1) = (x + ay + b)^2$.

20. $3z^3 = (a^2 - 9)x + 3ay + 3b$.

21. $(a^2 + 1)(1 - z^2) = (x + ay + b)^2$.

22. $z^2 = (x + ay + b)^2 + a^2$.

23. $az - 1 = be^{x + ay}$.

24. $z = \frac{2}{3}(x + a)^{3/2} - \frac{2}{3}(a - y)^{3/2} + b$.

25. $z = a^2(x + y) + \frac{x^2}{2} - \frac{4a}{3}x^{3/2} + b$.

26. $2z = ax^2 + \frac{y^2}{a} + b$.

27. $z = \frac{2}{3}(x + a)^{3/2} + \frac{2}{3}(y - a)^{3/2} + b$.

28. $2z = (x + a)^2 + (y - a)^2 + b$.

29. $z = a(x - y) - (\cos x + \cos + y) + b$.

30. $(2z + ay^2 - 2b)^2 = 4a(1 - x^2)$.

31. $z = -\cos x + ax + \frac{\sin y}{a} + b.$ **32.** $z = a \log x - a \log \sec y - y + b.$

33. $z^{3/2} = (x + a)^{3/2} + (y + a)^{3/2} + b.$ **34.** $z = a\sqrt{(x + y)} + \sqrt{(1 - a^2)}\sqrt{(x - y)} + b.$

35. $z = \frac{1}{2} a \log (x^2 + y^2) + \sqrt{(1 - a^2)} \tan^{-1} \frac{y}{x} + b.$

36. $(4 + a^2)(\log z)^2 = (x^2 + a \log y + b)^2.$

37. $z = \frac{a^2}{2} \sin h^{-1} \frac{x}{a} + \frac{x}{2}\sqrt{(a^2 + x^2)} + \frac{y}{2}\sqrt{(y^2 - a^2)} - \frac{a^2}{2} \cosh^{-1} \frac{x}{a} + b.$

38. $z = 2ax\, e^y + 2a^2 e^{2y} + b.$ **39.** $\sqrt{(1 + a)}\, z = \sqrt{(ax)} + \sqrt{y} + b.$

40. $z^2\sqrt{1 + a^2} = \pm 2\,(ay + \log x) + b.$

41. $z^{3/2} + (x + a)^{3/2} + (y + a)^{3/2} + b\,\phi\,(xy + yz + zx)$

Exercises 2(d) Page 145

1. $F\left(\frac{x - y}{z}, \frac{xy}{z}\right) = 0.$ **2.** $F(xy, x - \log z) = 0$

3. $\log (x^2 + y^2 + z^2 + 2xy) - 2x = F(x + y)$ **4.** $F\left(xy - z^2, \frac{x}{y}\right) = 0.$

5. $f\left(\frac{1}{x} - \frac{1}{y}, \frac{1}{y} - \frac{1}{z}\right) = 0$ **6.** $F(x - z, y - z) = 0$

7. $F\left(\frac{a - x}{b - y}, \frac{b - y}{c - z}\right) = 0.$ **8.** $\sqrt{x} - \sqrt{y} = f(\sqrt{y} - \sqrt{z})$

9. $(x - y) = f(x - z)$ **10.** $z = x^n \phi\left(\frac{y}{x}\right).$

11. $F(x^3 - y^3, x^2 - z^2) = 0$ **12.** $xyz = F(x^2 + y^2 - 2z).$

13. $\frac{1}{x} + \frac{1}{y} = f\left(\frac{1}{y} + \frac{1}{z}\right).$ **14.** $x^2 + y^2 = f(yz - y^2)$

15. $f\left(\frac{\cos y}{\cos x}, \frac{\cos z}{\cos y}\right) = 0.$ **16.** $\sqrt{y} - \sqrt{z} = f(\sqrt{x} - \sqrt{y}).$

17. $z(y - x) = \phi\left(\frac{x + y + 2}{z}\right)$ **18.** $x^2 - y^2 - z^2 = F(y^2 - x^2 + 2xy).$

23. $x^2 - z^2 = \phi\left(\frac{x + y + z}{y}\right).$ **24.** $x^2 + y^2 + z^2 = f(xyz)$

25. $\log (x^2 + y^2 + z^2 + 2xy) - zx = f(x + y)$

Exercises 2(e) Page 164

1. $z = Q_1(y - 2x) + Q_2(y - \frac{1}{2}x).$ **2.** $z = f_1(y - 2x) + f_2(y + x).$

3. $z = Q_1(y - 3x) + x\,Q_2(y - 3x)$ 4. $z = \varphi_1(y + x) + x\,\varphi_2(y + x) + x^2\varphi_3(y + x)$

5. $z = f_1(y + 2x) + xf_2(y + 2x) + f_3(y - 3x).$ **6.** $y = f(x - a\,t) + \varphi(x + a\,t).$

7. $z = \varphi_1(y - x) + \varphi_2(x + 2y) + \frac{x^2}{4}$.

8. $z = \varphi_1(y + x) + x\,\varphi_2(y + x) + \varphi_3(y + 2x) - \frac{x^2}{2}e^{y + x} - \frac{1}{36}e^{y - 2x} + x\,e^{y + 2x}$.

9. $z = \varphi_1(2y - 3x) + x\,\varphi_2(2y - 3x) + \frac{x^2}{8}e^{3x - 2y}$.

10. $z = f_1(y + x) + f_2(y - x) - \frac{1}{3}e^{x + 2y}$.

11. $z = \varphi_1(y + x) + x\varphi_2(y + x) + e^{x + 2y}$.

12. $z = f_1(y) + f_2(y + 2x) + \frac{1}{4}e^{2x} + \frac{x^4 y}{12} + \frac{x^5}{30}$

13. $z = f_1(y) + xf_2(y) + f_3(y + 2x) + \frac{1}{4}e^{2x} + \frac{1}{20}x^5 y + \frac{1}{60}x^6$.

14. $z = \varphi_1(y + x) + \varphi_2(y - 5x) + \frac{x^3}{6} - \frac{y^4}{60}$.

15. $z = \varphi_1(y - x) + x\phi_2(y - x) + \frac{x^4 y}{12} - \frac{x^5}{30}$.

16. $z = f_1(y - 2x) + f_2(y + 3x) + \frac{x^4}{24} + \frac{x^3 y}{6}$.

17. $z = \phi_1(y - 2x) + \phi_2(y - x) + \frac{x^2 y}{2} - \frac{x^3}{3}$.

18. $z = f(y - x) + \phi(y - 2x) + F(y + 3x) + \frac{x}{4}\cos(x - y) + \frac{5}{72}x^6$

$$+ \frac{1}{60}x^5(1 + 21y) + \frac{1}{24}x^4 y^2 + \frac{x^3 y^3}{6}.$$

19. $z = f_1(y + x) + f_2(y - x) + xf_3(y - x) + \frac{1}{9}e^{2x + y} - \frac{1}{4}x\cos(x + y)$.

20. $z = \phi_1(y) + \phi_2(y + x) + \frac{1}{2}\cos(x + 2y) - \frac{1}{6}\cos(x - 2y)$

21. $z = \phi_1(y + 2x) + x\phi_2(y + 2x) + \frac{x^2}{2}e^{2x + y}$.

22. $z = \phi_1(y + 3x) + x\phi_2(y + 3x) + x^2(3x + y)$

23. $z = \phi_1(y + 2x) + \phi_2(y - 3x) + \frac{x}{5}\sin(2x + y)$.

24. $z = \phi_1(y + x) + x\phi_2(y + x) - \sin x$.

25. $z = \phi_1(y) + \phi_2(x + y) + \frac{1}{3}(\sin x\cos 2y + 2\cos x\sin 2y)$.

26. $z = \phi(y - 2x) + f(y + 2x) + F(y + 3x) + \frac{1}{4}\sin(2x + y)$.

27. $z = \phi_1(y + 2x) + \phi_2(y - x) + ye^x$.

28. $z = \phi_1(y-5x) + \phi_2\{y-x) - \frac{1}{10}xy^2 - \frac{2}{75}y^2 - \frac{1}{60}y^4$

29. $z = \phi_1(y+2x) + \phi_2(y-2x) - \frac{x}{4}\cos(2x+y) + \frac{1}{6}\sin(2x+y)$

30. $z = \phi_1(y+x) + x\phi_2(y-x) + \phi_3(y+x) - \frac{3x}{4}\sin x + y.$

31. $u = \frac{1}{2}[\log(x^2+y^2)]^2 + 2\left(\tan^{-1}\frac{y}{x}\right)^2 + f(x+iy) + g(x-iy)$

32. $z = f(y-x) + x\phi(y-x) + F(y+x) - \frac{3x}{4}\sin(x+y).$

33. $z = \phi_1(y-x) + \phi_2(y+3x) + \phi_3(y-2x)$

$$+\frac{x}{4}\cos(y-x) + \frac{x^5}{60} + \frac{x^4y^2}{24} + \frac{x^3y^3}{6} + \frac{7x^5y}{20} + \frac{5x^6}{72}.$$

34. $z = \phi_1(x+y) + x\phi_2(x+y) + \frac{x^2}{2}e^{x+y}[1+\cos(x+y)]$

Exercises 2(f) Page 170

1. $z = e^x\phi(y-x) + e^{-x}f(y) - \frac{x}{2}e^{-x}.$

2. $z = f_1(y-x) + e^{2x}f_2(y-x) - \frac{x}{2}e^{x-y}.$

$$-\frac{1}{2}\left(\frac{x^3y}{3} + \frac{x^2y}{2} + \frac{x}{2} + \frac{x^2}{2} - \frac{x^4}{12}\right).$$

3. $z = f_1(y) + e^x f_2(y-x) + e^{2x}f_3(y-3x)$

$$+\frac{1}{72}(e^{2x+3y}) + \frac{1}{2}\left(\frac{x^2y}{2} + \frac{3xy}{2} + 7x + \frac{5x^2}{4}\right).$$

4. $z = f_1(x) + e^{3y}f_2(x-2y) + \frac{3}{50}[4\cos(3x-2y) + 3\sin(3x-2y)].$

5. $z = \phi_1(x-2y) + e^{3y}\phi_2(y+x).$

6. $z = \phi_1(y-x) + e^{-2x}\phi_2(y+2x) - \frac{1}{10}e^{2x+3y}$

$$-\frac{1}{6}\cos(2x+y) + \frac{x}{24}(6xy - 2x^2 + 9x - 6y - 12).$$

7. $z = e^x\phi_1(y) + e^{-x}\phi_2(y+x) + \frac{1}{2}\sin(x+2y) - xe^y.$

8. $z = \phi_1(xy) + \phi_2\left(\frac{x}{y}\right).$

9. $z = f(y) + e^{-x}\phi(x+y) + \frac{1}{3}x^3 + xy^2 - x^2 + 6x$

10. $z = e^x f(y) + e^{-x}\phi(x+y) - xe^y + \frac{1}{2}\sin(x+2y)$

11. $z = e^x\phi_1(x+y) + e^{2x}\phi_2(x+y) + \frac{1}{20}e^{y-2x}.$

12. $z = e^x\phi(y) + e^{-x}f(y+x) + \frac{1}{2}\sin(x+2y) + \frac{1}{2}e^{y-x}.$

13. $Z = e^{x}\,\phi_1(y) + e^{-x}\,\phi_2(x+y) + \frac{1}{2}\sin(x+2y)$

(BOUNDARY VALUE PROBLEMS)

Exercises 3(a) Page 175

1. $z = \left\{ Ae^{[1+\sqrt{(1+k)}]x} + Be^{[1-\sqrt{(1+k)}]x} \right\} e^{-ky}$.

2. $z = \frac{1}{\sqrt{2}}\sinh x\sqrt{2} + e^{-3y}\sin x.$ **3.** $z = ke^{4cx^3}\cdot e^{-3cy^4}$.

4. $z = (Ae^{x\sqrt{k}} + Be^{-x\sqrt{k}})\,e^{2ky}$ if $k > 0$.

$z = (A\sin x\sqrt{-k} + B\cos x\sqrt{-k})e^{2ky}$ if $k < 0$

$z = Ax + B$ if $k = 0$.

6. $z = 6e^{-3x-2y}$ **7.** $u = 4e^{-3x-2t}$ **8.** $u = 4e^{-x+\frac{3}{2}y}$

Exercises 3(b) Page 200

1. (a) $y = y_0\sin\frac{2\pi x}{l}\cos\frac{2\pi at}{l}$.

(b) $y = k\,\text{Sin}\frac{\pi x}{l}\,\text{Cos}\frac{\pi at}{l} - k\,\text{Sin}\frac{2\pi x}{l}\,\text{Cos}\frac{2\pi at}{l}$

2. $y(x,t) = \frac{v_0 l}{\pi a}\sin\frac{\pi x}{l}\sin\frac{\pi at}{l}$.

4. $y(x,t) = \frac{24l^2}{\pi^3}\sum_{n=1}^{\infty}\frac{1}{(2n-1)^3}\sin\frac{(2n-1)\pi x}{l}\cos\frac{(2n-1)\pi at}{l}$.

6. $y(x,t) = \frac{8l^2}{\pi^3}\sum_{n=1}^{\infty}\frac{1}{(2n-1)^3}\sin\frac{(2n-1)\pi x}{l}\cos\frac{(2n-1)\pi at}{l}$

where $l = 60$ and $a^2m = 2000$ and m = mass per unit length of the string.

9. $y(x,t) = \sum_{n=1}^{\infty}\frac{(-1)^{n+1}8p}{(2n-1)^2\pi^2}\sin\frac{(2n-1)\pi x}{a}\cos\frac{(2n-l)\pi ct}{a}$.

10. $y(x,t) = \frac{9d\sqrt{3}}{2\pi^2}\left[\sin\frac{\pi x}{l}\cos\frac{\pi at}{l} - \frac{1}{2^2}\sin\frac{2\pi x}{l}\cos\frac{2\pi at}{l} + \frac{1}{4^2}\sin\frac{4\pi x}{l}\cos\frac{4\pi at}{l} - \cdots\right]$.

Exercises 3(c) Page 225

1. $\theta(x,t) = \frac{4}{\pi}\left[e^{-a^2t}\sin x - \frac{1}{3^2}e^{-3^2a^2t}\sin 3x + \frac{1}{5^2}e^{-5^2a^2t}\sin 5x - \cdots\right]$

2. $u(x,t) = \frac{4l}{\pi^2}\left[\frac{1}{1^2}\sin\frac{\pi x}{l}e^{-\frac{a^2\pi^2 t}{l^2}} - \frac{1}{3^2}\sin\frac{3\pi x}{l}e^{-\frac{3^2\pi^2 a^2 t}{l^2}} + \ldots\right]$.

3. $T(x,t) = \frac{a_0}{2} + \sum_{n=1}^{\infty} a_n \cos\frac{n\pi x}{l}\exp(-\alpha^2 n^2\pi^2 t/l^2)$.

4. $\theta(x,t) = \frac{2l}{\pi}\left[\sum_{n=1}^{\infty}\frac{(-1)^{n+1}}{n}\sin\frac{n\pi x}{l}e^{-\frac{a^2 n^2\pi^2 t}{l^2}}\right]$.

5. $u(x,t) = \frac{8}{\pi}\sum_{n=1}^{\infty}\frac{1}{(2n-1)^3}\,u(x,t) = \frac{8}{\pi}\sum_{n=1}\frac{1}{(2n-1)^3}\sin(2n-1)\,xe^{-\alpha^2(2n-1)^2 t}$.

7. $u(x,t) = \frac{l}{2} - \frac{4l}{\pi^2}\sum_{n=0}^{\infty}\frac{1}{(2n+1)^2}\cos\frac{(2n+1)\pi x}{l}\cdot e^{-\frac{a^2(2n+1)^2\pi^2 t}{l^2}}$

8. $u(x,t) = \frac{l^2}{6} - \frac{l^2}{\pi^2}\sum_{n=1}^{\infty}\frac{1}{n^2}\cos\frac{2n\pi x}{l}\operatorname{Exp}[-4\alpha^2 n^2\pi^2 t/l^2]$.

9. $u(x,t) = \frac{2}{\pi} - \frac{4}{\pi}\sum_{n=1}^{\infty}\frac{1}{(4n^2-1)}\cos 2nx\cdot e^{-4n^2 a^2 t}$.

12. $u(x,t) = 50 - \frac{400}{\pi^2}\sum_{n=1}^{\infty}\frac{1}{(2n-1)^2}\cos\frac{(2n-1)\pi x}{100}e^{-\frac{a^2\pi^2(2n-1)^2 t}{100^2}}$

13. (*a*) $u(x,t) = \frac{200}{\pi}\sum_{n=1}^{\infty}(-1)^{n+1}\sin\frac{n\pi x}{l}\operatorname{Exp}[-n^2\pi^2\alpha^2 t/l^2]$.

(*b*) $u(x,t) = 25 + \frac{50x}{l} - \frac{50}{\pi}\sum_{n=1}^{\infty}\frac{1}{n}\sin\frac{2n\pi x}{l}\operatorname{Exp}[-4n^2\pi^2\alpha^2 t/l^2]$.

14. $u(x,t) = \frac{3}{4}e^{-kt}\sin x - \frac{1}{4}e^{-9kt}\sin 3x$.

15. $u(x,t) = 90 - 30x - \frac{80}{\pi}\sum_{n=1}^{\infty}\frac{1}{n}\sin\frac{n\pi x}{5}e^{-\frac{n^2\pi^2 a^2 t}{25}}$

16. $u(x,t) = 20x + 20 + \frac{20}{\pi}\sum_{1}^{\infty}\frac{1}{n}(1 - 6\cos n\pi)\sin\frac{n\pi x}{10}e^{-\frac{\alpha^2 n^2\pi^2 t}{100}}$.

17. $u(x,t) = 50 - 4x - \frac{60}{\pi}\sum_{n=1}^{\infty}\frac{1}{n}\sin\frac{n\pi x}{5}e^{-\frac{\alpha^2 n^2\pi^2 t}{25}}$

18. $u(x, t) = -\frac{5x}{4} - 50 + \frac{100}{\pi} \sum_{1}^{\infty} \frac{1}{n} (2 \cos n\pi - 1) \sin \frac{n\pi x}{40} e^{-\frac{n^2\pi^2\alpha^2 t}{1600}}$

19. $u(x, t) = 50 + \frac{100x}{l} - \frac{200}{\pi} \sum_{n=1}^{\infty} \frac{1}{(2n-1)} \sin \frac{(2n-1)\pi x}{l} e^{-\frac{\alpha^2 (2n-1)^2 \pi^2 t}{l^2}}$

20. $u(x, t) = 30 + x + \frac{20}{\pi} \sum_{n=1}^{\infty} (3 + 2 \cos n\pi) \cdot \frac{1}{n} \sin \frac{n\pi x}{50} \cdot \exp\left(-\frac{0{\cdot}15 n^2 \pi^2 t}{2500}\right).$

21. $u(x, t) = \frac{2x}{3} + 40 - \frac{40}{\pi} \sum_{n=1}^{\infty} \frac{1}{n} \sin \frac{n\pi x}{5} e^{-\frac{\alpha^2 n^2 \pi^2 t}{225}}.$

22. $u(x, t) = 50 - \frac{400}{\pi^2} \sum_{1}^{\infty} \frac{1}{(2n-1)^2} \cos \frac{(2n-1)\pi x}{50} e^{-\frac{\alpha^2 \pi^2 (2n-1)^2 t}{2500}}.$

24. $\theta(x, t) = \frac{8T}{\pi^2} \sum_{n=1}^{\infty} \frac{(-1)^n}{(2n+1)^2} \sin \frac{(2n+1)\pi x}{l} e^{-\frac{(2n+1)^2 \pi^2 kt}{l^2}}.$

25. $u(x, t) = \frac{100}{\pi} \sum_{n=1}^{\infty} \frac{1}{n} \sin \frac{2n\pi x}{l} e^{-\frac{\alpha^2 4 n^2 \pi^2 t}{l^2}}.$

26. $u(x, t) = 50 + \sum_{1}^{\infty} \left[\frac{(-1)^{n+1}\, 800}{(2n-1)^2 \pi^2} - \frac{200}{(2n-1)\pi} \right) \sin \frac{(2n-1)\pi x}{2l} e^{-\frac{(2n-1)^2 \alpha^2 \pi^2 t}{4l^2}}$

Exercises 3(e) Page 249

5. $u(x, t) = \frac{400}{\pi} \sum_{n=1}^{\infty} \frac{\sin \frac{(2n-1)\pi x}{100} \sinh \frac{(2n-1)\pi y}{100}}{(2n-1) \sinh (2n-1)\pi}$

$u(50, 50) = \frac{400}{\pi} \sum_{n=1}^{\infty} \frac{1}{(2n-1) \sinh (2n-1)\pi} \sin \frac{(2n-1)\pi}{2} \sinh \frac{(2n-1)\pi}{2}.$

6. $T(x, y) = \frac{3200}{\pi^3} \sum_{n=1}^{\infty} \frac{1}{(2n-1)^3} \sin \frac{(2n-1)\pi x}{10} \cdot e^{-\frac{(2n-1)\pi y}{10}}$

7. $T(x, y) = \frac{160}{\pi} \sum_{n=1}^{\infty} \frac{1}{(2n-1)} \sin \frac{(2n-1)\pi x}{30} e^{-\frac{(2n-1)\pi y}{30}}$

9. $$u(x, y) = \frac{4u_0}{\pi} \sum_{n=1}^{\infty} \frac{\sin (2n-1) \pi x/100 \cdot \sinh (2n-1)\pi y/100}{(2n-1) \sinh (2n-1) \pi}.$$

$$u(50, 50) = \frac{2u_0}{\pi}\left[\operatorname{sech} \frac{\pi}{2} - \frac{1}{3}\operatorname{sech} \frac{3\pi}{2} + \frac{1}{5} \operatorname{sech} \frac{5\pi}{2} - \cdots \right].$$

10. $$u(x, y) = \frac{3200}{\pi^3} \sum_{n=1}^{\infty} \frac{\sin \frac{(2n-1)\pi x}{20} \sinh \frac{(2n-1)\pi y}{20}}{(2n-1)^3 \sinh (2n-1)\pi}.$$

11. $$u(x, y) = \sum_{n=1}^{\infty} \left(C_n \cosh \frac{n\pi y}{a} + D_n \sinh \frac{n\pi y}{a} \right) \sin \frac{n\pi x}{a}$$

where $C_n = \frac{2}{a} \int_0^a f(x) \sin \frac{n\pi x}{a}\, dx$ and $D_n = -C_n \tanh \frac{n\pi b}{a}$.

12. $$u(x, y) = \frac{200}{\pi} \sum_{n=1}^{\infty} \frac{1}{n} (1 - \cos n\pi) \sin \frac{n\pi x}{a} \left(\cos \frac{n\pi y}{a} - \sinh \frac{n\pi y}{a} \tanh \frac{n\pi b}{a} \right).$$

13. $$u(x, y) = 3 \sin \frac{3\pi x}{a} \sinh \frac{3\pi (b-y)}{a} \operatorname{cosech} \frac{3\pi b}{a} + 5 \sin \frac{4\pi x}{a} \sinh \frac{4\pi (b-y)}{a} \operatorname{cosech} \frac{4\pi b}{a}.$$

16. $$u(x, y) = \frac{8a^2}{\pi^3} \sum_{n=1}^{\infty} \frac{\sin \frac{(2n-1)\pi x}{a} \sinh \frac{(2n-1)\pi (b-y)}{a}}{(2n-1)^3 \sinh \frac{(2n-1)\pi b}{a}}.$$

18. $$u(x, y) = \frac{30}{\pi} \sum_{1}^{\infty} \frac{(-1)^{n+1}}{n} \sin \frac{n\pi x}{5} e^{-\frac{n\pi y}{5}}$$

21. $$u(x, y) = \frac{\pi}{3}(\pi - y) + 4 \sum_{n=1}^{\infty} \frac{(-1)^n \sinh n(\pi - y) \cos nx}{n^2 \sinh n\pi}.$$

Exercises 3(f) Page 266

2. $$u(r, \theta) = \frac{800}{\pi^2}\left[\left(\frac{r}{10}\right) \sin \theta - \frac{1}{3^2}\left(\frac{r}{10}\right)^3 \sin 3\theta + \frac{1}{5^2}\left(\frac{r}{10}\right)^5 \sin 5\theta - \cdots \right].$$

3. $$u(r, \theta) = \frac{200}{\pi} \sum_{n=1}^{\infty} \frac{(-1)^{n+1}}{(2n-1)^2} \left(\frac{r}{10}\right)^{2n-1} \sin (2n-1)\theta.$$

4. $u(r, \theta) = \frac{8k}{\pi} \sum_{n=1}^{\infty} \frac{1}{(2n-1)^3} \left(\frac{r}{a}\right)^{2n-1} \sin(2n-1)\theta$.

5. $u(r, \theta) = \frac{400}{\pi^2} \sum_{n=1}^{\infty} \left[\frac{1}{n^2} \sin \frac{n\pi}{2} \left(\frac{r}{a}\right)^n \sin n\theta \right]$.

6. $u(r, \theta) = 50 + \frac{200}{\pi} \sum_{n=1}^{\infty} \frac{1}{(2n-1)} \left(\frac{r}{a}\right)^{2n-1} \sin(2n-1)\theta$.

7. $u(r, \theta) = \frac{2}{\pi} \sum_{n=1, 3, 5 \cdots} \frac{1}{n^3} \left(\frac{a}{r}\right)^{2n} \left(\frac{r^{4n} - b^{4n}}{a^{4n} - b^{4n}}\right) \sin 2n\theta$.

8. $u(r, \theta) = \frac{800}{9\pi} \sum \frac{1}{(2n-1)^3} \left(\frac{r}{a}\right)^{\frac{2n-1}{3}} \cdot \sin\left(\frac{2n-1}{3}\right)\theta$.

9. $V = \sum_n \left(\frac{r}{a}\right)^n C_n \cos n\theta$.

10. $u(r, \theta) = \frac{400}{\pi} \sum_{n=1, 3, 5 \cdots}^{\infty} \frac{1}{n^3} \left(\frac{r}{10}\right)^{2n} \sin 2n\theta$; 45·7°.

Exercises 3(g) Page 272

1. $T = 2\left(r - \frac{100}{r}\right) \sin\theta + \frac{1}{2}\left(\frac{400}{r} - r\right) \cos\theta$.

2. $u(r, \theta) = 4\left(r - \frac{1}{r}\right) \sin\theta + 4\left(r + \frac{1}{r}\right) \cos\theta$.

CHAPTER V

THE Z-TRANSFORMS

The *Z*-transform plays an important rôle in the communication engineering and control engineering. It is used in the analysis and representation of discrete-time linear shift invariance system. The application of *Z*-transform in discrete-time systems is similar to that of the Laplace Transform in continuous-time systems. When continuous signals are sampled, discrete-time functions arise. We shall see something about the sampler and holding devices.

Sampler and holding devices : The important element of a discrete-time system is the sampler. In a sampler, a switch closes to admit an input signal in every T seconds. A sampler is a conversion device which converts a continuous signal into a sequence of pulses occuring at the sampling instants $0, T, 2T, \ldots$; here T is called the sampling period. If two signals give equal values at the sampling instants, then they give the same sampled signal.

A holding device is one which converts the sampled signal into a continuous signal, which is very close to the signal applied to the sampler. The *simplest holding device* is one which converts the sampled signal into one which is constant between two consecutive sampling instants.

Definition of the *Z*-Transform

Let $\{x(n)\}$ be a sequence defined for $n = 0, \pm 1, \pm 2, \pm 3, \ldots$. Then the *two sided Z-transform* of the sequence $x(n)$ is defined as

$$Z\{x(n)\} = X(z) = \sum_{n=-\infty}^{\infty} x(n) \cdot z^{-n} \qquad \ldots(1)$$

where z is a complex variable in general.

If $\{x(n)\}$ is a causal sequence, *i.e.*, $x(n) = 0$ for $n < 0$, then the *Z*-transform reduces to *one-sided Z-transform* and its definition is

$$Z\{x(n)\} = X(z) = \sum_{n=0}^{\infty} x(n) \cdot z^{-n} \qquad \ldots(2)$$

The infinite series on the right hand side of (1) or (2) will be convergent only for certain values of z depending on the sequence $x(n)$. The *inverse* z-transform of $Z\{x(n)\} = X(z)$ is defined as

$$Z^{-1}[X(z)] = \{x(n)\}$$

Symbols : Sometimes we write $Z\{x(n)\} = Z[\{x(n)\}]$ and $\{x(n)\}$ as $x(n)$ (which is a sequence and not a single data).

The double braces { } are used for a sequence.

Unit Sample Sequence and Unit Step Sequence

The *unit sample sequence* $\delta(n)$ is defined as the sequence with values

$$\delta(n) = \begin{cases} 1 & \text{for } n = 0 \\ 0 & \text{for } n \neq 0 \end{cases} \qquad ...(3)$$

The *unit step sequence* $u(n)$ has values

$$u(n) = \begin{cases} 1 & \text{for } n \geq 0 \\ 0 & \text{for } n < 0 \end{cases} \qquad ...(4)$$

The above two sequences are related as

$$u(n) = \sum_{m=-\infty}^{n} \delta(m) \qquad ...(5)$$

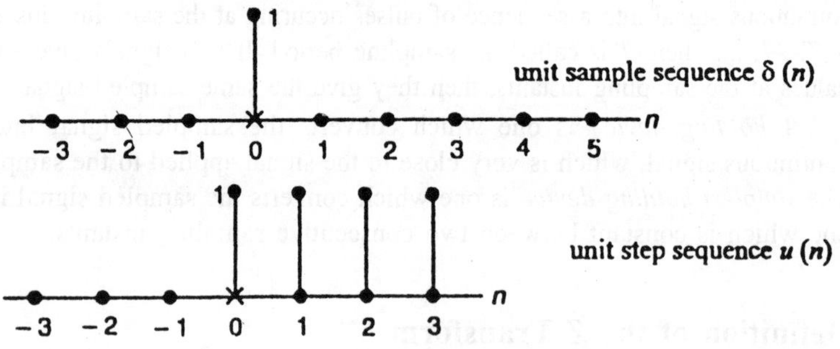

and $$\delta(n) = u(n) - u(n-1) \qquad ...(6)$$

we have $$\delta(n-k) = 1 \text{ if } k = n$$
$$= 0 \text{ if } k \neq n \qquad ...(7)$$

and $$u(n-k) = 1 \text{ for } (n-k) \geq 0$$
$$= 0 \text{ for } (n-k) < 0 \qquad ...(8)$$

Also, $$x(n) = \sum_{m=-\infty}^{\infty} x(m)\,\delta(n-m)$$

In discrete-time systems, the input signal $f(t)$ is sampled at discrete instants 0, T, $2T$, ...nT, ... where T is the sampling period. For such functions, the Z-transform (one sided) becomes

$$Z[f(t)] = F(z) = \sum_{n=0}^{\infty} f(nT)\, z^{-n} \qquad ...(9)$$

Def. *Therefore*, if $f(t)$ is a function defined for discrete values of t where $t = nT$, $n = 0, 1, 2, 3, ...$ to ∞, T being the sampling period, then Z-transform of $f(t)$ is defined as

$$Z[f(t)] = \sum_{n=0}^{\infty} f(t)\, z^{-n} = \sum_{n=0}^{\infty} f(nT)\, z^{-n} \qquad \text{...(10)}$$

The continuous function $f(t)$ and discrete function $f(t) = f(nT)$, $n = 0, 1, 2, \ldots \infty$ are given below for understanding by way of graph.

Notation : $Z[f(t)] = F(z)$

Convergence of series

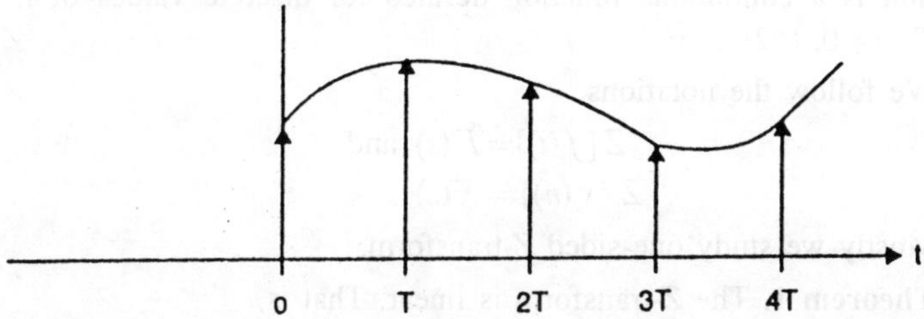

The series of one-sided z-transform, namely, $\sum_{n=0}^{\infty} x(n)\, z^{-n}$ is convergent, by Cauchy's ratio test,

if $$\underset{n \to \infty}{\text{Lt}} \left| \frac{x(n+1) \cdot z^{-n-1}}{x(n)\, z^{-n}} \right| < 1$$

That is,

if $$\underset{n \to \infty}{\text{Lt}} \left| \frac{x(n+1)}{x(n)} \right| |z|^{-1} < 1$$

i.e., $$\underset{n \to \infty}{\text{Lt}} \left| \frac{x(n+1)}{x(n)} \right| < |z|$$

Hence the region of convergence is outside the circle of radius R where

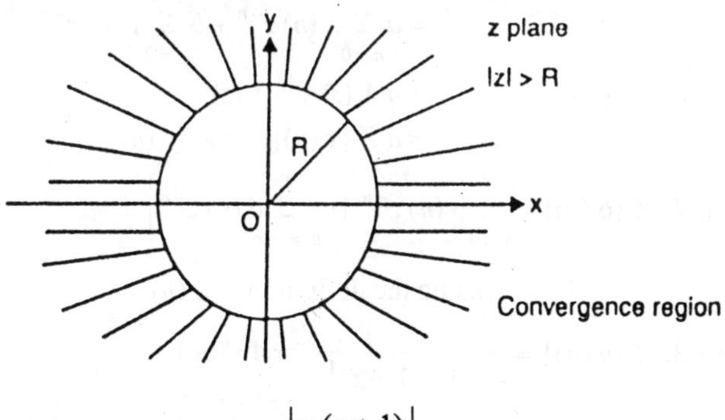

H
e

$$|z| > \underset{n \to \infty}{\text{Lt}} \left| \frac{x(n+1)}{x(n)} \right| = R.$$

re R is called the radius of convergence of the series $\sum_{n=0}^{\infty} x(n)\, z^{-n}$.

Properties and Theorems of Z-Transform

In what follows, we use $x(n)$ if $\{x(n)\}$ is given and use $f(t)$ if the given function is a continuous function defined for discrete values of t, say, $t = nT,\ n = 0, 1, 2, \ldots \infty$.

We follow the notations

$$Z[f(t)] = F(z) \text{ and}$$
$$Z\{x(n)\} = X(z)$$

Mostly we study one-sided Z-transform.

Theorem 1. The Z-transform is linear. That is,

$$Z[a f(t) + b\, g(t)] = a\, Z[f(t)] + b\, Z[g(t)]$$

or $$Z[a\{x(n)\} + b\{y(n)\}] = a\, Z\{x(n)\} + b\, Z\{y(n)\}$$

Proof is similar to both cases.

$$\begin{aligned} Z[a f(t) + b\, g(t)] &= \sum_{n=0}^{\infty} [a f(nT) + b\, g(nT)]\, z^{-n} \\ &= a \sum_{n=0}^{\infty} f(nT)\, z^{-n} + b \sum_{n=0}^{\infty} g(nT)\, z^{-n} \\ &= a\, F(z) + b\, G(z). \\ &= a\, Z[f(t)] + b\, Z[g(t)] \end{aligned}$$

Or

$$\begin{aligned} Z[a\{x(n)\} + b\{y(n)\}] &= \sum_{n=0}^{\infty} [ax(n) + by(n)]\, z^{-n} \\ &= a \sum_{n=0}^{\infty} x(n)\, z^{-n} + b \sum_{n=0}^{\infty} y(n)\, z^{-n} \\ &= a\, X(z) + b\, Y(z) \\ &= a\, Z\{x(n)\} + b\, Z\{y(n)\} \end{aligned}$$

Theorem 2. $Z\{\delta(n)\} = \sum_{n=0}^{\infty} \delta(n)\, z^{-n} \left(\text{or } \sum_{n=-\infty}^{\infty} \delta(n)\, z^{-n}\right)$

$= 1$ using the definition of $\delta(n)$

Theorem 3. $Z\{u(n)\} = \dfrac{z}{z-1} = \dfrac{1}{1-z^{-1}}$ for $|z| > 1$.

Proof : $Z\{u(n)\} = \sum_{n=0}^{\infty} u(n)\, z^{-n} = \sum_{n=0}^{\infty} z^{-n} = \sum_{n=0}^{\infty} \left(\dfrac{1}{z}\right)^n$

$$= \frac{1}{1 - z^{-1}} = \frac{z}{z-1} \text{ if } |z^{-1}| < 1 \text{ i.e., } |z| > 1$$

(Here $u(n)$ is the unit step sequence already defined)

Theorem 4. $Z[a^n f(t)] = F\left(\frac{z}{a}\right)$

or $$Z[\{a^n x(n)\}] = X\left(\frac{z}{a}\right)$$

Proof : $$Z[a^n f(t)] = \sum_{n=0}^{\infty} a^n f(nT) z^{-n}$$

$$= \sum_{n=0}^{\infty} f(nT) \cdot \left(\frac{z}{a}\right)^{-n} = F\left(\frac{z}{a}\right)$$

Or

$$Z[\{a^n x(n)\}] = \sum_{n=0}^{\infty} a^n x(n) z^{-n}$$

$$= \sum_{n=0}^{\infty} x(n) \left(\frac{z}{a}\right)^{-n} = X\left(\frac{z}{a}\right)$$

Theorem 5. $Z\{a^n u(n)\} = \frac{z}{z-a}$ if $|z| > a$.

Proof : $$Z\{a^n u(n)\} = \sum_{n=0}^{\infty} a^n z^{-n}$$

$$= \sum_{n=0}^{\infty} (az^{-1})^n = \frac{1}{1 - az^{-1}} \text{ if } |az^{-1}| < 1$$

$$= \frac{z}{z-a} \text{ if } |z| > |a|$$

Theorem 6. $Z[n f(t)] = -z \frac{dF(z)}{dz}$

Proof : $F(z) = Z[f(t)] = \sum_{n=0}^{\infty} f(nT) z^{-n}$

$$\frac{dF(z)}{dz} = \sum_{n=0}^{\infty} -n f(nT) z^{-n-1}$$

$$z\frac{dF}{dz} = -\sum_{n=0}^{\infty} n f(nT) z^{-n} = -Z[n f(t)]$$

$$\therefore \quad Z[n f(t)] = -z \frac{dF(z)}{dz}$$

Theorem 7. $Z[f(t+kT)] = Z[f((n+k)T)]$

$$= z^k \left[F(z) - f(0.T) - \frac{f(1.T)}{z} - \frac{f(2.T)}{z^2} - \dots - \frac{f((k-1)T)}{z^{k-1}} \right]$$

Proof : $Z[f(k+n)T] = \sum_{n=0}^{\infty} f((n+k)T)\, z^{-n}$

$$= z^k \sum_{n=0}^{\infty} f((n+k)T)\, z^{-(n+k)}$$

$$= z^k \sum_{m=k}^{\infty} f(mT)\, z^{-m} \quad \text{where } m = n+k$$

$$= z^k \left[\sum_{m=0}^{\infty} f(mT)\, z^{-m} - \sum_{m=0}^{k-1} f(mT)\, z^{-m} \right]$$

$$= z^k \left[F(Z) - f(0) - \frac{f(T)}{z} - \frac{f(2T)}{z^2} - \frac{f(3T)}{z^3} - \dots - \frac{f((k-1)T)}{z^{k-1}} \right]$$

Note : If $f((n+k)T)$ is denoted by f_{n+k}, then

$Z[f(t+kT)] = Z[f_{n+k}]$

$$= z^k \left[F(z) - f_0 - \frac{f_1}{z} - \frac{f_2}{z^2} - \dots - \frac{f_{k-1}}{z^{k-1}} \right]$$

Z-Transform of Standard Functions

Find the Z-Transform of sequence $\{x(n)\}$ or $\{f_n\}$ where $x(n)$ is given by

(*i*) $x(n) = k$ (*ii*) $x(n) = (-1)^n$ (*iii*) $x(n) = a^n$

(*iv*) $x(n) = n$ (*v*) $x(n) = na^n$ (*vi*) $x(n) = \sin n\theta$

(*vii*) $x(n) = \cos n\theta$ (*viii*) $x(n) = r^n \sin n\theta$ (*ix*) $x(n) = r^n \cos n\theta$

(*x*) $x(n) = n(n-1)$ (*ix*) $x(n) = n^2$

Solution. (*i*) $Z(k) = \sum_{n=0}^{\infty} kz^{-n} = k\left[1 + \frac{1}{z} + \frac{1}{z^2} + \dots \text{ to } \infty\right]$

$$= k\frac{1}{1-\frac{1}{z}} = \frac{kz}{z-1} \quad \text{if } |z| > 1$$

Cor. $Z(1) = \dfrac{z}{z-1}$ if $|z| > 1$.

(*ii*) $Z\{(-1)^n\} = \sum_{n=0}^{\infty} (-1)^n z^{-n} = \sum_{n=0}^{\infty} (-z)^{-n}$

$$= \frac{1}{1+\frac{1}{z}} \text{ if } |z| > 1 = \frac{z}{z+1} \text{ if } |z| > 1$$

(*iii*) $Z\{a^n\} = \sum_{n=0}^{\infty} a^n z^{-n} = \sum_{n=0}^{\infty} \left(\frac{a}{z}\right)^n$

$$= \frac{1}{1-\frac{a}{z}} \text{ if } |z| > |a| = \frac{z}{z-a} \text{ if } |z| > |a|$$

(*iv*) $Z\{(n)\} = \sum_{n=0}^{\infty} nz^{-n} = -z\frac{d}{dz}[Z(1)]$

$$= -z\frac{d}{dz}\left(\frac{z}{z-1}\right) = \frac{z}{(z-1)^2}$$

(*v*) $z\{na^n\} = -z\frac{d}{dz}\left[\frac{z}{z-a}\right] = \frac{az}{(z-a)^2}$

(*vi*), (*vii*) $Z\{\sin n\theta\}$ **and** $Z\{\cos n\theta\}$

We have proved $Z\{a^n\} = \frac{z}{z-a}$ if $|z| > |a|$

Taking $a = e^{i\theta}$,

$$Z\{e^{in\theta}\} = \frac{z}{z-e^{i\theta}}$$

$$Z\{\cos n\theta + i\sin n\theta\} = \frac{z(z-\cos\theta + i\sin\theta)}{(z-\cos\theta - i\sin\theta)(z-\cos\theta + i\sin\theta)}$$

$$= \frac{z(z-\cos\theta + i\sin\theta)}{(z-\cos\theta)^2 + \sin^2\theta}$$

Equating real and imaginary parts,

$$Z(\cos n\theta) = \frac{z(z-\cos\theta)}{z^2 - 2z\cos\theta + 1} \text{ if } |z| > 1$$

and

$$Z\{\sin n\theta\} = \frac{z\sin\theta}{z^2 - 2z\cos\theta + 1} \text{ if } |z| > 1$$

(*viii*) $Z\{r^n \cos n\theta\} = \frac{z(z - r\cos\theta)}{z^2 - 2zr\cos\theta + r^2}$ if $|z| > |r|$

(*ix*) $Z\{r^n \sin n\theta\} = \frac{zr\sin\theta}{z^2 - 2zr\cos\theta + r^2}$ if $|z| > |r|$

Taking $a = re^{i\theta}$ in $Z\{a^n\}$

$$Z\{r^n e^{in\theta}\} = \frac{z}{z - re^{i\theta}}$$

$$Z\{r^n(\cos n\theta + i\sin n\theta)\} = \frac{\left(\frac{z}{r}\right)}{\left(\frac{z}{r}\right) - e^{i\theta}}$$

$$= \frac{\left(\frac{z}{r}\right)\left[\frac{z}{r} - \cos\theta + i\sin\theta\right]}{\left[\left(\frac{z}{r}\right) - \cos\theta\right]^2 + \sin^2\theta}$$

Equating real and imaginary parts,

$$Z\{r^n \cos n\theta\} = \frac{z(z - r\cos\theta)}{z^2 - 2zr\cos\theta + r^2}$$

$$Z\{r^n \sin n\theta\} = \frac{zr\sin\theta}{z^2 - 2zr\cos\theta + r^2} \text{ if } |z| > |r|$$

(x) $Z(n^2) = Z(n.n) = -z\dfrac{d}{dz}\left(\dfrac{z}{(z-1)^2}\right)$

$$= \frac{z(z+1)}{(z-1)^3}$$

(xi) $Z[\{n(n-1)\}] = Z[\{n^2 - n\}]$

$$= Z\{n^2\} - z(n) = \frac{z(z+1)}{(z-1)^3} - \frac{z}{(z-1)^2}$$

$$= \frac{z^2 + z - z(z-1)}{(z-1)^3} = \frac{2z}{(z-1)^3}$$

Standard Results

Find the Z-transform of $f(t)$ where $f(t)$ is given by

(i) t (ii) e^{-at} (iii) e^{at}

(iv) $\sin \omega t$ (v) $\cos \omega t$ (vi) t^k

(i) $$Z(t) = \sum_{n=0}^{\infty} (nT) z^{-n} = T \sum_{n=0}^{\infty} n z^{-n}$$

$$= T \cdot (-z)\frac{d}{dz}[Z\{1\}] = -Tz\frac{d}{dz}\left(\frac{z}{z-1}\right)$$

$$= \frac{Tz}{(z-1)^2}$$

ii) $$Z(e^{-at}) = \sum_{n=0}^{\infty} e^{-anT} z^{-n} = \sum_{n=0}^{\infty} (e^{-aT})^n z^{-n}$$

$$= \frac{z}{z - e^{-aT}} \text{ using the result of } Z\{a^n\} \text{ if } |z| > |e^{-aT}|$$

(iii) $$Z(e^{at}) = \sum_{n=0}^{\infty} e^{anT} z^{-n} = \sum_{n=0} (e^{at}) z^{-n}$$

$$= \frac{z}{z - e^{aT}} \text{ if } |z| > |e^{aT}|$$

(iv) $$Z[\sin \omega t] = \sum_{n=0}^{\infty} \sin n\omega T . z^{-n}$$

$$= Z\{\sin n\theta\} \text{ where } \theta = \omega T$$

$$= \frac{z \sin \omega T}{z^2 - 2z \cos \omega T + 1} \text{ if } |z| > 1$$

(v) $$Z(\cos \omega t) = \sum_{n=0}^{\infty} \cos n\omega T . z^{-n}$$

$$= Z\{\cos n\theta\} \text{ where } \theta = \omega T$$

$$= \frac{z(z - \cos \omega T)}{z^2 - 2z \cos \omega T + 1}$$

or $$Z[\cos \omega t] = Z\left[\frac{e^{i\omega t} + e^{-i\omega t}}{2}\right]$$

$$= \frac{1}{2}\left[\frac{z}{z - e^{i\omega T}} + \frac{z}{z - e^{-i\omega T}}\right]$$

$$= \frac{1}{2} \cdot z \left[\frac{2z - (e^{i\omega T} + e^{-i\omega T})}{(z - e^{i\omega T})(z - e^{-i\omega T})}\right]$$

$$= \frac{1}{2} \cdot z \left[\frac{2z - 2\cos \omega T}{z^2 - (e^{i\omega T} + e^{-i\omega T}) z + 1}\right]$$

$$= \frac{z(z - \cos \omega T)}{z^2 - 2z \cos \omega T + 1}$$

Similarly, $$Z[\sin \omega t] = Z \cdot \left[\frac{e^{i\omega t} - e^{-i\omega t}}{2i}\right]$$

$$= \frac{1}{2i}\left[\frac{z}{z - e^{i\omega T}} - \frac{z}{z - e^{-i\omega T}}\right]$$

$$= \frac{z}{2i}\left[\frac{(e^{i\omega T} - e^{-i\omega T})}{z^2 - z\,(e^{i\omega T} + e^{-i\omega T}) + 1}\right]$$

$$= \frac{z \sin \omega T}{z^2 - 2z \cos \omega T + 1}$$

$$Z\,(t^k) = -\,Tz \frac{d}{dz}\,[Z\,(t^{k-1})]$$

(*vi*) $$Z\,(t^k) = \sum_{n=0}^{\infty} (nT)^k\, z^{-n}$$

$$= Tz \sum_{n=0}^{\infty} n^k\, T^{k-1}\, z^{-(n+1)}$$

$$= Tz \sum_{n=0}^{\infty} (nT)^{k-1} \cdot nz^{-(n+1)} \qquad \text{...(1)}$$

Similarly, $$Z\,(t^{k-1}) = \sum_{n=0}^{\infty} (nT)^{k-1}\, z^{-n}$$

$$\frac{d}{dz}\,[Z\,(t^{k-1})] = \sum_{n=0}^{\infty} (nT)^{k-1}\,(-\,n)\, z^{-(n+1)}$$

$$= -\sum_{n=0}^{\infty} (nT)^{k-1}\, n.\, z^{-(n+1)} \qquad \text{...(2)}$$

Using (2) in (1), we get

$$Z\,(t^k) = -\,Tz \frac{d}{dz}\,[Z\,(t^{k-1})] \qquad \text{...(3)}$$

Setting $k = 1, 2, 3, \ldots..$ we get $Z\,(t),\ Z\,(t^2), \ldots..$

$$Z\,(t) = -\,Tz \frac{d}{dz}\,[z\,(1)]$$

$$= -\,Tz \frac{d}{dz}\left(\frac{z}{z-1}\right)$$

$$Z\,(t) = \frac{Tz}{(z-1)^2} \qquad \text{...(4)}$$

$$Z\,(t^2) = -\,Tz \frac{d}{dz}\,Z\,(t) = -\,Tz \frac{d}{dz}\left[\frac{Tz}{(z-1)^2}\right]$$

$$Z\,(t^2) = \frac{T^2 z\,(z+1)}{(z-1)^3} \qquad \text{...(5)}$$

Shifting Theorem I

If $\quad Z\,[f\,(t)] = F\,(z)$, then $Z\,[e^{-at} f\,(t)] = F\,[ze^{aT}]$

Proof : $Z[e^{-at}f(t)] = \sum_{n=0}^{\infty} e^{-anT} f(nT)\, z^{-n}$

$$= \sum_{n=0}^{\infty} f(nT)\,(ze^{aT})^{-n}$$

$$= F[ze^{aT}] \text{ since } F(z) = \sum_{n=0}^{\infty} f(nT)\, z^{-n}$$

i.e., $\quad Z[e^{-at}f(t)] = Z[f(t)]_{z \to ze^{aT}}$

$$= F(z) \text{ where } z \to ze^{aT}$$

Deductions : Using shifting theorem, derive the following results (which are already proved by using definition).

1.

$$Z(e^{-at}) = Z(e^{-at}.\,1)$$

$$= [Z(1)]_{z \to ze^{aT}}$$

$$= \left(\frac{z}{z-1}\right)_{z \to ze^{aT}} = \frac{ze^{aT}}{ze^{aT}-1}$$

$$Z(e^{-at}) = \frac{z}{z-e^{-aT}}$$

2.

$$Z[e^{-at} \cdot t] = [z(t)]_{z \to ze^{aT}}$$

$$= \left[\frac{Tz}{(z-1)^2}\right]_{z \to ze^{aT}}$$

$$= \frac{Tz\,e^{aT}}{(ze^{aT}-1)^2}$$

$$Z(te^{-at}) = \frac{Tz\,e^{-aT}}{(z-e^{-aT})^2}$$

3.

$$Z[e^{-iat}] = Z[e^{-at} \cdot 1]$$

$$Z[\cos at - i \sin at] = [z(1)]_{z \to ze^{iaT}}$$

$$= \left(\frac{z}{z-1}\right)_{z \to ze^{iaT}}$$

$$= \frac{ze^{iaT}}{ze^{iaT}-1} = \frac{z}{z-e^{-iaT}}$$

$$= \frac{z(z-e^{iaT})}{(z-e^{-iaT})(z-e^{iaT})}$$

$$= \frac{z[z-(\cos aT + i \sin aT)]}{z^2 - z(e^{iaT}+e^{-iaT})+1}$$

$$= \frac{z[(z-\cos aT) - i \sin aT]}{z^2 - 2z\cos aT + 1}$$

Equating real and imaginary parts,

$$Z[\cos at] = \frac{z(z - \cos aT)}{z^2 - 2z\cos aT + 1}$$

and
$$Z[\sin at] = \frac{z \sin aT}{z^2 - 2z\cos aT + 1}$$

4. $$Z[e^{-at}\cos bt] = Z(\cos bt)_{z \to ze^{aT}}$$

$$= \left[\frac{z(z - \cos bT)}{z^2 - 2z\cos bT + 1}\right]_{z \to ze^{aT}}$$

$$= \frac{ze^{aT}[ze^{aT} - \cos bT]}{z^2 e^{2aT} - 2z\, e^{aT}\cos bT + 1}$$

5. $$Z[e^{-at}\sin bt] = [Z(\sin bt)]_{z \to ze^{aT}}$$

$$= \frac{ze^{aT}\sin bT}{z^2 e^{2aT} - 2z\, e^{aT}\cos bT + 1}$$

Initial Value Theorem

If $Z[f(t)] = F(z)$, then $f(0) = \underset{z \to \infty}{\text{Lt}}\, F(z)$

Proof : $F(z) = Z[f(t)] = \sum_{n=0}^{\infty} f(nT)\, z^{-n}$

$$= f(0.T) + \frac{f(1.T)}{z} + \frac{f(2.T)}{z^2} + \ldots \text{ to } \infty$$

$$= f(0) + \frac{1}{z} f(T) + \frac{1}{z^2} f(2T) + \ldots \text{ to } \infty$$

Taking limit as $z \to \infty$

$$\underset{z \to \infty}{\text{Lt}}\, F(z) = f(0)$$

Example. If $F(z) = \dfrac{z(z - \cos aT)}{z^2 - 2z\cos aT + 1}$, find $f(0)$.

$f(0) = \underset{z \to \infty}{\text{Lt}}\, F(z) = 1$ by L'Hospital's rule.

Shifting Theorem II

If $Z[f(t)] = F(z)$ then $Z[f(t + T)] = z[F(z) - f(0)]$

$$\textbf{Proof : } Z[f(t+T)] = \sum_{n=0}^{\infty} f(nT+T)\, z^{-n}$$

$$= \sum_{n=0}^{\infty} f((n+1)\,T)\,.\,z^{-n}$$

$$= z \sum_{n=0}^{\infty} f[(n+1)\,T]\, z^{-(n+1)}$$

$$= z \sum_{k=1}^{\infty} f(kT)\, z^{-k}$$

$$= z\left[\sum_{k=0}^{\infty} f(kT)\, z^{-k} - f(0)\right]$$

$$= z\,[F(z) - f(0)]$$

Final Value Theorem

If $\quad Z[f(t)] = F(z)$ then $\underset{t\to\infty}{\text{Lt}}\, f(t) = \underset{z\to 1}{\text{Lt}}\,(z-1)\,F(z)$

$$\textbf{Proof : } Z[f(t+T) - f(t)] = \sum_{n=0}^{\infty} [f(nT+T) - f(nT)]\, z^{-n}$$

$$Z[f(t+T) - Z[f(t)] = \sum_{n=0}^{\infty} [f(nT+T) - f(nT)]\, z^{-n}$$

$$zF(z) - zf(0) - F(z) = \sum_{n=0}^{\infty} [f(nT+T) - f(nT)]\, z^{-n}$$

Taking limit as $z \to 1$,

$$\underset{z\to 1}{\text{Lt}}\,(z-1)\,F(z) - f(0) = \underset{z\to 1}{\text{Lt}} \sum_{n=0}^{\infty} [f(nT+T) - f(nT)]\, z^{-n}$$

$$= \sum_{n=0}^{\infty} [f(nT+T) - f(nT)]$$

$$= \underset{n\to\infty}{\text{Lt}}\, [f(T) - f(0) + f(2T) - f(T) + \ldots + f((n+1)\,T) - f(nT)]$$

$$= \underset{n\to\infty}{\text{Lt}}\, f((n+1)\,T) - f(0)$$

$$= f(\infty) - f(0)$$

$$\therefore \qquad f(\infty) = \underset{t\to\infty}{\text{Lt}}\, f(t) = \underset{z\to 1}{\text{Lt}}\,(z-1)\,F(z)$$

Example. *If* $F(z) = \dfrac{z}{z - e^{-T}}$, *find* $\underset{t\to\infty}{\text{Lt}}\, f(t)$.

Solution. $\underset{z \to 1}{\text{Lt}} (z-1) \cdot \dfrac{z}{z-e^{-T}} = \sum_{n=0}^{\infty} \underset{t \to \infty}{\text{Lt}} f(t)$

$$0 = \underset{t \to \infty}{\text{Lt}} f(t)$$

Example. *If* $F(z) = \dfrac{10z}{(z-1)(z-2)}$, *find* $f(0)$.

Solution. $f(0) = \underset{z \to \infty}{\text{Lt}} F(z) = \underset{z \to \infty}{\text{Lt}} \dfrac{10}{2z-3} = 0$

Example 1. *Find the Z-transform of*

(*i*) $\left\{\dfrac{1}{n}\right\}$ (*ii*) $\left\{\cos \dfrac{n\pi}{2}\right\}$ (*iii*) $\left\{\dfrac{1}{n(n+1)}\right\}, n \geq 1$

Solution. (*i*) $Z\left\{\dfrac{1}{n}\right\} = \sum_{n=1}^{\infty} \dfrac{1}{n} z^{-n}$

$$= \frac{1}{z} + \frac{1}{2z^2} + \frac{1}{3z^3} + \dots \text{ to } \infty$$

$$= -\log\left(1 - \frac{1}{z}\right) \text{ if } \left|\frac{1}{z}\right| < 1$$

$$= \log\left(\frac{z}{z-1}\right) \text{ if } |z| > 1$$

(*ii*) $Z\left\{\cos \dfrac{n\pi}{2}\right\} = \sum_{n=0}^{\infty} \cos \dfrac{n\pi}{2} \cdot z^{-n}$

$$= 1 - \frac{1}{z^2} + \frac{1}{z^4} - \dots \text{ to } \infty$$

$$= \left(1 + \frac{1}{z^2}\right)^{-1} = \frac{z^2}{z^2+1} \text{ if } |z| > 1$$

(*iii*) Let $\dfrac{1}{n(n+1)} = \dfrac{A}{n} + \dfrac{B}{n+1}$

$$= \frac{1}{n} - \frac{1}{n+1}$$

$$Z\left[\left\{\frac{1}{n(n+1)}\right\}\right] = \sum_{n=1}^{\infty} \frac{1}{n} \cdot z^{-n} - \sum_{n=1}^{\infty} \frac{1}{n+1} \cdot z^{-n}$$

$$= \log \frac{z}{z-1} - \left[\frac{1}{2z} + \frac{1}{3z^2} + \cdots\right]$$

$$= \log \frac{z}{z-1} - z\left[\frac{1}{2}\left(\frac{1}{z}\right)^2 + \frac{1}{3}\left(\frac{1}{z}\right)^3 + \cdots\right]$$

$$= \log \frac{z}{z-1} + z \log\left(1 - \frac{1}{z}\right) + 1$$

$$= (z-1) \log\left(\frac{z-1}{z}\right) + 1$$

Shift Property. (Shift of a causal sequence to the right)

$Z\{x(n-m)\} = z^{-m} X(z)$ where $x(n)$ is a causal sequence and m is a positive integer.

Proof : $Z\{x(n-m)\} = \sum_{n=0}^{\infty} x(n-m) z^{-n}$

$$= \sum_{p=-m}^{\infty} x(p) z^{-(m+p)}$$

$$= \sum_{p=0}^{\infty} x(p) z^{-p} \cdot z^{-m}$$

$$= z^{-m} X(z)$$

$\therefore \qquad Z\{x(n-m)\} = z^{-m} X(z)$

Cor. $Z^{-1}[z^{-m} X(z)] = \{x(n-m)\} = [Z^{-1} X(z)]_{n \to n-m}$

Example 2. *Find* $Z^{-1}\left[\dfrac{1}{z-\frac{1}{2}}\right]$

Solution. $Z^{-1}\left[\dfrac{1}{z-\frac{1}{2}}\right] = Z^{-1}\left[z^{-1}\left(\dfrac{z}{z-\frac{1}{2}}\right)\right]$

$$= Z^{-1}\left(\frac{z}{z-\frac{1}{2}}\right)_{n \to n-1}$$

$$= \left(\frac{1}{2}\right)^{n-1} \text{ or } \left(\frac{1}{2}\right)^{n-1} u(n-1)$$

Example 3. *Evaluate* $Z^{-1}\left(\dfrac{1}{z+1}\right)$ *given* $Z^{-1}\left(\dfrac{z}{z+1}\right) = (-1)^n$.

Solution. $Z^{-1}\left[\dfrac{1}{z+1}\right] = Z^{-1}\left[z^{-1} \cdot \dfrac{z}{z+1}\right]$

$$= Z^{-1}\left(\frac{z}{z+1}\right)_{n \to n-1}$$

$$= (-1)^{n-1}, \quad n = 1, 2, 3, \ldots..$$

Example 4. *Prove* $Z\{a^n \cos n\theta\, u(n)\} = \dfrac{z(z - a\cos\theta)}{z^2 - 2az\cos\theta + a^2}$.

$$Z\{a^n \sin n\theta\, u(n)\} = \frac{az\sin\theta}{z^2 - 2az\cos\theta + a^2}.$$

Solution. These are same as $Z\{r^n \cos n\theta\}$ and $Z\{r^n \sin n\theta\}$ (which are worked out earlier)

Example 5. *Find the Z-transform of*

(*i*) $u(n-1)$ (*ii*) $3^n\, \delta(n-1)$ (*iii*) $\cos\dfrac{n\pi}{2}\, u(n)$ (*iv*) $\delta(n-k)$

Solution. (*i*) $Z[\{u(n-1)\}] = \sum\limits_{n=1}^{\infty} 1 \,.\, z^{-n}$

$$= \frac{1}{z} + \frac{1}{z^2} + \frac{1}{z^3} + \ldots \text{ to } \infty$$

$$= \frac{1}{z}\left[\frac{1}{1 - \frac{1}{z}}\right] \text{ if } \left|\frac{1}{z}\right| < 1$$

$$= \frac{1}{z-1} \text{ if } |z| > 1$$

(*ii*) $Z[3^n\, \delta(n-1)] = \sum\limits_{n=0}^{\infty} 3^n\, \delta(n-1)\, z^{-n} = \dfrac{3}{z}$

(*iii*) This problem is same as $Z\left[\cos\dfrac{n\pi}{2}\right]$

See Example 1.

(*iv*) $Z[\delta(n-k)] = \sum\limits_{n=0}^{\infty} \delta(n-k)\, z^{-n}$

$$= \frac{1}{z^k}, \text{ if } k \text{ is a positive integer.}$$

$\therefore\ Z[\delta(n-1)] = \dfrac{1}{z}$

Example 6. *Find the Z-transform of*

(*i*) $a^n \cos n\pi$ (*ii*) $e^{-t}\, t^2$ (*iii*) $e^t \sin 2t$ (*iv*) $e^{3t} \cos t$

(*v*) $e^{-2t}\, t^3$ (*vi*) e^{3t+7} (*vii*) e^{at+b}

Solution. (*i*) $Z(a^n \cos n\pi) = \sum_{n=0}^{\infty} a^n \cos n\pi \,.\, z^{-n}$

$$= 1 - \frac{a}{z} + \left(\frac{a}{z}\right)^2 - \left(\frac{a}{z}\right)^3 + \ldots \text{ to } \infty$$

$$= \left(1 + \frac{a}{z}\right)^{-1} \text{ if } \left|\frac{a}{z}\right| < 1$$

$$= \frac{z}{z+a} \text{ if } |z| > |a|$$

(*ii*) $$Z(e^{-t} t^2) = [Z(t^2)]_{z \to ze^T}$$

$$= \left[\frac{T^2 z(z+1)}{(z-1)^3}\right]_{z \to ze^T}$$

$$= \frac{T^2 . ze^T (ze^T + 1)}{(ze^T - 1)^3}$$

(*iii*) $$Z[e^t \sin 2t] = Z[\sin 2t]_{z \to ze^{-T}}$$

$$= \left[\frac{z \sin 2T}{z^2 - 2z \cos 2T + 1}\right] \text{ where } z \to ze^{-T}$$

$$= \frac{ze^{-T} \sin 2T}{z^2 e^{-2T} - 2ze^{-T} \cos 2T + 1}$$

(*iv*) $$Z[e^{3t} \cos t] = [Z(\cos t)]_{z \to ze^{-3T}}$$

$$= \left[\frac{z(z - \cos T)}{z^2 - 2z \cos T + 1}\right] \text{ where } z \to ze^{-3T}$$

$$= \frac{ze^{-3T}(ze^{-3T} - \cos T)}{z^2 e^{-6T} - 2ze^{-3T} \cos T + 1}$$

(*v*) $$Z[e^{-2t} t^3] = [z(t^3)]_{z \to ze^{2T}}$$

$$= \left[\frac{T^3 z(1 + 4z + z^2)}{(z-1)^4}\right]_{z \to ze^{2T}}$$

$$= \frac{T^3 ze^{2T}[1 + 4ze^{2T} + z^2 e^{4T}]}{(ze^{2T} - 1)^4}$$

(*vi*) $$Z[e^{3t+7}] = e^7 Z(e^{3t}) = e^7 \cdot \frac{z}{z - e^{3T}}$$

(*vii*) $$Z[e^{at+b}] = e^b . Z(e^{at})$$

$$= e^b \cdot \frac{z}{z - e^{aT}} \text{ if } |z| > |e^{aT}|$$

Example 7. *Find the Z-transforms of*

(*i*) $e^{2(t+T)}$ (*ii*) $\sin(t+T)$ (*iii*) $(t+T)\,e^{-(t+T)}$

Solution. (*i*) $Z[e^{2(t+T)}] = Z[f(t+T)]$ where $f(t) = e^{2t}$

$$= z[F(z) - f(0)]$$

$$= z\left[\frac{z}{z-e^{2T}} - 1\right]$$

$$= \frac{z \cdot e^{2T}}{z-e^{2T}}, \quad \text{using shifting theorem II.}$$

Or

$$Z[e^{2(t+T)}] = e^{2T} Z(e^{2t})$$

$$= e^{2T} \frac{z}{z-e^{2T}}$$

(*ii*) $Z[\sin(t+T)] = Z[f(t+T)]$ where $f(t) = \sin t$

$$= z[F(z) - f(0)]$$

$$= z\left[\frac{z\sin T}{z^2 - 2z\cos T + 1} - 0\right]$$

$$= \frac{z^2 \sin T}{z^2 - 2z\cos T + 1}$$

Or

$$Z[\sin(t+T)] = Z[\sin t\cos T + \cos t\sin T]$$

$$= \cos T\left[\frac{z\sin T}{z^2 - 2z\cos T + 1}\right] + \sin T\left[\frac{z(z-\cos T)}{z^2 - 2z\cos T + 1}\right]$$

$$= \frac{z^2 \sin T}{z^2 - 2z\cos T + 1}$$

(*iii*) $Z[(t+T)\,e^{-(t+T)}] = e^{-T} Z[(t+T)\,e^{-t}]$

$$= e^{-T}[Z(te^{-t}) + TZ(e^{-t})]$$

$$= e^{-T}\left[\frac{Tze^T}{(ze^T-1)^2} + \frac{Tz}{z-e^{-T}}\right]$$

$$= e^{-T}\left[\frac{Tze^T}{(ze^T-1)^2} + \frac{Te^Tz}{ze^T-1}\right]$$

$$= e^{-T} \cdot Tze^T\left[\frac{ze^T}{(ze^T-1)^2}\right] = \frac{Tz^2e^T}{(ze^T-1)^2}$$

Or

$$Z[(t+T)\,e^{-(t+T)}] = Z[f(t+T)] \text{ where } f(t) = te^{-t}$$

$$= z\,[F(z) - f(0)]$$

$$= z\left[\frac{Tze^{T}}{(ze^{T}-1)^2} - 0\right]$$

$$= \frac{Tz^2\,e^{T}}{(ze^{T}-1)^2}$$

Example 8. *Find the Z-transforms of*

(*i*) ab^n, $(a \neq 0,\ b \neq 0)$

(*ii*) $x(n) = n,\ n \geq 0$
$\quad = 0,\ n < 0$

(*iii*) $\dfrac{(n+1)(n+2)}{2}$

(*iv*) $x(n) = 1$ *for* $n = k$
$\quad = 0$ *otherwise.*

Solution. (*i*) $Z[\{ab^n\}] = \sum\limits_{n=0}^{\infty} ab^n z^{-n}$

$$= a \sum_{n=0}^{\infty} \left(\frac{b}{z}\right)^n$$

$$= a\left[1 + \frac{b}{z} + \left(\frac{b}{z}\right)^2 + \ldots \text{ to } \infty\right]$$

$$= a \cdot \frac{1}{1 - \dfrac{b}{z}} \text{ if } \left|\frac{b}{z}\right| < 1$$

$$= \frac{az}{z-b} \text{ if } |z| > |b|$$

(*ii*) $\quad z[x(n)] = \sum\limits_{n=0}^{\infty} n \,.\, z^{-n}$

$$= \frac{1}{z} + \frac{2}{z^2} + \frac{3}{z^3} + \ldots \text{ to } \infty$$

$$= \frac{1}{z}\left[1 + 2\left(\frac{1}{z}\right) + 3\left(\frac{1}{z}\right)^2 + \ldots\ldots\right]$$

$$= \frac{1}{z}\left(1 - \frac{1}{z}\right)^{-2}$$

$$= \frac{1}{z}\left(\frac{z-1}{z}\right)^{-2} \text{ if } \left|\frac{1}{z}\right| < 1$$

$$= \frac{z}{(z-1)^2} \text{ if } |z| > 1$$

(*iii*) $$Z\left[\frac{(n+1)(n+2)}{2}\right] = \frac{1}{2}[z(n^2) + 3z(n) + z(2)]$$

$$= \frac{1}{2}\left[\frac{z(z+1)}{(z-1)^3} + \frac{3z}{(z-1)^2} + \frac{2z}{(z-1)}\right] \text{ if } |z| > 1$$

(*iv*) $$Z[\{x(n)\}] = \sum_{n=0}^{\infty} x(n) z^{-n}$$

$$= z^{-k} = \frac{1}{z^k}$$

Example 9. *Find the Z-transform of*

(*i*) $x(n) = 0$ *if* $n > 0$

$\quad = 1$ *if* $n \leq 0$

(*ii*) $x(n) = \dfrac{a^n}{n!}$ *for* $n \geq 0$

$\quad = 0$ *otherwise.*

Solution. (*i*) $$Z[\{x(n)\}] = \sum_{n=-\infty}^{\infty} x(n) z^{-n} = \sum_{n=-\infty}^{0} z^{-n}$$

$$= 1 + z + z^2 + \ldots z^n + \ldots \text{ to } \infty$$

$$= \frac{1}{1-z} \text{ if } |z| < 1$$

(*ii*) $$Z[\{x(n)\}] = \sum_{n=0}^{\infty} \frac{a^n}{n!} z^{-n} = \sum_{n=0}^{\infty} \frac{(az^{-1})^n}{n!} = e^{az^{-1}} = e^{\frac{a}{z}}$$

Convolution of Sequences

The convolution of two sequences $\{x(n)\}$ and $\{y(n)\}$ is defined as

$$w(n) = \sum_{k=-\infty}^{\infty} x(k)\, y(n-k)$$

[**Note** : If it is one sided (right sided) sequence, take $x(k) = 0$, $y(k) = 0$ for $k < 0$]

Convolution Theorem

If $w(n)$ is the convolution of two sequences $x(n)$ and $y(n)$, then

$$Z[w(n)] = W(z) = Z[x(n)] \,.\, Z[y(n)] = X(z)\, Y(z)$$

Proof : $W(z) = Z[w(n)]$

$$= Z\left[\sum_{k=-\infty}^{\infty} x(k)\, y(n-k)\right]$$

$$= \sum_{n=-\infty}^{\infty}\left[\sum_{k=-\infty}^{\infty} x(k)\, y(n-k)\right] z^{-n}$$

$$= \sum_{k=-\infty}^{\infty} x(k)\left(\sum_{n=-\infty}^{\infty} y(n-k)\, z^{-n}\right)$$

by change of order of summation

$$= \sum_{k=-\infty}^{\infty} x(k)\left[\sum_{m=-\infty}^{\infty} y(m)\, z^{-(m+k)}\right] \qquad \text{putting } n-k=m$$

$$= \sum_{k=-\infty}^{\infty} x(k)\, z^{-k}\left[\sum_{m=-\infty}^{\infty} y(m)\, z^{-m}\right]$$

$$= \sum_{k=-\infty}^{\infty} x(k)\, z^{-k} \cdot Y(z)$$

$$= Y(z) \sum_{k=-\infty}^{\infty} x(k)\, z^{-k}$$

$$= Y(z)\, X(z) = X(z)\, Y(z)$$

Note : The result is true only for those values of z inside the region of convergence.

Another Form of Convolution Theorem

If $F(z)$ and $G(z)$ are one sided *Z*-transforms of $f(t)$ and $g(t)$ respectively, then the *Z*-transform of $\sum_{k=0}^{n} f(kT)\, g(nT-kT)$ is $F(z) \cdot G(z)$.

Proof : (We are dealing here with one sided *Z*-transform)

$$F(z) = \sum_{i=0}^{\infty} f(iT)\, z^{-i}$$

$$G(z) = \sum_{n=0}^{\infty} g(nT)\, z^{-n}$$

$$\therefore \quad F(z)\, G(z) = \left(\sum_{i=0}^{\infty} f(iT)\, z^{-i}\right)\left(\sum_{n=0}^{\infty} g(nT)\, z^{-n}\right)$$

$$= \sum_{n=0}^{\infty}\sum_{i=0}^{\infty} [f(iT)\, g(nT)\, z^{-i} \cdot z^{-n}]$$

$$= \sum_{n=0}^{\infty}\left(\sum_{k=0}^{n} f(kT)\, g((n-k)\,T)\right) z^{-n}$$

llecting the coefficient of z^{-n} in double sigma summation.

$$= \sum_{n=0}^{\infty} C(n) z^{-n} \text{ where } C(n) = \sum_{k=0}^{n} f(kT)\, g((n-k)T)$$

$$= Z(C(n))$$

$$= Z\text{-transform of } \sum_{k=0}^{n} f(kT)\, g((n-k)T)$$

Note : By symmetry, Z-transform of $\sum_{k=0}^{n} g(kT) f((n-k)T)$ is also the same value $F(z) \,.\, G(z)$.

Example 10. *Find the Z-transform of the convolution of* $x(n) = u(n)$ *and* $y(n) = a^n u(n)$.

Solution. $X(z) = \sum_{n=0}^{\infty} z^{-n} = \dfrac{z}{z-1}$ if $|z| > 1$

$$Y(z) = \sum_{n=0}^{\infty} a^n z^{-n} = \frac{1}{1-\frac{a}{z}} = \frac{z}{z-a} \text{ if } |z| > |a|$$

The Z-transform of the convolution of $x(n)$ and $y(n)$ is

$$W(z) = \frac{z}{z-1} \cdot \frac{z}{z-a} = \frac{z^2}{(z-1)(z-a)} \text{ if } |z| > \max(|a|, 1)$$

Example 11. *Find the Z-transform of the convolution of* $x(n) = a^n u(n)$, *and* $y(n) = b^n u(n)$.

Solution. $X(z) = \dfrac{z}{z-a}$ if $|z| > |a|$

$$Y(z) = \frac{z}{z-b} \text{ if } |z| > |b|$$

$W(z) = Z$- transform of convolution of $x(n)$ and $y(n)$.

$$= \frac{z^2}{(z-a)(z-b)} \text{ if } |z| > \max(|a|, |b|)$$

Note : $Z[x(n)\, y(n)] = \sum_{n=0}^{\infty} x(n)\, y(n)\, z^{-n}$

$$= \sum_{n=0}^{\infty} (ab)^n \,.\, z^{-n} = \frac{1}{1-abz^{-1}} = \frac{z}{z-ab} \text{ if } |z| > |ab|$$

The convolution of two causal sequences $x(n)$ and $y(n)$, we define

$$\{x(n)\} * \{y(n)\} = \sum_{k=0}^{n} x(n-k)\, y(k)$$

Convolution Theorem

If $f(n)$ and $g(n)$ are two causal sequences,

$$Z[\{f(n) * g(n)] = Z\{f(n)\} \cdot Z\{g(n)\} = F(z) \cdot G(z)$$

Proof : $$F(z) \cdot G(z) = \left[\sum_{n=0}^{\infty} f(n) z^{-n}\right]\left[\sum_{n=0}^{\infty} g(n) z^{-n}\right]$$

$$= \sum_{n=0}^{\infty}\left[\sum_{k=0}^{n} f(k) g(n-k)\right] z^{-n}$$

$$= \sum_{n=0}^{\infty}\left[\sum_{k=0}^{n} f(n-k) g(k)\right] z^{-n} \qquad \text{...(1)}$$

By definition,

$$Z[f(n) * g(n)] = \sum_{n=0}^{\infty}\left[\sum_{k=0}^{n} f(n-k) g(k)\right] z^{-n} \qquad \text{...(2)}$$

From (1) and (2),

$$Z[f(n) * g(n)] = F(z) G(z) = Z[f(n)] \cdot Z[g(n)]$$

Cor. $Z^{-1}[F(z) \cdot G(z)] = f(n) * g(n)$

$$= \sum_{k=0}^{n} f(n-k) g(k) = Z^{-1}(F(z)) * Z^{-1}(G(z))$$

Example 12. *Find the inverse Z-transform of* $X(Z) = \dfrac{z^2}{\left(z-\frac{1}{2}\right)\left(z-\frac{1}{4}\right)}$ *using convolution theorem.*

Solution. $$Z^{-1}\left[\frac{z^2}{\left(z-\frac{1}{2}\right)\left(z-\frac{1}{4}\right)}\right] = Z^{-1}\left[\frac{z}{z-\frac{1}{2}}\right] * Z^{-1}\left(\frac{z}{z-\frac{1}{4}}\right)$$

$$= \left(\frac{1}{2}\right)^n * \left(\frac{1}{4}\right)^n = \sum_{k=0}^{n}\left(\frac{1}{2}\right)^{n-k}\left(\frac{1}{4}\right)^k$$

$$= \left(\frac{1}{2}\right)^n \sum_{k=0}^{n} \frac{\left(\frac{1}{4}\right)^k}{\left(\frac{1}{2}\right)^k} = \left(\frac{1}{2}\right)^n \sum_{k=0}^{n}\left(\frac{1}{2}\right)^k$$

$$= \left(\frac{1}{2}\right)^n\left[1 + \frac{1}{2} + \left(\frac{1}{2}\right)^2 + \dots \left(\frac{1}{2}\right)^n\right]$$

$$= \left(\frac{1}{2}\right)^n \left[\frac{1-\left(\frac{1}{2}\right)^{n+1}}{1-\frac{1}{2}}\right]$$

$$x(n) = \left(\frac{1}{2}\right)^{n-1}\left[1-\left(\frac{1}{2}\right)^{n+1}\right] = \left(\frac{1}{2}\right)^{n-1} - \left(\frac{1}{2}\right)^{2n}$$

[**Note :** We will do the same problem by partial fraction method later.]

Example 13. *Find the inverse Z-transform of* $\frac{z^2}{(z-a)^2}$ *using convolution theorem.*

Solution. $Z^{-1}\left[\frac{z^2}{(z-a)^2}\right] = Z^{-1}\left[\frac{z}{z-a}\cdot\frac{z}{z-a}\right]$

$$= Z^{-1}\left(\frac{z}{z-a}\right) * Z^{-1}\left(\frac{z}{z-a}\right) = a^n * a^n$$

$$= \sum_{k=0}^{n} a^{n-k}\cdot a^k = \sum_{k=0}^{n} a^n = (n+1)\,a^n$$

or $$= (n+1)\,a^n\,u(n)$$

TABLE OF Z-TRANSFORMS

$x(n)$	$X(z)$
1	$\frac{z}{z-1}$
$(-1)^n$	$\frac{z}{z+1}$
$a^n\,u(n)$	$\frac{z}{z-a}$ if $\lvert z\rvert > \lvert a\rvert$
$u(n-m)$	$z^{-m}\cdot\frac{z}{z-1}$
n	$\frac{z}{(z-1)^2}$
n^2	$\frac{z^2+z}{(z-1)^3}$
$n(n-1)$	$\frac{2z}{(z-1)^3}$
$n^{(k)}$	$\frac{k!\,z}{(z-1)^{k+1}}$
$a^n\;u(n-1)$	$\frac{a}{z-a}$

$u(n-1)$	$\dfrac{1}{z-1}$
$n\,a^n$	$\dfrac{az}{(z-a)^2}$
$u(n)\cos n\theta$	$\dfrac{z(z-\cos\theta)}{z^2-2z\cos\theta+1}$
$u(n)\sin n\theta$	$\dfrac{z\sin\theta}{z^2-2z\cos\theta+1}$
$r^n\cos n\theta$	$\dfrac{z(z-r\cos\theta)}{z^2-2zr\cos\theta+r^2}$
$r^n\sin n\theta$	$\dfrac{rz\sin\theta}{z^2-2zr\cos\theta+r^2}$
$a^n\,x(n)$	$X\left(\dfrac{z}{a}\right)$
$n\,x(n)$	$\dfrac{1}{z}\dfrac{dX(z)}{dz^{-1}}$

$x(n)$ or $f(t)$	$X(z)$ or $F(z)$
$\dfrac{1}{n}$	$\log\left(\dfrac{z}{z-1}\right)$ if $\lvert z\rvert > 1$
$\delta(n)$	1
$\delta(n-k)$	$\dfrac{1}{z^k}$
$a^n\,u(n)$	$\dfrac{z}{z-a}$
$a^n\cos n\theta\,.\,u(n)$	$\dfrac{z(z-a\cos\theta)}{z^2-2az\cos\theta+a^2}$
$a^n\sin n\theta\,.\,u(n)$	$\dfrac{az\sin\theta}{z^2-2az\cos\theta+a^2}$
$n\,a^n\,u(n)$	$\dfrac{az}{(z-a)^2}$
$(n+1)\,a^n\,u(n)$	$\dfrac{z^2}{(z-a)^2}$
$n(n-1)\,a^n\,u(n)$	$\dfrac{2a^2z}{(z-a)^3}$
$ax(n)+by(n)$	$aX(z)+bY(z)$
$x(n-m)$	$z^{-m}X(z)$
$x(n) * y(n)$	$X(z)\,Y(z)$

$u(n)$	$\dfrac{z}{z-1}$ if $\|z\| > 1$
$Z(t^k)$	$-Tz\dfrac{d}{dz}[Z(t^{k-1})]$
$Z(t)$	$\dfrac{Tz}{(z-1)^2}$
$Z(t^2)$	$\dfrac{T^2 z(z+1)}{(z-1)^3}$
$Z(t^3)$	$\dfrac{T^3 z(1+4z-z^2)}{(z-1)^4}$
$a^n \cos\dfrac{n\pi}{2}$	$\dfrac{z^2}{z^2+a^2}$
$a^n \sin\dfrac{n\pi}{2}$	$\dfrac{az}{z^2+a^2}$
$a^n f(t)$	$F\left(\dfrac{z}{a}\right)$
$n f(nT) = n f(t)$	$-z\dfrac{d}{dz}F(z)$
k	$\dfrac{kz}{z-1}$ if $\|z\| > 1$
e^{-at}	$\dfrac{z}{z-e^{-aT}}$ if $\|z\| > \|e^{-aT}\|$
e^{at}	$\dfrac{z}{z-e^{aT}}$ if $\|z\| > \|e^{aT}\|$
$\sin \omega t$	$\dfrac{z\sin\omega T}{z^2-2z\cos\omega T+1}$ if $\|z\| < 1$
$\cos \omega t$	$\dfrac{z(z-\cos\omega T)}{z^2-2z\cos\omega T+1}$

Theorems. $Z[e^{-at} f(t)] = F(ze^{aT})$

$$Z[f((n+1)T)] = Z[f(t+T)] = z[F(z) - f(0)]$$

$$Z[f(t+kT)] = Z[f_{n+k}] = z^k\left[F(z) - f(0) - \frac{f(T)}{z} - \frac{f(2T)}{z^2} - \frac{f(3T)}{z^3} - \dots \frac{f((k-1)T)}{z^{k-1}}\right]$$

$$Z[f_{n+2}] = z^2\left[F(z) - f(0) - \frac{f(T)}{z}\right]$$

TABLE OF INVERSE *Z*-TRANSFORM

$X(z)$	$x(n) = Z^{-1}[X(z)]$
$\frac{z}{z-1}$	1
$\frac{z}{z+1}$	$(-1)^n$
$\frac{z}{z-a}$	$a^n u(n)$
$z^{-m} \cdot \frac{z}{z-1}$	$u(n-m)$
$\frac{z}{(z-1)^2}$	n
$\frac{z^2+z}{(z-1)^3}$	n^2
$\frac{2z}{(z-1)^2}$	$n(n-1)$
$\frac{1}{z-1}$	$u(n-1)$
$\frac{az}{(z-a)^2}$	na^n
$X\left(\frac{z}{a}\right)$	$a^n x(n)$
$\frac{z(z-r\cos\theta)}{z^2-2zr\cos\theta+r^2}$	$r^n \cos n\theta$
$\frac{rz\sin\theta}{z^2-2zr\cos\theta+r^2}$	$r^n \sin n\theta$
$\frac{z^2}{z^2+a^2}$	$a^n \cos\frac{n\pi}{2}$
$\frac{az}{z^2+a^2}$	$a^n \sin\frac{n\pi}{2}$
$\frac{1}{z-a}$	$a^{n-1} u(n-1)$
	etc. See the other table also for more functions.

Inverse Z-Transform

Method I. Long Division Method :

Since Z-transform is defined by the series $X(z) = \sum_{n=0}^{\infty} x(n) z^{-n}$ (one sided), to find the inverse Z-transform $[x(n) = Z^{-1}[X(z)]$ of $X(z)$, expand $X(z)$ in the proper power series and collect the coefficient of z^{-n} to get $x(n)$. Long division method is useful to get $x(n)$.

Example 1. *Find inverse Z-transform of*

(*i*) $\dfrac{10z}{(z-1)(z-2)}$ (*ii*) $\dfrac{(z+2)z}{z^2+2z+4}$

(*iii*) $\dfrac{2z(z^2-1)}{(z^2+1)^2}$ (*iv*) $\dfrac{z}{z^2+7z+10}$

Solution. (*i*) $X(z) = \dfrac{10z}{(z-1)(z-2)} = \dfrac{10z}{z^2-3z+2} = \dfrac{10z^{-1}}{1-3z^{-1}+2z^{-2}}$

By actual division,

$$\begin{array}{r|l} & 10z^{-1} + 30z^{-2} + 70z^{-3} + \ldots \\ \hline 1 - 3z^{-1} + 2z^{-2} & 10z^{-1} \\ & \underline{10z^{-1} - 30z^{-2} + 20z^{-3}} \\ & \quad 30z^{-2} - 20z^{-3} \\ & \quad \underline{30z^{-2} - 90z^{-3} + 60z^{-4}} \\ & \qquad 70z^{-3} - 60z^{-4} \\ & \qquad \underline{70z^{-3} - 210z^{-4} + 40z^{-5}} \\ & \qquad\quad + 150z^{-4} - 140z^{-5} \end{array}$$

Comparing the quotient with $\sum_{n=0}^{\infty} x(n) z^{-n}$

$$= x(0) + x(1) z^{-1} + x(2) z^{-2} + \ldots..$$

We get the sequence $x(n)$,

$$x(0) = 0, \ x(1) = 10, \ x(2) = 30, \ x(3) = 70, \ldots..$$

i.e., we can get $x(n) = 10(2^n - 1), \ n = 0, 1, 2, \ldots..$

(*ii*) $X(z) = \dfrac{z^2+2z}{z^2+2z+4} = \dfrac{1+2z^{-1}}{1+2z^{-1}+4z^{-2}}$

By actual division,

$$
\begin{array}{r}
1 - 4z^{-2} + 8z^{-3} - 32z^{-5} + \ldots \\
1 + 2z^{-1} + 4z^{-2} \,\big)\, 1 + 2z^{-1} \qquad\qquad\qquad \\
1 + 2z^{-1} + 4z^{-2} \qquad\qquad \\
\hline
- 4z^{-2} \qquad\qquad \\
- 4z^{-2} - 8z^{-3} - 16z^{-4} \\
\hline
8z^{-3} + 16z^{-4} \\
8z^{-3} + 16z^{-4} + 32z^{-5} \\
\hline
- 32z^{-5}
\end{array}
$$

Therefore $x(0) = 1,\ x(1) = 0,\ x(2) = -4$

$x(3) = 8,\ x(4) = 0,\ x(5) = -32, \ldots..$

The sequence is $1, 0, -4, 8, 0, -32, \ldots..$

(*iii*) $$X(z) = \frac{2z(z^2-1)}{(z^2+1)^2} = \frac{2z^3 - 2z}{z^4 + 2z^2 + 1} = \frac{2z^{-1} - 2z^{-3}}{1 + 2z^{-2} + z^{-4}}$$

By actual division, this is equal to

$$X(z) = 2z^{-1} - 6z^{-3} + 10z^{-5} - 14z^{-7} + \ldots..$$

$\therefore$ $x(0) = 0,\ x(1) = 2,\ x(2) = 0,\ x(3) = -6,$

$x(4) = 0,\ x(5) = 10,\ x(6) = 0.$

In general $x(n) = 2n \sin \dfrac{n\pi}{2},\ n = 0, 1, 2, \ldots..$

(*iv*) $$X(z) = \frac{z}{z^2 + 7z + 10} = \frac{z^{-1}}{1 + 7z^{-1} + 10z^{-2}}$$

By actual division,

$$X(z) = z^{-1} - 7z^{-2} + 39z^{-3} - 203z^{-4} + \ldots..$$

$\therefore$ $x(0) = 0,\ x(1) = 1,\ x(2) = -7,\ x(3) = 39,\ x(4) = -203, \ldots..$

Example 2. *Find the Z-inverse of* $X(z) = \dfrac{(1 - e^{-aT})z}{(z-1)(z - e^{-aT})}$.

Solution. $X(z) = \dfrac{(1 - e^{-aT})z}{(z-1)(z - e^{-aT})} = \dfrac{(1 - e^{-aT})z^{-1}}{1 - (1 + e^{-aT})z^{-1} + e^{-aT}z^{-2}}$

By actual division,

$$X(z) = (1 - e^{-aT})z^{-1} + (1 - e^{-2aT})z^{-2} + [(1 - e^{-aT}) + (1 + e^{-aT})(1 - e^{-2aT})]z^{-3} + \ldots$$

$\therefore$ $x(0) = 0,\ x(1) = 1 - e^{-aT}$

$x(2) = (1 - e^{-2aT})$

$x(3) = (1 - e^{-aT}) + (1 + e^{-aT})(1 - e^{-2aT})$

....

Method 2. (By Partial Fraction Method)

Example 3. *Find the inverse Z-transform of*

(*i*) $\dfrac{z}{z^2+7z+10}$ (*ii*) $\dfrac{z^2+z}{(z-1)(z^2+1)}$ (*iii*) $\dfrac{kTz}{(z-1+kT)(z-1)}$

(*iv*) $\dfrac{z^2}{\left(z-\frac{1}{2}\right)\left(z-\frac{1}{4}\right)}$ (*v*) $\dfrac{z}{(z-1)^2(z+1)}$

Solution. (*i*) Let $X(z) = \dfrac{z}{z^2+7z+10}$

$$\therefore \quad \frac{X(z)}{z} = \frac{1}{(z+5)(z+2)} = \frac{A}{z+5} + \frac{B}{z+2} = -\frac{1}{3}\cdot\frac{1}{z+5} + \frac{1}{3}\cdot\frac{1}{z+2}$$

$$\therefore \quad X(z) = \frac{1}{3}\left[\frac{z}{z+2} - \frac{z}{z+5}\right]$$

$$x(n) = Z^{-1}\left[\frac{1}{3}\left(\frac{z}{z+2} - \frac{z}{z+5}\right)\right]$$

$$= \frac{1}{3}[(-2)^n - (-5)^n], \; n = 0, 1, 2, \ldots\ldots \text{ since } Z(a^n) = \frac{z}{z-a}.$$

(*ii*) Let $X(z) = \dfrac{z(z+1)}{(z-1)(z^2+1)}$

$$\frac{X(z)}{z} = \frac{z+1}{(z-1)(z^2+1)} = \frac{z+1}{(z-1)(z-i)(z+i)}$$

$$= \frac{A}{z-1} + \frac{B}{z-i} + \frac{C}{z+i} = \frac{1}{z-1} - \frac{1}{2}\cdot\frac{1}{z-i} - \frac{1}{2}\cdot\frac{1}{z+i}$$

$$\therefore \quad X(z) = \frac{z}{z-1} - \frac{1}{2}\cdot\frac{z}{z-i} - \frac{1}{2}\cdot\frac{z}{z+i}$$

$$x(n) = 1 - \frac{1}{2}(i)^n - \frac{1}{2}(-i)^n.$$

(*iii*) $X(z) = \dfrac{kTz}{(z-1+kT)(z-1)}$

$$\frac{X(z)}{z} = \frac{kT}{(z+kT-1)(z-1)} = \frac{A}{(z+kT-1)} + \frac{B}{z-1}$$

$$= \frac{1}{z-1} - \frac{1}{z-(1-kT)}$$

$$\therefore \quad X(z) = \frac{z}{z-1} - \frac{z}{z-(1-kT)}$$

$$\therefore \quad x(n) = Z^{-1}[\text{RHS}]$$

$$= Z^{-1}\left[\frac{z}{z-1}\right] - Z^{-1}\frac{z}{z-(1-kT)}$$

$$x(n) = 1 - (1-kT)^n,\ n = 0, 1, 2, ...$$

(*iv*) Let $\quad X(z) = \dfrac{z^2}{\left(z-\frac{1}{2}\right)\left(z-\frac{1}{4}\right)}$

$$\frac{X(z)}{z} = \frac{z}{\left(z-\frac{1}{2}\right)\left(z-\frac{1}{4}\right)} = \frac{A}{z-\frac{1}{2}} + \frac{B}{z-\frac{1}{4}} = \frac{2}{z-\frac{1}{2}} - \frac{1}{z-\frac{1}{4}}$$

$$X(z) = 2\cdot\frac{z}{z-\frac{1}{2}} - \frac{z}{z-\frac{1}{4}}$$

$$X(n) = 2Z^{-1}\left[\frac{z}{z-\frac{1}{2}}\right] - Z^{-1}\left[\frac{z}{z-\frac{1}{4}}\right] = 2\cdot\left(\frac{1}{2}\right)^n - \left(\frac{1}{4}\right)^n,\ n = 0, 1, 2, ...$$

$$= \left[2\left(\frac{1}{2}\right)^n - \left(\frac{1}{4}\right)^n\right] u(n)$$

(*v*) Let $X(z) = \dfrac{z}{(z-1)^2(z+1)}$

$$\frac{X(z)}{z} = \frac{1}{(z-1)^2(z+1)} = \frac{\frac{1}{4}}{z+1} - \frac{\frac{1}{4}}{z-1} + \frac{\frac{1}{2}}{(z-1)^2}$$

$$X(z) = \frac{1}{4}\cdot\frac{z}{z+1} - \frac{1}{4}\cdot\frac{z}{z-1} + \frac{1}{2}\cdot\frac{z}{(z-1)^2}$$

$\therefore \quad x(n) = Z^{-1}$ [RHS]

$$= \frac{1}{4}(-1)^n - \frac{1}{4} + \frac{1}{2}n,\ n = 0, 1, 2, ...$$

Method 3. (Inverse of *Z*-transform by Inverse integral method)

By using the theory of complex variables, it can be shown that the inverse *Z*-transform is given by

$$x(n) = \frac{1}{2\pi i}\oint_c X(z)\cdot z^{n-1}\,dz$$

where *c* is the circle (may be even closed contour) which contains all the isolated singularities of $X(z)$ and containing the origin of the *z*-plane, in the region of convergence.

By Cauchy's Residue theorem, $x(n) = \Sigma R$ where ΣR is the sum of the residues of $X(z) \cdot z^{n-1}$ at the isolated singularities of $X(z) \cdot z^{n-1}$.

Example 4. *Find the inverse Z-transform of*

(*i*) $\dfrac{z}{(z-1)(z-2)}$ (*ii*) $\dfrac{z(z^2-1)}{(z^2+1)^2}$ *using residues.*

Solution. (*i*) Let $X(z) = \dfrac{z}{(z-1)(z-2)}$

$$X(z) \cdot z^{n-1} = \frac{z^n}{(z-1)(z-2)}$$

has simple poles at $z = 1$ and $z = 2$.

$z = 0$ is not a singularity of $X(z) \cdot z^{n-1}$ since $n = 0, 1, 2, \ldots$

$$\left.\text{Residue of } X(z) \cdot z^{n-1} \text{ at } z = 1\right\} = \underset{z \to 1}{\text{Lt}}\ (z-1) \cdot \frac{z^n}{(z-1)(z-2)} = -1$$

$$\left.\text{Residue of } X(z) \cdot z^{n-1} \text{ at } z = 2\right\} = \underset{z \to 2}{\text{Lt}}\ (z-2) \cdot \frac{z^n}{(z-1)(z-2)} = 2^n$$

$\therefore \quad x(n) = \Sigma R = 2^n - 1, \; n = 0, 1, 2, \ldots$

(*ii*) Let $X(z) = \dfrac{z(z^2-1)}{(z^2+1)^2}$

$$f(z) = X(z) \cdot z^{n-1} = \frac{z^n (z^2-1)}{(z^2+1)^2}$$

poles of $X(z) \cdot z^{n-1}$ are $z = \pm i$, each pole of order 2; $z = 0$ is not a singularity.

R_1 = Residue of $f(z)$ at $z = i$

$$= \underset{z \to i}{\text{Lt}}\ \frac{d}{dz}\left[(z-i)^2 \cdot X(z) \cdot z^{n-1}\right]$$

$$= \underset{z \to i}{\text{Lt}}\ \frac{d}{dz}\left[\frac{z^n (z^2-1)}{(z+i)^2}\right]$$

$$= \underset{z \to i}{\text{Lt}}\ \frac{(z+i)^2 \left[z^n \cdot 2z + nz^{n-1}(z^2-1)\right] - z^n (z^2-1) \cdot 2(z+i)}{(z+i)^4}$$

$$= \frac{n}{2} i^{n-1}$$

Similarly, R_2 = Residue of $f(z)$ at $z = -i$ (pole of order 2)

$$= \frac{n}{2} (-i)^{n-1}$$

$$\therefore \quad x(n) = \frac{n}{2}\left[(i)^{n-1} + (-i)^{n-1}\right], \; n = 0, 1, 2, \ldots$$

Example 5. *Find* $Z^{-1}\left[\dfrac{z}{z^2-2z+2}\right]$ *by Residue method.*

Solution. $X(z) = \dfrac{z}{(z^2-2z+2)}$

$$f(z) = X(z)\cdot z^{n-1} = \frac{z^n}{z^2-2z+2}$$

The poles of $f(z)$ are $z = 1 \pm i$, each simple pole.

R_1 = Residue of $f(z)$ at $z = 1+i = \dfrac{P(a)}{Q'(a)} = \dfrac{(1+i)^n}{2(1+i)-2} = \dfrac{1}{2i}(1+i)^n$

R_2 = Residue of $f(z)$ at $z = 1-i = -\dfrac{1}{2i}(1-i)^n$

$$\therefore \quad x(n) = \frac{1}{2i}\left[(1+i)^n - (1-i)^n\right]$$

$$= \frac{1}{2i}\left[(\sqrt{2})^n\left\{\cos\frac{n\pi}{4} + i\sin\frac{n\pi}{4}\right\} - (\sqrt{2})^n\left\{\cos\frac{n\pi}{4} - i\sin\frac{n\pi}{4}\right\}\right]$$

$$x(n) = (\sqrt{2})^n \sin\frac{n\pi}{4}, \; n = 0, 1, 2, \ldots$$

Example 6. *Find* $Z^{-1}\left[\dfrac{z^2-3z}{(z+2)(z-5)}\right]$.

Solution. Let $X(z) = \dfrac{z^2-3z}{(z+2)(z-5)}$

The simple poles of $f(z) = X(z)\cdot z^{n-1}$ are $z = -2, 5$.

Residue of $f(z)$ at $z = -2$ is $\dfrac{5}{7}(-2)^n$

Residue of $f(z)$ at $z = 5$ is $\dfrac{2}{7}5^n$

$$\therefore \quad x(n) = \frac{1}{7}\left[2(5)^n + 5(-2)^n\right], \; n = 0, 1, 2, \ldots$$

Example 7. *Find* $Z^{-1}(X(z))$ *where* $X(z) = \dfrac{4z^2-2z}{z^3-5z^2+8z-4}$.

Solution. $f(z) = X(z)\cdot z^{n-1}$ has poles given by

$$z^3 - 5z^2 + 8z - 4 = 0$$

i.e., $z = 1$ is a simple pole and

$z = 2$ is a pole of order 2.

Residue of $f(z)$ at $z = 1$ is $\underset{z\to 1}{\text{Lt}}\,(z-1)f(z)$

$$= \underset{z \to 1}{\text{Lt}} \frac{(4z^2 - 2z)\, z^{n-1}}{(z-2)^2} = 2$$

Residue of $f(z)$ at $z = 2$ is $2^n (3n - 2)$

$\therefore \quad x(n) = 2 + 2^n (3n - 2), \; n = 0, 1, 2, \ldots$

Example 8. *Find the inverse Z-transform of* $\dfrac{z(z+1)}{(z-1)^3}$.

Solution. Let $f(z) = X(z) \cdot z^{n-1} = \dfrac{z^n (z+1)}{(z-1)^3}$

$f(z)$ has only one pole at $z = 1$ of order 3.

Residue of $f(z)$ at $z = 1$ is

$$= \frac{1}{2!} \underset{z \to 1}{\text{Lt}} \frac{d^2}{dz^2} [z(z+1) \,.\, z^{n-1}]$$

$$= \frac{1}{2} \underset{z \to 1}{\text{Lt}} [(n+1)\, nz^{n-1} + n(n-1)\, z^{n-2}]$$

$$= \frac{1}{2} [n^2 + n + n^2 - n] = n^2$$

$\therefore \quad x(n) = n^2, \text{ for } n = 0, 1, 2, \ldots$

Example 9 . If $X(z) = (z-1) \log \left(1 - \dfrac{1}{z}\right) + 1$, find $x(n)$

Solution . $x(n) = Z^{-1} \left[(z-1) \log \left(1 - \dfrac{1}{z}\right) + 1 \right]$

$$= Z^{-1} \left[(z-1) \left(-\frac{1}{z} - \frac{1}{2z^2} - \frac{1}{3z^3} \ldots\right) + 1 \right]$$

$$= Z^{-1} \left[\left(-1 - \frac{1}{2z} - \frac{1}{3z^2} - \ldots\right) + \left(\frac{1}{z} + \frac{1}{2z^2} + \ldots\right) + 1 \right]$$

$$= Z^{-1} \left[\left(1 - \frac{1}{2}\right) \frac{1}{z} + \left(\frac{1}{2} - \frac{1}{3}\right) \frac{1}{z^2} + \ldots \right]$$

$$= Z^{-1} \left[\frac{1}{1.2} \frac{1}{z} + \frac{1}{2.3} \frac{1}{z^2} + \ldots \frac{1}{n(n+1)} \cdot \frac{1}{z^n} + \ldots \right]$$

$$= Z^{-1} \sum_{n=1}^{\infty} \frac{1}{n(n+1)} \cdot \frac{1}{z^n}$$

$$= \begin{cases} 0 & \text{if } n = 0 \\ \dfrac{1}{n(n+1)}, & \text{if } n \geq 1 \end{cases}$$

Differentiation

Let $Z[\{x(n)\}] = X(z)$.

An infinite series can be differentiated term by term within its region of convergence. $X(z)$ may be treated as a function of z^{-1}.

$$X(z) = \sum_{n=0}^{\infty} x(n) \cdot z^{-n} = \sum_{n=0}^{\infty} x(n)(z^{-1})^n$$

Differentiate both sides w.r.t. z^{-1}.

$$\frac{dX(z)}{dz^{-1}} = \sum_{n=0}^{\infty} n\,x(n) \cdot (z^{-1})^{n-1} \qquad ...(1)$$

$$z^{-1}\frac{dX(z)}{dz^{-1}} = \sum_{n=0}^{\infty} n\,(x(n)\,z^{-n}$$

$$= Z[nx(n)]$$

$$\boldsymbol{Z[nx(n)] = z^{-1}\frac{dX}{dz^{-1}}} \qquad ...(2)$$

Differentiating (1) w.r.t. to z^{-1} again,

$$\frac{d^2X(z)}{d(z^{-1})^2} = \sum_{n=0}^{\infty} n(n-1)\,x(n)\,(z^{-1})^{n-2}$$

$$z^{-2}\frac{d^2X(z)}{d(z^{-1})^2} = \sum_{n=0}^{\infty} n(n-1)\,x(n)\,z^{-n}$$

$$= Z[n(n-1)\,x(n)]$$

$$\therefore \qquad \boldsymbol{Z[n(n-1)\,x(n)] = z^{-2}\frac{d^2X(z)}{d(z^{-1})^2}} \qquad ...(3)$$

Example 10. *Find the Z-transform of*

(*i*) $na^n u(n)$ (*ii*) $n(n-1)\,a^n u(n)$

Solution. Take $x(n) = a^n u(n)$ in (2) and (3)

(*i*)
$$Z[na^n u(n)] = z^{-1}\frac{d}{d(z^{-1})}\left(\frac{z}{z-a}\right)$$

$$= z^{-1}\frac{d}{d(z^{-1})}[1-az^{-1}]^{-1}$$

$$= z^{-1}\frac{a}{(1-az^{-1})^2} = \frac{az^{-1}}{(1-az^{-1})^2}$$

(*ii*)
$$Z[n(n-1)\,a^n u(n)] = z^{-2}\frac{d^2[1-az^{-1}]^{-1}}{d(z^{-1})^2} = \frac{2a^2z^2}{(1-az^{-1})^3}$$

Setting $a = 1$,

$$Z[nu(n)] = \frac{z^{-1}}{(1-z^{-1})^2}$$

$$Z[n(n-1)u(n)] = \frac{2z^{-2}}{(1-z^{-1})^2}$$

Example 11. *If* $X(z) = (1 - az^{-1})^{-2}$*, find* $x(n)$.

Solution.
$$X(z) = \frac{1}{(1-az^{-1})^2} = \frac{1 - az^{-1} + az^{-1}}{(1-az^{-1})^2}$$

$$= \frac{1}{(1-az^{-1})} + \frac{az^{-1}}{(1-az^{-1})^2}$$

$$= \frac{z}{z-a} + \frac{az}{(z-a)^2}$$

$$x(n) = Z^{-1}[\text{RHS}]$$
$$= a^n u(n) + na^n u(n)$$
$$= (n+1)a^n u(n)$$

$$\therefore \quad Z[(n+1)a^n u(n)] = \frac{z^2}{(z-a)^2} = (1-az^{-1})^{-2}$$

Applications to Solving Finite Difference Equations

Z-transform can be employed in solving difference equation remembering $Z[x(n-m)] = z^{-m} X(z)$ and the expansions of
$Z[f_{n+k}]$, $Z[f_{n+1}]$, $Z[f_{n+2}]$ etc.

$$Z[y(n+2)] = z^2 Y(z) - z^2 y(0) - zy(1)$$
$$Z[y(n+1)] = zY(z) - zy(0) \text{ where } Y(z) = Z[y(n)]$$

Example 1. *Solve :* $y_{n+2} - 4y_{n+1} + 4y_n = 0$ *given* $y_0 = 1$ *and* $y_1 = 0$.

Solution. Taking Z-transform on both sides of the difference equation, we get

$$Z(y_{n+2}) - 4Z(y_{n+1}) + 4Z(y_n) = Z(0)$$

$$z^2\left[Y(z) - y_0 - \frac{y_1}{z}\right] - 4[z(Y(z) - y_0)] + 4Y(z) = 0$$

$$(z^2 - 4z + 4)Y(z) = z^2 - 4z$$

$$Y(z) = \frac{z^2 - 4z}{z^2 - 4z + 4} = \frac{z^2 - 4z}{(z-2)^2}$$

$Y(z)$ has pole at $z = 2$ (of order 2)

$\therefore \quad y(n) = R$ where R is residue of $Y(z) \cdot z^{n-1}$

$$= \underset{z \to 2}{\text{Lt}} \frac{d}{dz} [(z^2 - 4z) z^{n-1}]$$

$$= \underset{z \to 2}{\text{Lt}} [(n+1) z^n - 4nz^{n-1}]$$

$$= (n+1) 2^n - 4n \, 2^{n-1}$$

$$= 2^n [n + 1 - 2n]$$

$$y(n) = 2^n (1 - n), \; n = 0, 1, 2, \ldots$$

Example 2. *Solve* $y_{n+2} + 6y_{n+1} + 9y_n = 2^n$ *given* $y_0 = y_1 = 0$.

Solution. Taking Z-transform on both sides of the difference equation,

$$Z[y_{n+2}] + 6Z[y_{n+1}] + 9Z[y_n] = Z(2^n)$$

$$[z^2 Y(z) - z^2 y_0 - zy_1] + 6[z(Y(z) - y_0)] + 9Y(z) = \frac{z}{z-2}$$

$$(z^2 + 6z + 9) Y(z) = \frac{z}{z-2}$$

$$\therefore \quad Y(z) = \frac{z}{(z-2)(z+3)^2}$$

$$\frac{Y(z)}{z} = \frac{1}{(z-2)(z+3)^2} = \frac{A}{z-2} + \frac{B}{z+3} + \frac{C}{(z+3)^2}$$

$$= \frac{1}{25} \cdot \frac{1}{z-2} - \frac{1}{25} \cdot \frac{1}{z+3} - \frac{1}{5} \cdot \frac{1}{(z+3)^2}$$

$$Y(z) = \frac{1}{25} \cdot \frac{z}{z-2} - \frac{1}{25} \cdot \frac{z}{z+3} - \frac{1}{5} \cdot \frac{z}{(z+3)^2}$$

Taking inverse Z-transform,

$$y(n) = \frac{1}{25} \cdot 2^n - \frac{1}{25} (-3)^n - \frac{1}{5} n (-3)^{n-1}$$

since

$$Z[n(-3)^n] = \frac{-3z}{(z+3)^2}$$

$$y(n) = \frac{1}{25} \left[2^n - (-3)^n + \frac{5}{3} n (-3)^n \right]$$

Example 3. *Solve* $x(n+2) - 3x(n+1) + 2x(n) = 0$, *given* $x(0) = 0$, $x(1) = 1$.

Solution. Taking Z-transform of the equation given,

$$Z[x(n+2)] - Z[3x(n+1)] + 2Z[x(n)] = 0$$

$$z^2 X(z) - z^2 x(0) - zx(1) - 3[z(X(z) - x(0))] + 2X(z) = 0$$

$$(z^2 - 3z + 2) X(z) = z$$

$$\therefore \quad \frac{X(z)}{z} = \frac{1}{(z-1)(z-2)} = \frac{1}{z-2} - \frac{1}{z-1}$$

$$\therefore \quad X(z) = \frac{z}{z-2} - \frac{z}{z-1}$$

Taking inverse Z-transform,

$$x(n) = 2^n - 1, \text{ where } n = 0, 1, 2, \ldots$$

Example 4. *Solve $x(k+2) - 3x(k+1) + 2x(k) = u(k)$ given $x(k) = 0$ for $k \le 0$ and $u(0) = 1$, $u(k) = 0$ for $k < 0$ and $k > 0$.*

Solution. Taking Z-transform on both sides of the given equation,

$$Z[x(k+2)] - 3Z[x(k+1)] + 2Z[x(k)] = Z[u(k)]$$

$$[z^2 X(z) - z^2 x(0) - zx(1)] - 3[zX(z) - zx(0)] + 2X(z) = Z[u(k)] \qquad \ldots(1)$$

$x(0) = 0$, $x(-1) = 0$; putting $k = -1$ in the given equation,

$$x(1) - 3x(0) + 2x(-1) = u(-1)$$

$$\therefore \quad x(1) = 0 \text{ since } u(-1) = 0$$

$$Z[u(k)] = \sum_{k=0}^{\infty} u(k) z^{-k} = 1$$

Equation (1) reduces to,

$$(z^2 - 3z + 2) X(z) = 1$$

$$X(z) = \frac{1}{(z-1)(z-2)}$$

Poles of $X(z) . z^{k-1}$ are $z = 1$, $z = 2$, for $k = 1, 2, \ldots$

Residue of $f(z) = X(z) z^{k-1}$ at $z = 1$ is

$$\underset{z \to 1}{\text{Lt}} (z-1) \frac{z^{k-1}}{(z-1)(z-2)} = -1$$

Residue of $f(z)$ at $z = 2$ is

$$\underset{z \to 2}{\text{Lt}} (z-2) \frac{z^{k-1}}{(z-1)(z-2)} = 2^{k-1}$$

$$\therefore \quad x(k) = \Sigma R = 2^{k-1} - 1, \text{ for } k = 1, 2, 3, \ldots$$

Example 5. *Solve : $y(n) - ay(n-1) = u(n)$.*

Solution. Taking Z-transform on both sides

$$Z[y(n)] - aZ[y(n-1)] = Z[u(n)]$$

$$Y(z) - a.z^{-1} Y(z) = \frac{z}{z-1}$$

$$(1 - az^{-1}) Y(z) = \frac{z}{z-1}$$

$$Y(z) = \frac{z^2}{(z-1)(z-a)}$$

$$Y(z) = \frac{1}{1-a}\left[\frac{z}{z-1} - \frac{az}{z-a}\right]$$

Taking inverse Z-transform,

$$y(n) = \frac{1}{(1-a)}\,[1 - a^{n+1}\,u(n)]$$

Example 6. *Solve : $y(n) - y(n-1) = u(n) + u(n-1)$.*

Solution. Taking Z-transform on both sides,

$$Z[y(n)] - Z[y(n-1)] = Z[u(n)] + Z[u(n-1)]$$

$$Y(z) - z^{-1}\,Y(z) = \frac{z}{z-1} + z^{-1}\cdot\frac{z}{z-1}$$

$$(1 - z^{-1})\,Y(z) = \frac{z+1}{z-1}$$

$$Y(z) = \frac{z(z+1)}{(z-1)^2}$$

$$\frac{Y(z)}{z} = \frac{z+1}{(z-1)^2} = \frac{z-1+2}{(z-1)^2} = \frac{1}{z-1} + \frac{2}{(z-1)^2}$$

$$Y(z) = \frac{z}{z-1} + \frac{2z}{(z-1)^2}$$

$$y(n) = Z^{-1}\,[\text{RHS}] = 1 + 2n$$

Example 7. *Solve $x(n+1) - 2x(n) = 1$, given $x(0) = 0$.*

Solution. Taking Z-transform on both sides,

$$Z[x(n+1)] - 2Z[x(n)] = Z[1]$$

$$z[X(z) - x(0)] - 2\,.\,X(z) = \frac{z}{z-1}$$

$$(z-2)\,X(z) = \frac{z}{z-1}$$

$$\therefore \quad X(z) = \frac{z}{(z-1)(z-2)}$$

$$\frac{X(z)}{z} = \frac{1}{(z-1)(z-2)} = \frac{1}{z-2} - \frac{1}{z-1}$$

$$\therefore \quad X(z) = \frac{z}{(z-2)} - \frac{z}{(z-1)}$$

$$x(n) = Z^{-1}\,[\text{RHS}]$$

$$x(n) = 2^n - 1$$

Example 8. *Solve $y_{n+2} + y_n = 2$, given $y_0 = y_1 = 0$.*

Solution. $Z[y_{n+2}] + Z[y_n] = Z(2)$

$$z^2Y(z) - z^2y(0) - zy(1) + Y(z) = 2 \cdot \frac{z}{z-1}$$

$$(z^2+1)\,Y(z) = 2 \cdot \frac{z}{z-1}$$

$$Y(z) = 2 \cdot \frac{z}{(z-1)(z^2+1)} = \frac{z}{z-1} - \frac{z^2+z}{z^2+1}$$

$$y(n) = Z^{-1}\left(\frac{z}{z-1}\right) - Z^{-1}\left[\frac{z(z+1)}{z^2+1}\right]$$

$$= 1 - \left[\cos\frac{n\pi}{2} + \sin\frac{n\pi}{2}\right] u(n)$$

$$\left[\textbf{Hint.}\ \frac{Y(z)}{z} = \frac{2}{(z-1)(z^2+1)} = \frac{A}{z-1} + \frac{Bz+C}{z^2+1} = \frac{1}{z-1} - \frac{z+1}{z^2+1}\right.$$

$$\left.\text{Hence } Y(z) = \frac{z}{z-1} - \frac{(z^2+z)}{z^2+1}\right]$$

Example 9. *Solve $y_{n+2} - 4y_n = 0$ using Z-transform.*

Solution. Here, the conditions y_0 and y_1 are not given.

$\therefore$ Take $y_0 = A$, $y_1 = B$.

$$\therefore \quad Z[y_{n+2}] - 4Z[y_n] = 0$$

$$z^2Y(z) - z^2y(0) - zy(1) - 4Y(z) = 0$$

$$(z^2-4)\,Y(z) = Az^2 + Bz$$

$$Y(z) = \frac{Az^2}{z^2-4} + \frac{Bz}{z^2-4}$$

$$= \frac{A}{2}\left[\frac{z}{z-2} + \frac{z}{z+2}\right] + \frac{B}{4}\left[\frac{z}{z-2} - \frac{z}{z+2}\right]$$

$$y(n) = \frac{A}{2}[2^n + (-2)^n] + \frac{B}{4}[2^n - (-2)^n]$$

$$= \left(\frac{A}{2} + \frac{B}{4}\right)2^n + \left(\frac{A}{2} - \frac{B}{4}\right)(-2)^n$$

$$y(n) = C.2^n + D(-2)^n.$$

Example 10. *Using Z-transform, solve the simultaneous equations,*

$$x_{n+1} - y_n = 1 \qquad \text{given } x_0 = 0,\ y_0 = -1$$
$$y_{n+1} + x_n = 0$$

Solution. Taking Z-transforms

$$z[X(z)-x(0)]-Y(z)=\frac{z}{z-1}$$

$$zX(z)-Y(z)=\frac{z}{z-1} \qquad ...(1)$$

Similarly, $z[Y(z)-y(0)]+X(z)=0$

$$zY(z)+X(z)=-z \qquad ...(2)$$

Solving for $X(z)$ and $Y(z)$ from (1) and (2),

$$X(z)=\frac{z}{(z-1)(z^2+1)}$$

$$\therefore \quad x(n)=Z^{-1}[\text{RHS}]$$

$$=\frac{1}{2}\left[1-\cos\frac{n\pi}{2}-\sin\frac{n\pi}{2}\right] \qquad \text{(Refer Example 8)}$$

From given equation,

$$y_{n+1}=-x_n=\frac{1}{2}\left[\cos\frac{n\pi}{2}+\sin\frac{n\pi}{2}-1\right]$$

$$\therefore \quad y_n=\frac{1}{2}\left[\cos(n-1)\frac{\pi}{2}+\sin(n-1)\frac{\pi}{2}-1\right]$$

$$=\frac{1}{2}\left[\cos\frac{n\pi}{2}\cos\frac{\pi}{2}+\sin\frac{n\pi}{2}\sin\frac{\pi}{2}\right.$$

$$\left.+\sin\frac{n\pi}{2}\cos\frac{\pi}{2}-\cos\frac{n\pi}{2}\sin\frac{\pi}{2}-1\right]$$

$$y_n=\frac{1}{2}\left[\sin\frac{n\pi}{2}-\cos\frac{n\pi}{2}-1\right]$$

$$x_n=\frac{1}{2}\left[1-\cos\frac{n\pi}{2}-\sin\frac{n\pi}{2}\right]$$

EXERCISE 1

1. Find the Z-transform (one sided) of the sequences $\{x(n)\}$ where $x(n)$ is

(*i*) $\left(\frac{1}{3}\right)^n u(n)$ (*ii*) $\delta(n-2)$ (*iii*) $u(n-3)+\left(\frac{1}{2}\right)^n+2^n$

(*iv*) $(-1)^n u(n)$ (*v*) $2^n \sin\frac{n\pi}{2}$ (*vi*) $3^n \cos\frac{n\pi}{2}$

(*vii*) $n.2^n$ (*viii*) $2^n u(n)$ (*ix*) $n\,u(n), n^2 u(n)$

(*x*) $u(n-2)$ (*xi*) $u(n)\cos\frac{n\pi}{4}$ (*xii*) $u(n)\sin\frac{n\pi}{4}$

(*xiii*) an^2+bn+c; (Give region of convergence)

(*xiv*) $2^n u(n-1)$ (*xv*) $u(n-k)$ (*xvi*) $2^n u(n-k)$

2. Find the Z-transform of

(i) e^{2t+3} (ii) e^{-3t-7} (iii) $\sin 2t$ (iv) $\cos 3t$

(v) $\sin^2 3t$ (vi) $\cos^2 t$ (vii) $\sin^3 t$ (viii) $\cos^3 t$

(ix) $n.2^n$ (x) $n \sin n\theta$ (xi) $n \cos n\theta$

3. Find the inverse Z-transform of (assume $x(n)$ is a causal sequence) by any method; if possible do by more than one method.

(i) $\dfrac{z^2+z}{(z-1)^2}$ (ii) $\dfrac{1+2z^{-1}}{1-z^{-1}}$ (iii) $\dfrac{z}{(z-1)(z-2)}$

(iv) $\dfrac{z^2+z}{(z-1)^3}$ (v) $\dfrac{z^2+2z}{(z-1)(z-2)(z-3)}$ (vi) $\dfrac{4z^{-1}}{(1-z^{-1})^2}$

(vii) $\dfrac{z^2}{(z-a)(z-b)}$ (viii) $\dfrac{z}{(z-1)(z-2)(z-3)}$ (ix) $\dfrac{z^2-z}{(z+1)^2}$

(x) $\dfrac{z^2}{(z-1)^2}$ (xi) $\dfrac{4-8z^{-1}+6z^{-2}}{(1-2z^{-1})^2(1+z^{-1})}$

(xii) $\dfrac{3+\frac{11}{2}z^{-1}+7z^{-2}}{\left(1-\frac{1}{2}z^{-1}\right)(1+2z^{-1}+4z^{-2})}$ (xiii) $\dfrac{z^2}{(z+2)(z^2+4)}$

(xiv) $\dfrac{z}{z^2+11z+30}$ (xv) $\dfrac{2z^2+4z}{(z-2)^3}$ (xvi) $\dfrac{5z}{(2-z)(3z-1)}$

(xvii) $\dfrac{5z}{(2z-1)(z-3)}$ (xviii) $\dfrac{z(z^2-z+2)}{(z+1)(z-1)^2}$

4. Find the Z-transforms of

(i) $e^{-3t}\cdot t$ (ii) $e^{-t}\cdot t^2$ (iii) $e^{-at}\sin\omega t$

(iv) $e^{-t}\cos 2t$ (v) at^2+bt+c

5. Find the Z-transforms of the convolutions of $x(n) * y(n)$

(i) $2^n * n$ (ii) $n(n-1) * 3^n$ (iii) $3^n * \cos n\theta$

(iv) $2^n * \sin n\theta$ (v) $\cos\dfrac{n\pi}{2} * \sin\dfrac{n\pi}{2}$

6. Using convolution theorem, find the inverse Z-transform of

(i) $\dfrac{z^2}{(z-a)(z-b)}$ (ii) $\dfrac{1}{\left(1-\frac{1}{2}z^{-1}\right)\left(1-\frac{1}{4}z^{-1}\right)}$

7. Find (i) $a^n u(n) * a^n u(n)$ (ii) $3^n * 3^n$

8. Find the inverse Z-transform of

$$\frac{3-\frac{5}{6}z^{-1}}{\left(1-\frac{1}{4}z^{-1}\right)\left(\frac{1}{3}z^{-1}\right)} \text{ if } |z| > \frac{1}{3}$$

9. Find the Z-transform of $f(n) = -u(-n-1)$ where $u(n)$ is the unit step sequence.

10. Solve the difference equations, using Z-transforms :

(*i*) $x(n+2) - 3x(n+1) - 10x(n) = 0$, given $x(0) = 1, x(1) = 0$

(*ii*) $y(k+2) - 5y(k+1) + 6y(k) = 6k$ if $y(0) = y(1) = 0$

(*iii*) $y(n+2) - 5y(n+1) + 6y(n) = 4^n$ given $y(0) = 0,\ y(1) = 1$

(*iv*) $y_{n+2} + 3y_{n+1} + 2y_n = 0$ if $y_0 = 0,\ y_1 = 1$

(*v*) $f(n) + 3f(n-1) - 4f(n-2) = 0, n \geq 2$ given that $f(0) = 3$ and $f(1) = -2$

(*vi*) $f(n) - 2f(n-1) - 3f(n-2) = 0, n \geq 0$ given that $f(-1) = \frac{-5}{3}$ and $f(-2) = \frac{19}{9}$

(*vii*) $x(k+2) + 2x(k+1) + x(k) = u(k)$ where $x(0) = 0,\ x(1) = 0$ and $u(k) = k$ for $k = 0, 1, 2, \ldots$

(*viii*) $y_{n+2} - 4y_{n+1} + 3y_n = 0$ given $y_0 = 2,\ y_1 = 4$

(*ix*) $y_{n+2} - 4y_{n+1} + 4y_n = \pi$, if $y_0 = y_1 = 0$.

(*x*) $y_{n+1} - 3y_n = 0$ where $y_0 = 1$

(*xi*) $y_{n+3} + y_{n+2} - 8y_{n+1} - 12y_n = 0$ given $y_0 = 1,\ y_1 = y_2 = 0$

(*xii*) $u_{x+2} + u_x = 5.2^x$ given $u_0 = 1,\ u_1 = 0$

(*xiii*) $y_{x+2} + y_x = xa^x$

(*xiv*) $y_{n+2} - 7y_{n+1} + 12y_n = 2^n$ given $y_0 = y_1 = 0$

(*xv*) $y_{n+3} - 3y_{n+1} + 2y_n = 0$ given $y_1 = 0, y_2 = 8$ and $y_3 = -8$

(*xvi*) $y_{x+2} - 4y_{x+1} + 4y_x = 0$ if $y_0 = 1,\ y_1 = 2$

(*xvii*) $y_{n+2} - 4y_n = 2^n$ given $y_0 = 0,\ y_1 = 0$

(*xviii*) $x(n+2) - 5x(n+1) + 6x(n) = 5^n$ given $x(0) = x(1) = 0$

(*xix*) $y(n) - ay(n-1) = u(n)$

(*xx*) $y_{n+2} - 5y_{n+1} + 6y_n = 36$ given $y_0 = y_1 = 0$

(*xxi*) $x_{n+1} = 7x_n + 10y_n$

$y_{n+1} = x_n + 4y_n$ given $x_0 = 3,\ y_0 = 2$

(*xxii*) $a_{n+1} = a_n + \alpha b_n$

$b_{n+1} = b_n + \alpha a_n$ given $a_0 = 0,\ b_0 = 1.$

Find $\frac{a_n}{b_n}$ as a function of α and n.

(*xxiii*) $y_{n+2} - 2\cos\alpha\, y_{n+1} + \alpha\, y_n = 0$ given $y_0 = 1,\ y_1 = \cos\alpha$

(*xxiv*) $y_{n+2} - 2\cos\alpha\, y_{n+1} + y_n = 0$ if $y_0 = 0,\ y_1 = 1$

(*xxv*) $y_{n+2} - y_n = 2^n$ if $y_0 = 0,\ y_1 = 1$

(xxvi) $y(n+2) = y(n+1) + y(n)$ if $y(0) = 0,\ y(1) = 1$

(xxvii) $y_{n+1} + y_n = 1$ if $y_0 = 1$

(xxviii) $y_{n+1} - 3y_n = 2^n$ if $y_0 = 1$

(xxix) $y_{n+2} + 2y_{n+1} + y_n = n$ if $y_0 = y_1 = 0$

(xxx) $y_{n+2} - 4y_{n+1} + 4y_n = 0$ given $y_0 = 1,\ y_1 = 0$

EXERCISE 2
(Short Answer Questions)

1. $u(n) - u(n-1)$ is ....
2. Radius of convergence of $\sum_{n=0}^{\infty} x(n) z^{-n}$ is ....
3. $Z[a^n u(n)]$ exists only if ....
4. Z-transform of $nx(n)$ is ....
5. Z-transform of $n(n-1)u(n)$ is ....
6. Define the sampler.
7. Define a holding device.
8. Define two sided Z-transform.
9. Define one sided Z-transform.
10. Define unit step sequence and unit sample sequence.
11. Define $u(n-k)$.
12. $Z[n f(t)] =$ ....
13. $Z\{(-1)^n\} =$ ....
14. If $x(n) = n$, then $Z[x(n)]$ is ....

Write True or False

15. $Z\left[\sin \frac{n\pi}{2}\right] = \frac{z}{z^2+1}$
16. $Z\left[\cos \frac{n\pi}{2}\right] = \frac{z^2}{z^2+1}$
17. $\delta(n-k) = \frac{1}{z^k}$
18. $Z\left[\frac{1}{n}\right] = \log\left(\frac{z}{z-1}\right)$ if $|z| > 1$
19. $Z^{-1}\left(\frac{z}{z-1}\right)$ if $|z| > 1$ is ....
20. State convolution theorem for causal sequences.
21. $Z^{-1}\left[\frac{z}{(z-3)(z-4)}\right]$ is ....

22. If $Z[x(n)] = X(z)$ then $Z[nx(n)]$ ....

23. $Z^{-1}\left[\frac{z^2}{z^2+1}\right]$ is .... 24 . $Z^{-1}\left[\frac{z}{z^2+1}\right]$ is ....

25. Solve : $y_{n+1} - 2y_n = 1$ given $y_0 = 0$

26. Find the Residue of $\frac{z^n(z+1)}{(z-1)^3}$

27. Find the Residue of $\frac{z^n(z-4)}{(z-2)^2}$

28. State initial value theorem.

29. State final value theorem.

30. $Z[n(n-1)x(n)]$ is ....

ANSWERS

(Z-Transforms)

EXERCISE 1

1. (i) $\frac{3z}{3z-1}$ (ii) $\frac{1}{z^2}$ (iii) $\frac{1}{z^2(z-1)} + \frac{2z}{2z-1} + \frac{z}{z-2}$

(iv) $\frac{z}{z+1}$ (v) $\frac{2z}{z^2+4}$ (vi) $\frac{z^2}{z^2+9}$

(vii) $\frac{2z}{(z-2)^2}$ (viii) $\frac{z}{z-2}$ (ix) $\frac{z}{(z-1)^2}, \frac{z(z+1)}{(z-1)^3}$

(x) $\frac{1}{z(z-1)}$ (xi) $\frac{z(\sqrt{2}z-1)}{\sqrt{2}[z^2-\sqrt{2}z+1]}$ (xii) $\frac{z}{\sqrt{2}[z^2-\sqrt{2}z+1]}$

(xiii) $\frac{az(z+1)}{(z-1)^3} + \frac{bz}{(z-1)^2} + \frac{cz}{z-1}$ (xiv) $\frac{2}{z-2}$

(xv) $z^{-k} \cdot \frac{z}{z-1}$ (xvi) $\frac{\left(\frac{z}{2}\right)^{-k} \cdot \left(\frac{z}{2}\right)}{\left(\frac{z}{2}\right) - 1}$

2. (i) $e^3 \cdot \frac{z}{z-e^{2T}}$ (ii) $e^{-7} \cdot \frac{z}{z-e^{-3T}}$ (iii) $\frac{z\sin 2T}{z^2 - 2z\cos 2T + 1}$

(iv) $\frac{z(z-\cos 3T)}{z^2 - 2z\cos 3T + 1}$ (v) $\frac{1}{2} \cdot \frac{z}{z-1} - \frac{1}{2}\left[\frac{z(z-\cos 6T)}{z^2 - 2z\cos 6T + 1}\right]$

(vi) $\frac{1}{2}\frac{z}{z-1} + \frac{1}{2}\left[\frac{z(z-\cos 2T)}{z^2 - 2z\cos 2T + 1}\right]$

(vii) $\frac{3}{4}\left[\frac{z\sin T}{z^2-2z\cos T+1}\right]-\frac{1}{4}\left[\frac{z\sin 3T}{z^2-2z\cos 3T+1}\right]$

(viii) $\frac{3}{4}\left[\frac{z(z-\cos T)}{z^2-2z\cos T+1}\right]+\frac{1}{4}\left[\frac{z(z-\cos 3T)}{z^2-2z\cos 3T+1}\right]$

(ix) $\frac{2z}{(z-2)^2}$ (x) $z^{-1}\frac{d}{dz^{-1}}\left[\frac{\sin\theta \,.\, z^{-1}}{1-2z^{-1}\cos\theta+z^{-2}}\right]$

(xi) $z^{-1}\frac{d}{dz^{-1}}\left[\frac{1-\cos\theta \,.\, z^{-1}}{1-2\cos\theta \,.\, z^{-1}+z^{-2}}\right]$

3. (i) $2n+1$ (ii) $1+2u(n-1)$ (iii) $(2^n-1)\,u(n)$

(iv) $\frac{t^2}{T^2}$ (v) $\frac{3}{2}-4.2^n+\frac{5}{2}\cdot 3^n$ (vi) $2n(n-1)\,u(n)$

(vii) $\frac{1}{ab}[a^{n+1}-b^{n+1}]$ (viii) $\frac{1}{2}-\frac{1}{2}2^n+\frac{1}{2}\cdot 3^n$

(ix) $(2n+1)(-1)^n$ (x) $(n+1)\,u(n)$ (xi) $2(-1)^n+2^n(n+2)$

(xii) $2\left(\frac{1}{2}\right)^n+2^n\left[\cos\frac{2\pi n}{3}+\frac{1}{\sqrt{3}}\sin\frac{2\pi n}{3}\right]$

(xiii) $\frac{(-2)^{n+1}}{8}+\frac{(2i)^{n+1}}{8i(1+i)}-\frac{(2i)^{n+1}}{2i(1-i)}$

(xiv) $(-5)^n-(-6)^n$ (xv) $n^2 2^n$ (xvi) $\left(\frac{1}{3}\right)^n-2^n$

(xvii) $3^n-\left(\frac{1}{2}\right)^n$ (xviii) $(-1)^n+\frac{1}{2}(2n-1),\ n=0,1,2,\ldots$

4. (i) $\frac{T\,.\,ze^{3T}}{(ze^{3T}-1)^2}$ (ii) $\frac{T^2\,.\,ze^T(ze^T+1)}{(ze^T-1)^3}$ (iii) $\frac{ze^{aT}\sin\omega t}{z^2e^{2aT}-2ze^{aT}\cos\omega T+1}$

(iv) $\frac{ze^T(ze^T-\cos 2T)}{z^2e^{2T}-2ze^T\cos 2T+1}$ (v) $\frac{aT^2z(z+1)}{(z-1)^3}+\frac{bTz}{(z-1)^2}+\frac{cz}{z-1}$

5. (i) $\frac{z^2}{(z-2)(z-1)^2}$ (ii) $\frac{2z^2}{(z-1)^2(z-3)}$ (iii) $\frac{z^2(z-\cos\theta)}{(z-3)(z^2-2z\cos\theta+1)}$

(iv) $\frac{z^2\sin\theta}{(z-2)(z^2-2z\cos\theta+1)}$ (v) $\frac{z^3}{(z^2+1)^2}$

6. (i) $\frac{1}{a-b}[a^{n+1}-b^{n+1}]$ (ii) $\frac{2}{3}\left(\frac{1}{2}\right)^n+\frac{1}{3}\left(-\frac{1}{4}\right)^n$

7. (i) $(1+n)\,a^n$ (ii) $(n+1)\,3^n$

8. $\left(\frac{1}{4}\right)^n-2\left(\frac{1}{3}\right)^n$ **9.** $\frac{z}{z-1}$ if $|z|<1$

10. (*i*) $x(n) = \frac{2}{7} \cdot (5)^n + \frac{5}{7}(-2)^n, \; n = 0, 1, 2, \ldots$

(*ii*) $y(k) = \frac{1}{12} 6^k - \frac{1}{3} \cdot 3^k + \frac{1}{4} . 2^k, \; k = 0, 1, 2, \ldots$

(*iii*) $y(n) = \frac{1}{2}(4^n - 2^n), \; n = 0, 1, 2, \ldots$

(*iv*) $y_n = (-1)^n (-2)^n, \; n = 0, 1, 2, \ldots$ (*v*) $f(n) = \frac{1}{5} \cdot 4^n + \frac{14}{5}(-1)^n$

(*vi*) $f(n) = 3^n + 2(-1)^n$ (*vii*) $x(k) = \frac{k+1}{4} + \frac{(-1)^k}{4}(1-k), \; k = 0, 1, 2, \ldots$

(*viii*) $y_n = 1 + 3^n$ (*ix*) $y_n = \pi\left[1 + 2^n\left(\frac{n}{2} - 1\right)\right]$

(*x*) $y(n) = 3^n$ (*xi*) $y_n = \frac{4}{25} \cdot 3 + \frac{1}{25}(21 - 15n)(-2)^n$

(*xii*) $u_x = 2^x - 2 \sin \frac{\pi x}{2}$ (*xiii*) $y_x = A \cos \frac{\pi x}{2} + B \sin \frac{\pi x}{2} + \frac{a^x}{1+a^2}\left(x - \frac{2a^2}{1+a^2}\right)$

(*xiv*) $y_n = (-1)\, 3^n + \frac{1}{2} \cdot 4^n + 2^{n-1}$ (*xv*) $y_n = \frac{8}{3} + \frac{4}{3}(-2)^n$ (*xvi*) $y_x = 2^x$

(*xvii*) $y_n = \frac{1}{16}[(-2)^n - 2^n + n \,.\, 2^{n+1}]$ (*xviii*) $x(n) = \frac{1}{3} \cdot 2^n - \frac{1}{2} \cdot 3^n + \frac{5^n}{6}$

(*xix*) $y(n) = \frac{1}{1-a}(1 - a^{n+1})$ (*xx*) $y_n = 18 \cdot 3^n - 36 \cdot 2^n + 18$

(*xxi*) $x_n = 5.9^n - 2.2^n; \; y_n = 9^n + 2^n$ (*xxii*) $\tan(n \tan^{-1} \alpha)$

(*xxiii*) $y_n = \cos n\alpha$ (*xxiv*) $y_n = \frac{\sin n\alpha}{\sin \alpha}$

(*xxv*) $y_n = \frac{1}{3}[2^n - (-1)^n]$ (*xxvi*) $y_n = \frac{1}{\sqrt{5}}\left[\left(\frac{\sqrt{5}+1}{2}\right)^n - \left(\frac{1-\sqrt{5}}{2}\right)^n\right]$

(*xxvii*) $y_n = \frac{1}{2}[1 - (-1)^n]$ (*xxviii*) $y_n = 2.3^n - 2^n$

(*xxix*) $y_n = \frac{n-1}{4}[1 - (-1)^n]$ (*xxx*) $y_n = 2^n(1-n)$

EXERCISE 2

1. $\delta(n)$ **2.** $\underset{n \to \infty}{\text{Lt}} \left|\frac{x(n+1)}{x(n)}\right|$ **3.** $|z| > |a|$ **4.** $z^{-1} \frac{dX}{dz^{-1}}$ **5.** $\frac{2z^{-1}}{(1-z^{-1})^3}$

11. $\begin{cases} = 1 & \text{if } n = k \\ = 0 & \text{if } n \neq k \end{cases}$ **12.** $-z \frac{dF(z)}{dz}$ 13. $\frac{z}{z+1}$ if $|z| > 1$ **14.** $\frac{z}{(z-1)^2}$ **15** to

1 8. True **19.** $u(n)$ **21.** $4^n - 3^n$ **22.** $z^{-1} \frac{dX}{dz^{-1}}$ **23.** $\cos \frac{n\pi}{2}$ **24.** $\sin \frac{n\pi}{2}$

25. $2^n - 1$ **26.** n^2 **27.** $2^n(1-n)$ **30.** $z^{-2} \frac{d^2X}{d(z^{-1})^2}$

CHAPTER VI

LINEAR ALGEBRA

Revision : THE GROUP

Definition : A *group* is a non-empty set G of elements together with a *binary* operation * satisfying the following axioms or postulates.

P_1 : **Closure property.** For all $a, b \in G$, $a * b \in G$. *i.e.,* the set G is closed with respect to the composition $*$.

P_2 : **Associative law.** For all $a, b, c \in G$,

$$(a * b) * c = a * (b * c)$$

i.e., the operation * obeys associative law.

P_3 : **Existence of identity element.** There exists *an element* $e \in G$ such that

$$a * e = e * a = a \text{ all for } a \in G.$$

i.e., the binary composition admits identity element.

P_4 : **Existence of inverse element :** For every $a \in G$, there exists an element $a^{-1} \in G$ such that

$$a * a^{-1} = a^{-1} * a = e$$

where e is *identity element* or *unit element* of G.

Definition : Abelian group. A group $(G, *)$ is said to be an **Abelian group** or commutative group if the group operation $*$ satisfies the commutative law. That is, *for every* $a, b, \in G$, $a * b = b * a$.

Note. 1. If the first postulate only is satisfied then G is called a *groupoid or an algebraic structure or a quasi group.*

2. If the first two postulates are satisfied, then G is called a *semi group.*

3. If the first three postulates are satisfied, then G is called a *monoid.*

4. If all the four postulates are satisfied, then G is called a *group.*

Properties of groups (without proof)

Let $(G, *)$ be a group. Then,

1. the identity element e is unique
2. for every element $a \in G$, its inverse a^{-1} is unique
3. for all $a, b, c \in G$,

$$a * b = a * c \Rightarrow b = c$$
$$b * a = c * a \Rightarrow b = c$$

4. for $a, b \in G$, the equations

$a * x = b$ and $y * a = b$ have unique solutions in G for x and y.

5. for $a, b \in G$,

$(a * b)^{-1} = b^{-1} * a^{-1}$

THE RING

A non empty set R is said to be an *associative ring* if, in R, there are two defined operations '+' and '.' such that for all $a, b, c \in R$.

I : $p_1 : a + b \in R.$

$p_2 : a + b = b + a$

$p_3 : (a + b) + c = a + (b + c)$

p_4 : There exists an element 0 in R, such that $0 + a = a + 0 = a$ for every a in R

p_5 : There exists an element $-a$ in R such that

$$a + (-a) = (-a) + a = 0$$

II : $p_6 : \forall\, a, b \text{ in } R, a \,.\, b \in R$

$p_7 : \forall\, a, b, c \in R, (a \,.\, b) \,.\, c = a \,.\, (b \,.\, c)$

III : p_8 : Distributive laws.

For all $a, b, c \in R$,

$a \,.\, (b + c) = a \,.\, b + a \,.\, c$ (dot is distributive over

and $(b + c) \,.\, a = b \,.\, a + c \,.\, a$ plus – both left and right)

These eight postulates must be satisfied in R, called *associative Ring*.

Note : 1. Axioms P_1 to P_5 state that R is an abelian group under the operation '+'. (addition)

2. Axioms P_6 and P_7 insist that R must be closed under an associative operation '.'. (multiplication)

3. Axiom P_8 serves to interrelate the two operations. That is, . is distributive over +, both right and left.

Unit element: If there exists an element $1 \in R$ such that $1 \,.\, a = a \,.\, 1 = a$ $\forall\, a \in R$, then R is a ring with *unity* or unit element 1.

Commutative ring : If the operation '.' is such that $a \,.\, b = b \,.\, a \;\forall\, a, b \in R$, than R is said to be *a commutative ring.*

Zero division : If R is a commutative ring then $a \neq 0$ and $a \in R$, is said to be a *zero divisor* if there exists an element $b \in R$, $b \neq 0$ such that $a \,.\, b = 0$.

Integral Domain : A commutative ring is an integral domain if it has *no zero divisors.*

Division ring : A ring is said to be a *division ring* if its non zero elements form a group under multiplication.

Notation. A ring is denoted by $(R, +, .)$

THE FIELD

A *field* is a *commutative division ring.*

A finite integral domain is a *field.*

In other words,

Definition. A ring $(R, +, .)$ which has at least two elements is called a *field* if

(*i*) it is a commutative ring with unity and

(*ii*) all the non-zero elements are invertable w.r.t the multiplication.

i.e., $\forall\ a \neq 0$, and $a \in R$,

There exists $b \in R$ such that $a \cdot b = b \cdot a = 1$.

Stating in other words (R has at least two elements)

(1) $(R, +)$ is an abelian group

(2) $(R - \{0\}, .)$ is an abelian group

(3) distributive law of . over + holds good (both left and right)

Examples : (Prove yourself)

1. The set of integers with ordinary addition (+) and multiplication (.) as the compositions forms a ring'.

2. The set of numbers of the form $a + b\sqrt{3}$ where a, b are integers forms a ring under '+' and '.'.

3. The set of 2×2 matrices is a ring w.r.t addition and multiplication of matrices.

Note: 1. In a field 0 and 1 are distinct elements.

2. A *subfield S* of a field $(F, +, .)$ is a subset of F which is also field under + and . (same laws of compositions)

Fields (Examples) (Prove yourself)

1. The set of all real numbers R is a field w.r.t '+' and '.'

2. The set of all rational number Q is a field w.r.t + and .

3. $(C, +, .)$ is a field where C is the set of all complex numbers.

4. The set of positive integers with addition and multiplication modulo p, p being prime, is a finite field with p elements.

VECTOR SPACE

Definition: Let $(F, +, .)$ be a given field and V be a non-empty set having defined in it one internal and one external composition $\oplus$ and $*$. Then the given set V is said to be a *vector space* or *a linear space* over the field F, denoted by $V(F)$, iff the following postulates are satisfied.

V_1 : The algebraic structure (V, $\oplus$) is an abelian group.

That is, for all $x, y, z, \in V$.

(1) $x \oplus y \in V$

(2) $(x \oplus y) \oplus z = x \oplus (y \oplus z)$

(3) $\exists$ an element $\overline{0} \in V$, such that $\forall x \in V,\ x \oplus 0 = 0 \oplus x = x$

(4) $\forall x \in V$, $\exists$ an element $-x \in V$, such that $x \oplus (-x) = (-x) \oplus x = \overline{0}$

(5) $x \oplus y = y \oplus x \ \forall\ x, y, z, \in V$

V_2 : (6) Scalar multiplication is closed.

i.e., $\forall\ \alpha \in F, x \in V, \alpha * x \in V$ ($\alpha * x$ is a vector)

(7) The scalar multiplication is associative.

i.e., For $\alpha, \beta \in F$, and $x \in V$

$(\alpha . \beta) * x = \alpha * (\beta * x)$

(8) For unit scalar $1 \in F$,

$1 * x = x \ \forall\ x \in V$

V_3 : (9) Scalar multiplication is distributive over vector addition.

$\alpha * (x \oplus y) = (\alpha * x) \oplus (\alpha * y)$ for $\alpha \in$ F, $x, y \in V$.

(10) Multiplication by vector is distributive over scalar addition.

$(\alpha + \beta) * x = (\alpha * x) \oplus (\beta * x)$, $\alpha, \beta \in F$ and $x \in V$.

Note: 1. The elements of V are called *vectors*.

2. The elements of the field F are called *scalars*.

3. V_1 defines the additive structure of the system

4. V_2 defines the multiplicative structure of the system.

5. V_3 expresses the connection between the two structures.

Null space: The vector space which has only one element $\overline{0}$, the additive identity of the vector space, is called a *null space* or *zero vector space* and is denoted by $\{\overline{0}\}$.

Note: 1. 1 does not belong to V but it belong to F.

2. **Normally,** we take $(R, +, .)$ or $(C, +, .)$ as fields. In that case, real numbers or complex numbers are scalars. In our course, mostly we adopt this.

3. We use α, β, γ .... for scalars and x, y, z .... for vectors.

Examples 1. *A field F is a vector space over F or over any subfield S of F.*

i.e., F(F) or *F(S)* is a vector space.

2. *C(R), C(C), R(R)*, are vector spaces. (where C is the complex field and R is the real field)

Let us prove one of the above namely $C(R)$ is a vector space.

To Prove V_1: Here + and . of the field are ordinary + and .

The $\oplus$ and $*$ of the space C are also ordinary + and . of C.

$(C, \oplus) = (C, +)$ is an abelian group

This is true from group ideas. We accept this is true.

$V_2 : \alpha, \beta \in F$ means $\alpha, \beta \in R$

$x, y \in V$ means $x, y \in C$.

i.e, x, y, z are complex numbers and, α, β are real numbers

$\alpha * x = \alpha . x =$ (real no) . (complex no) = complex no $\in C$. So $\alpha * x \in C$

$(\alpha . \beta) * x = (\alpha . \beta) . x = \alpha . (\beta . x) = \alpha * (\beta * x)$ is true. Also $1 \in R$. Hence

$1 * x = 1 . x = x$ is true $\forall\, x \in C$

$V_3 : \alpha * (x \oplus y) = \alpha . (x + y) = \alpha . x + \alpha . y = (\alpha * x) \oplus (\alpha * y)$ is true

$(\alpha + \beta) * x = (\alpha + \beta) . x = \alpha . x + \beta . x = \alpha * x \oplus \beta * x$

All the ten axioms of the vector space are verified to be true. Hence C(R) is vector space (C is space and R is the field here)

Example 2. *Is R (C) a vector space* ?

Here R is the space and C is the field.

$\alpha, \beta, \in C$ and $x, y \in R$.

V_1 is true here. (you can try).

$\alpha * x \neq$ (complex no) . (real no) = complex no $\notin$ R

$\alpha * x \in$ field and not to space.

This postulates is violated.

Hence it is not a vector space.

Example 3. *Show that V, the set of all ordered n-tuples of the elements of any field F for a fixed $n \in N$, is a vector space.*

Proof : We will verify the axioms of a vector space

(*i*) Let $\bar{x} = (x_1, x_2, \ldots\ldots\ldots x_n)$, $\bar{y} = (y_1, y_2, \ldots\ldots\ldots\ldots y_n)$ be two vectors of V. Here $x_i, y_i \in F$

We define $\bar{x} \oplus \bar{y} = (x_1, x_2, \ldots\ldots\ldots x_n) \oplus (y_1, y_2, \ldots\ldots\ldots\ldots y_n)$

$$= (x_1 + y_1, x_2 + y_2, \ldots\ldots\ldots\ldots x_n + y_n)$$

$x_i + y_i \in F$ since $x_i, y_i \in F$.

Hence $\bar{x} \oplus \bar{y} \in V$.

Addition of vectors is binary. *i.e.,*

(*ii*) Associativity :

$$\bar{x} \oplus (\bar{y} \oplus \bar{z}) = (x_1, x_2, \ldots\ldots x_n) \oplus [(y_1, y_2, \ldots\ldots y_n) \oplus (z_1, z_2, \ldots\ldots z_n)]$$

$= (x_1, x_2, \ldots\ldots x_n) \oplus (y_1 + z_1, y_2 + z_2, \ldots\ldots y_n + z_n)$

$= (x_1 + y_1 + z_1, x_2 + y_2 + z_2, \ldots\ldots\ldots x_n + y_n + z_n)$

Similarly $(\overline{x} \oplus \overline{y}) \oplus \overline{z} = (x_1 + y_1 + z_1\ x_2 + y_2 + z_2 + \ldots\ldots\ldots + x_n + y_n + z_n)$

Hence $\overline{x} \oplus (\overline{y} \oplus \overline{z}) = (\overline{x} \oplus \overline{y}) \oplus \overline{z}$

(iii) $\forall\ \overline{x} = (x_1, x_2, \ldots\ldots x_n), \exists\ \overline{0} = (0, 0, \ldots\ldots\ldots 0) \in V$

Such that $\overline{x} \oplus \overline{0} = (x_1 + 0, x_2 + 0, \ldots\ldots\ x_n + 0) = (x_1, x_2, \ldots\ldots\ x_n) = \overline{x}$

Also $\overline{0} \oplus \overline{x} = \overline{x}$

(iv) $\forall\ \overline{x} = (x_1, x_2, \ldots\ldots x_n), -\overline{x} = (-x_1, -x_2, \ldots\ldots -x_n) \in V$

and $\overline{x} + (-\overline{x}) = \overline{0}$ and $(-\overline{x}) + \overline{x} = \overline{0}$

(v) Also $\overline{x} \oplus \overline{y} = \overline{y} \oplus \overline{x}$ is true

Hence $(V, \oplus)$ is an abelian group.

(vi) Let $\alpha, \beta \in F$

Define $\alpha * \overline{x} = \alpha\ .\ (x_1, x_2, \ldots\ x_n) = (\alpha x_1, \alpha x_2, \ldots\ldots \alpha x_n) \in V$

since $\alpha_1, x_i \in F$ and

so $\alpha x_i \in F$

$\therefore \quad \alpha \in F,\ \overline{x} \in v \Rightarrow \alpha * \overline{x} \in v$

(vii) $\alpha * (\beta * x) = \alpha * (\beta x_1, \beta x_2 \ldots\ldots\ \beta x_n)$

$= (\alpha\beta x_1, \alpha\beta x_2 \ldots\ldots\ \alpha\beta x_n)$

$= (\alpha\ .\ \beta) * (x_1, x_2 \ldots\ldots x_n)$

$= (\alpha\ .\ \beta) * \overline{x}$

(viii) $\forall\ \overline{x} \in$ V, and $1 \in F$

$1 * \overline{x} = 1 * (x_1, x_2 \ldots\ldots\ldots\ x_n)$

$= (1.x_1, 1.x_2 \ldots\ldots\ldots\ 1.x_n)$

$= (x_1, x_2 \ldots\ldots\ x_n)$

$= \overline{x}$

(ix) $\forall\ \overline{x}, \overline{y} \in V, \alpha \in F,$

$\alpha * (\overline{x} \oplus \overline{y}) = \alpha * (x_1 + y_1, x_2 + y_2, \ldots\ldots\ x_n + y_n)$

$= (\alpha\, x_1 + \alpha\, y_1, \alpha\, x_2 + \alpha\, y_2, \ldots\ldots\ \alpha\, x_n + \alpha\, y_n)$

$= (\alpha\, x_1\ , \alpha\, x_2 \ldots\ldots\ \alpha\, x_n) \oplus (\alpha\, y_1\ , \alpha\, y_2 \ldots\ldots\ \alpha\, y_n)$

$= \alpha * \overline{x}\ \oplus \alpha * \overline{y}$

(x) $(\alpha + \beta) * \overline{x} = ((\alpha + \beta)x_1, (\alpha + \beta)x_2, \ldots\ldots\ldots\ (\alpha + \beta)x_n)$

$$= (\alpha x_1 + \beta x_1, \alpha x_2 + \beta x_2, ... \alpha x_n + \beta x_n)$$
$$= (\alpha x_1, \alpha x_2, ... \alpha x_n) + (\beta x_1, \beta x_2, ... \beta x_n)$$
$$= \alpha * \bar{x} \oplus \beta * \bar{y}$$

Hence $V(F)$ is a vector space.

EXERCISE 6 (a)

1. Let F be a field and let V be the set of all ordered pairs (x, y) where $x, y \in F$.
   Define $(x_1, y_1) \oplus (x_2, y_2) = (x_1 + x_2, y_1 + y_2)$ and $\alpha (x, y) = (\alpha x, y)$ where $\alpha \in F$.
   Prove $V(F)$ is not a vector space.
2. $V = \{(a, b) : a, b \in R\}$
   Define $(a, b) \oplus (c, d) = (o, b + d)$
   $$\alpha * (a, b) = (\alpha a, \alpha b)$$
   Prove $V(R)$ is not a vector space.
3. In problem 2, define $\alpha * (a, b) = (\alpha a, b), \alpha \in R$. Prove $V(R)$ is not a vector space.
4. Prove $C(C)$, $C(R)$, $R(R)$ are vector spaces, where addition and scalar multiplication of vectors are ordinary addition and multiplication.
5. Show that the set of all $m \times n$ matrices with elements as real numbers is a vector space over the field R with usual operations of matrices.
6. If A is the set of all real valued continuous functions defined in $[0, 1]$, show that A is a vector space over R with addition and scalar multiplication defined as follows.
   $$(f + g)(x) = f(x) + g(x) \ \forall \ f, g \in A$$
   $$(\alpha f)(x) = \alpha f(x), \ \forall \ f \in A, \alpha \in R.$$
7. Let $P(x)$ denote the set of all polynomials in one variable x over a field F. Then, prove $P(x)$ is a vector space over F with addition defined as addition of polynomials and scalar multiplication defined as the product of the polynomials by an element of F.
8. R^n is the set of n-tuples of real numbers. Prove R^n is a real vector space.
9. C^n is the set of all n-tuples of complex numbers. Define $x \oplus y = (x_1 + y_1, x_2 + y_2, ... x_n + y_n)$ and $\alpha * x = (\alpha x_1, \alpha x_2, ... \alpha x_n)$ as usual. Prove C^n is a vector space over C.
10. Taking $n = 1$ in problem 9, Prove $C(C)$ is a vector space.
11. V consists of only zero vector with scalar multiplication defined by $\alpha * \bar{0} = \bar{0}$ for all $\alpha \in R$. Prove $V(R)$ is a vector space.

LINEAR INDEPENDENCE AND DEPENDENCE

Definition: If V be a vector space over the field F, then the vectors x_1, x_2, ... $x_n \in V$ are said to be *linearly dependent* over F, if there exist elements $\alpha_1, \alpha_2 ....\alpha_n \in F$, not all of them zero, that is, at least one of the $a_i's$ is not equal to zero, such that $\sum_{i=1}^{n} \alpha_i x_i = \overline{0}$

Otherwise, the vectors are said to be *linearly independent.*

In other words, if $\sum_{i=1}^{n} \alpha_i x_i = \overline{0}$ implies each $\alpha_i = 0$

Then $x_1, x_2, ... x_n$ are *linearly independent* and at least one α_i is not zero, then $x_1, x_2 ......x_n$ are *linearly dependent.*

Note. 1. The set $S = \{x_1, x_2, ... x_n\}$ is linearly independent or linearly dependent according as the vectors are linearly independent or linearly dependent.

2. Any subset of a linearly independent set is linearly independent.

Proof. Let $x_1, x_2, ... x_n$ be linearly independent.

$$\text{Therefore, } \sum_{i=1}^{n} \alpha_i x_i = 0 \Rightarrow \alpha_1 = \alpha_2 = ... \alpha_n = 0 \qquad ...(1)$$

To prove $x_1, x_2 ....x_m$ are linearly independent, $m \leq n$.
The order of the elements is immaterial.

$$\sum_{i=1}^{m} \alpha_i x_i = 0 \Rightarrow \sum_{i=1}^{m} \alpha_i x_i + \sum_{i=m+1}^{n} 0.x_i = 0$$

i.e., $\sum_{i=1}^{n} \alpha_i x_i = 0$ where $\alpha_i = 0$, for $i = m + 1, ... n$

By (1), each $\alpha_l = 0$

$\therefore \alpha_1, \alpha_2, ... \alpha_m = 0$

Hence $x_1, x_2, ... x_m$ are linearly independent.

3. Any superset of a *linearly dependent set* is *linearly dependent.*

4. Any subset of a vector space is either linearly independent or linearly dependent.

5. A set containing zero vector only is linearly dependent.

6. Any set of a vector space containing only a non-zero vector is linearly independent.

7. A set containing one of the vectors as zero vector is linearly dependent.
8. The set of non-zero vectors $x_1, x_2, \ldots\ldots x_n$ is linearlydependent iff some x_k, $2 \le k \le n$, is a linear combination of the preceding ones.
9. If x is a linear combination of $x_1, x_2, \ldots\ldots x_n$ then $\{x, x_1, x_2, \ldots\ldots x_n\}$ is linearly dependent.

Example 1. *Determine whether the following vectors are linearly dependent or not.*

(*i*) $x = (4, 3, -1), y = (1, 1, 5)$ in R^3

(*ii*) $x = (1, 2, 3, 4); y = (2, 4, 6, 8)$ in R^4

(*iii*) $x = \begin{pmatrix} 1 & -2 & 4 \\ 3 & 0 & -1 \end{pmatrix}, y = \begin{pmatrix} 2 & -4 & 8 \\ 6 & 0 & -2 \end{pmatrix}$

(*iv*) $x = t^3 + 3t + 4, y = t^3 + 4t + 3$

Proof. When two vectors are given, they are linearly dependent if one is a scalar multiple of the other. Hence

(*i*) $x \ne ky$ $\therefore$ x, y are linearly independent

(*ii*) $y = 2x$ $\therefore$ x, y are linearly dependent

(*iii*) $y = 2x$; x, y are linearly dependent

(*iv*) x, y are linearly independent.

Example 2. *Determine whether the following vectors in V_3 (R) or R^3 are linearly dependent or not.*

(*i*) $x = (1, 2, 3), y = (4, 1, 5), z = (-4, 6, 2)$

(*ii*) $(1, 2, -3), (1, -3, 2), (2, -1, 5)$

(*iii*) $(3, 0, -3), (-1, 1, 2), (4, 2, -2), (2, 1, 1)$

Proof. (*i*) Let $\alpha, \beta, \gamma \in R$ (scalars)

$$\alpha x + \beta y + \gamma z = 0$$

$$\Rightarrow \quad \alpha (1, 2, 3) + \beta (4, 1, 5) + \gamma (-4, 6, 2) = (0, 0, 0)$$

$$\therefore \quad \alpha + 4\beta - 4\gamma = 0 \quad \ldots(1) \qquad (1) + (2) \text{ gives } (3)$$

$$2\alpha + \beta + 6\gamma = 0 \quad \ldots(2) \qquad \text{Solve } \alpha + 4\beta - 4\gamma = 0$$

$$3\alpha + 5\beta + 2\gamma = 0 \quad \ldots(3) \qquad 2\alpha + \beta + 6\gamma = 0$$

Hence $\alpha = -4, \beta = 2, \gamma = 1$

Hence the vectors are linearly dependent.

(*ii*) Write the vectors as row vectors of a matrix.

$$A = \begin{pmatrix} 1 & 2 & -3 \\ 1 & -3 & 2 \\ 2 & -1 & 5 \end{pmatrix}$$

$$\sim \begin{pmatrix} 1 & 2 & -3 \\ 0 & -5 & 5 \\ 0 & -5 & 11 \end{pmatrix}$$

$$\sim \begin{pmatrix} 1 & 2 & -3 \\ 0 & -5 & 5 \\ 0 & 0 & 6 \end{pmatrix}$$

$R(A) = 3$ and the three vectors are linear independent

(*iii*) There are 4 vectors in R^3. Hence they are linearly dependent.

Example 3. *Prove that the vectors (1, 0, 0), (0, 1, 0) (0, 0, 1) and (1, 1, 1) are linearly dependent. Also prove any three vectors are linearly independent.*

Proof. There are 4 vectors in R^3.

Hence they are linearly dependent.

OR $\alpha(1, 0, 0) + \beta(0, 1, 0) + \gamma(0, 0, 1) + \delta(1, 1, 1) = (0, 0, 0)$

$\Rightarrow \quad \alpha + \delta = 0, \beta + \delta = 0, \gamma + \delta = 0$

$\Rightarrow \quad \alpha = \beta = \gamma = 1, \delta = -1$

Hence the vectors are linearly dependent.

Also (1, 0, 0), (0, 1, 0), (0, 0, 1) are linearly independent.

Example 4. *If x, y, z are linearly independent vectors in V(F) prove (i) x + y, y + z, z + x are linearly independent and (ii) x–y, y – z, z – x are linearly dependent.*

Proof: $\alpha(x + y) + \beta(y + z) + \gamma(z + x) = 0$

$\Rightarrow \quad (\alpha + \gamma)x + (\alpha + \beta)y + (\beta + \gamma)z = 0$

$\Rightarrow \quad \alpha + \gamma = 0, \alpha + \beta = 0, \beta + \gamma = 0 \qquad \ldots(1)$

Adding $2(\alpha + \beta + \gamma) = 0$

$\therefore \quad \alpha + \beta + \gamma = 0$

using (1), $\alpha = 0, \beta = 0, \gamma = 0$

Hence $x + y, y + z, z + x$ are linearly independent.

(*ii*) $\alpha(x - y) + \beta(y - z) + \gamma(z - x) = 0$

$\Rightarrow \alpha = 1, \beta = 1, \gamma = 1$

Hence $x - y, y - z, z - x$ are linearly dependent.

Example 5. *If V(R) is a vector space of 2 × 3 matrices over R, show that*

$$A = \begin{pmatrix} 2 & 1 & -1 \\ 3 & -2 & 4 \end{pmatrix}, B = \begin{pmatrix} 1 & 1 & -3 \\ -2 & 0 & 5 \end{pmatrix}, C = \begin{pmatrix} 4 & -1 & 2 \\ 1 & -2 & 3 \end{pmatrix}$$

in V(R) are linearly independent.

Proof. $\alpha A + \beta B + \gamma C = 0$

$$\Rightarrow \alpha \begin{pmatrix} 2 & 1 & -1 \\ 3 & -2 & 4 \end{pmatrix} + \beta \begin{pmatrix} 1 & 1 & -3 \\ -2 & 0 & 5 \end{pmatrix} + \gamma \begin{pmatrix} 4 & -1 & 2 \\ 1 & -2 & 3 \end{pmatrix} = 0$$

$\therefore$ $2\alpha + \beta + 4\gamma = 0$ ... (1) $3\alpha - 2\beta + \gamma = 0$ ... (4)

$\alpha + \beta - \gamma = 0$ ...(2) $-2\alpha \; . - 2\gamma = 0$ ... (5)

$-\alpha - 3\beta + 2\gamma = 0$ ...(3) $4\alpha + 5\beta + 3\gamma = 0$ ... (6)

Solving we get the only solution $\alpha = 0, \beta = 0, \gamma = 0$

Hence they are linearly independent.

Example 6. *Prove (1, 0, 0) (0, 1, 0) and (0, 0, 1) are linearly independent in R^3. Find three sets of three linearly independent vectors from your answer.*

$$\alpha(1, 0, 0) + \beta(0, 1, 0) + \gamma(0, 0, 1) = 0$$

$$\Rightarrow \alpha = 0, \beta = 0, \gamma = 0$$

Hence the vectors are linearly independent.

$x + y, y + z, z + x$ are also linearly independent.

Hence (1, 0, 0) + (0, 1, 0), (0, 1, 0) + (0, 0, 1), (0, 0, 1) + (1, 0, 0)

i.e., (1, 1, 0), (0, 1, 1) and (1, 0, 1) are linearly independent.

(1, 2, 1), (1, 1, 2) and (2, 1, 1) are linearly independent.

Also (2, 3, 3), (3, 2, 3) and (3, 3, 2) are linearly independent.

Example 7. *Show that the vectors (1, 2, 1, 0), (1, 3, 1, 2) (4, 2, 1, 0) and (6, 1, 0, 1) are linearly independent in R^4.*

Proof. Write vectors as row vectors of a matrix A.

$$A = \begin{pmatrix} 1 & 2 & 1 & 0 \\ 1 & 3 & 1 & 2 \\ 4 & 2 & 1 & 0 \\ 6 & 1 & 0 & 1 \end{pmatrix}$$

$$\sim \begin{pmatrix} 1 & 2 & 1 & 0 \\ 0 & 1 & 0 & 2 \\ 0 & -6 & -3 & 0 \\ 0 & -11 & -6 & 1 \end{pmatrix}$$

$$\sim \begin{pmatrix} 1 & 2 & 1 & 0 \\ 0 & 1 & 0 & 2 \\ 0 & 0 & -3 & 12 \\ 0 & 0 & -6 & 23 \end{pmatrix}$$

$$\sim \begin{pmatrix} 1 & 2 & 1 & 0 \\ 0 & 1 & 0 & 2 \\ 0 & 0 & -3 & 12 \\ 0 & 0 & 0 & -1 \end{pmatrix}$$

$R(A) = 4$. Hence the four vectors are linearly independent.

Example 8. *Express (2, –1, 4) as a linear combination of (4, 0, 12) and (0, 1, 2).*

Proof. Let $(2, -1, 4) = \alpha (4, 0, 12) + \beta (0, 1, 2)$

Hence $2\alpha = 2, \beta = -1, 12\alpha + 2\beta = 4$

$\alpha = ½, \beta = -1$ satisfy the three equations.

$$\therefore \quad (2, -1, 4) = \frac{1}{2} (4, 0, 12) - (0, 1, 2).$$

EXERCISE 6 (b)

1. Show that the following vectors are linearly dependent. Find their relationship in each case.
   (*i*) (1, 2, 0), (2, 3, 0) and (8, 13, 0) of R^3
   (*ii*) (2, 3, –1, –1), (1, –1, –2, –4), (3, 1, 3, –2) and (6, 3, 0, –7) of R^4
   (*iii*) (1, 2, –1, 3), (0, –2, 1, –1), and (2, 2, –1, 5) of R^4
   (*iv*) (1, 2, 3), (3, –2, 1) and (1, –6, –5) of $V_3 (R)$
2. Show that the vectors are linearly independent in each case
   (*i*) (1, 1, 1, 1), (1, –1, 1, 1), (1, 1, –1, 1) and (1, 1, 1, –1) of R^4
   (*ii*) (1, 2, 3) and (2, –2, 0) of R^3
   (*iii*) (1, 0, 0, 0), (0, 1, 0, 0), (0, 0, 1, 0) and (0, 0, 0, 1) of $V_4 (R)$.
   (*iv*) (0, 2, 1, –1), (1, –1, –1, 3) and (7, 2, –7, 4) of R^4.
3. If x, y, z are linearly independent, prove $x + y, x - y, x - 2y + z$ are also linearly independent.
4. Express (x, y, z) as a linear combination of the vectors (1, 1, 0), (0, 1, 1), (1, 0, 1).

BASES AND DIMENSION

Definition : A *basis* in a vector space $V(F)$ is a set B of *linearly independent vectors* of $V(F)$ such that every vector in V is a linear combination of elements of B.

In other words, a basis for a vector space V over a field F, is a linearly independent set of vectors from V which *spans V.*

If a basis for $V(F)$ contains only a finite number of vectors, then $V(F)$ is *finite dimensional.*

Definition: The number of elements in any basis of a finite dimensional vector space $V(F)$ is called the *dimension* of the vector space and is denoted as dim V.

Theorem. *If $\{x_1, x_2, ... x_n\}$ is a basis of $V(F)$, then any vector y of $V(F)$ can be expressed uniquely as a combination of the vectors of the basis.*

Let $y = \sum_{i=1}^{n} \alpha_i x_i$

Now we have to prove the uniqueness of the expression.

So, let $y = \sum_{i=1}^{n} \beta_i x_i$, if not unique.

$$\therefore \quad \sum_{i=1}^{n} \alpha_i x_i = \sum_{i=1}^{n} \beta_i x_i$$

$$\text{Hence} \quad \sum_{i=1}^{n} (\alpha_i - \beta_i) x_i = 0$$

Hence $\alpha_i - \beta_i = 0$ for $i = 1, 2, \ldots\ldots. n$ since $\{x_1, x_2, \ldots\ldots. x_n\}$ is linearly independent.

So, $\quad \alpha_i = \beta_i$

$$\text{Hence} \quad y = \sum_{i=1}^{n} \alpha_i x_i \text{ is unique.}$$

Note 2. A set of vectors having zero vector as one of the vectors cannot be a basis.

Note 3. If $\{x_1, x_2, \ldots.. x_n\}$ is a basis and if $x \in V(F)$ and $x = \sum_{i=1}^{n} \alpha_i x_i$. then the set of scalars $\{\alpha_1, \alpha_2, \ldots.. \alpha_n\}$ can be taken as coordinates of the vector x corresponding to the basis set as coordinate system.

Note 4. There exists a basis for each finite dimensional vector space.

Theorem: *Each set of (n + 1) or more vectors in an n-dimensional vector space V(F) is linearly dependent.*

Cor: In an n-dimensional vector space $V(F)$, any linearly independent set of n vectors is a basis of $V(F)$.

Theorem : *If V(F) is finite dimensional vector space of dimension n and if W is any subspace of V then W is also finite dimensional and dim W ≤ dim V.*

Example 1. *Show that S = {(1, 0, 0, 0), (0, 1, 0, 0), (0, 0, 1,0), (0, 0, 0, 1)} is a basis of V_4 (R) or R^4.*

Firstly, let us prove S is linearly independent.

$$\alpha_1 (1, 0, 0, 0) + \alpha_2 (0, 1, 0, 0) + \alpha_3 (0, 0, 1, 0) + \alpha_4 (0, 0, 0, 1) = 0$$

$$\Rightarrow \quad (\alpha_1, \alpha_2, \alpha_3, \alpha_4) = (0, 0, 0, 0)$$

Hence $\alpha_1 = 0, \alpha_2 = 0, \alpha_3 = 0, \alpha_4 = 0$

Therefore S is linearly by independent.

Next, we should prove any vector of R^4 is a linear combination of vectors of S.

Let $(x_1, x_2, x_3, x_4) \in R^4$

Then $(x_1, x_2, x_3, x_4) = x_1 (1, 0, 0, 0) + x_2 (0, 1, 0, 0) + x_3 (0, 0, 1, 0)$
$+ x_4 (0, 0, 0, 1)$

Therefore S spans R^4.

Hence S is a basis of R^4.

Example 2. *If $V(R) = P_n$, be vector space of polynomials in t over the field of real, of degree $\leq n$, show that $S = \{1, t, t^2, ... t^n\}$ is a basis of $V(R)$.*

Proof. $V(R) = P_n = \{p(t) : p(t) = a_0 + a_1 t + a_2 t^2 + ... + a_n t^n\}$
for $a_i \in R$

Let α_i's be scalars such that

$$\alpha_0 . 1 + \alpha_1 . t + \alpha_2 . t^2 + ...+ \alpha_n . t^n = 0 \text{ (zero polynomial)}$$

$$\Leftrightarrow \quad \alpha_0 = 0, \alpha_1 = 0, ... \alpha_n = 0.$$

So, S in linearly independent.

To Prove: S spans $V(F)$.

Let $\quad b_0 + b_1 t + b_2 t^2 + ... + b_n t^n \in V(F)$

Then $b_0 + b_1 t + ... + b_n t^n = b_1 .1 + b_2 . t + b_3 . t^2 + ... + b_n . t^n$
= a linear combination of S.

Hence S is a basis of P_n.

Example 3. *Prove $S = \{(1, 0, 0), (1, 1, 0), (1, 1, 1)\}$ is a basis of R^3. Hence find the coordinates of the vector (6, 5, 3) w.r.t. this basis. What is the dimension of the space ?*

Proof. $\alpha_1 (1, 0, 0) + \alpha_2 (1, 1, 0) + \alpha_3 (1, 1, 1) = 0 \Rightarrow$

$$\alpha_1 + \alpha_2 + \alpha_3 = 0$$
$$\alpha_2 + \alpha_3 = 0$$
$$\alpha_3 = 0$$

Going backwards, $\alpha_3 = 0, \alpha_2 = 0, \alpha_1 = 0$

Hence S is linearly independent in R^3.

Therefore S spans R^3.

S is a basis of R^3.

$(6, 5, 3) = a_1 (1, 0, 0) + a_2 (1, 1, 0) + a_3 (1, 1, 1)$

$$\therefore \quad a_1 + a_2 + a_3 = 6$$
$$a_2 + a_3 = 5$$
$$a_3 = 3$$

Hence $a_2 = 2$ and $a_1 = 1$

Coordinates of the vector (6, 5, 3) are (1, 2, 3); the dimension of R^3 is three.

Example 4. *Show that $B = \{(1, 0, 0), (1, 1, 0), (1, 1, 1)\}$ is a basis of V_3 (C). Hence find the coordinates of the vector (3 + 4i, 6i, 3 + 7i) in V_3 (C) w.r.t. B.*

Proof. $\alpha_1 (1, 0, 0) + \alpha_2 (1, 1, 0) + \alpha_3 (1, 1, 1) = (0, 0, 0)$ implies

$$\alpha_1 + \alpha_2 + \alpha_3 = 0$$
$$\alpha_2 + \alpha_3 = 0$$
$$\alpha_3 = 0.$$

Hence, solving, $\alpha_3 = 0, \alpha_2 = 0, \alpha_1 = 0$.

Therefore B is linearly independent. $V_3 (C)$ is of dimension 3 and B has 3 vectors. Hence B is a basis of $V_3 (C)$.

$$(3 + 4 i, 6i , 3 + 7i) = a_1 (1, 1, 0) + a_2 (1, 1, 0) + a_3 (1, 1, 1)$$

$$\Rightarrow \quad a_1 + a_2 + a_3 = 3 + 4 i$$
$$a_2 + a_3 = 6i$$
$$a_3 = 3 + 7i$$

$$\therefore \quad a_3 = 3 + 7i, a_2 = -3 - i, a_1 = 3 - 2i$$

Coordinates are $(3 - 2i, -3 - i, 3 + 7i)$

Example 5. *Show that the vectors* $\alpha_1 = (1, 0, -1)$, $\alpha_2 = (1, 2, 1)$ *and* $\alpha_3 = (0, -3, 2)$ *form a basis of* R^3. *Express each of the standard basis vector as a linear combination of* $\alpha_1, \alpha_2, \alpha_3$.

Proof. We can easily prove $\{\alpha_1, \alpha_2, \alpha_3\}$ is linearly independent and they form a basis.

Let $(1, 0, 0)= \alpha(1, 0, -1) + \beta(1, 2, 1) + r(0, -3, 2)$

Then $\alpha = \frac{7}{10}, \beta = \frac{3}{10}, r = \frac{1}{5}$

Coordinates of $(1, 0, 0)$ w.r.t the basis $\{\alpha_1, \alpha_2, \alpha_3\}$ are

$$\left(\frac{7}{10}, \frac{3}{10}, \frac{1}{5}\right)$$

Similarly, we can find the coordinates of $(0, 1, 0)$ and $(0, 0, 1)$.

Example 6. *Prove* $B = \{(1, 0, 0), (0, 1, 0), (0, 0, 1)\}$ *is a standard basis of* R^3. *Find the coordinates of the vector (6, 5, 3) w.r.t this standard basis.*

Proof. Let us prove B is linearly independent

$$\alpha_1(1, 0, 0) + \alpha_2(0, 1, 0) + \alpha_3(0, 0, 1) = \bar{0} \text{ implies}$$
$$(\alpha_1, \alpha_2, \alpha_3) = (0, 0, 0)$$

Hence $\alpha_1 = 0, \alpha_2 = 0, \alpha_3 = 0$

Further, B is a linearly independent set of three vectors in a three diamensional vector space R^3.

Hence B is a basis of R^3 (called standard basis of R^3)

$$(6, 5, 3) = 6(1, 0 , 0) + 5(0, 1, 0) + 3(0, 0, 1)$$

Hence the coordinates of the vector $(6, 5, 3)$ are $(6, 5, 3)$

Example 7. *(a, b) and (c, d) are two vectors which form a basis of R^2. Prove $ad - bc \neq 0$.*

Proof. Since (a, b) and (c, d) are linearly independent, the rank of $\begin{pmatrix} a & b \\ c & d \end{pmatrix}$ is 2 and hence $\begin{vmatrix} a & b \\ c & d \end{vmatrix} \neq 0$ *i.e* $ad - bc \neq 0$

EXERCISE 6 (c)

1. Prove that the following sets of vectors are basis for $V_3(R)$.
   (*a*) {(5, 5, 6), (6, 5, 5), (5, 6, 5)}
   (*b*) {(2, 3, 3), (3, 2, 3), (3, 3, 2)}
   (*c*) {(1, 2, 1), (1, 1, 2), (2, 1, 1)}
   (*d*) {(2, –3, 1), (0, 1, 2), (1, 1, –2)}
2. Prove {(1, 0, 0, 0), (0, 1, 0, 0), (0, 0, 1, 0) (0, 0, 0, 1)} constitute a basis for R^4. Also find the coordinates of the vector (*a, b, c, d*) with respect to this basis. What is the dimension this space ?

INNER PRODUCT SPACES

General Definition : If F be either the field of real numbers or the field of complex numbers and $V(F)$ be a vector space, then an *inner product* in $V(F)$ is a numerical valued function which assigns to each *ordered pair of vectors* x, y in V, a scalar (x, y) in F, such that

$$I_1 : \overline{(x, y)} = (y, x) \qquad \text{(conjugate symmetry)}$$

$$I_2 : (\alpha x + \beta y, z) = \alpha(x, z) + \beta\,(y, z) \text{ where } x, y, z \in V \text{ and } \alpha, \beta \in F \qquad \text{(linearity)}$$

$$I_3 : (x, x) \geq 0 \text{ and } (x, x) = 0 \Leftrightarrow 0$$

(The above is the general definition of an inner product.)

The vector space $V(F)$ with an *inner product* is called an *inner product space.*

Norm : We define norm of vector x or length of vector x as

$$\| x \| = \sqrt{(x, x)}.$$

We read $\| x \|$ as norm of x.

Unit vector. A vector x is a unit vector if $\| x \| = 1$

Note 1 : In the case of vector space over real field,

$$\overline{(x, y)} = (x, y) = (y, x)$$

Note 2. $\overline{(\Sigma a_i)} = \Sigma \overline{a}_i$

Note 3. $\overline{ab} = \overline{a}\,\overline{b}$

Note 4. $\overline{(\overline{a})} = a$

Note 5. If the filed is the complex field, the inner product is *complex inner product.*

Note 6. If the field is the real field, the inner product is *real inner product.*

Note 7. $\| x \|$ is real and ≥ 0.

Example 1. *In V_3 (C), define an inner product*

$$(x, y) = \sum_{i=1}^{n} a_i \overline{b_i} \text{ if } x = (a_1, a_2, \ldots a_n) \text{ and }$$

$$y = (b_1, b_2, \ldots b_n)$$

$$\text{and} \quad a_i, b_i \in C.$$

Prove this is really an inner product in V_3 (C).

Proof. $(x, y) = \sum_{i=1}^{n} a_i \overline{b_i}$ is a complex number $\in C$.

This is a scalar valued function. we will prove the three condition I_1, I_2, I_3.

(*i*) $\overline{(x, y)} = \overline{\Sigma a_i \overline{b_i}} = \Sigma \bar{a}_c b_i = \sum b_i \bar{a}_i = (y, x)$

(*ii*) $(\alpha x, + \beta y, z) = \Sigma (\alpha a_i + \beta b_i) \bar{c}_i$

$$= \alpha \Sigma a_i \bar{c}_i + \beta \Sigma b_i \bar{c}_i$$

$$= \alpha (x, z) + \beta (y, z)$$

(*iii*) $(x, x) = \Sigma a_i \bar{a}_i = \Sigma | a_i |^2 \geq 0$

If $(x, x) = 0$, then $\Sigma | a_i |^2 = 0$ and $| a_i | = 0$ for each i and hence $a_i = 0$ and then $x = 0$.

Conversely, if $x = 0$, the $(x, x) = (0, 0) = 0$

The axioms of inner product are verified to be true. Hence with this definition as inner product, V_3 (*C*) is an inner product space.

This is called *standard inner product* in V_3 (*C*).

Note 1. If $x = (a_1, a_2, \ldots\ldots a_n)$, $y = (b_1, b_2 \ldots b_n)$ in V_3 (*R*), then the definition $(x, y) = \Sigma a_i b_i$ gives an inner product.

This is standard inner product in V_3 (*R*).

Definition. A real inner product space is called an *Euclidean* space and a complex inner product space is called *a unitary space.*

Note. 2. If $x = (x_1, x_2, \ldots. x_n)$ is a complex vector in V_n (*C*), then

$$\| x \| = \sqrt{| x_1 |^2 + | x_2 |^2 + \ldots\ldots + | x_n |^2}$$

Example 2. *Prove every finite dimensional vector space over the field of reals or complex numbers is an inner product space with respect to the standard definition of inner product.*

Proof. Follow the pattern of the proof in example 1.

Example 3. *If one of The vectors x or y is zero. The inner product is zero. That is* $(\bar{0}, y) = 0$ or $(x, \bar{0}) = 0$.

Also prove $(x, \alpha y) = \bar{\alpha}(x, y)$, and $(z, \alpha x + \beta y) = \bar{\alpha}(z, x) + \bar{\beta}(z, y)$

Proof. $(\bar{0}, y) = (0.\bar{0}, y) = 0(\bar{0}, y) = 0$ since $(\alpha x, y) = \alpha(x, y)$

$$(x, \alpha y) = \overline{(\alpha y, x)} = \overline{\alpha(y, x)} = \bar{\alpha}\,\overline{(y, x)} = \bar{\alpha}(x, y)$$

$$(z, \alpha x + \beta y) = \overline{(\alpha x + \beta y, z)} = \overline{\alpha(x, z) + \beta(y, z)}$$

$$= \bar{\alpha}\,\overline{(x, z)} + \bar{\beta}\,\overline{(y, z)}$$

$$= \bar{\alpha}(z, x) + \bar{\beta}(z, y)$$

NORMED VECTOR SPACE

Definition : A vector space $V(F)$ which assigns to each vector x in V a real number $\|x\|$ (read norm x) such that

(*i*) $\|x\| \geq 0$ and $\|x\| = 0 \Leftrightarrow x = 0$ for each x of V

(*ii*) $\|\alpha x\| = |\alpha|\,\|x\|$ for $\alpha \in F, x \in V$ and

(*iii*) $\|x + y\| \leq \|x\| + \|y\|$ for $x, y \in V$ is called a normed vector space.

Theorem: *Every inner product space is a normed vector space.*

Proof. Let $V(F)$ be an inner product space.

Let us verify the postulates of normed vector space.

(*i*) $\|x\| = \sqrt{(x, x)} = \sqrt{\text{a non-negative number}}$ = a non-negative number.

$\therefore \quad \|x\| \geq 0.$

$(x, x) = 0 \Rightarrow x = 0$

$\|x\|^2 = 0 \Rightarrow x = 0$

$\therefore \quad \|x\| = 0 \Rightarrow x = 0$

If $x = 0$, then $(x, x) = 0$

$\Rightarrow \quad \|x\|^2 = 0$ *i.e.,* $\|x\| = 0$

(*ii*) $\|\alpha x\|^2 = (\alpha x, \alpha x) = \alpha\bar{\alpha}(x, x) = |\alpha|^2\,\|x\|^2$

$\therefore \quad \|\alpha x\| = |\alpha|\,\|x\|$

(*iii*) $\|x + y\|^2 = (x + y, x + y)$

$$= (x, x + y) + (y, x + y)$$

$$= \overline{(x + y, x)} + \overline{(x + y, y)}$$

$$= \overline{(x, x)} + \overline{(y, x)} + \overline{(x, y)} + \overline{(y, y)}$$

$= (x, x) + (x, y) + \overline{(x, y)} + (y, y)$

$= \| x \|^2 + 2 \text{ Real } (x, y) + \| y \|^2$

$\leq \| x \|^2 + 2 \ | (x, y) | + \| y \|^2$ since Real $(x, y) \leq | (x, y) |$

$\leq \| x \|^2 + 2 \| x \| \| y \| + \| y \|^2$ by Schwarz inequality

$\leq (\| x \| + \| y \|)^2$ (see next pages)

$\therefore \quad \| x + y \| \leq \| x \| + \| y \|$

The inner product space satisfies all the conditions of normed vector space. Hence, it is a normed vector space

Note 1. Combining the above theorem and worked example 2, we get that :

"Every finite dimesional vector space is a normed vector space". The converse is not true.

Example 4. *If* $\alpha_1, \beta \in F$ *and* $x, y, z \in V$ *where V(F) is a normed vector space, prove the following.*

(*i*) $\| \alpha x + \beta y \|^2 = | \alpha |^2 \| x \|^2 + \alpha\bar{\beta} (x, y) + \bar{\alpha}\beta (y, x) + | \beta |^2 \| y \|^2$

(*ii*) $\| x + y \|^2 + \| x - y \|^2 = 2(\| x \|^2 + \| y \|^2)$

(*iii*) $(x, y) = 0 \Leftrightarrow \|\alpha x + \beta y \|^2 = |\alpha|^2 \| x \|^2 + |\beta|^2 \| y \|^2$

(*iv*) $\| x \| = \| y \| \Rightarrow (x - y, x + y) = 0$ *in real inner product space*

(*v*) *If* $(x, y) = 0$, *then* $\|x + y\|^2 = \| x \|^2 + \| y \|^2$

(*vi*) $4 (x, y) = \| x + y \|^2 - \| x - y \|^2 + i \| x + iy \|^2 - i \| x - iy \|^2$.

Proof. (*i*) $\| \alpha x + \beta y \|^2 = (\alpha x + \beta y, \alpha x + \beta y)$

$= \alpha (x, \alpha x + \beta y) + \beta (y, \alpha x \ \beta y)$

$= \alpha \overline{(\alpha x + \beta y, x)} + \beta \overline{(\alpha x + \beta y, y)}$

$= \alpha\bar{\alpha} \overline{(x, x)} + \alpha\bar{\beta} \overline{(y, x)} + \beta\bar{\alpha} \overline{(x, y)} + \beta\bar{\beta} \overline{(y, y)}$

$= \alpha\bar{\alpha} (x, \bar{x}) + \alpha\bar{\beta} (x, y) + \beta\bar{\alpha} (y, x) + \beta\bar{\beta} (y, y)$

$= | \alpha |^2 \| x \|^2 + \alpha\bar{\beta} (x, y) + \bar{\alpha}\beta (y, x) + | \beta |^2 \| y \|^2$

(*ii*) Putting $\alpha = 1, \beta = 1$ in the above result

$\| x + y \|^2 = \| x \|^2 + (x, y) + (y, x) + \| y \|^2$

Put $\alpha = 1, \beta = -1$ in the above result,

$\| x - y \|^2 = \| x \|^2 - (x, y) - (y, x) + \| y \|^2$

Hence $\| x + y \|^2 + \| x - y \|^2 = 2 (\| x \|^2 + \| y\|^2)$

(*iii*) In the result of (*i*), put $(x, y) = 0. (y, x) = 0$

$\| \alpha x + \beta y \|^2 = | \alpha |^2 \| x \|^2 + | \beta |^2 . \| y \|^2$

(iv) $(x-y, x+y) = (x, x+y) - (y, x+y)$

$= (x, x) + (x, y) - (y, x) - (y, y)$

$= \| x \|^2 - \| y \|^2 \because (x, y) = (y, x)$ in real product space

$= 0$ since $\| x \| = \| y \|$

(v) If $(x, y) = 0$ use the result in (ii). Then,

$\| x+y \|^2 = \| x \|^2 + \| y \|^2$

(vi) $\| x+iy \|^2 = \|x\|^2 + \bar{i}\ (x, y) + i\,(y, x) + i.\bar{i}\ \| y \|^2$

$i \| x+iy \|^2 = i \| x \|^2 + (x, y) - (y, x) + i \| y \|^2$

Similarly, $-i \| x-iy \|^2 = -i \| x \|^2 + (x, y) - (y, x) - i \| y \|^2$

$i \| x+iy \|^2 - i \| x-iy \|^2 = 2\,(x, y) - 2\,(y, x)$ ... (i)

From the result in (ii),

$\| x+y \|^2 - \| x-y \|^2 = 2\,(x, y) + 2\,(y, x)$ ... (ii)

Adding (i) and (ii)

$4\,(x, y) = \| x+y \|^2 - \| x-y \|^2 + i \| x+iy \|^2 - i \|x-iy\|^2$

Example 5. *If V(C) be the vector space of all continuous complex valued functions on the real interval* $a \le t \le b$, *and* $(f, g) = \int_a^b f(t)\,\overline{g(t)}\,dt, f(t), g(t) \in V$. *Prove this is an inner product in V.*

Proof. From the difinition of (f, g) it is scalar valued.

(i) $\overline{(f, g)} = \overline{\int_a^b f(t)\, g(t)\, dt}$

$= \int_a^b \overline{f(t)}\, g(t)\, dt$

$= (g, f)$

(ii) $\alpha f(t) + \beta\, g(t) \in$ V.

$(\alpha f + \beta g, h) = \int_a^b (\alpha f + \beta g)\bar{h}\, dt$

$= \alpha \int_a^b f\,\bar{h}\, dt + \beta \int_a^b g\bar{h}\, dt, h(t) \in V$

$= \alpha\,(f, h) + \beta\,(g, h)$

(iii) $(f, f) = \int_a^b f(t)\,\overline{f(t)}\, dt$

$= \int_a^b | f(t) |^2\, dt \ge 0$

If $f = 0$, $(f, f) = 0$ and the converse

The above definition of (f, g) is really an inner product.

Example 6. *$x, y \in V_2$ (C) and $x = (a_1, a_2)$, $y = (b_1, b_2)$, $z = (c_1, c_2)$. Verify whether the following defines an inner product, if* (*i*) $(x, y) = 2a_1 \bar{b}_1 + a_1 \bar{b}_2 + a_2 \bar{b}_1 + a_2 \bar{b}_2$

(*ii*) $(x, y) = a_1 \bar{b}_1 + (a_1 + a_2)\overline{(b_1 + b_2)}$

(*iii*) $(x, y) = a_1 b_1 - a_2 b_1 - a_1 b_2 + 4a_2 b_2$.

This is left as an exercise to the reader.

Theorem : Schwarz's inequality: *If x, y are vectors in an inner product space, then prove* $|(x, y)| \le \|x\|\,\|y\|$

Proof. Let λ be real and $x, y \in V(F)$; $\lambda \neq 0$

$$\|x + \lambda y\|^2 \ge 0$$

$$\|x\|^2 + \lambda^2 \|y\|^2 + \lambda\,[(x, y) + (y, x)] \ge 0$$

$$\lambda^2 \|y\|^2 + \lambda\,[(x, y) + \overline{(x, y)}] + \|x\|^2 \ge 0$$

i.e, $\lambda^2 \|y\|^2 + \lambda.[2 \text{ Real part of } (x, y)] + \|x\|^2 \ge 0$

i.e., $\lambda^2 \|y\|^2 + 2\lambda\,|(x, y)| + \|x\|^2 \ge 0$ since $Rl\,(x, y) \le |(x, y)|$

This is a quadratic expression in λ and $\lambda^2 > 0$

$\therefore$ "$b^2 - 4ac$" ≤ 0

i.e., $4\,|(x, y)|^2 - 4\,\|y\|^2 \|x\|^2 \le 0$

i.e., $|(x, y)|^2 \le \|x\|^2 \|y\|^2$

Hence, $|(x, y)| \le \|x\|\,\|y\|$

Cauchy's inequality: *If $V_n(C)$ is a unitary space, for any two vectors $x = (a_1, a_2, \ldots a_n)$, $y = (b_1, b_2, \ldots b_n)$ define $(x, y) = \sum_{i=1}^{n} a_i \bar{b}_i$; then by the above inequality.*

$$|\Sigma a_i \bar{b}_i|^2 \le (\Sigma a_i \bar{a}_i)(\Sigma b_i \bar{b}_i)$$

i.e., $|\Sigma a_i \bar{b}_i|^2 \le (\Sigma |a_i|^2)(\Sigma |b_i|^2)$

In the case of Euclidean space (real field)

$$(a_1 b_1 + a_2 b_2 + a_n b_n)^2 \le (a_1^2 + a_2^2 + \ldots\ldots + a_n^3)(b_1^2 + b_2^2 + \ldots\ldots. + b_n^2)$$

ORTHOGONALITY

Definition : If x and y are two vectors in an inner product space $V(F)$, then x and y are orthogonal if $(x, y) = 0$

i.e., x is orthogonal to y if $(x, y) = 0$

If x is orthogonal to y, then y is orthogonal to x because of symmetry

$$(x, y) = 0 \Leftrightarrow (y, x)$$

Note 1. Every vector is orthogonal to zero vector since $(x, \bar{0}) = 0$ $\forall\, x \in V$

2. Zero vector is orthogonal to itself
3. The angle between two vectors, θ, is given by

$$\cos\theta = \frac{(x, y)}{\|x\|\|y\|}$$

If $(x, y) = 0$, $\theta = 90°$

Orthogonal set : If S be a set of vectors in an inner product space $V(F)$, then S is called an orthogonal set if any two distinct vectors in S are orthogonal *i.e.*, $(x_i, x_j) = 0, i \neq j$

Orthonormal set : In an orthogonal set S if every vector is a unit vector, then the set is an orthonormal set. In other words, S is orthonormal if

(*i*) $(x, y) = 0$ for $x \neq y$

(*ii*) $(x, y) = 1$ for $x = y$ when $x, y \in S$.

Theorem : *An orthogonal set of non-zero vectors in an inner product space V (F) is linearly independent.*

Proof : Let $S = \{x_1, x_2, ... x_m\}$ be an orthogonal set of non-zero vectors $\therefore$ $(x_i, x_j) = 0$ for $i \neq j$

Let $\sum_{i=1}^{m} \alpha_i x_i = 0$ We have to prove each $\alpha_i = 0$

$$\left(\sum_{i=1}^{m} \alpha_i x_i, x_j\right) = (0, x_j) = 0$$

i.e., $\sum_{i=1}^{m} a_i (x_i, x_j) = 0$

i.e., $\alpha_j (x_j, x_j) = 0 \because (x_e, x_j) = 0, i \neq j$

$\therefore$ $\alpha_j = 0$ since $x_j \neq 0, j = 1, 2, ... m$.

$\therefore$ S is linearly independent

Note 1: Let $x \neq 0$ and $x \in V(F)$, inner product space; $\|x\| \neq 0$, $\frac{x}{\|x\|} \in V$

$A = \left\{\frac{x}{\|x\|}\right\}$ is an orthonormal singleton set for $\left(\frac{x}{\|x\|}, \frac{x}{\|x\|}\right)$

$= \frac{1}{\|x\|^2}(x, x) = 1$

Example 7. *In R^3, prove $B = \{(1, 0, 0), (0, 1, 0), (0, 0, 1)\}$ is an orthonormal set under usual standard inner product. Also find the angle between (1, 2, 3) and (0, 2, 5)*

Proof. We define $(x, y) = \Sigma a_i b_i$ as inner product in R^3

Let the vector be denoted by x, y, z.

$(x, y) = 1.0 + 0.1 + 0.0 = 0$

$(y, z) = 0.0 + 1.0 + 0.1 = 0$

$(x, z) = 0.1 + 0.1 + 0.1 = 0$

$\| x \|^2 = (x, x) = 1.1 = 1 \therefore \| x \| = 1$

Similarly $\| y \| = 1, \| z \| = 1$

Hence B is an orthonormal set

$$\cos\theta = \frac{(x,y)}{\|x\|\|y\|};\ (x, y) = \sum a_i b_i = 1.0 + 2.2 + 3.5 = 19$$

$$= \frac{19}{\sqrt{1+4+9}\,\sqrt{4+25}} \quad \text{since } \|x\| = \sqrt{\sum ai^2}$$

$$= \frac{19}{\sqrt{14}\,\sqrt{29}}$$

$$\theta = \cos^{-1}\left(\frac{19}{\sqrt{14}\,\sqrt{29}}\right)$$

Example 8. *Find the norm of the vector $\upsilon = (2, -3, 4)$ and normalize it.*

Proof. $(\upsilon, \upsilon) = \| \upsilon \|^2 = 2^2 + (-3)^2 + 4^2 = 29$

$$\| \upsilon \| = \sqrt{29}$$

Normalized vector is $= \dfrac{\upsilon}{\|\upsilon\|} = \left(\dfrac{2}{\sqrt{29}}, \dfrac{-3}{\sqrt{29}}, \dfrac{4}{\sqrt{29}}\right)$

GRAM-SCHMIDT ORTHOGONALIZATION PROCESS

Theorem: Every finite dimensional inner product space has an orthonormal basis.

Proof. Let V be an inner product space and $\{\beta_1, \beta_2, \ldots\ldots\beta_n\}$ be a basis for V. From this, we shall obtain an orthogonal basis $\{\alpha_1, \alpha_2, \ldots\ldots\alpha_n\}$ by means of construction known as *Gram-Schmidt orthogonalization* process.

First, let $\alpha_1 = \beta_1$. Then $\alpha_1 \neq 0$.

Let $\quad \alpha_2 = \beta_2 - \dfrac{(\beta_2, \alpha_1)}{\|\alpha_1\|^2} . \alpha_1$

Since, β_1, β_2 are linearly independent, $\alpha_2 \neq 0$, for

if $\alpha_2 = 0$, then β_2 depends upon $\alpha_1 = \beta_1$ which contradicts.

Also $(\alpha_2, \alpha_1) = 0$ by direct computation, for

$$(\alpha_2, \alpha_1) = (\beta_2, \beta_1) - \frac{(\beta_2, \alpha_1)}{\|\alpha_1\|^2}(\alpha_1, \alpha_1)$$

$$= (\beta_2, \alpha_1) - (\beta_2, \alpha_1) = 0.$$

$\therefore \quad \{\alpha_1, \alpha_2\}$ is an orthogonal set

Next, let $\alpha_3 = \beta_3 - \frac{(\beta_3, \alpha_1)}{\| \alpha_1 \|^2} \alpha_1 - \frac{(\beta_3, \alpha_2)}{\| \alpha_2 \|^2} . \alpha_2$

Again $\alpha_3 \neq 0$; otherwise, if $\alpha_3 = 0$, then

β_3 is a linear combination of β_1 and β_2 which is untrue.

$$\text{Further, } (\alpha_3, \alpha_1) = (\beta_3, \alpha_1) - \frac{(\beta_3, \alpha_1)}{\| \alpha_1 \|^2}(\alpha_1, \alpha_1) - \frac{(\beta_3, \alpha_2)}{\| \alpha_2 \|^2}(\alpha_2, \alpha_1)$$

$$= (\beta_3, \alpha_1) - (\beta_3, \alpha_1) - 0$$

$$= 0$$

Similarly $(\alpha_3, \alpha_2) = 0$

Hence $\{\alpha_1, \alpha_2, \alpha_3\}$ is an orthogonal set.

Now, suppose we have constructed non-zero orthogonal vectors $\alpha_1, \alpha_2, \ldots\ldots \alpha_k$ in such a way that α_j is β_j minus some linear combination of $\beta_1, \beta_2 \ldots\ldots \beta_{j-1}$ for $2 \leq j \leq k$.

Let $\qquad \alpha_{k+1} = \beta_{k+1} - \sum_{j=1}^{k} \frac{(\beta_{k+1}, \alpha_j)}{\| \alpha_j \|^2} \alpha_j$

Then $\qquad \alpha_{k+1} \neq 0$ and

$$(\alpha_{k+1}, \alpha_c) = (\beta_{k+1}, \alpha_i) - \sum_{j=1}^{k} \frac{(\beta_{k+1}, \alpha_j)}{\| \alpha_j \|^2}(\alpha_j, \alpha_i)$$

$$= (\beta_{k+1}, \alpha_i) - (\beta_{k+1}, \alpha_i)$$

$$= 0 \qquad \text{for } 1 \leq i \leq k.$$

Thus α_{k+1} is orthogonal to each of the vectors $\alpha_1, \alpha_2, \ldots\ldots\ldots \alpha_k$.

Suppose $\alpha_{k+1} = 0$ Then β_{k+1} is a linear combination of $\alpha_1, \alpha_2, \ldots\ldots\ldots \alpha_k$ and hence of $\beta_1, \beta_2, \ldots\ldots\ldots \beta_k$ which is not true. Therefore $\alpha_{k+1} \neq 0$.

Continuing like this, ultimately, we obtain a non zero orthogonal set $\{\alpha_1, \alpha_2, \ldots\ldots \alpha_n\}$ containing n district vectors. Since the set is orthogonal, it is independent and has n vectors and the dimension of V is n. Hence it is an orthogonal basis.

To obtain orthonormal basis, divide each vector by its norm.

Also $\qquad \left(\frac{\alpha_i}{\| \alpha_i \|}, \frac{\alpha_j}{\| \alpha_j \|} \right) = 0, \; i \neq j$

and $\qquad \left(\frac{\alpha_i}{\| \alpha_1 \|}, \frac{\alpha_i}{\| \alpha_c \|} \right) = \frac{1}{\| \alpha_c \|^2} (\alpha_i, \alpha_i) = 1$

Therefore, the set $\left\{ \frac{\alpha_1}{\| \alpha_1 \|}, \frac{\alpha_2}{\| \alpha_2 \|}, \ldots\ldots\ldots, \frac{\alpha_n}{\| \alpha_n \|} \right\}$ is the orthonormal set.

Example 9. *$B = \{(1, 1,, 1), (0, 1, 1), (0, 0, 1)\}$ is a linear basis of V_3 (R). Transform B into an orthonormal basis of V_3 (R) by Gram-Schmidt orthogonal ization process (Standard inner product).*

Solution. Let $\{\alpha_1, \alpha_2, \alpha_3\}$ be the required basis and $\{\beta_1, \beta_2, \beta_3\}$ be the given linear basis.

Take $\alpha_1 = \beta_1 = (1, 1, 1)$.

$$\text{Select } \alpha_2 = \beta_2 - \frac{(\beta_2, \alpha_1)}{\|\alpha_1\|^2} . \alpha_1$$

$$= (0, 1, 1) - \frac{(1.0 + 1.1 + 1.1)}{(1 + 1 + 1)} (1, 1, 1)$$

$$= (0, 1, 1) - \frac{2}{3} (1, 1, 1)$$

$$= \left(\frac{-2}{3}, \frac{1}{3}, \frac{1}{3}\right)$$

$$\text{Select } \alpha_3 = \beta_3 - \frac{(\beta_3, \alpha_2)}{\|\alpha_2\|^2} . \alpha^2 - \frac{(\beta_3, \alpha_1)}{\|\alpha_1\|^2} \alpha_1$$

$$= (0, 0, 1) - \frac{\left(1.\frac{1}{3}\right)}{\left(\frac{4}{9} + \frac{1}{9} + \frac{1}{9}\right)} \left(-\frac{2}{3}, \frac{1}{3}, \frac{1}{3}\right) - \frac{(1.1)}{(1 + 1 + 1)} . (1, 1, 1)$$

$$= (0, 0, 1) + \left(\frac{1}{3}, -\frac{1}{6}, -\frac{1}{6}\right) - \left(\frac{1}{3}, \frac{1}{3}, \frac{1}{3}\right)$$

$$= \left(0, -\frac{1}{2}, \frac{1}{2}\right)$$

$$\|\alpha_1\| = \sqrt{3}, \|\alpha_2\| = \sqrt{\frac{2}{3}}, \|\alpha_3\| = \frac{1}{\sqrt{2}}$$

$$\text{Hence, orthonormal basis is } \left(\frac{\alpha_1}{\|\alpha_1\|}, \frac{\alpha_2}{\|\alpha_2\|}, \frac{\alpha_3}{\|\alpha_3\|}\right)$$

$$\text{i.e., } \left\{\left(\frac{1}{\sqrt{3}}, \frac{1}{\sqrt{3}}, \frac{1}{\sqrt{3}}\right), \left(-\frac{2}{\sqrt{6}}, \frac{1}{\sqrt{6}}, \frac{1}{\sqrt{6}}\right), \left(0, -\frac{1}{\sqrt{2}}, \frac{1}{\sqrt{2}}\right)\right\}$$

Example 10. *Apply Gram-Schmidt orthogonalization process to the vectors $\beta_1 = (1, 0, 1)$, $\beta_2 = (1, 0, -1)$, $\beta_3 = (0_3, 3, 4)$ so that the linear basis*

$\{\beta_1, \beta_2, \beta_3\}$ becomes an orthonormal basis for $V_3(R)$ with the standard inner product.

Solution. Let $(\alpha_1, \alpha_2, \alpha_3)$ be the required orthonormal basis. We take $(x, y) = \Sigma\, a_i\, b_i$ as inner product.

Take $\alpha_1 = \beta_1 = (1, 0, 1)$

$$\text{Let} \quad \alpha_2 = \beta_2 - \frac{(\beta_2, \alpha_1)}{\|\alpha_1\|^2}\alpha_1$$

$$= (1, 0, -1) - \frac{(1.1 - 1.1)}{\|\alpha_1\|^2}\alpha_1 = (1, 0, -1)$$

$$\alpha_3 = \beta_3 - \frac{(\beta_3, \alpha_1)}{\|\alpha_1\|^2}\alpha_1 - \frac{(\beta_3, \alpha_2)}{\|\alpha_2\|^2}\alpha_2 \qquad \text{.... (1)}$$

$(\beta_3, \alpha_1) = (\beta_3, \beta_1) = 1.0 + 0.3 + 1.4 = 4$

$(\beta_3, \alpha_2) = 0.1 + 3.0 + 4.\ (-1) = -4$

$\|\alpha_1\|^2 = 1 + 0 + 1 = 2\ ;\ \|\alpha_1\| = \sqrt{2}$

$\|\alpha_2\|^2 = 1 + 0 + 1 = 2\ ;\ \|\alpha_2\| = \sqrt{2}$

Now use in (1).

$$\alpha_3 = (0, 3, 4) - \frac{4}{2}(1, 0, 1) - \frac{(-4)}{2}.\ (1, 0, -1)$$

$$= (0, 3, 4) - (2, 0, 2) + (2, 0, -2)$$

$$= (0, 3, 0)$$

$$\|\alpha_3\| = 3$$

$$\frac{\alpha_1}{\|\alpha_1\|} = \left(\frac{1}{\sqrt{2}}, 0, \frac{1}{\sqrt{2}}\right); \frac{\alpha_2}{\|\alpha_2\|} = \left(\frac{1}{\sqrt{2}}, 0, -\frac{1}{\sqrt{2}}\right)$$

$$\frac{\alpha_3}{\|\alpha_3\|} = (0, 1, 0)$$

Hence, the orthonormal basis is

$$B = \left\{\left(\frac{1}{\sqrt{2}}, 0, \frac{1}{\sqrt{2}}\right), \left(\frac{1}{\sqrt{2}}, 0, -\frac{1}{\sqrt{2}}\right), (0, 1, 0)\right\}$$

Example 11. *The vector space P_2 containing polynomials of at most second degree in t has the inner product $(p(t), q(t)) = \int_0^1 p(t)\,\overline{q(t)}\,dt$ for p, $q \in P_2$. Taking $B = \{1, t, t^2\}$ as a basis, convert to orthonormal basis.*

Let $\{\alpha_1, \alpha_2, \alpha_3\}$ be the required basis

$$\beta_1 = 1, \beta_2 = t, \beta_3 = t^2.$$

$$\text{Let} \qquad \alpha_1 = \beta_1 = 1.$$

and $$\alpha_2 = \beta_2 - \frac{(\beta_2, \alpha_1)}{\| \alpha_1 \|^2} \alpha_1 \quad \text{...(1)}$$

$$(\beta_2, \alpha_1) = \int_0^1 t \,.1 \, dt = \frac{1}{2}$$

$$\| \alpha_1 \|^2 = (\alpha_1, \alpha_1) = \int_0^1 1 \, dt = (t)_0^1 = 1$$

$\therefore$ $$\| \alpha_1 \| = 1$$

Use in (1).

$$\alpha_2 = t - \frac{\left(\frac{1}{2}\right)}{1} .1 = t - \frac{1}{2}$$

$$\alpha_3 = \beta_3 - \frac{(\beta_3, \alpha_1)}{\| \alpha_1 \|^2} \alpha_1 - \frac{(\beta_3, \alpha_2)}{\| \alpha_2 \|^2} \alpha_2 \quad \text{...(2)}$$

$$(\beta_3, \alpha_1) = \int_0^1 t^2 \,.1 \, dt = \frac{1}{3}$$

$$(\beta_3, \alpha_2) = \int_0^1 t^2 \left(t - \frac{1}{2}\right) dt$$

$$= \int_0^1 \left(t^3 - \frac{1}{2} t^2\right) dt$$

$$= \frac{1}{4} - \frac{1}{6} = \frac{1}{12}$$

$$\| \alpha_2 \|^2 = (\alpha_2, \alpha_2) = \int_0^1 \left(t - \frac{1}{2}\right)^2 dt$$

$$= \left(\frac{t^3}{3} - \frac{t^2}{2} + \frac{1}{4} t\right)_0^1$$

$$= \frac{1}{3} - \frac{1}{2} + \frac{1}{4} = \frac{1}{12}$$

$$\| \alpha_2 \| = \frac{1}{\sqrt{12}}$$

Now use in (2).

$$\alpha_3 = t^2 - \frac{\left(\frac{1}{3}\right)}{1} \times 1 - \frac{\left(\frac{1}{12}\right)}{\frac{1}{12}} \cdot \left(t - \frac{1}{2}\right)$$

$$= t^2 - \frac{1}{3} - t + \frac{1}{2} = t^2 - t + \frac{1}{6}$$

Hence $\{\alpha_1, \alpha_2, \alpha_3\}$ is

$$\left\{1, t-\frac{1}{2},\ t^2-t+\frac{1}{6}\right\}$$

$$\|\alpha_3\|^2 = (\alpha_3, \alpha_3)$$

$$= \int_0^1 \left(t^2 - t + \frac{1}{6}\right)^2 dt$$

$$= \int_0^1 \left(t^4 + t^2 + \frac{1}{36} - 2t^3 - \frac{1}{3}t + \frac{1}{3}t^2\right) dt$$

$$= \frac{1}{5} + \frac{1}{3} + \frac{1}{36} - \frac{1}{2} - \frac{1}{6} + \frac{1}{9}$$

$$= \frac{36 + 60 + 5 - 90 - 30 + 20}{180}$$

$$= \frac{1}{180}$$

$$\|\alpha_3\| = \frac{1}{\sqrt{180}} = \frac{1}{3\sqrt{20}}$$

Hence, orthonormal basis is

$$\left\{\frac{\alpha_1}{\|\alpha_1\|}, \frac{\alpha_2}{\|\alpha_2\|}, \frac{\alpha_3}{\|\alpha_3\|}\right\}$$

i.e., $\left\{1, \sqrt{12}\left(t - \frac{1}{2}\right), 3\sqrt{20}\left(t^2 - t + \frac{1}{6}\right)\right\}$

Example 12. *If x, y are orthogonal unit vectors what is the distance between x and y ?*

Solution: $\delta(x, y) = \|x - y\|$

$$\|x - y\|^2 = (x - y, x - y)$$

$$= (x, x) - (x, y) - (y, x) + (y, y)$$

$$= \|x\|^2 - 0 - 0 + \|y\|^2 \text{ since } x, y \text{ are orthogonal}$$

$$= 1 + 1$$

$$= 2 \text{ since } x, y \text{ are unit vectors}$$

$$\therefore \quad \delta(x, y) = \sqrt{2}$$

EXERCISE 6 (d)

1. Obtain an orthonormal basis of R^3 with standard inner product starting from the linear basis.

(*a*) $B = \{(1, -1, 3), (0, 1, -1), (0, 3, -2)\}$

(*b*) $B = \{(1, 1, 1), (1, 2, 3), (2, 3, 8)\}$

(*c*) $B = \{(1, 2, 1), (1, 1, 2), (2, 1, 1)\}$

2. Let V be a space of polynomials in a variable x over the reals of degree 2 or less. We define an inner product

$$(p, q) = \int_{-1}^{1} P(x)\, q(x)\, dx,\ p, q \in V.$$

Find an orthonormal basis starting from the basis $\{1, t, t^2\}$.

3. V is a vector space of $R[x]$ of polynomials of degree at most 3. Define an inner product.

$(f, g) = \int_0^1 f(t)\, g(t)\, dt$. Apply Gram-Smidt orthogonalization process to the basis $\{1, t, t^2, t^3\}$.

4. If x and y are linearly dependent, prove
$|(x, y)| = \|x\|\, \|y\|$.

ANSWERS 4(d)

1. (*a*) $\left\{\frac{1}{\sqrt{11}}(1,-1,3), \frac{1}{\sqrt{66}}(4,7,1), \frac{1}{\sqrt{6}}(-2,1,1)\right\}$

(*b*) $\left\{\left(\frac{1}{\sqrt{3}}, \frac{1}{\sqrt{3}}, \frac{1}{\sqrt{3}}\right), \left(-\frac{1}{\sqrt{2}}, 0, \frac{1}{\sqrt{2}}\right), \left(\frac{1}{\sqrt{6}}, \frac{-2}{\sqrt{6}}, \frac{1}{\sqrt{6}}\right)\right\}$

(*c*) $\left\{\left(\frac{1}{\sqrt{6}}, \frac{2}{\sqrt{6}}, \frac{1}{\sqrt{6}}\right), \sqrt{\frac{6}{11}}\left(\frac{1}{6}, \frac{-4}{6}, \frac{7}{6}\right), \frac{11}{\sqrt{176}}\left(\frac{12}{11}, \frac{-4}{11}, \frac{-4}{11}\right)\right\}$

2. $\left\{\frac{1}{\sqrt{2}}, \sqrt{\frac{3}{2}}\,.t, \sqrt{\frac{45}{8}}\left(t^2 - \frac{1}{3}\right)\right\}$

MODEL QUESTION PAPER – I

Time: 3 Hrs Max. Marks : 100

Answer all Questions

PART-A (10 × 2 = 20 Marks)

1. Form a partial differential equation by eliminating the arbitrary function ϕ from $z = (x + y)\,\phi\,(x^2 - y^2)$.
2. Find the complete integral of $q = 2px$.
3. Find the half range sine series for $f(x) = 2$ in $0 < x < 4$.
4. If the cosine series for $f(x) = x \sin x$ for $0 < x < \pi$ is given by

$$x \text{ sine } x = 1 - \frac{1}{2}\cos x - 2\sum_{n=2}^{\infty}\frac{(-1)^n}{n^2-1} \cos nx, \text{ show that}$$

$$1+2\left[\frac{1}{1.3} - \frac{1}{3.5} + \frac{1}{5.7} - \ldots\ldots\ldots\ldots\right] = \frac{\pi}{2}$$

5. Classify the partial differential equation

$$(1 - x^2)z_{xx} - 2xyz_{xy} + (1 - y^2)z_{yy} + xz_x + 3x^2yz_y - 2z = 0.$$

6. The steady state temperature distribution is considered in a square plate with sides $x = 0$, $y = 0$, $x = a$ and $y = a$. The edge $y = 0$ is kept at a constant temperature T and the other three edges are insulated. The same state is continued subsequently. Express the problem mathematically.
7. Define the sampler.
8. Find Z-transform of $\{n\}$.
9. If Fourier transform of $f(x)$ is $F(s)$, prove that the Fourier transform of

$$f(x) \cos ax \text{ is } \frac{1}{2}[F(s-a) + F(s+a)].$$

10. Find the Fourier cosine integral representation of $f(x) = \begin{cases} 1, & 0 < x < 1 \\ 0 & x > 1 \end{cases}$.

PART –B (5 × 16 = 80)

Question No. 11 has no choice; Questions 12 to 15 have one choice (either–or type) each.

11. (*i*) Expand in Fourier series of periodicity

$$2n \text{ of } f(x) = \begin{cases} x & \text{if } 0 < x < \pi \\ 2\pi - x & \text{if } \pi < x < 2\pi \end{cases} \qquad (8)$$

(ii) Find the half-range cosine series for the function $f(x) = x$, $0 < x < \pi$ and hence deduce the sum of the series $\sum_{n=0}^{\infty} \frac{1}{(2n-1)^4}$ (8)

12. (a) (i) Find the complete solution and singular solution of $z = px + qy + p^2 - q^2$. (8)

(ii) Find the general solution of $x(z^2 - y^2)p + y(x^2 - z^2)q = z(y^2 - x^2)$ (8)

OR

(b) (i) Solve : $(D^2 - 4D'^2)z = \cos 2x$ coa $3y$. (8)

(ii) Solve : $[(D + D' - 1)(D + 2D' - 3)]z = 4 + 3x + 6y + e^{x+y}$. (8)

13. (a) A taut string of length L is fastened at both ends. The midpoint of the string is taken to a height of b and then released from rest in this position. Find the displacement of the string at any time t. (16)

OR

(b) A rod 30 cm long, has its ends A and B at 20°C and 80°C respectively, until steady state conditions prevail. The temperature at the end B is then suddenly reduced to 60°C and at the end A is raised to 40°C and maintained so. Find the resulting temperature $u(x,t)$. (16)

14. (a) (i) Find z-transform of t and e^{at}. (8)

(ii) Find inverse z-transform of $\frac{4z^2 - 2z}{z^3 - 5z^2 + 8z - 4}$ by Residue method. (8)

OR

(b) (i) Find z-transform of $\{a^n\}$. (8)

(ii) Solve : $y_{n+2} - 5y_{n+1} + 6y_n = 36$ (8)

given $y_0 = y_1 = 0$

15. (a) (i) Find the Fourier transform of $f(x) =$ $f(x) = \begin{cases} 1 - |x| & \text{for } |x| \le 1 \\ 0 & \text{for } |x| > 1 \end{cases}$ (6)

Hence evaluate the following integral :

(ii) $\int_0^{\infty} \left(\frac{\sin x}{x}\right)^2 dx$ (5)

(*iii*) $\int_0^\infty \left(\frac{\sin x}{x}\right)^4 dx$ (5)

OR

(*b*) (*i*) Find the Fourier sine and cosine transform of e^{-2x}. (6)

Hence find the value of the following integrals :

(*ii*) $\int_0^\infty \frac{dx}{(x^2+4)^2}$ (5)

(*iii*) $\int_0^\infty \frac{x^2\, dx}{(x^2+4)^2}$ (5)

MODEL QUESTION PAPER – II

Time: 3 Hrs *Max. Marks : 100*

Answer all the Questions

PART–A (10 × 2 = 20 Marks)

1. Find the partial differential equation of the family of spheres having their centres on the line $x = y = z$.

2. Solve $\frac{\partial^3 z}{\partial x^3} - 2\frac{\partial^3 z}{\partial x^2 \partial y^2} - 4\frac{\partial^3 z}{\partial x \partial y^2} + 8\frac{\partial^3 z}{\partial y^3} = 0.$

3. Find the constant term in the Fourier series corresponding to $f(x) = \cos^2 x$ expressed in the interval $(-\pi, \pi)$.

4. To which value, the half range sine series corresponding to $f(x) = x^2$ expressed in the interval (0,2) converges at $x = 2$?

5. Classify the following partial differential equations :

(a) $\frac{\partial^2 u}{\partial x^2} = \frac{\partial^2 u}{\partial y^2}$

(b) $\frac{\partial^2 u}{\partial x dy} = \left(\frac{\partial u}{\partial x}\right)\left(\frac{\partial u}{\partial y}\right) + xy$

6. Write any two solutions of the Laplace equation $u_{xx} + u_{yy} = 0$ involving exponential terms in x or y.

7. State initial value theorem on z-transform.

8. Find z-transform of e^{at+b}.

9. Write the Fourier transform pair.

10. If $F_c(s)$ is the Fourier cosine transform of $f(x)$, prove that the Fourier cosine transform of $f(ax)$ is $\frac{1}{a}F_c\left(\frac{s}{a}\right)$

PART B — (5 ×16 = 80 marks)

11. (*i*) Find the Fourier series expansion of the periodic function $f(x)$ of period $2l$ defined by

$$f(x) = l + x, \; -l \le x \le 0$$
$$= l - x, \; 0 \le x \le l$$

Deduce that $\sum_{1}^{\infty} \frac{1}{(2n-1)^2} = \frac{\pi^2}{8}$

(*ii*) Find the half range sine series of $f(x) = x \cos x$ in $(0, \pi)$.

12.(*a*) (*i*) Form the differential equation by eliminating the arbitrary functions f and g in $z = f(x^3 + 2y) + g(x^3 - 2y)$.

(*ii*) Solve :

$$(y - xz)p + (yz - x)q = (x + y)(x - y).$$

OR

(*b*) (*i*) Find the singular solution of $z = px + qy + \sqrt{p^2 + q^2 + 16}$

(*ii*) Solve : $(D^2 - DD' - 30\,D'^2)\,z = xy + e^{6x+y}$.

13. (*a*) A tightly stretched string of length l has its ends fastened at $x = 0$ and $x = 1$. The mid point of the string is then taken to a height h and then released from rest in that position. Obtain an expression for the displacement of the string at any subsequent time.

OR

(*b*) The boundary value problem governing the steady-state temperature distribution in a flat, thin, square plate is given by

$$\frac{\partial^2 u}{\partial x^2} + \frac{\partial^2 u}{\partial y^2} = 0,\ 0 < x < a, 0 < y < a,$$

$$u(x,0) = 0, u(x,a) = 4\sin^3\left(\frac{\pi x}{a}\right), 0 < x < a,$$

$$u\,(0, y) = 0,\ u\,(a\,, y) = 0,\ 0 < y < a.$$

Find the steady-state temperature distribution in the plate.

14. (*a*) (*i*) Find the inverse *z*-transform of $\dfrac{z^2}{\left(z-\frac{1}{2}\right)\left(z-\frac{1}{4}\right)}$

(*ii*) Solve $x(n + 1) - 2x\,(n) = 1$ if $x(0) = 0$

OR

(*b*)(*i*) Find *z*-transform of $\left\{\cos\dfrac{n\pi}{2}\right\}$.

(*ii*) Find the inverse z-transform of $\dfrac{z^2 + z}{(z-1)(z^2+1)}$.

15.(*a*)(*i*) Find the Fourier transform of $f(x) = 1$ for $|x| < 1$

$= 0$ otherwise.

Hence prove that $\displaystyle\int_0^\infty \frac{\sin x}{x}\,dx = \int_0^\infty \frac{\sin^2 x}{x^2}\,dx = \frac{\pi}{2}$

(*ii*) Find the Fourier sine transform

$$f(x) = \sin x, 0 < x \le \pi$$
$$= 0, \pi \le x < \infty$$

OR

(*b*) (*i*) Find the Fourier cosine transform of e^{-4x}. Deduce that

$$\int_0^\infty \frac{\cos 2x}{x^2+16}\,dx = \frac{\pi}{8}e^{-8} \quad\text{and}\quad \int_0^\infty \frac{x\sin 2x}{x^2+16}\,dx = \frac{\pi}{2}e^{-8}$$

(*ii*) State and prove convolution theorem for Fourier transforms.

MODEL QUESTION PAPER – III

Time: 3 Hrs *Max. Marks : 100*

Answer all the Questions

PART–A (10 × 2 = 20 Marks)

1. Find the complete integral of $p + q = pq$ where $p = \frac{\partial z}{\partial x}$ and $q = \frac{\partial z}{\partial y}$.

2. Solve $\left(D^3 - 3DD'^2 + 2D'^3\right)z = 0.$

3. If $f(x) = x^2 + x$ is expressed as a Fourier series in the interval $(-2, 2)$, to which value this series converges at $x = 2$?

4. If the fourier series corresponding to $f(x) = x$ in the interval $(0, 2\pi)$ is $\frac{a_0}{2} + \sum_{1}^{\infty}\{a_n \cos nx + b_n \sin nx\}$, without finding the values of $a_0\, a_n,\, b_n$ find the value of $\frac{a_0^2}{2} + \sum_{1}^{\infty}\left(a_n^2 + b_n^2\right)$.

5. Classify the following second order partial differential equations :

(*a*) $4\frac{\partial^2 u}{\partial x^2} + 4\frac{\partial^2 u}{\partial x \partial y} + \frac{\partial^2 u}{\partial y^2} - 6\frac{\partial u}{\partial x} - 8\frac{\partial u}{\partial y} - 16u = 0$

(*b*) $\frac{\partial^2 u}{\partial x^2} + \frac{\partial^2 u}{\partial y^2} = \left(\frac{\partial u}{\partial x}\right)^2 + \left(\frac{\partial u}{\partial y}\right)^2$

6. Write any two solutions of the Laplace equation obtained by the method of separation of variables.

7. Find z-transform of $\{(-1)^n\}$.

8. State final value theorem on *z*-Transform.

9. If $F(s)$ is the Fourier transform of $f(x)$, write the formula for the Fourier transform of $f(x)\cos(ax)$ in terms in F.

10. State the convolution theorem for Fourier transforms.

PART B—(5 × 16 = 80 marks)

11. (*i*) Find inverse of *z*-transform of $\frac{z^2}{(z-a)^2}$ using convolution theorem.

(*ii*) Using *z*-transform

Solve : $y(n) - ay(n-1) = u(n)$

12. (*a*) (*i*) Solve : $z = 1 + p^2 + q^2$.

(*ii*) Solve : $(y - z)p - (2x + y)\, q = 2x + z$.

OR

(*b*) (*i*) Form the partial differential equation by eliminating the arbitrary functions f and g in $z = x^2 f(y) + y^2 g(x)$.

(*ii*) Solve : $(D^2 - DD' - 20D'^2)\, z = e^{5x+y} + \sin(4x - y)$.

13. (*a*) (*i*) Obtain the Fourier series for $f(x) = 1 + x + x^2$ in $(-\pi, \pi)$. Deduce that $\frac{1}{1^2} + \frac{1}{2^2} + \frac{1}{3^2} + \ldots\ldots\ldots \infty = \frac{\pi^2}{6}$.

(*ii*) Obtain the constant term and the first harmonic in the Fourier series expansion for $f(x)$ where $f(x)$ is given in the following table :

x:	0	1	2	3	4	5	6	7	8	9	10	11
$f(x)$:	18.0	18.7	17.6	15.0	11.6	8.3	6.0	5.3	6.4	9.0	12.4	15.7

OR

(*b*) (*i*) Expand the function $f(x) = x \sin x$ as a Fourier series in the interval $-\pi \le x \le \pi$.

(*ii*) Obtain the half range cosine series for $f(x) = (x - 2)^2$ in the interval $0 < x < 2$. Deduce that $\sum_{1}^{\infty} \frac{1}{(2n-1)^2} = \frac{\pi^2}{8}$.

14. (*a*) A tightly stretched flexible string has its ends fixed at $x = 0$ and $x = l$. At time $t = 0$, the string is given a shape defined by $f(x) = kx^2 (l - x)$, where k is a constant, and then released from rest. Find the displacement of any point x of the string at any time $t > 0$.

OR

(*b*) The ends A and B of a rod l cm long have the temperatures 40°C and 90°C until steady state prevails. The temperature at A is suddenly raised to 90°C and at the same time that at B is lowered to 40°C. Find the temperature distribution in the rod at time t. Also show that the temperature at the mid point of the rod remains unaltered for all time, regardless of the material of the rod.

15. (*a*) (*i*) Find the Fourier transform of $e^{-a|x|}$ if a > 0. Deduce that

$$\int_0^{\infty} \frac{1}{(x^2 + a^2)^2}\, dx = \frac{\pi}{4a^3} \text{ if } a > 0.$$

(*ii*) Find the Fourier sine transform of $xe^{-x^2/2}$.

OR

(*b*) (*i*) Find the Fourier cosine transform of

$f(x) = 1 - x^2, 0 < x < 1$

$= 0$, otherwise.

Hence prove that $\int_0^{\infty} \frac{\sin x - x\cos x}{x^3} \cos\left(\frac{x}{2}\right) dx = \frac{3\pi}{16}$.

(*ii*) Derive the Parseval's identity for Fourier transforms.

MODEL QUESTION PAPER – IV

Time : Three hours *Maximum : 100 marks*

Answer All questions.

Part — A (10 × 2 = 20 marks)

1. Obtain partial differential equation by eliminating arbitrary constants a and b from $(x-a)^2+(y-b)^2+z^2=1$.

2. Find the general solution of $4\frac{\partial^2 z}{\partial x^2}-12\frac{\partial^2 z}{\partial x\,\partial y}+9\frac{\partial^2 z}{\partial y^2}=0$.

3. State Dirichlet's conditions for a given function to expand in Fourier series.

4. If the Fourier series of the function $f(x)=x+x^2$ in the interval $-\pi<x<\pi$ is

$$\frac{\pi^2}{3}+\sum_{n=1}^{\infty}(-1)^n\left[\frac{4}{n^2}\cos nx-\frac{2}{n}\sin nx\right],$$

then find the value of the infinite series $\frac{1}{1^2}+\frac{1}{2^2}+\frac{1}{3^2}+\ldots$.

5. Classify the following partial differential equations

(a) $y^2u_{xx}-2xy\,u_{xy}+x^2u_{yy}+2u_x-3u=0$

(b) $y^2u_{xx}+u_{yy}+u_x^2+u_y^2+7=0$.

6. An insulated rod of length 60 cm has its ends at A and B maintained at 20°C and 80°C respectively. Find the steady state solution of rod.

7. State convolution theorem on Fourier Transform.

8. Find the z-transform of $u(n)$.

9. Find the z-transform of $\left\{\frac{1}{n}\right\}$.

10. If $F_s(s)$ is the Fourier sine transform of $f(x)$, show that

$$F_s[f(x)\cos ax]=\frac{1}{2}[F_s(s+a)+F_s(s-a)].$$

Part — B (5 × 16 = 80 marks)

11. Obtain Fourier series for $f(x)$ of period $2l$ and defined as follows

$$f(x) = l - x \text{ in } 0 < 0 \le l$$
$$= 0 \quad \text{in } l \le x \le 2l$$

Hence deduce that

$$1 - \frac{1}{3} + \frac{1}{5} - \frac{1}{7} + \ldots.. = \frac{\pi}{4}$$

$$\frac{1}{1^2} + \frac{1}{3^2} + \frac{1}{5^2} + \ldots. = \frac{\pi^2}{8}. \quad (16)$$

12. (*a*) (*i*) Find the singular integral of the partial differential equation $z = px + qy + p^2 - q^2$. (6)

(*ii*) Solve : $(D^2 + 4DD' - 5D'^2)z = 3e^{2x-y} + \sin(x - 2y)$. (10)

OR

(*b*) (*i*) Find the general solution of

$$(3z - 4y)p + (4x - 2z)q = 2y - 3x$$

(*ii*) Solve : $(D^2 - 2DD' + D'^2 - 3D + 3D' + 2)z = (e^{3x} + 2e^{-2y})^2$ (10)

13. (*a*) A tightly stretched string with fixed end points $x = 0$ and $x = l$ is initially at rest in its equilibrium position. If it is set vibrating by giving each point a velocity $kx(l - x)$. Find the displacement of the string at any time. (16)

OR

(*b*) A rectangular plate with insulated surface is 10 cm wide so long compared to its width that it may be considered infinite length. If the temperature along short edge $y = 0$ is given $u(x, 0) = 8\sin\frac{\pi x}{10}$ when $0 < x < 10$, while the two long edges $x = 0$ and $x = 10$ as well as the other short edge, are kept at 0°C, find the steady state temperature function $u(x, y)$. (16)

14. (*a*) Find the Fourier transform of $f(x)$ given by

$$f(x) = 1 \text{ for } |x| < 2$$
$$= 0 \text{ for } |x| > 2$$

and hence evaluate $\int_0^\infty \frac{\sin x}{x} dx$ and $\int_0^\infty \left(\frac{\sin x}{x}\right)^2 dx$. (16)

OR

(*b*) Find Fourier sine and cosine transform of e^{-x} and hence find the Fourier sine transform of $\frac{x}{1+x^2}$ and Fourier cosine transform of $\frac{1}{1+x^2}$. (16)

15. (*a*) (*i*) Find the inverse *z*-transform of $\frac{z}{z^2+2z+2}$.

(*ii*) Solve : $x(n+2)-3x(n+1)+2x(n)=0$ given $x(0) = 0$, $x(1) = 1$ using *z*-transform.

OR

(*b*) (*i*) Find the inverse transform of $\frac{z}{z^2+7z+10}$.

(*ii*) Solve : $y_{n+2}+6y_{n+1}+9y_n=2^n$ if $y_0 = y_1 = 0$. (16)

MODEL QUESTION PAPER – V

Time : Three hours *Maximum : 100 marks*

Answer All questions.

Part —A (10 × 2 = 20 marks)

1. Eliminate the arbitrary function f from $z = f\left(\frac{xy}{z}\right)$ and form the partial differential equation.

2. Find the complete integral of $p + q = pq$.

3. Find a Fourier sine series for the function $f(x) = 1;\ 0 < x < \pi$.

4. If the Fourier series for the function

$$f(x) = 0;\ 0 < x < \pi$$
$$= \sin x;\ \pi < x < 2\pi$$

is $f(x) = -\frac{1}{\pi} + \frac{2}{\pi}\left[\frac{\cos 2x}{1.3} + \frac{\cos 4x}{3.5} + \frac{\cos 6x}{5.7} + \ldots\right] + \frac{1}{2}\sin x$

deduce that $\frac{1}{1.3} - \frac{1}{3.5} + \frac{1}{5.7} - \ldots\infty = \frac{\pi - 2}{4}$.

5. Classify the partial differential equation $u_{xx} + xu_{yy} = 0$.

6. A rod 30 cm long has its ends A and B kept at 20°C and 80°C respectively until steady state conditions prevail. Find the steady state temperature in the rod.

7. Define the sampler.

8. Define unit step sequence and unit sample sequence.

9. State change of scale property on Fourier Transform.

10. Find the Fourier transform of $f(x)$ if

$$f(x) = \begin{cases} 1;\ |x| < a \\ 0;\ |x| > a > 0. \end{cases}$$

Part — B (5 × 16 = 80 marks)

11. (*i*) Solve $x(y - z)p + y(z - x)q = z(x - y)$

(*ii*) Solve $(D^3 - 7DD'^2 - 6D'^3)z = \sin(x + 2y) + e^{2x+y}$.

12. (*a*) (*i*) Determine the Fourier expansion of $f(x) = x$ in the interval $-\pi < x < \pi$.

(*ii*) Find the half range cosine series for $\sin x$ in $(0, \pi)$.

OR

(*b*) (*i*) Obtain the Fourier series for the function

$$f(x) = \pi x;\ 0 \le x \le 1$$
$$= \pi(2 - x);\ 1 \le x \le 2.$$

(*ii*) Find the Fourier series of period 2π for the function

$$f(x) = \begin{cases} 1 \text{ in } (0,\pi) \\ 2 \text{ in } (\pi, 2\pi) \end{cases}$$

and hence find the sum of the series $\frac{1}{1^2}+\frac{1}{3^2}+\frac{1}{5^2}+...\infty$.

13. (*a*) A string is stretched between two fixed points at a distance $2l$ apart and the points of the string are given initial velocities v where

$$v = \frac{cx}{l} \text{ in } 0 < x < l$$
$$= \frac{c}{l}(2l - x) \text{ in } l < x < 2l$$

x being the distance from one end point. Find the displacement of the string at any subsequent time.

OR

(*b*) Find the solution of the one dimensional diffusion equation satisfying the boundary conditions :

(*i*) u is bounded as $t \to \infty$

(*ii*) $\left[\frac{\partial u}{\partial x}\right]_{x=0} = 0$ for all t

(*iii*) $\left[\frac{\partial u}{\partial x}\right]_{x=a} = 0$ for all t

(*iv*) $u(x, 0) = x(a - x), 0 < x < a.$

14. (*a*) (*i*) Find the z-transform of te^{-3t}.

(*ii*) Using convolution theorem, find inverse of z-trasnform of $\frac{z^2}{(z-a)(z-b)}$.

OR

(*b*) (*i*) Solve: $y_{n+2} - 4y_{n+1} + 4y = \pi$ if $y_0 = y_1 = 0$.

(*ii*) Find the z-Transform of $na^n u(n)$.

15. (*a*) (*i*) Show that the Fourier transform of

$$f(x)=\begin{cases} a^2-x^2; & |x|<a \\ 0; & |x|>a>0 \end{cases}$$

is

$$2\sqrt{\frac{2}{\pi}}\left(\frac{\sin\lambda a-\lambda a\cos\lambda a}{\lambda^3}\right).$$

Hence deduce $\displaystyle\int_0^\infty \frac{\sin t - t\cos t}{t^3}\,dt=\frac{\pi}{4}.$

(*ii*) Find the Fourier sine and cosine transform of

$$f(x)=\begin{cases} x; & 0<x<1 \\ 2-x; & 1<x<2 \\ 0; & x>2 \end{cases}$$

OR

(*b*) (*i*) If $\overline{f}(\lambda)$ is the Fourier transform of $f(x)$, find the Fourier transform of $f(x-a)$ and $f(ax)$.

(*ii*) Verify Parseval's theorem of Fourier transform for the function

$$f(x)=\begin{cases} 0; & x<0 \\ e^{-x}; & x>0. \end{cases}$$

SOLVED QUESTION PAPERS

B.E./ B.Tech. Degree Examination, April/May 2011
Third Semester
Civil Engineering
MA 231 – Mathematics –III
(Common to all Other Branches)
(Regulation 2001)

Time : Three Hours *Maximum : 100 Marks*

Answer all Questions
PART– A (10 × 2 = 20 Marks)

1. Find the complete integral of $pq + p + q = 0$.
2. Form the partial differential equation by eliminating the arbitrary function from $z = f(x^2 + y^2 + z^2)$.
3. State Dirichlet conditions for the existence of the Fourier series of $f(x)$.
4. State the Parseval's theorem for the Fourier series.
5. Prove that $F[f(x-a)] = e^{ias} F(s)$.
6. Find the Fourier sine transform of $f(x) = e^{-x}, x > 0$.
7. Classify the partial differential equation $f_{xx} + 2f_{xy} + 4f_{yy} = 0, x > 0, y > 0$.
8. Write down all possible solutions of one dimensional wave equation.
9. Find the Laplace transform of $e^{-t} \sin t \cos t$.
10. Find the inverse Laplace transform of $\dfrac{s}{s^2+2s+3}$.

PART–B (5 × 16 = 80 Marks)

11. (*a*) (*i*) Find the singular integral of $z = px + qy + p^2 + pq + q^2$.

(*ii*) Solve $(D^2 - DD' - 30D'^2)\, z = xy + e^{6x+y}$.

OR

(*b*) (*i*) Find the general solution of
$$x(z^2 - y^2)\, p + y(x^2 - z^2)\, q = z(y^2 - x^2).$$

(*ii*) Find the general solution of $p^2 + q^2 = x^2 + y^2$.

12. (*a*) (*i*) Find the Fourier series of $f(x) = (\pi - x)^2$ in $(0, 2\pi)$.

(*ii*) Find the half range cosine series of $f(x) = x(l - x), 0 < x < l$.

OR

(*b*) (*i*) Find the Fourier series of $f(x) = x^2, -\pi < x < \pi$. Hence show that
$$\frac{1}{1^4}+\frac{1}{2^4}+\frac{1}{3^4}+\frac{1}{4^4}+.. = \frac{\pi^4}{90}.$$

(*ii*) Find the Fourier series of $f(x)$ as far as the second harmonics from the given data :

x	0	$\pi/6$	$\pi/3$	$\pi/2$	$2\pi/3$	$5\pi/6$	π
$f(x)$	2.34	2.20	1.60	0.83	0.51	0.88	1.19

13. (*a*) Find the Fourier transform of $f(x) = \begin{cases} 1-x^2, & if \ |x| \le 1 \\ 0, & if \ |x| > 1 \end{cases}$. Hence

evaluate $\int_0^\infty \frac{\sin x - x\cos x}{x^3} \cos\frac{x}{2}\, dx$ and $\int_0^\infty \left(\frac{\sin t - t\cos t}{t^3}\right)^2 dt.$

OR

(*b*) Find the Fourier sine transforms of $f(x) = e^{-ax}$, $x > 0$, $a > 0$ and $g(x) = e^{-bx}$, $x > 0$, $b > 0$. Hence evaluate $\int_0^\infty \frac{x^2 dx}{(x^2+a^2)(x^2+b^2)}$.

4. (*a*) A tightly stretched string of length l has its ends fastened at $x = 0$ and $x - l$. At time $t = 0$ the string was released from rest with the displacement $f(x) = kx(l-x)$. Find the displacement at a distance x from one end at any time t.

OR

(*b*) A square plate is bounded by the lines $x = 0$, $y = 0$, $x = 10$, $y = 10$. Its faces are insulated. The temperature along the upper horizontal edge is given by $u(x, 10) = x(10 - x)$, $0 < x < 10$, while the other edges are kept at 0°C. Find the temperature distribution in the plate.

15. (*a*) (*i*) Find the Laplace transform of $f(t) = \begin{cases} E, & 0 \le t \le \frac{T}{2} \\ -E, & T/2 \le t \le T \end{cases}$ and $f(t+T) = f(t)$.

(*ii*) Using Laplace transform, solve $y'' - 3y' + 2y = e^{3x}$, where $y(0) = 0$, $y'(0) = 0$.

OR

(*b*) (*i*) Find the inverse Laplace transform of $\frac{s^2 - s + 2}{s(s+2)(s-3)}$

(*ii*) Using convolution theorem, find the inverse Laplace transform of $\frac{s^2}{(s^2+a^2)^2}$.

Answers for MA 231 – Mathematics – III
Third Semester – May 2011
PART–A (10 × 2 = 20 Marks)

1. This is of the form $F(p, q) = 0$.

Hence complete integral is $z = ax + by + c$ where $ab + a + b = 0$.

$$\therefore b = \frac{-a}{1+a}$$

$$\text{Hence, } z = ax - \frac{a}{1+a}y + c$$

2. $z = f(x^2 + y^2 + z^2)$

$$\frac{\partial z}{\partial x} = p = f'(x^2 + y^2 + z^2) \times (2x + 2zp)$$

$$\frac{\partial z}{\partial y} = q = f'(x^2 + y^2 + z^2) \times (2y + 2zq)$$

$$\text{Dividing, } \frac{p}{q} = \frac{x + pz}{y + qz}$$

$$py + pqz = qx + pqz$$

$\therefore\ py - qx = 0$ is the required answer.

3. Dirichlet's conditions

(*i*) $f(x)$ is single valued and finite in $(c, c + 2\pi)$

(*ii*) $f(x)$ is continuous or piece-wise continuous with finite number of finite discontinuities in $(c, c + 2\pi)$

(*iii*) $f(x)$ has a finite number of maxima or minima in $(c, c + 2\pi)$

4. Parseval's theorem

$$\text{If } f(x) = \frac{a_0}{2} + \sum_{n=1}^{\infty}(a_n \cos nx + b_n \sin nx)$$

$$\text{Then } (2\pi)\left[\frac{a_0^2}{4} + \sum \frac{1}{2}(a_n^2 + b_n^2)\right] = \int_c^{c+2\pi} [f(x)]^2\, dx.$$

5. $$F[f(x-a)] = \frac{1}{\sqrt{2\pi}} \int_{-\infty}^{\infty} f(x-a)\, e^{isx}\, dx$$

$$= \frac{1}{\sqrt{2\pi}} \int_{-\infty}^{\infty} f(t)\, e^{is(a+b)}\, dt, \text{ where } t = x - a$$

$$= e^{ias} \cdot \frac{1}{\sqrt{2\pi}} \int_{-\infty}^{\infty} f(t)\, e^{ist}\, dt = e^{ias} F[f(x)] = e^{ias} F(s)$$

6. $$F_s(e^{-x}) = \sqrt{\frac{2}{\pi}} \int_0^{\infty} e^{-x} \sin sx\, dx$$

$$= \sqrt{\frac{2}{\pi}} \left[\frac{e^{-x}}{1+s^2}(-\sin sx - s\cos sx)\right]_0^{\infty} = \sqrt{\frac{2}{\pi}} \cdot \frac{s}{1+s^2}$$

7. $f_{xx} + 2f_{xy} + 4f_{yy} = 0$

$$A = 1, B = 2, C = 4$$

$$B^2 - 4AC = 4 - 16 = -12 = -\text{ve}$$

Hence, it is elliptic at every point of the domain.

8. $\dfrac{\partial^2 y}{\partial t^2} = a^2 \dfrac{\partial^2 y}{\partial x^2}$ wave equation.

The solutions are

$$y(x, t) = (A e^{\lambda x} + B e^{-\lambda x})(C e^{\lambda at} + D e^{-\lambda at})$$

Or $y(x, t) = (A \cos \lambda x + B \sin \lambda x)(C \cos \lambda at + D \sin \lambda at)$

Or $y(x, t) = (Ax + B)(Ct + D)$

9. $$L[e^{-t} \sin t \cos t] = \frac{1}{2} L\left[e^{-t} \sin 2t\right] = \frac{1}{2}\left[\frac{2}{s^2 + 4}\right]_{s \to s+1}$$

$$= \frac{1}{(s+1)^2 + 4}$$

10. $$L^{-1}\left[\frac{s}{s^2 + 2s + 3}\right] = L^{-1} \frac{(s+1) - 1}{(s+1)^2 + (\sqrt{2})^2}$$

$$= e^{-t}.L^{-1} \frac{s - 1}{s^2 + (\sqrt{2})^2}$$

$$= e^{-t}\left[\cos \sqrt{2}t - \frac{1}{\sqrt{2}} \sin \sqrt{2}t\right]$$

PART–B (5 × 16 = 80 Marks)

11. (a) (i) $z = px + qy + p^2 + pq + q^2$

This is of the form $z = px + qy + f(p, q)$

Hence complete integral is $z = ax + by + a^2 + ab + b^2$ ...(1)

where a, b are arbitrary constants.

To find, singular integral, differentiate (1) w.r.t a and b partially.

$$0 = x + 2a + b$$

$$0 = y + 2b + a$$

i.e., $$2a + b = -x$$

$$a + 2b = -y$$

$\therefore$ $$3a = y - 2x$$

$$a = \frac{y - 2x}{3}$$

$$b = -x - 2a = -x - \frac{2}{3}(y - 2x) = \frac{x}{3} - \frac{2y}{3}$$

Substituting (1)

$$z = \left(\frac{y-2x}{3}\right)x + \left(\frac{x-2y}{3}\right)y + \frac{(y-2x)^2}{9} + \frac{(x-2y)^2}{9} + \frac{(y-2x)(x-2y)}{9}$$

$$9z = 3xy - 3x^2 - 3y^2$$

$\therefore\ 3z = xy - x^2 - y^2$ is the singular solution.

(*ii*) $(D^2 - DD' - 30D'^2)\, z = xy + e^{6x+y}$

(printing mistake in question paper)

Auxiliary equation is

$m^2 - m - 30 = 0$

$(m-6)(m+5) = 0$

$\therefore\ m = 6, -5$

C.F. is $\phi_1(y+6x) + \phi_2(y-5x)$ ....(1)

$$\text{PI}_1 = \frac{xy}{D^2 - DD' - 30D'^2} = \frac{1}{D^2}\frac{xy}{\left(1 - \frac{D'}{D} - \frac{30D'^2}{D^2}\right)}$$

$$= \frac{1}{D^2}\left[1 - \left(\frac{D'}{D} + 30\frac{D'^2}{D^2}\right)\right]^{-1}(xy)$$

$$= \frac{1}{D^2}\left[1 + \frac{D'}{D} + \frac{30D'^2}{D^2} + \left(\frac{D'}{D} + \frac{30D'^2}{D^2}\right)^2 + \ldots\right](xy)$$

$$= \frac{1}{D^2}\left[1 + \frac{D'}{D} + \ldots\right](xy)$$

$$= \left(\frac{1}{D^2} + \frac{D'}{D^3}\right)(xy)$$

$$= \frac{x^3 y}{6} + \frac{x^4}{24}$$

$$PI_2 = \frac{e^{6x+y}}{D^2 - DD' - 30D'^2} = \frac{e^{6x+y}}{(D-6D')(D+5D')}$$

$$= \frac{1}{11}\frac{e^{6x+y}}{D-6D'} = \frac{1}{11}.x\, e^{6x+y}$$

Hence, general solution is

$$z = \phi_1(y+6x) + \phi_2(y-5x) + \frac{x^3 y}{6} + \frac{x^4}{24} + \frac{1}{11}x\, e^{6x+y}$$

11. (*b*) (*i*) $x(z^2 - y^2)\,p + y(x^2 - z^2)\,q = z(y^2 - x^2)$ (Text P. 136)

The subsidiary equations are

$$\frac{dx}{x(z^2-y^2)} = \frac{dy}{y(x^2-z^2)} = \frac{dz}{z(y^2-x^2)} \quad ...(1)$$

Taking the two sets of multipliers *x, y, z* and $\frac{1}{x}, \frac{1}{y}, \frac{1}{z}$ each ratio of (1) equals

$$\frac{xdx+ydy+zdz}{\sum x^2(z^2-y^2)} = \frac{\frac{1}{x}dx+\frac{1}{y}dy+\frac{1}{z}dz}{\sum(z^2-y^2)}$$

$$= \frac{xdx+ydy+zdz}{0} = \frac{\frac{1}{x}dx+\frac{1}{y}dy+\frac{1}{z}dz}{0}$$

This implies $xdx + ydy + zdz = 0$ and $\frac{dx}{x}+\frac{dy}{y}+\frac{dz}{z}=0$

Integrating, $x^2 + y^2 + z^2 = a$ and $\log x + \log y + \log z = b$ *i.e.,* $xyz = k$

Hence, the general solution is

$$\phi(x^2 + y^2 + z^2, xyz) = 0.$$

(*ii*) $p^2 + q^2 = x^2 + y^2$ (Text P. 125)

$$p^2 - x^2 = y^2 - q^2 = a^2 \text{ (say)}$$

$$\therefore \quad p = \sqrt{a^2+x^2},\ q = \sqrt{y^2-a^2}$$

$$dz = pdx + qdy$$

$$= \sqrt{a^2+x^2}\ dx + \sqrt{y^2-a^2}\ dy$$

Integrating,

$$z = \frac{a^2}{2}\sinh^{-1}\frac{x}{a} + \frac{x}{2}\sqrt{a^2+x^2} + \frac{y}{2}\sqrt{y^2-a^2} - \frac{a^2}{2}\cosh^{-1}\frac{y}{a} + b$$

This is the complete solution and not general solution. (Question is wrongly worded)

12. (*a*) (*i*) Let $f(x) = (\pi - x)^2 = \frac{a_0}{2} + \sum_{n=1}^{\infty}(a_n \cos nx + b_n \sin nx)$ (Text P. 28)

where $a_0 = \frac{1}{\pi}\int_0^{2\pi}(\pi - x)^2 dx$

$$= \frac{1}{\pi}\left[\frac{(\pi-x)^3}{(-3)}\right]_0^{2\pi}$$

$$= -\frac{1}{3\pi}\left[-\pi^3-\pi^3\right] = \frac{2\pi^2}{3}$$

$$a_n = \frac{1}{\pi}\int_0^{2\pi} (\pi - x)^2 \cos n\,x dx$$

$$= \frac{1}{\pi}\left[(\pi - x)^2\left(\frac{\sin nx}{n}\right) + 2(\pi - x)\left(-\frac{\cos nx}{n^2}\right) + 2\left(-\frac{\sin nx}{n^3}\right)\right]_0^{2\pi}$$

$$= \frac{1}{\pi}\left[\frac{2\pi}{n^2} + \frac{2\pi}{n^2}\right] = \frac{4}{n^2}$$

$$b_n = \frac{1}{\pi}\int_0^{2\pi} (\pi - x)^2 \sin nx\, dx$$

$$= \frac{1}{\pi}\left[(\pi - x)^2\left(-\frac{\cos nx}{n}\right) + 2\,(\pi - x)\left(-\frac{\sin nx}{n^2}\right) + 2\left(\frac{\cos nx}{n^3}\right)\right]_0^{2\pi}$$

$$= \frac{1}{\pi}\left[-\frac{\pi^2}{n} + \frac{\pi^2}{n} + \frac{2}{n^3}(1-1)\right] = 0$$

Hence, $f(x) = (\pi - x)^2 = \dfrac{\pi^2}{3} + 4\displaystyle\sum_{n=1}^{\infty} \frac{1}{n^2}\cos nx$

(*ii*) $f(x) = (lx - x^2), \qquad 0 < x < l.$

Let $f(x) = \dfrac{a_0}{2} + \displaystyle\sum_{n=1}^{\infty} a_n \cos\frac{n\pi x}{e}$

$$a_0 = \frac{2}{l}\int_0^l (lx - x^2)\, dx = \frac{2}{l}\left[\frac{lx^2}{2} - \frac{x^3}{3}\right]_0^l$$

$$= \frac{2}{l}\left[\frac{l^3}{6}\right] = \frac{l^2}{3}$$

$$a_n = \frac{2}{l}\int_0^l (lx - x^2)\cos\frac{n\pi x}{l}\, dx$$

$$= \frac{2}{l}\left[(lx - x^2)\left(\frac{\sin\frac{n\pi x}{l}}{\frac{n\pi}{l}}\right) - (l - 2x)\left(-\frac{\cos\frac{n\pi x}{l}}{\frac{n^2\pi^2}{l^2}}\right) + (-2)\left(-\frac{\sin\frac{n\pi x}{l}}{\frac{n^3\pi^3}{l^3}}\right)\right]_0^l$$

$$= \frac{2}{l}\left[-\frac{l^3}{n^2\pi^2}\cos n\pi - \frac{l^3}{n^2\pi^2}\right] = \frac{-2l^2}{n^2\pi^2}[1 + (-1)^n]$$

$= 0$ if n is odd

$= \frac{-4l^2}{n^2\pi^2}$ if n is even

$$\therefore\ f(x) = \frac{l^2}{6} - \frac{4l^2}{\pi^2} \sum_{n=2,4,6....}^{\infty} \frac{1}{n^2} \cos\frac{n\pi x}{l}$$

12. (*b*) (*i*) $f(x) = x^2, -\pi < x < \pi$ (Text P. 40, 63)

$f(x)$ is even in $(-\pi, \pi)$

Hence $b_n = 0$ in its Fourier series

$$\therefore\ f(x) = \frac{a_0}{2} + \sum_{1}^{\infty} a_n \cos nx$$

$$a_0 = \frac{2}{\pi}\int_0^{\pi} x^2 dx = \frac{2\pi^2}{3}$$

$$a_n = \frac{2}{\pi}\int_0^{\pi} x^2 \cos nx\, dx$$

$$= \frac{2}{\pi}\left[(x^2)\left(\frac{\sin nx}{n}\right) - (2x)\left(-\frac{\cos nx}{n^2}\right) + (2)\left(-\frac{\sin nx}{n^3}\right)\right]_0^{\pi}$$

$$= \frac{2}{\pi}\left[\frac{2\pi}{n^2}(-1)^n\right] = \frac{4}{n^2}(-1)^n$$

$$\therefore\ f(x) = \frac{\pi^2}{3} + 4\sum_{n=1}^{\infty} \frac{(-1)^n}{n^2} \cos nx$$

Here $a_0 = \frac{2\pi^2}{3}$, $a_n = \frac{4(-1)^n}{n^2}$

Using Parsival's theorem

$$(\text{range})\left[\frac{a_0^2}{4} + \frac{1}{2}\Sigma a_n^2\right] = \int_{-\pi}^{\pi} (f(x))^2 dx = \int_{-\pi}^{\pi} x^4 dx$$

$$2\pi\left[\frac{\pi^4}{9} + \frac{1}{2}\Sigma\frac{16}{n^4}\right] = 2\frac{\pi^5}{5}$$

$$\frac{\pi^4}{9} + 8\Sigma\frac{1}{n^4} = \frac{\pi^4}{5}$$

$$8\Sigma\frac{1}{n^4} = \frac{\pi^4}{5} - \frac{\pi^4}{9} = \frac{4}{45}\pi^4$$

$$\therefore\ \sum_{n=1}^{\infty}\frac{1}{n^4} = \frac{\pi^4}{90}.$$

(*ii*) We can omit the last value of $f(x)$, since $f(x)$ is given at $x = 0$

Let $\theta = 2x$; As x varies from 0 to π, θ varies from 0 to 2π with an increase of $\frac{2\pi}{6} = \frac{\pi}{3}$.

Let $f(x) = F(\theta) = \frac{a_0}{2} + a_1 \cos\theta + a_2 \cos 2\theta + b_1 \sin\theta + b_2 \sin 2\theta$

θ	$y = F(\theta)$	$\cos\theta$	$\cos 2\theta$	$\sin\theta$	$\sin 2\theta$	$y\cos\theta$	$y\cos 2\theta$	$y\sin\theta$	$y\sin 2\theta$
0	2.34	1	1	0	0	2.34	2.34	0	0
$\frac{\pi}{3}$	2.20	0.5	–0.5	0.866	0.866	1.10	–1.1	1.91	1.91
$\frac{2\pi}{3}$	1.60	–0.5	–0.5	0.866	–0.866	–0.80	–0.8	1.39	–1.39
π	0.83	–1	1	0	0	–0.83	0.83	0	0
$\frac{4\pi}{3}$	0.51	–0.5	–0.5	–0.866	0.866	–0.25	–0.25	–0.43	0.43
$\frac{5\pi}{3}$	0.88	0.5	–0.5	–0.866	–0.866	0.44	–0.44	–0.76	–0.76
Σ	8.36					2.00	0.58	2.11	0.19

$$y = \frac{a_0}{2} + a_1 \cos\theta + a_2 \cos 2\theta + b_1 \sin\theta + b_2 \sin 2\theta.$$

$$a_0 = \frac{2}{6}\Sigma f(x) = \frac{8.36}{3} = 2.79 \qquad b_1 = \frac{2}{6}(2.11) = 0.70$$

$$a_1 = \frac{2}{6}(2.0) = 0.67 \qquad b_2 = \frac{2}{6}(0.19) = 0.06$$

$$a_2 = \frac{2}{6}(0.58) = 0.19$$

$\therefore\ f(x) = F(\theta) = 1.39 + 0.67\cos\theta + 0.19\cos 2\theta + 0.7\sin\theta + 0.06\sin 2\theta$
where $\theta = 2x$.

13. (*a*) $F\{f(x)\} = \frac{1}{\sqrt{2\pi}}\int_{-\infty}^{\infty} e^{isx} f(x)\,dx$

$$= \frac{1}{\sqrt{2\pi}}\int_{-1}^{1} e^{isx}(1-x^2)\,dx$$

$$= \frac{1}{\sqrt{2\pi}}\left[(1-x^2)\left(\frac{e^{isx}}{is}\right) - (-2x)\left(\frac{e^{isx}}{i^2 s^2}\right) + (-2)\left(\frac{e^{isx}}{i^3 s^3}\right)\right]_{-1}^{1}$$

$$= \frac{1}{\sqrt{2\pi}}\left[-\frac{2}{s^2}\left(e^{is} + e^{-is}\right) - \frac{-2i}{s^3}\left(e^{is} - e^{-is}\right)\right]$$

$$= -\frac{2}{s^3}\cdot\frac{1}{\sqrt{2\pi}}[2s\cos s - 2\sin s]$$

$$= -\frac{4}{\sqrt{2\pi}}\left[\frac{s\cos s - \sin s}{s^3}\right] \qquad \text{...(1)}$$

Using inversion formula,

$$f(x) = \frac{1}{\sqrt{2\pi}}\int_{-\infty}^{\infty} \frac{-4}{\sqrt{2\pi}}\left(\frac{s\cos s - \sin s}{s^3}\right)e^{-isx}\,ds$$

$$= \frac{-2}{\pi}\int_{-\infty}^{\infty} \frac{s\cos s - \sin s}{s^3}\, e^{-isx}\,ds$$

$$\int_{-\infty}^{\infty} \frac{s\cos s - \sin s}{s^3}(\cos sx - i\sin sx)\,ds = -\frac{\pi}{2}(1-x^2) \text{ if } |x| < 1$$

$$= 0 \text{ if } |x| > 1 \qquad \text{...(2)}$$

Put $x = 1/2$ in (2) and equate real parts,

$$\int_{-\infty}^{\infty} \frac{s\cos s - \sin s}{s^3} \cos\frac{s}{2}\, ds = -\frac{\pi}{2}\left(1 - \frac{1}{4}\right) = \frac{-3\pi}{8}$$

$$2\int_{0}^{\infty} \frac{s\cos s - \sin s}{s^3} \cos\frac{s}{2}\, ds = -\frac{3\pi}{8} \qquad (\because \text{ integrand is even})$$

$$\int_{0}^{\infty} \frac{s\cos s - \sin s}{s^3} \cos\frac{s}{2}\, ds = -\frac{3\pi}{16}$$

$$\int_{0}^{\infty} \frac{\sin s - s\cos s}{s^3} \cos\frac{s}{2}\, ds = -\frac{3\pi}{16} \qquad \text{...(3)}$$

Put $x = 0$ in (2), we get

$$\int_{-\infty}^{\infty} \frac{s\cos s - \sin s}{s^3}\, ds = -\frac{\pi}{2}$$

$$\therefore \int_{0}^{\infty} \frac{\sin s - s\cos s}{s^3}\, ds = \frac{\pi}{4} \qquad \text{(change } s \text{ to } t\text{)}$$

13. (*b*) $F_s(e^{-ax}) = \sqrt{\frac{2}{\pi}} \int_0^{\infty} e^{-ax} \sin sx \, dx$

$$= \sqrt{\frac{2}{\pi}} \left[\frac{e^{-ax}}{a^2+s^2}(-a\sin sx - s\cos sx)\right]_0^{\infty}$$

$$= \sqrt{\frac{2}{\pi}} \cdot \frac{s}{a^2+s^2}, \; a > 0$$

$$F_s(e^{-bx}) = \sqrt{\frac{2}{\pi}} \cdot \frac{s}{b^2+s^2}, \; b > 0$$

Using $\int_0^{\infty} f(x)g(x)\,dx = \int_0^{\infty} F_s(s)\,G_s(s)\,ds$

$$\int_0^{\infty} e^{-ax} \cdot e^{-bx}\, dx = \frac{2}{\pi}\int_0^{\infty} \frac{s^2}{(a^2+s^2)(b^2+s^2)}\, ds$$

$$\therefore \int_0^{\infty} e^{-(a+b)x}\, dx = \frac{2}{\pi}\int_0^{\infty} \frac{x^2}{(a^2+x^2)(b^2+x^2)}\, dx$$

$$\left[\frac{e^{-(a+b)x}}{-(a+b)}\right]_0^{\infty} = \frac{2}{\pi}\int_0^{\infty} \frac{x^2}{(a^2+x^2)(b^2+x^2)}\, dx$$

$$= \frac{2}{\pi} \int_0^\infty \frac{x^2}{(a^2+x^2)(b^2+x^2)}\, dx = \frac{1}{a+b}$$

$$\therefore \int_0^\infty \frac{x^2}{(a^2+x^2)(b^2+x^2)}\, dx = \frac{\pi}{2(a+b)}.$$

14. (*a*) Refer Text P 194.

Also Refer to Question 14 (*a*) under MA 2211 question paper.

14. (*b*) Boundary conditions are

$u(0, y) = 0\ ;\ 0 \le y \le 10$ ...(i)

$u(x, 0) = 0\ ;\ 0 \le x \le 10$ ...(ii)

$u(10, y) = 0;\ 0 \le y \le 10$ ...(iii)

$u(x, 10) = 10x - x^2;\ 0 \le x \le 10$ ...(iv)

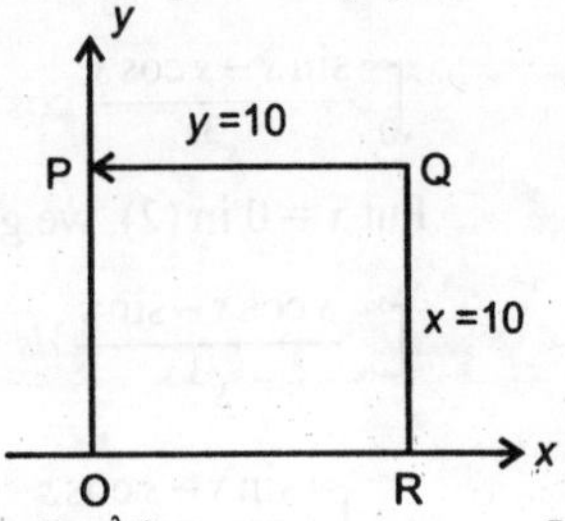

Take $u(x, y) = (A \cos \lambda x + B \sin \lambda x)(Ce^{\lambda y} + De^{-\lambda y})$ I

as the solution of Laplace equation (Steady state)

Using $u(0, y) = 0$, we get $A = 0$

Using $u(10, y) = 0$, we get

$B \sin 10\lambda$ (variable) $= 0$

$\therefore \quad B \ne 0, \quad \lambda = \frac{n\pi}{10}, \quad n = 1, 2, 3, ...$

Using $u(x, 0) = 0$, we get

$B \sin \lambda x\, (C + D) = 0$

$\therefore \quad D = -C$

Hence I becomes,

$$u(x, y) = B \sin \frac{n\pi x}{10} \times c\,(e^{\lambda y} - e^{-\lambda y})$$

$$= B_n \sin \frac{n\pi x}{10} \sinh \lambda y$$

$$= B_n \sin \frac{n\pi x}{10} \sinh \frac{n\pi y}{10}$$

Most general solution is

$$u(x, y) = \sum_{n=1}^{\infty} B_n \sin \frac{n\pi x}{10} \sinh \frac{n\pi y}{10} \qquad ...\text{II}$$

Using boundary condition (iv) in II,

$$u(x, 10) = \sum_{1}^{\infty} B_n \sin \frac{n\pi x}{10} \cdot \sinh n\pi = 10x - x^2.$$

$$\therefore\ B_n \sinh n\pi = b_n = \frac{2}{10}\int_0^{10}(10x-x^2)\sin\frac{n\pi x}{10}\,dx$$

$$= \frac{1}{5}\left[(10x-x^2)\left(\frac{-\cos\frac{n\pi x}{10}}{\frac{n\pi}{10}}\right)-(10-2x)\left(\frac{-\sin\frac{n\pi x}{10}}{\frac{n^2\pi^2}{100}}\right)+(-2)\left(\frac{\cos\frac{n\pi x}{10}}{\frac{n^3\pi^3}{10^3}}\right)\right]_0^{10}$$

$$= \frac{1}{5}\left[\frac{-2000}{n^3\pi^3}\left((-1)^n-1\right)\right] = \frac{400}{n^3\pi^3}\,[1-(-1)^n].$$

$= 0$ for n even

$= \dfrac{800}{n^3\pi^3}$ for n odd

Hence, $B_n = \begin{cases} 0 \text{ for } n \text{ even} \\ \dfrac{800}{n^3\pi^3 \sin h n\pi} \text{ for } n \text{ odd}, \end{cases}$ put in II

$$u(x,t) = \sum_{n=1,3,5,\ldots}^{\infty} \frac{800}{n^3\pi^3 \sin h\, n\pi}\cdot \sin\frac{n\pi x}{10}\sin h\frac{n\pi y}{10}$$

15. (*a*) (*i*) Since $f(t)$ is periodic of period T, (Text P. 781)

$$L[f(t)] = \frac{1}{1-e^{-sT}}\int_0^T e^{-st} f(t)\,dt$$

$$= \frac{E}{1-e^{-sT}}\left[\int_0^{T/2} e^{-st}dt - \int_{T/2}^{T} e^{-st}dt\right]$$

$$= \frac{E}{1-e^{-sT}}\left[\left(\frac{e^{-st}}{-s}\right)_0^{\frac{T}{2}} - \left(\frac{e^{-st}}{-s}\right)_{\frac{T}{2}}^{T}\right]$$

$$= \frac{E}{s(1-e^{-sT})}\left[e^{-sT} - e^{\frac{-sT}{2}} - e^{\frac{-sT}{2}} + 1\right]$$

$$= \frac{E}{s\,(1-e^{-sT})}\left(1-e^{-\frac{sT}{2}}\right)^2$$

(*ii*) $y'' - 3y' + 2y = e^{3x}, \quad y(0) = 0, y'(0) = 0$ (Text P. 746)

Taking Laplace Transform,

$L(y'') - 3L(y') + 2L(y) = L(e^{3x})$

$$s^2\bar{y} - sy(0) - y'(0) - 3[s\bar{y} - y(0)] + 2\bar{y} = \frac{1}{s-3}$$

Using $y(0) = 0, y'(0) = 0$, we get

$$\bar{y}\,(s^2 - 3s + 2) = \frac{1}{s-3}$$

$$\bar{y} = \frac{1}{(s-1)(s-2)(s-3)} = \frac{A}{s-1} + \frac{B}{s-2} + \frac{C}{s-3}$$

$$= \frac{\frac{1}{2}}{s-1} - \frac{1}{s-2} + \frac{\frac{1}{2}}{s-3}$$

$\therefore \quad y = L^{-1}$ (RHS)

$$y = \frac{1}{2}e^{x} - e^{2x} + \frac{1}{2}e^{3x}$$

15. (*b*) (*i*) $L^{-1}\left[\frac{s^2-s+2}{s\,(s+2)(s-3)}\right] = ?$ (Text P. 725)

Let $\frac{s^2-s+2}{s(s+2)(s-3)} = \frac{A}{s} + \frac{B}{s+2} + \frac{C}{s-3}$

$$A = \left(\frac{s^2-s+2}{(s+2)(s-3)}\right)_{s=0} = \frac{2}{-6} = \frac{-1}{3}$$

$$B = \left(\frac{s^2-s+2}{s(s-3)}\right)_{s=-2} = \frac{8}{10} = \frac{4}{5}$$

$$C = \left(\frac{s^2-s+2}{s(s+2)}\right)_{s=3} = \frac{8}{15}$$

$$\therefore \quad L^{-1}\frac{s^2-s+2}{s(s+2)(s-3)} = L^{-1}\left[\frac{-1/3}{s}+\frac{4/5}{s+2}+\frac{8/15}{s-3}\right]$$

$$= \frac{-1}{3}+\frac{4}{5}e^{-2t}+\frac{8}{15}e^{3t}$$

(*ii*) $L^{-1}\frac{s^2}{(s^2+a^2)^2} = ?$ (Text P. 787)

$$L^{-1}\frac{s^2}{(s^2+a^2)^2} = L^{-1}\frac{s}{s^2+a^2} * L^{-1}\frac{s}{s^2+a^2}$$

$$= \cos at * \cos at$$

$$= \int_0^t \cos au . \cos a(t-u)\,du$$

$$= \frac{1}{2}\int_0^t [\cos at + \cos(2au - at)]\,du$$

$$= \frac{1}{2}\left[u\cos at + \frac{\sin(2au-at)}{2a}\right]_0^t$$

$$= \frac{1}{2}\left[t\cos at + \frac{1}{2a}(\sin at + \sin at)\right]$$

$$= \frac{1}{2a}(\sin at + at\cos at)$$

B.E./ B.Tech. Degree Examination, April/May 2011
Regulation 2008
Third Semester
Common to all branches
MA 2211 Transforms and Partial Differential Equations

Time : Three Hours *Maximum : 100 Marks*

Answer ALL Questions

PART–A (10 × 2 = 20 Marks)

1. Give the expression for the Fourier Series coefficient b_n for the function $f(x)$ defined in $(-2, 2)$.
2. Without finding the values of a_0, a_n and b_n, the Fourier coefficients of Fourier series, for the funciton $F(x) = x^2$ in the interval $(0, \pi)$ find the value of $\left[\frac{a_0^2}{2} + \sum_{n=1}^{\infty}(a_n^2 + b_n^2)\right]$.
3. State and prove the change of scale property of Fourier Transform.
4. If $F_C(s)$ is the Fourier cosine transform of $f(x)$, prove that the Fourier cosine transform of $f(ax)$ is $\frac{1}{a}F_C\left(\frac{s}{a}\right)$.
5. Form the partial differential equation by eliminating the arbitrary constants a and b from $z = (x^2 + a)(y^2 + b)$.
6. Solve the equation $(D - D')^3 z = 0$.
7. A rod 40 cm long with insulated sides has its ends A and B kept at 20°C and 60°C respectively. Find the steady state temperature at a location 15 cm from A.
8. Write down the three possible solutions of Laplace equation in two dimensions.
9. Find the Z-transform of a^n.
10. What advantage is gained when Z-transform is used to solve difference equation?

PART–B (5 × 16 = 80 Marks)

11. (*a*) (*i*) Exapand $f(x) = x(2\pi - \pi)$ as Fourier series in $(0, 2\pi)$ and hence deduce that the sum of $\frac{1}{1^2} + \frac{1}{2^2} + \frac{1}{3^2} + \frac{1}{4^2} + \ldots\ldots$

(*ii*) Obtain the Fourier series for the function $f(x)$ given by

$$f(x) = \begin{cases} 1 - x, & -\pi < x < 0 \\ 1 + x, & 0 < x < \pi \end{cases}$$

Hence deduce that $\frac{1}{1^2}+\frac{1}{3^2}+\frac{1}{5^2}+....\frac{\pi^2}{8}$

OR

11. (*b*) (*i*) Obtain the sine series for $f(x)=\begin{cases} x \text{ in } 0 \le x \le \frac{l}{2} \\ (l-x) \text{ in } \frac{l}{2} \le x \le l \end{cases}$

(*ii*) Find the Fourier series up to second harmonic for $y=f(x)$ from the following values.

x	0	π/3	2π/3	π	4π/3	5π/3	2π
y	1.0	1.4	1.9	1.7	1.5	1.2	1.0

12. (*a*) (*i*) Find the Fourier transform of $f(x)=\begin{cases} 1-x^2 \text{ if } |x|<1 \\ 0 \text{ if } |x|>1 \end{cases}$

Hence evaluate $\int_0^{\infty}\left(\frac{x\cos x-\sin x}{x^3}\right)\cos\frac{x}{2}dx.$

(*ii*) Find the Fourier transform of $f(x)$ given by

$$f(x)=\begin{cases} 1 \text{ for } |x|<a \\ 0 \text{ for } |x|>a>0 \end{cases}$$

and using Parseval's identity prove that $\int_0^{\infty}\left(\frac{\sin t}{t}\right)^2 dt=\frac{\pi}{2}.$

OR

12. (*b*) (*i*) Find the Fourier sine transform of $f(x)=\begin{cases} x, & 0<x<1 \\ 2-x, & 1<x<2 \\ 0, & x<2 \end{cases}$

(*ii*) Evaluate $\int_0^{\infty}\frac{dx}{(x^2+a^2)(x^2+b^2)}$ using Fourier cosine tansforms of e^{-ax} and e^{-bx}.

13. (*a*) (*i*) Form the partial differential equation by eliminating arbitrary functions f and ϕ from $z=f(x+ct)+\phi(x-ct)$.

(*ii*) Solve the partial differential equation

$$(mz-ny)\,p+(nx-lz)\,q=ly-mx.$$

OR

13. (*b*) (*i*) Solve $(D^2-D'^2)\,z=e^{x-y}\sin(2x+3y)$.

(*ii*) Solve $(D^2-3\,DD'+2\,D'^2+2\,D-2D')\,z=x+y+\sin(2x+y)$.

14. (*a*) A uniform string is stretched and fastened to two points 'l' apart. Motion is started by displacing the string into the form of the curve $y = kx(l - x)$ and then released from this position at time $t = 0$. Derive the expression for the desplacment of any point of the string at a distance x from one end at time t.

OR

14. (*b*) A retangular plate with insulated surfaces is 20 cm wide and so long compared to its width that it may be considered infinite in length without introducing an appreciable error. If the temperature of the short edge $x = 0$ is given by

$$u = \begin{cases} 10y, & \text{for } 0 \le y \le 10 \\ 10(20 - y), & \text{for } 10 \le y \le 20 \end{cases}$$

and the two long edges as well as the other short edge are kept at 0°C. Find the steady state temperature distribution in the plate.

15. (*a*) (*i*) Using convolution theorem, find inverse Z-transform of $\dfrac{z^2}{(z-1)(z-3)}$.

(*ii*) Find the Z-transforms of $a^n \cos n\theta$ and $e^{-at} \sin bt$

OR

15. (*b*) (*i*) Solve the difference equation $y(n+3) - 3y(n+1) + 2y(n) = 0$, given that $y(0) = 4$, $y(1) = 0$ and $y(2) = 8$.

(*ii*) Derive the difference equation from $y_n = (A + Bn)(-3)^n$.

Answers for MA 2211 – Transforms and Partial Differential Equations – April 2011

Third Semester

PART–A (10 × 2 = 20 Marks)

1. $b_n = \dfrac{1}{2}\displaystyle\int_{-2}^{2} f(x).\sin\frac{n\pi x}{2}\, dx.$

2. $\dfrac{a_0^2}{2} + \displaystyle\sum_{n=1}^{\infty}(a_n^2 + b_n^2) = 2\bar{y}^2 = \frac{2}{\pi}\int_0^{\pi} x^4\, dx = \frac{2}{\pi}\left(\frac{x^5}{5}\right)_0^{\pi} = \frac{2}{5}\pi^4$

3. If $F\{f(x)\} = \bar{F}(s)$ then $F\{f(ax)\} = \dfrac{1}{|a|}\bar{F}\left(\dfrac{s}{a}\right), a \ne 0.$

4. $F_C\{f(x)\} = \sqrt{\dfrac{2}{\pi}}\displaystyle\int_0^{\infty} f(x)\cos sx\, dx$

$$F_C\{f(ax)\} = \sqrt{\frac{2}{\pi}}\int_0^{\infty} f(ax)\cos sx\, dx$$

$$= \sqrt{\frac{2}{\pi}}\int_0^{\infty} f(u)\cos\left(\frac{us}{a}\right)\frac{du}{a} \qquad \text{where } ax = u \text{ and } a > 0$$

$$= \frac{1}{a}\sqrt{\frac{2}{\pi}}\int_0^\infty f(u)\cos\left\{\left(\frac{s}{a}\right)u\right\}du = \frac{1}{a}F_C\left(\frac{s}{a}\right)$$

5. $Z = (x^2 + a)(y^2 + b)$ (Text P. 99)

$$\frac{\partial z}{\partial x} = 2x(y^2 + b)$$

$$\frac{\partial z}{\partial y} = (x^2 + a).\,2y$$

$$\frac{\partial z}{\partial x}\cdot\frac{\partial z}{\partial y} = 4xy(x^2 + a)(y^2 + b^2) = 4xyz \text{ i.e., } pq = 4xyz.$$

6. $(D - D')^3 z = 0$

Put $D = m$, $D' = 1$. Then auxiliary equation is $(m - 1)^3 = 0$

$m = 1, 1, 1$

General solution is

$$z = f(y + x) + x\,\phi(y + x) + x^2\,\psi(y + x)$$

7. Steady state solution is

$$u(x) = ax + b \quad \text{...(1)}$$

Given, $u(0) = 20$ and $u(40) = 60$. Using in (1),

$$b = 20;\ u(40) = 40a + 20 = 60 \therefore a = 1$$

Hence, $u(x) = x + 20$

$$u(15) = 35°C$$

8. $\nabla^2 u = 0$ is Laplace equation. Then

$$u(x, y) = (Ae^{\lambda x} + Be^{-\lambda x})(C\cos\lambda y + D\sin\lambda y) \quad \text{...(1)}$$

$$u(x, y) = (A\cos\lambda x + B\sin\lambda x)(Ce^{\lambda y} + De^{-\lambda y}) \quad \text{...(2)}$$

$$u(x, y) = (Ax + B)(Cy + D) \quad \text{...(3)}$$

9. $$Z(a^n) = \sum_{n=0}^{\infty} a^n z^{-n} = \sum_{n=0}^{\infty}\left(\frac{a}{z}\right)^n$$

$$= \frac{1}{1 - \frac{a}{z}} \text{ if } |z| > |a| = \frac{z}{z - a} \text{ if } |z| > |a|$$

10. The difference equation is reduced to rational integral functions and its inverse z transform becomes simple. Arbitrary constants do not occur in the solution. This is the main advantage in using the transform.

PART–B (5 × 16 = 80 marks)

11. (*a*) (*i*) Let $f(x) = \frac{a_0}{2} + \sum_{n=1}^{\infty}(a_n\cos nx + b_n\sin nx)$

$$\text{where } a_0 = \frac{1}{\pi}\int_0^{2\pi}(2\pi x - x^2)\,dx = \frac{1}{\pi}\left[\pi x^2 - \frac{x^3}{3}\right]_0^{2\pi}$$

$$= \frac{1}{\pi}\left[4\pi^3 - \frac{8\pi^3}{3}\right] = \frac{4}{3}\pi^2$$

$$a_n = \frac{1}{\pi}\int_0^{2\pi}(2\pi x - x^2)\cdot\cos nx\,dx$$

$$= \frac{1}{\pi}\left[(2\pi x - x^2)\left(\frac{\sin nx}{n}\right) - (2\pi - 2x)\left(\frac{-\cos nx}{n^2}\right) + (-2)\left(\frac{-\sin nx}{n^3}\right)\right]_0^{2\pi}$$

$$= \frac{1}{\pi}\left[2\pi\left(-\frac{1}{n^2} - \frac{1}{n^2}\right)\right] = -\frac{4}{n^2}$$

$$b_n = \frac{1}{\pi}\int_0^{2\pi}(2\pi x - x^2)\sin nx\,dx$$

$$= \frac{1}{\pi}\left[(2\pi x - x^2)\left(\frac{-\cos nx}{n^2}\right) - (2\pi - 2x)\left(\frac{-\sin nx}{n^2}\right) + (-2)\left(\frac{\cos nx}{n^3}\right)\right]_0^{2\pi}$$

$$= -\frac{2}{\pi n^3}(1-1) = 0$$

$$\therefore\ f(x) = x(2\pi - x) = \frac{4}{6}\pi^2 - \sum_{n=1}^{\infty}\frac{4}{n^2}\cos nx$$

At $x = 0$, value of Fourier series is $= \dfrac{f(0) + f(2\pi)}{2}$

$$\therefore\ \frac{2}{3}\pi^2 - \sum_{n=1}^{\infty}\frac{4}{n^2} = \frac{0+0}{2} = 0$$

$$\therefore\ \sum_{1}^{\infty}\frac{1}{n^2} = \frac{\pi^2}{6}$$

(*ii*) $f(x) = \begin{cases} 1 - x, -\pi < x < 0 \\ 1 + x, 0 < x < \pi \end{cases}$

$f(x)$ is evidently even in $(-\pi, \pi)$

$\therefore\ b_n = 0$ in its Fourier series.

Let $f(x) = \dfrac{a_0}{2} + \displaystyle\sum_{n=1}^{\infty} a_n \cos nx$

where $a_0 = \frac{1}{\pi}\int_{-\pi}^{\pi} f(x)dx = \frac{2}{\pi}\int_0^{\pi}(1+x)\,dx = \frac{2}{\pi}\left[x+\frac{x^2}{2}\right]_0^{\pi} = 2+\pi$

$$a_n = \frac{1}{\pi}\int_{-\pi}^{\pi} f(x)\cos nx\,dx = \frac{2}{\pi}\int_0^{\pi}(1+x)\cos nx\,dx$$

$$= \frac{2}{\pi}\left[(1+x)\left(\frac{\sin nx}{n}\right)-(1)\left(\frac{-\cos nx}{n^2}\right)\right]_0^{\pi} = \frac{2}{\pi}\left[\frac{1}{n^2}\left((-1)^n-1\right)\right]$$

$$= \begin{cases} 0 \text{ if } n \text{ even} \\ \dfrac{-4}{\pi n^2} \text{ if } n \text{ odd} \end{cases}$$

$$\therefore\ f(x) = \frac{2+\pi}{2} - \sum_{n=1,3,5..}^{\infty}\frac{4}{\pi n^2}\cos nx$$

Putting $x = 0$

$$\frac{2+\pi}{2} - \sum_{n=1,3,5..}^{\infty}\frac{4}{\pi n^2} = f(0) = 1$$

$$\frac{\pi}{2}+1-\frac{4}{\pi}\left[\frac{1}{1^2}+\frac{1}{3^2}+\frac{1}{5^2}+....\right] = 1$$

$$\therefore\ \frac{1}{1^2}+\frac{1}{3^2}+\frac{1}{5^2}+\text{...to}\ \infty = \frac{\pi^2}{8}$$

11. (*b*) (*i*) $f(x) = \begin{cases} x \text{ in } 0 \le x \le l/2 \\ l-x \text{ in } l/2 \le x \le l \end{cases}$

$$\text{Let } f(x) = \sum_{n=1}^{\infty} b_n \sin\frac{n\pi x}{l}$$

$$b_n = \frac{2}{l}\int_0^l f(x)\sin\frac{n\pi x}{l}\,dx$$

$$= \frac{2}{l}\left[\int_0^{l/2} x\sin\frac{n\pi x}{l}\,dx + \int_{l/2}^{l}(l-x)\sin\frac{n\pi x}{l}\,dx\right]$$

$$= \frac{2}{l}\left[\left\{(x)\left(\frac{-\cos\frac{n\pi x}{l}}{\frac{n\pi}{l}}\right)-(1)\left(\frac{-\sin\frac{n\pi x}{l}}{\frac{n^2l^2}{l^2}}\right)\right\}_0^{l/2}\right.$$

$$\left.+\left\{(l-x)\left(\frac{-\cos\frac{n\pi x}{l}}{\frac{n\pi}{l}}\right)-(1)\left(\frac{-\sin\frac{n\pi x}{l}}{\frac{n^2l^2}{l^2}}\right)\right\}_0^{l/2}\right]$$

$$= \frac{2}{l}\left[-\frac{l^2}{2n\pi}\cos\frac{n\pi}{2}+\frac{l^2}{n^2\pi^2}\sin\frac{n\pi}{2}+\frac{l^2}{2n\pi}\cos\frac{n\pi}{2}+\frac{l^2}{n^2\pi^2}\sin\frac{n\pi}{2}\right]$$

$$= \frac{4l}{n^2\pi^2}\sin\frac{n\pi}{2}$$

Hence, $f(x) = \frac{4l}{\pi^2}\sum_{n=1}^{\infty}\frac{1}{n^2}\sin\frac{n\pi}{2}\sin\frac{n\pi x}{l}$

$$= \frac{4l}{\pi^2}\left[\frac{1}{1^2}\sin\frac{\pi x}{l}-\frac{1}{3^2}\sin\frac{3\pi x}{l}+\frac{1}{5^2}\sin\frac{5\pi x}{l}\text{....to }\infty\right]$$

(*ii*) (See Text P. 86)

We will exclude the last point $x = 2\pi$

x	$f(x)$	$\cos x$	$\sin x$	$\cos 2x$	$\sin 2x$
0	1.0	1	0	1	0
$\frac{\pi}{3}$	1.4	0.5	0.866	–0.5	0.866
$\frac{2\pi}{3}$	1.9	–0.5	0.866	–0.5	–0.866
π	1.7	–1	0	1	0
$\frac{4\pi}{3}$	1.5	–0.5	–0.866	–0.5	0.866
$\frac{5\pi}{3}$	1.2	0.5	–0.866	–0.5	–0.866

$$a_0 = \frac{2}{6}\sum f(x) = \frac{1}{3}(1.0 + 1.4 + 1.9 + 1.7 + 1.5 + 1.2) = 2.9$$

$$a_1 = \frac{2}{6}\sum f(x)\cos x = \frac{1}{3}(1 + 0.7 - 0.95 - 1.7 - 0.75 + 0.6) = -0.37$$

$$a_2 = \frac{2}{6}\sum f(x)\cos 2x = -0.1$$

$$b_1 = \frac{2}{6}\sum f(x)\sin x = 0.17$$

$$b_2 = \frac{2}{6}\sum f(x)\sin 2x = -0.06$$

$\therefore f(x) = 1.45 - 0.37\cos x - 0.1\cos 2x + 0.17\sin x - 0.06\sin 2x$

12. (*a*) (*i*) (See Text P. 281)

$$F[f(x)] = \frac{1}{\sqrt{2\pi}} \int_{-\infty}^{\infty} e^{isx} f(x)\, dx = \frac{1}{\sqrt{2\pi}} \int_{-1}^{1} e^{isx} (1-x^2)\, dx$$

$$= \frac{1}{\sqrt{2\pi}} \left[(1-x^2)\left(\frac{e^{isx}}{is}\right) - (-2x)\left(\frac{e^{isx}}{i^2 s^2}\right) + (-2)\left(\frac{e^{isx}}{i^3 s^3}\right) \right]_{-1}^{1}$$

$$= \frac{1}{\sqrt{2\pi}} \left[\frac{-2}{s^2}(e^{is} + e^{-is}) - \frac{2i}{s^3}(e^{is} + e^{-is}) \right]$$

$$= \frac{-2}{s^3} \cdot \frac{1}{\sqrt{2\pi}} [2s\cos s - 2\sin s] = \frac{-4}{\sqrt{2\pi}} \left[\frac{s\cos s - \sin s}{s^3} \right] \qquad ...(1)$$

Using inversion, $f(x) = \frac{1}{\sqrt{2\pi}} \int_{-\infty}^{\infty} \frac{-4}{\sqrt{2\pi}} \left(\frac{s\cos s - \sin s}{s^3} \right) e^{-isx}\, ds$

$$= \frac{-2}{\pi} \int_{-\infty}^{\infty} \frac{s\cos s - \sin s}{s^3} e^{-isx}\, ds$$

$$\int_{-\infty}^{\infty} \left(\frac{s\cos s - \sin s}{s^3} \right) (\cos sx - i\sin sx)\, ds = \frac{-\pi}{2}(1 - x^2) \text{ if } |x| < 1$$

$$= 0 \text{ if } |x| > 1$$

Put $x = \frac{1}{2}$ and equate real parts

$$\int_{-\infty}^{\infty} \frac{s\cos s - \sin s}{s^3} \cos\frac{s}{2}\, ds = \frac{-\pi}{2}\left(1 - \frac{1}{4}\right) = \frac{-3\pi}{8}$$

$$\therefore \quad 2\int_{-\infty}^{\infty} \frac{s\cos s - \sin s}{s^3} \cos\frac{s}{2}\, ds = \frac{-\pi}{2}\left(\frac{3}{4}\right)$$

$$\therefore \quad \int_{-\infty}^{\infty} \left(\frac{s\cos s - \sin s}{s^3} \right) \cos\frac{s}{2}\, ds = \frac{-3\pi}{16}.$$

Change s to x to get the result.

(*ii*) $f(x) = \begin{cases} 1 \text{ for } |x| < a \\ 0 \text{ for } |x| > a > 0. \end{cases}$ (Refer Text P. 283)

$$F[f(x)] = \frac{1}{\sqrt{2\pi}} \int_{-\infty}^{\infty} f(x)\, e^{isx}\, dx$$

$$= \frac{1}{\sqrt{2\pi}} \int_{-a}^{a} e^{isx}\, dx = \frac{1}{\sqrt{2\pi}} \left(\frac{e^{isx}}{is} \right)_{-a}^{a}$$

$$= \frac{1}{\sqrt{2\pi}} \cdot \frac{1}{is}\left(e^{isa} - e^{-isa}\right) = \frac{1}{\sqrt{2\pi}} \cdot \frac{1}{is}\,(2i \sin sa)$$

$$= \sqrt{\frac{2}{\pi}} \cdot \frac{\sin as}{s} \qquad \text{....(1)}$$

Using parseval's identity,

$$\int_{-\infty}^{\infty} |f(x)|^2 dx = \int_{-\infty}^{\infty} |F(s)|^2\, ds \qquad \text{(Refer Text P. 286)}$$

$$\therefore \int_{-a}^{a} 1\, dx = \int_{-\infty}^{\infty} \frac{2}{\pi} \frac{\sin^2 as}{s^2}\, ds$$

$$2a = \frac{2}{\pi}\int_{-\infty}^{\infty} \left(\frac{\sin as}{s}\right)^2 ds$$

Putting as $= t$, $ds = \dfrac{dt}{a}$

$$\int_{-\infty}^{\infty} \left(\frac{\sin t}{t}\right)^2 dt = \pi$$

$$2\int_{0}^{\infty} \left(\frac{\sin t}{t}\right)^2 dt = \pi$$

$$\therefore \int_{0}^{\infty} \left(\frac{\sin t}{t}\right)^2 dt = \frac{\pi}{2}$$

12. (*b*) (*i*) $f(x) = \begin{cases} x \text{ for } 0 < x < 1 \\ 2 - x \text{ for } 1 < x < 2 \\ 0 \text{ for } x > 2 \end{cases}$ (Text P. 303)

$$F_s(f(x)) = \sqrt{\frac{2}{\pi}} \int_0^{\infty} f(x) \sin sx\, dx$$

$$= \sqrt{\frac{2}{\pi}}\left[\int_0^1 x \sin sx\, dx + \int_1^2 (2-x)\sin sx\, dx\right]$$

$$= \sqrt{\frac{2}{\pi}}\left[\left\{(x)\left(\frac{-\cos sx}{s}\right) - (1)\left(\frac{-\sin sx}{s^2}\right)\right\}_0^1\right.$$

$$\left. + \left\{(2-x)\left(\frac{-\cos sx}{s}\right) - (-1)\left(\frac{-\sin sx}{s^2}\right)\right\}_1^2\right]$$

$$= \sqrt{\frac{2}{\pi}}\left[-\frac{1}{s}\cos s + \frac{1}{s^2}\sin s + \frac{1}{s}\cos s - \frac{\sin 2s}{s^2} + \frac{\sin s}{s^2}\right]$$

$$= \sqrt{\frac{2}{\pi}}\left[\frac{2}{s^2}\sin s - \frac{\sin 2s}{s^2}\right] = \frac{2\sqrt{2}\sin s}{\sqrt{\pi}s^2}(1-\cos s)$$

(*ii*) $\displaystyle\int_0^\infty \frac{dx}{(a^2+x^2)(x^2+b^2)}$ (Text P. 294)

$$F_c(e^{-ax}) = \sqrt{\frac{2}{\pi}}\int_0^\infty e^{-ax}\cos sx\, dx$$

$$= \sqrt{\frac{2}{\pi}}\left[\frac{e^{-ax}}{a^2+s^2}(-a\cos sx + s\sin sx)\right]_0^\infty$$

$$= \sqrt{\frac{2}{\pi}}\frac{a}{a^2+s^2}, \text{ if } a > 0$$

Similarly, $F_c(e^{-bx}) = \sqrt{\dfrac{2}{\pi}}\dfrac{b}{b^2+s^2}$, if $b > 0$

Using $\displaystyle\int_0^\infty f(x)\,g(x)\,dx = \int_0^\infty F_c(s)\,G_c(s)\,ds$

$$\int_0^\infty e^{-ax}\cdot e^{-bx}\,dx = \frac{2}{\pi}\int_0^\infty \frac{ab}{(a^2+s^2)(b^2+s^2)}\,ds$$

$$\left[\frac{e^{-(a+b)x}}{-(a+b)}\right]_0^\infty = \frac{2}{\pi}\int_0^\infty \frac{ab}{(a^2+x^2)(b^2+x^2)}\,dx$$

$$\frac{1}{a+b} = \text{R.H.S}$$

$$\therefore \int_0^\infty \frac{dx}{(a^2+x^2)(b^2+x^2)} = \frac{\pi}{2ab(a+b)}, \; a > 0, b > 0.$$ (Text P. 102)

13. (*a*) (*i*) $z = f(x+ct) + \phi(x-ct)$

$$\frac{\partial z}{\partial x} = p = f'(x+ct) + \phi'(x-ct)$$

$$\frac{\partial^2 z}{\partial x^2} = f''(x+ct) + \phi''(x-ct) \quad ...(1)$$

$$\frac{\partial z}{\partial t} = cf'(x+ct) - c\phi'(x-ct)$$

$$\frac{\partial^2 z}{\partial t^2} = c^2 f''(x+ct) + c^2 \phi''(x-ct)$$
$$= c^2 [f''(x+ct) + \phi''(x-ct)]$$

$$\frac{\partial^2 z}{\partial t^2} = c^2 \frac{\partial^2 z}{\partial x^2}$$

(*ii*) $(mz - ny)p + (nx - lz)q = ly - mx$ (Text P. 135)

Subsidiary equations are

$$\frac{dx}{mz-ny} = \frac{dy}{nx-lz} = \frac{dz}{ly-mz} \quad .....(1)$$

Using two sets of multipliers x, y, z and l, m, n each of the ratios in (1) equals

$$\frac{xdx+ydy+zdz}{x(mz-ny)+y(nx-lz)+z(ly-mx)} = \frac{ldx+mdy+ndz}{l(mz-ny)+m(nx-lx)+n(ly-mx)}$$

i.e.,
$$\frac{xdx+ydy+zdz}{0} = \frac{ldx+mdy+ndz}{0}$$

This implies, $xdx + ydy + zdz = 0$ and $ldx + mdy + ndz = 0$

Integrating, $x^2 + y^2 + z^2 = a$ and $lx + my + mz = b$

Hence, general solution is

$$\phi(x^2 + y^2 + z^2, lx + my + xz) = 0$$

13. (*b*) (*i*) $(D^2 - D'^2)Z = e^{x-y}\sin(2x+3y)$

Auxiliary equation is $m^2 - 1 = 0$, $m = \pm 1$

C.F. is $\phi(y+x) + f(y-x)$

$$\text{P.I.} = \frac{e^{x-y}\sin(2x+3y)}{D^2 - D'^2}$$

$$= e^{x-y} \times \frac{\sin(2x+3y)}{(D+1)^2 - (D'-1)^2}$$

$$= e^{x-y} \times \frac{\sin(2x+3y)}{D^2 - D'^2 + 2D + 2D'}$$

$$= e^{x-y} \times \frac{\sin(2x+3y)}{-4-(-9)+2(D+D')}$$

$$= e^{x-y} \frac{\sin(2x+3y)}{5+2(D+D')}$$

$$= e^{x-y}(5-2D-2D')\frac{\sin(2x+3y)}{25-4(D+D')^2}$$

$$= e^{x-y}(5-2D-2D')\frac{\sin(2x+3y)}{25-4(D^2+D'^2+2DD')}$$

$$= e^{x-y}(5-2D-2D')\frac{\sin(2x+3y)}{25-4(-4-9-12)}$$

$$= \frac{e^{x-y}}{125}(5-2D-2D')\sin(2x+3y)$$

$$= \frac{e^{x-y}}{125}[5\sin(2x+3y)-4\cos(2x+3y)-6\cos(2x+3y)]$$

$$= \frac{e^{x-y}}{25}[\sin(2x+3y)-2\cos(2x+3y)]$$

Hence, general solution is

$$Z = \phi(y+x)+f(y-x)+\frac{e^{x-y}}{25}[\sin(2x+3y)-2\cos(2x+3y)]$$

(*ii*) $D^2 - 3DD' + 2D'^2 + 2D - 2D')z = x + y + \sin(2x+y)$

i.e., $(D-2D'+2)(D-D')z = x + y + \sin(2x+y)$

Here, $m_1 = 2$, $c_1 = -2$. $m_2 = 1$, $c_2 = 0$

Hence C.F. is $e^{-2x}f(y+2x)+\phi(y+x)$

$$PI_1 = \frac{x+y}{(D-2D'+2)(D-D')} = \frac{x+y}{D\left(\frac{1-D'}{D}\right)(2)\left(1+\frac{D-2D'}{2}\right)}$$

$$= \frac{1}{2D}\left(1+\frac{D'}{D}\right)^{-1}\left(1+\frac{D-2D'}{2}\right)^{-1}(x+y)$$

$$= \frac{1}{2D}\left[1+\frac{D'}{D}+\frac{D'^2}{D^2}+\ldots\right]\left[1-\frac{D-2D'}{2}+\left(\frac{D-2D'}{2}\right)^2\ldots\right](x+y)$$

$$= \frac{1}{2D}\left[1+\frac{D'}{D}+\frac{D'^2}{D^2}+..\right]\left[1-\frac{D}{2}+D'+\frac{D^2}{4}-\frac{4DD'}{4}+D'^2\right](x-y)$$

$$= \frac{1}{2}\left[\frac{1}{D}+\frac{1}{2}+\frac{1}{2}\cdot\frac{D'}{D}+\frac{D}{4}-D'+\frac{D'}{D^2}\right](x-y)$$

$$= \frac{1}{2}\left[\frac{1}{D}-\frac{1}{2}+\frac{D'}{2D}+\frac{D}{4}-D'+\frac{D'}{D^2}\right](x+y)$$

$$= \frac{x^2}{2} + \frac{xy}{2} - \frac{y}{4} - \frac{3}{8}$$

$$PI_2 = \frac{\sin(2x+y)}{D^2 - 3DD' + 2D'^2 + 2D - 2D'}$$

$$= \frac{\sin(2x+y)}{-4+6-2+2D-2D'}$$

Replacing D^2 by $-a^2$, D'^2 by $-b^2$, DD' by $-ab$.

$$= \frac{\sin(2x+y)}{2(D-D')} = \frac{1}{2}\frac{(D+D')}{D^2-D'^2}\sin(2x+y)$$

$$= \frac{1}{2}\cdot\frac{(D+D')\sin(2x+y)}{-4+1}$$

$$= -\frac{1}{6}\,[\,2\cos(2x+y) + \cos(x+y)]$$

$$= -\frac{1}{2}\cos(2x+y)$$

∴ General solution is

$$Z = e^{-2x} f(y+2x) + \phi(y+x) + \frac{x^2}{2} + \frac{xy}{2} - \frac{y}{4} - \frac{3}{8} - \frac{1}{2}\cos(2x+y)$$

14. (*a*) (Refer Text P. 194)

Boundary conditions are

$$\left.\begin{aligned} y(0,t) &= 0 \\ y(l,t) &= 0 \end{aligned}\right\} t \geq 0 \qquad \begin{aligned} &...(1) \\ &...(2) \end{aligned}$$

$$\left.\begin{aligned} \left(\frac{\partial y}{\partial t}\right)_{t=0} &= 0 \qquad ...(3) \\ y(x,0) &= kx(l-x) \qquad ...(4) \end{aligned}\right\} 0 \leq x \leq l$$

The differential equation governing the displacment y is $\frac{\partial^2 y}{\partial t^2} = a^2 \frac{\partial^2 y}{\partial x^2}$

Solving and selecting the probable solution,

$y(x,t) = (A\cos\lambda x + B\sin\lambda x)(C\cos\lambda at + D\sin\lambda at)$ ...I

and using Boundary condition (1) and (2) we get

$A = 0$ and $\lambda = \frac{n\pi}{l}$, $n = 1, 2, 3$.........

$$\frac{\partial y}{\partial l} = (B\sin\lambda x)(-C\lambda a\sin\lambda at + D\lambda a\cos\lambda at)$$

Using $\dfrac{\partial y}{\partial l} = 0$ at $t = 0$, we get $D = 0$.

$$\therefore \quad y(x, t) = B_n \sin \frac{n\pi x}{l} \cdot \cos \frac{n\pi at}{l}, \; n = 1, 2, 3 \ldots\ldots\ldots$$

Most general solution is

$$y(x, t) = \sum_{n=1}^{\infty} B_n \sin \frac{n\pi x}{l} \cdot \cos \frac{n\pi at}{l} \qquad \ldots\text{II}$$

Using boundary condition (4) in

$$\sum_{n=1}^{\infty} B_n \sin \frac{n\pi x}{l} = kx(l - x) \qquad \ldots\text{III}$$

This is Fourier sine series.

$$\therefore B_n = b_n = \frac{2}{l} \int_0^l k(lx - x^2) \sin \frac{n\pi^2}{l} dx$$

$$= \frac{2k}{l} \left[(lx - x^2) \left(\frac{-\cos \frac{n\pi x}{l}}{\frac{n\pi}{l}} \right) - (l - 2x) \left(\frac{-\sin \frac{n\pi x}{l}}{\frac{n^2\pi^2}{l^2}} \right) + (-2) \left(-\frac{\cos \frac{n\pi x}{l}}{\frac{n^3\pi^3}{l^3}} \right) \right]_0^l$$

$$= \frac{4kl^2}{n^3\pi^3} [1 - (-1)^n] = 0 \text{ if } n \text{ even}$$

$$= \frac{8kl^2}{n^3\pi^3} \text{ if } n \text{ is odd}$$

Hence, using II, we get

$$y(x, t) = \frac{8kl^2}{\pi^3} \sum_{n=1}^{\infty} \frac{1}{(2n-1)^3} \sin \frac{(2n-1)\pi x}{l} \cdot \cos \frac{(2n-l)\pi at}{l}$$

14. (*b*) Take $u(x, y) = (A e^{\lambda x} + B e^{-\lambda x})(C \cos \lambda y + B \sin \lambda y)$ ...I

as the probable solution of Laplace equation.

Boundary conditions are

$$\left.\begin{aligned} u(x, 0) &= 0 \\ u(x, 20) &= 0 \end{aligned}\right\} 0 < x < \infty \qquad \begin{aligned} &\ldots(1) \\ &\ldots(2) \end{aligned}$$

$$u(\infty, y) = 0 \} \; 0 < y < 20$$

$$u(0, y) = \begin{cases} 10y &, 0 \le y \le 10 \\ 10(20 - y), & 10 \le y \le 20 \end{cases} \qquad \begin{aligned} &\ldots(3) \\ &\ldots(4) \end{aligned}$$

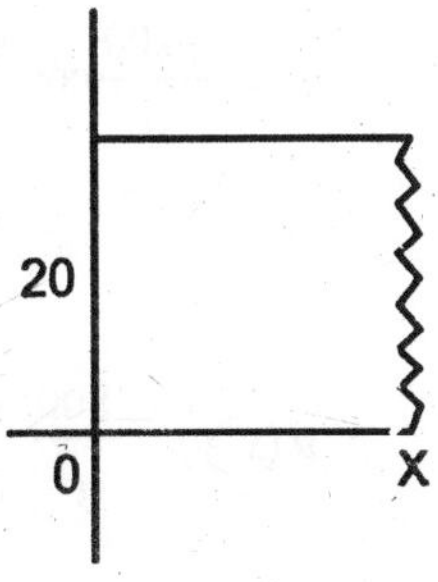

Using (1) in I, we get, $C = 0$

Using (2) $\sin 20\lambda = 0 \quad \therefore = \lambda = \frac{n\pi}{20}$, $n = 1, 2, 3.....$

Using (3) in I, we get $A = 0$; otherwise $u \to \infty$,

Hence, $u(x, y) = B_n e^{\frac{-n\pi x}{20}} \sin\frac{n\pi y}{20}$, $n = 1, 2, 3$ ......

Most general solution is

$$u(x, y) = \sum_{n=1}^{\infty} B_n e^{\frac{-n\pi x}{20}} \sin\frac{n\pi y}{20} \qquad ...II$$

Using boundary condition (4) in II

$$u(0, y) = \sum_{n=1}^{\infty} B_n \sin\frac{n\pi y}{20} = \begin{cases} 10y & \text{if } 0 \le y \le 10 \\ 10(20-y) & \text{if } 10 \le y \le 20 \end{cases}$$

$$\therefore\ B_n = b_n = \frac{2}{20}\left[\int_0^{10} 10y \cdot \sin\frac{n\pi y}{20}\, dy + \int_{10}^{20} 10(20-y)\sin\frac{n\pi y}{20}\, dy\right]$$

$$= \left[\left\{(y)\left(\frac{-\cos\frac{n\pi y}{20}}{\frac{n\pi}{20}}\right) - (1)\left(\frac{-\sin\frac{n\pi y}{20}}{\frac{n^2\pi^2}{400}}\right)\right\}_0^{10}\right.$$

$$\left. + \left\{(20-y)\left(\frac{-\cos\frac{n\pi y}{20}}{\frac{n\pi}{20}}\right) - (-1)\left(\frac{-\sin\frac{n\pi y}{20}}{\frac{n^2\pi^2}{400}}\right)\right\}_{10}^{20}\right]$$

$$= \frac{-200}{n\pi}\cos\frac{n\pi}{2} + \frac{400}{n^2\pi^2}\left(\sin\frac{n\pi}{2}\right) + \frac{200}{n\pi}\cos\frac{n\pi}{2} + \frac{400}{n^2\pi^2}\cdot\sin\frac{n\pi}{2}$$

$$= \frac{800}{n^2\pi^2}\cdot\sin\frac{n\pi}{2}$$

$$\therefore\ u(x, y) = \frac{800}{\pi^2}\sum_{n=1}^{\infty}\frac{1}{n^2}\sin\frac{n\pi}{2}\cdot e^{-\frac{n\pi x}{20}}\sin\frac{n\pi y}{20}$$

15. (*a*) (*i*) Find inverse Z transform of $\frac{Z^2}{(Z-1)(Z-3)}$, by convolution

$$Z^{-1}\left[\frac{z^2}{(z-1)(z-3)}\right] = Z^{-1}\left(\frac{z}{z-1}\right) * Z^{-1}\frac{z}{z-3}$$

$$= 1^n * 3^n \text{ if } |z| > 3$$

$$= \sum_{k=0}^{n} 1^{n-k} 3^k = 1 + 3 + 3^2 + ..3^n = \frac{3^{n+1}-1}{2}$$

(*ii*) $Z(a^n) = \frac{z}{z-a}$ if $|z| > |a|$

Take $a = re^{i\theta}$

$$Z\{r^n e^{in\theta}\} = \frac{z}{z - re^{i\theta}}$$

$$Z\{r^n \cos n\theta + ir^n \sin n\theta\} = \frac{\frac{z}{r}}{\left(\frac{z}{r} - \cos\theta - i\sin\theta\right)}$$

$$= \frac{z}{r} \frac{\left[\frac{z}{r} - \cos\theta + i\sin\theta\right]}{\left(\frac{z}{r} - \cos\theta\right)^2 - \sin^2\theta}$$

Taking real

$$Z(r^n \cos n\theta) = \frac{z(z - r\cos\theta)}{z^2 - 2zr\cos\theta + r^2}, \text{ if } |z| > |r|$$

Also, $Z\{e^{-at} \sin bt\} = Z(\sin bt)\, z \to ze^{at}$

$$= \frac{ze^{aT} \sin bT}{z^2 e^{2aT} - 2ze^{aT} \cos bt + 1}$$

15. (*b*) (*i*) $y_{n+3} - 3y_{n+1} + 2y_{n=0}$

$Z(y_{n+3}) - 3\,Z(y_{n+1}) + 2Z(y_n) = 0$

$[z^3 y(z) - 3^2 y(0) - zy(1)] - 3\,[z\,y(z) - z\,y(0)] + 2\,y(z) = 0$

Using $y(0) = 4$, $y(1) = 0$ and $y(2) = 8$, we get

$z^3 y(z) - 4z^2 - 3z\,y(3) + 12z + 2y(z) = 0$

$(z^3 - 3z + 2)\,y(z) = 4z^2 - 12z$

$$y(z) = \frac{4z^2 - 12z}{z^3 - 3z + 2} = \frac{4z^2 - 12z}{(z-1)(z^2 + z - 2)}$$

$$= \frac{4z^2 - 12z}{(z-1)^2 (z+2)} = z \frac{4z - 12}{(z-1)^2 (z+2)}$$

$$= z\left[\frac{A}{z-1} + \frac{B}{(z-1)^2} + \frac{C}{3+2}\right]$$

$$= z\left[\frac{20/9}{z-1} - \frac{8}{3} \cdot \frac{1}{(z-1)^2} - \frac{20}{9} \cdot \frac{1}{z+2}\right]$$

$$y(z) = \frac{20}{9}\left[\frac{z}{z-1}\right] - \frac{8}{3}\left(\frac{z}{(z-1)^2}\right) - \frac{20}{9} \cdot \frac{z}{z+2}$$

Taking inverse transform,

$$y(n) = \frac{20}{9} - \frac{8}{3} n - \frac{20}{9} \cdot (-2)^n$$

15. (*b*) (*ii*) $y_n = (A + Bn)(-3)^n$

$\therefore \; y_{n+1} = (A + Bn + B)(-3)^{n+1}$

$\therefore \; y_{n+2} = (A + Bn + 2B)(-3)^{n+2}$

$y_{n+2} + 6y_{n+1} + 9y_n$

$= (-3)^{n+2} [A + Bn + 2B - 2(A + Bn + B) + A + Bn]$

$= (-3)^{n+2} (0)$

$y_{n+2} + 6y_{n+1} + 9y_n = 0$

B.Tech. (PT) DEGREE EXAMINATION, DECEMBER 2013

First Semester

PMA201 - MATHEMATICS - III

Time: Three hours Max. Marks: 100

Answer ALL Questions

PART – A (10 × 2 = 20 Marks)

1. Determine the value of a_n in the Fourier series expansion of $f(x) = x^3$ in $-\pi < x < \pi$.
2. Expand $f(x) = 1$ in a sine series in $0 < x < \pi$.
3. Find the particular integral of $(D^2 - DD')z = \sin(x + y)$.
4. Find the complete integral of $q = 2px$.
5. The ends A and B of a rod of length 10 cm long have their temperature kept at 20°C and 70°C. Find the steady state temperature distribution on the rod.
6. In the wave equation $\frac{\partial^2 y}{\partial t^2} = C^2 \frac{\partial^2 y}{\partial x^2}$, what does C^2 stand for?
7. Find the Fourier cosine transform of

$$f(x) = \begin{cases} \cos x, 0 < x < a \\ 0, \quad x \geq a \end{cases}$$

8. If Fourier transform of $f(x) = F(s)$, then what is Fourier transform of $f(ax)$?
9. Write any two applications of Ψ^2 test.
10. Define Correlation and Regression.

PART – B (5 × 16 = 80 Marks)

11. (*a*) Find Fourier series to represent $f(x) = x - x^2$ from $x = -\pi$ to $x = \pi$ and hence deduce that

$$\frac{1}{1^2} - \frac{1}{2^2} + \frac{1}{3^2} - \frac{1}{4^2} + \ldots = \frac{\pi^2}{12}.$$

OR

(*b*) Table below gives the values of $f(x)$ at the specified values of x. Find the first three harmonic to represent $f(x)$ as a Fourier series.

x	0°	60°	120°	180°	240°	300°
y	1.98	1.30	1.05	1.30	–0.88	–0.25

12. (*a*) (*i*) Solve $(D^2 - 2DD')z = e^{2x} + x^3y$.

(*ii*) Find the general solution of the following linear partial differential equation $x^2(y - z)p + y^2(z - x)q = z^2(x - y)$.

(OR)

(*b*) (*i*) Solve $p(1 + q) = qz$.

(*ii*) Solve $(D^2 + DD' - 6D'^2)z = \cos(2x + y.)$

13. (*a*) A tightly stretched string with fixed end points $x = 0$ and $x = l$ is initially at rest in its equilibrium position. If it is set to vibrate by giving to each of its points a velocity $\lambda x(l - x)$. Find the displacement of the string at any distance x from one end at any time t.

(OR)

(*b*) The ends A and B of a rod 20 cm long have the temperature kept at 0°C and 20°C respectively until steady state conditions prevail. The temperature of the end B is then suddenly raised to 60°C and kept so, while that of the end A is kept at 0°C. Find the temperature $u(x, t)$.

14. (*a*) (*i*) Find the Fourier sine transform of $\dfrac{x}{x^2 + a^2}$.

(*ii*) Evaluate $\displaystyle\int_0^\infty \frac{dx}{(x^2 + a^2)(x^2 + b^2)}$ using Fourier transform method.

(OR)

(*b*) (*i*) Find the Fourier transform of $e^{-x^2/2}$

(*ii*) Find the Fourier cosine transform of $f(x)$ if $f(x) = \sin x$ in $0 < x < \pi$.

15. (*a*) (*i*) To verify whether a course in accounting improved performance, a similar test was given to 12 participants both before and after the course. The marks are

Before	44	40	61	52	32	44	70	41	67	72	53	72
After	53	38	69	57	46	39	73	48	73	74	60	78

Whether the course is useful or not?

(*ii*) The number of automobile accidents per week in a certain community are as follows: 12, 8, 20, 2, 14, 10, 15, 6, 9, 4. Are these frequencies in agreement with the belief that accident conditions were the same during the 10 week period?

(OR)

(*b*) (*i*) The means of two large samples 1000 and 2000 members are 67.5 inches and 68.0 inches respectively. Can the samples the regarded as drawn from the same population of standard deviation 2.5 inches.

(*ii*) Obtain two regression equations to the following data:

x	65	66	67	67	68	69	79	72
y	67	68	65	68	72	72	69	71

B.Tech. (PT) DEGREE EXAMINATION, DECEMBER 2012

First Semester

PMA211 - MATHEMATICS - III

Time: Three hours Max. Marks: 100

Answer ALL Questions

PART – A (10 × 2 = 20 Marks)

1. Write a_o, a_n in the expansion of $x + x^3$ as a Fourier series in $(-\pi, \pi)$.
2. Find the half range sine series for $f(x) = k$ in $0 < x < \pi$.
3. Form the P.D.E. by eliminating the arbitrary function from $z = f(x^2 - y^2)$.
4. Solve $(4D^2 - D'^2)z = 0$.
5. In wave equation $\frac{\partial^2 y}{\partial t^2} = c^2 \frac{\partial^2 y}{\partial x^2}$, what does c^2 stands for?
6. A rod of length 20 cm whose one end is kept at 30°C and the other end is kept at 70°C is maintained so until steady state prevails. Find the steady state temperature.
7. Write any two solution of the Laplace equation $u_{xx} + u_{yy}$ = involving exponential terms in x or y.
8. State two dimensional heat equation.
9. Find the Fourier sine transform of $3e^{-2x}$.
10. Prove that $F_c[f(ax)] = \frac{1}{a} F_c\left(\frac{s}{a}\right)$.

PART – B (5 x 16 = 80 Marks)

11. (*a*) Find the half range cosine series $f(x) = (\pi - x^2)$ in the interval $(0, \pi)$.

Hence find the sum of the series $\frac{1}{1^4} + \frac{1}{2^4} + \frac{1}{3^4} + ... + \infty$.

(OR)

(*b*) Find the Fourier series upto 3rd harmonic from the table.

x	0	$\frac{\pi}{3}$	$\frac{2\pi}{3}$	π	$\frac{4\pi}{3}$	$\frac{5\pi}{3}$	2π
y	1.0	1.4	1.9	1.7	1.5	1.2	1.0

12. (*a*) Solve: $(D^2 + DD' - 6D'^2)z = x^2y + e^{3x+y}$

(OR)

(*b*) Solve: $(3z-4y)p+(4x-2z)q=(2y-3x)$

13. (*a*) A string is stretched and fastened to two points l apart. Motion is started by displacing the string into the form $y = k(lx - x^2)$ from which it is released at time $t = 0$. Find the displacement of any point of the string at a distance x from one end at any time 't'.

(OR)

(*b*) A rod 30 cm long has its end A and B kept 20°C and 80°C respectively until steady state conditions prevail. The temperature at each end is then suddenly reduced to 0°C and kept so. Find the resulting temperature function $u(x, t)$.

14. (*a*) A rectangular plate with insulated surface is 10 cm wide so long compared to its width that it may be considered infinite length without introducing appreciable error. If the temperature at short edge $y = 0$ is given by

$$u = \begin{cases} 20x & \text{for } 0 \le x \le 5 \\ 20(10-x) & \text{for } 5 \le x \le 10 \end{cases}$$

and all the other three edges are kept 0°C. Find $u(x, y)$.

(OR)

(*b*) An infinite long metal plate in the form of an area is enclosed between the line $y = 0$ and $y = \pi$ for $x > 0$. The temperature is zero along the edges $y = 0$ and $y = \pi$ and at infinity. If the edge $x = 0$ is kept at a constant temperature T°C. Find $u(x, y)$.

$$\int_0^\infty \frac{dx}{(x^2+1)(x^2+4)}.$$

15. (*a*) Use Fourier transform method to evaluate

$$\int_0^\infty \frac{dx}{(x^2+1)(x^2+4)}.$$

(OR)

(*b*) Find the Fourier transform of the function $f(x)$ is defined by

$$f(x) = \begin{cases} 1-x^2 & \text{if } |x| < 1 \\ 0 & \text{if } |x| \ge 1 \end{cases}.$$ Hence prove that

$$\int_0^\infty \frac{\sin s - s\cos s}{s^3} \cos\left(\frac{s}{2}\right) ds = \frac{3\pi}{16}.$$

B.Tech.
DEGREE EXAMINATION, NOVEMBER 2013
Third Semester
MA0201 – MATHEMATICS - III

(For the candidates admitted from the academic year 2007–2008 to 2012–2013
(Use of statistical table is permitted)

Time: Three hours | Max. Marks: 100

Answer ALL Questions

PART – A (10 × 2 = 20 Marks)

1. State the Dirichlet's condition for a function $f(x)$ to be expanded as a Fourier series.
2. Find a_n for $f(x) = x \cos x$ in $(-\pi, \pi)$.
3. Form the partial differential equation by eliminating the arbitrary constants a and b from $Z = ax + by + ab$.
4. Find the particular integral of the equation $(D^2 + 2DD' + D'^2)z = e^{x-y}$.
5. Classify the following partial differential equation:

$$4u_{xx} + 4u_{xy} + u_{yy} + 2u_x - u_y = 0.$$

6. Write down the three mathematically possible solutions of one dimensional heat equation.
7. Prove that $F\{e^{iax}.f(x)\} = F(s + a)$
8. State Parseval's identity in Fourier transform.
9. The first four central moments of a distribution are 0, 2.5, 0.7 and 18.75. Find the Kurtosis.
10. Define Null and Alternate hypothesis.

PART – B (5 × 16 = 80 Marks)

11. (*a*) (*i*) Expand $f(x) = x^2$ in $-l < x < l$ in a Fourier series and hence deduce $\frac{1}{1^2} + \frac{1}{2^2} + \frac{1}{3^2} + \ldots\infty = \frac{\pi^2}{6}$. (10 Marks)

(*ii*) Find the half range sine series of the function $f(x) = \pi - x$ in $(0, \pi)$ and hence deduce $1 - \frac{1}{3} + \frac{1}{5} - \frac{1}{7} + \ldots\infty$. (6 Marks)

(OR)

(*b*) Find the first three harmonics of the Fourier series of $y = f(x)$ from the following data:

$x°$ (deg.)	0°	60°	120°	180°	240°	300°	360°
y	0	9.2	14.4	17.8	17.3	11.7	0

12. (*a*) (*i*) Solve: $Z = px + qy + p^2 + pq + q^2$. Find also singular solution.

(*ii*) Form the partial differential equation by eliminating the arbitrary function $f(x^2 + y^2 + z^2, ax + by + cz) = 0$.

(OR)

(*b*) (*i*) Solve: $x(z^2 - y^2)p + y(x^2 - z^2)q = z(y^2 - x^2)$.

(*ii*) Solve: $(D^2 + 4DD' - 5D'^2)z = xy + \sin(2x + 3y)$.

13. (a) A string is stretched and fastened to two points 'l' apart. Motion is started by displacing the string into the form $y = k(lx - x^2)$ and then released it from this position at time $t = 0$. Find the displacement of the point of the string at a distance 'x' from one end at any time 't'.

(OR)

(*b*) A rectangular plate with insulated surface is 10 cm wide and so long compared to its width that it may be considered infinite in length. The temperature at the short edge $y = 0$ is given by

$$u = \begin{cases} 20x, & 0 \le x \le 5 \\ 20(10 - x), & 5 \le x \le 10 \end{cases}$$ and all other three edges are kept at 0°C. Find the steady state temperature at any point in the plate.

14. (*a*) Find the Fourier transform of $f(x) = \begin{cases} 1 - x^2 & \text{if } |x| < 1 \\ 0 & \text{if } |x| > 1 \end{cases}$. Hence

evaluate $\int_0^\infty \left(\frac{x\cos x - \sin x}{x^3}\right) \cos\left(\frac{x}{2}\right) dx$.

(OR)

(*b*) Find the Fourier sine transform of e^{-ax} $(a > 0)$. Hence find

$F_S\{x.e^{-ax}\}$ and $F_S\left\{\frac{e^{-ax}}{x}\right\}$. Also deduce the value of $\int_0^\infty \frac{\sin sx}{x} dx$.

15. (*a*) Marks obtained by 10 students in Mathematics (x) and Statistics (y) are given below:

x	60	34	40	50	45	40	22	43	42	64
y	75	32	33	40	45	33	12	30	34	51

Find the correlation coefficient and the two regression lines. Also find y when $x = 55$.

(OR)

(*b*) (*i*) The average marks scored by 32 boys is 72 with a standard deviation of 8, while that for 36 girls is 70 with a standard deviation of 6. Test at 1% level of significance, whether the boys perform better than girls.

(*ii*) The theory predicts the proportion of beans in the four groups A, B, C and D should be 9:3:3:1. In an experiment among 1600 beans, the numbers in the four groups were 882, 313, 287 and 118. Does the experimental results support the theory?

NOTE

NOTE